Methods in Enzymology

Volume 329
REGULATORS AND EFFECTORS OF SMALL GTPases
Part E
GTPases Involved in Vesicular Traffic

METHODS IN ENZYMOLOGY

EDITORS-IN-CHIEF

John N. Abelson   Melvin I. Simon

DIVISION OF BIOLOGY
CALIFORNIA INSTITUTE OF TECHNOLOGY
PASADENA, CALIFORNIA

FOUNDING EDITORS

Sidney P. Colowick and Nathan O. Kaplan

Methods in Enzymology

Volume 329

Regulators and Effectors of Small GTPases

Part E
GTPases Involved in Vesicular Traffic

EDITED BY

W. E. Balch

THE SCRIPPS RESEARCH INSTITUTE
LA JOLLA, CALIFORNIA

Channing J. Der

LINEBERGER COMPREHENSIVE CANCER CENTER
THE UNIVERSITY OF NORTH CAROLINA AT CHAPEL HILL
CHAPEL HILL, NORTH CAROLINA

Alan Hall

UNIVERSITY COLLEGE LONDON, LONDON, ENGLAND

ACADEMIC PRESS

San Diego   London   Boston   New York   Sydney   Tokyo   Toronto

Academic Press
A Harcourt Science and Technology Company
525 B Street, Suite 1900, San Diego, California 92101-4495, USA
http://www.academicpress.com

Academic Press
Harcourt Place, 32 Jamestown Road, London NW1 7BY, UK
http://www.academicpress.com

International Standard Book Number: 0-12-182230-3

PRINTED IN THE UNITED STATES OF AMERICA
01 02 03 04 05 06 07 SB 9 8 7 6 5 4 3 2 1

Table of Contents

Section I. Rab GTPases

Section II. ADP-Ribosylation Factor (ARF) GTPases

Section III. Sar GTPases

Section IV. Dynamin GTPases

Section V. Septin GTPases

Contributors to Volume 329

Article numbers are in parentheses following the names of contributors.
Affiliations listed are current.

JOSEPH P. ALBANESI (51), *Department of Pharmacology, University of Texas Southwestern Medical Center, Dallas, Texas 75390-9041*

ŠTEFAN ALBERT (6), *Department of Molecular Genetics, Max Planck Institute for Biophysical Chemistry, Göttingen D-37070, Germany*

KIRILL ALEXANDROV (3), *Department of Physical Biochemistry, Max Planck Institute for Molecular Physiology, Dortmund 44202, Germany*

MEIR ARIDOR (45), *Department of Cell Biology, The Scripps Research Institute, La Jolla, California 92037*

LORRAINE M. ARON (23), *Monoclonal Antibody Facility, University of Georgia, Athens, Georgia 30602*

WILLIAM E. BALCH (1, 2, 25, 45), *Departments of Cell and Molecular Biology, The Scripps Research Institute, La Jolla, California 92037*

MANUEL A. BARBIERI (16), *Department of Cell Biology and Physiology, Washington University School of Medicine, St. Louis, Missouri 63110*

CHARLES BARLOWE (46), *Department of Biochemistry, Dartmouth Medical School, Hanover, New Hampshire 03755*

BARBARA BARYLKO (51), *Department of Pharmacology, University of Texas Southwestern Medical Center, Dallas, Texas 75390-9041*

CRESTINA L. BEITES (52), *Programme in Cell Biology, Hospital for Sick Children, Department of Biochemistry, University of Toronto, Toronto, Ontario, Canada M5G 1X8*

WILLIAM J. BELDEN (46), *Department of Biochemistry, Dartmouth Medical School, Hanover, New Hampshire 03755*

SOPHIE BÉRAUD-DUFOUR (25, 28), *Department of Molecular and Cell Biology, The Scripps Research Institute, La Jolla, California 92037*

KUN BI (38), *Department of Biochemistry, University of Texas Southwestern Medical Center, Dallas, Texas 75235*

DERK D. BINNS (51), *Department of Pharmacology, University of Texas Southwestern Medical Center, Dallas, Texas 75390-9041*

JAMES E. CASANOVA (23, 27), *Department of Cell Biology, University of Virginia Health Sciences Center, Charlottesville, Virginia 22908*

DAN CASSEL (33, 34), *Department of Biology, Technion-Israel Institute of Technology, Haifa 32000, Israel*

PHILIPPE CHAVRIER (29), *Institut Curie–Section Recherche, CNRS UMR 144, Paris Cedex 05, France*

WEI CHEN (18, 19), *National Center for Genome Resources, Santa Fe, New Mexico 87505*

SAVVAS CHRISTOFORIDIS (14), *Laboratory of Biological Chemistry, Medical School, University of Ioannina, Ioannina 45110, Greece*

SHAMSHAD COCKCROFT (38), *Department of Physiology, University College, London WC1E6JJ, United Kingdom*

EDNA CUKIERMAN (33), *Department of Biology, Technion-Israel Institute of Technology, Haifa 32000, Israel*

MICHAEL P. CZECH (30), *Program in Molecular Medicine and Department of Biochemistry and Molecular Biology, University of*

Massachusetts Medical School, Worcester, Massachusetts 01605

HANNA DAMKE (47), Department of Cell Biology, The Scripps Research Institute, La Jolla, California 92037

PIETRO DE CAMILLI (50), Department of Cell Biology, Howard Hughes Medical Institute, Yale University School of Medicine, New Haven, Connecticut 06510

MARIA ANTONIETTA DE MATTEIS (42), Department of Cell Biology and Oncology, Consorzio Mario Negri Sud, Santa Maria Imbaro, Chieti 66030, Italy

MAGDA DENEKA (13), Department of Cell Biology, Utrecht University School of Medicine, Utrecht 3584 CX, The Netherlands

JULIE G. DONALDSON (26), Laboratory of Cell Biology, National Heart, Lung, and Blood Institute, National Institutes of Health, Bethesda, Maryland 20892-0301

MATTHEW T. DRAKE (40), Department of Internal Medicine, Washington University School of Medicine, St. Louis, Missouri 63110

ROCKFORD K. DRAPER (39), Department of Molecular and Cell Biology, The University of Texas at Dallas, Richardson, Texas 75083-0688

LI-LIN DU (11), Department of Molecular Biophysics and Biochemistry, Yale University School of Medicine, New Haven, Connecticut 06520-8002

STEVEN DUNKELBARGER (12), Department of Biochemistry and Molecular Biology, Uniformed Services University of the Health Sciences, Bethesda, Maryland 20814

ARNAUD ECHARD (17), Laboratoire Mécanismes Moléculaires du Transport Intracellulaire, UMR CNRS 144, Institut Curie, Paris Cedex 05, France

AHMED EL MARJOU (17), Service des Protéines Recombinantes, UMR CNRS 144, Institut Curie, Paris Cedex 05, France

YAN FENG (19), Department of Chemistry and Cell Biology, Harvard Medical School, Boston, Massachusetts 02115

SUSAN FERRO-NOVICK (24), Department of Cell Biology, Boyer Center for Molecular Medicine, Howard Hughes Medical Institute, Yale University School of Medicine, New Haven, Connecticut 06510

MICHEL FRANCO (29), Institut de Pharmacologie, Moléculaire et Cellulaire, CNRS UPR 411, Valbonne 06650, France

SCOTT R. FRANK (27), DNAX, Palo Alto, California 94304

JOHANNA FURUHJELM (20), Institute of Biotechnology, PB56, University of Helsinki, Helsinki FIN 00014, Finland

THIERRY GALLI (21), Trafic Membranaire et Plasticité Neuronale, INSERM U536, Institut Curie, Paris Cedex 05, France

DIETER GALLWITZ (6), Department of Molecular Genetics, Max Planck Institute for Biophysical Chemistry, Göttingen D-37070, Germany

JAMES R. GOLDENRING (23), Institute for Molecular Medicine and Genetics, Departments of Medicine, Surgery, Cellular Biology and Anatomy, Medical College of Georgia and Augusta Veterans Affairs Medical Center, Augusta, Georgia 30912-3175

ROGER S. GOODY (3), Department of Physical Biochemistry, Max Planck Institute of Molecular Physiology, Dortmund 44202, Germany

BRUNO GOUD (17), Laboratoire Mécanismes Moléculaires du Transport Intracellulaire, UMR CNRS 144, Institut Curie, Paris Cedex 05, France

A. GUMUSBOGA (16), Department of Cell Biology and Physiology, Washington University School of Medicine, St. Louis, Missouri 63110

WEI GUO (12), Department of Cell Biology, Yale University School of Medicine, New Haven, Connecticut 06520-8002

HISANORI HORIUCHI (15), Department of Geriatric Medicine, Kyoto University Hospital, Kyoto City 606-01, Japan

TONGHUAN HU (39), Department of Molecular Biology, University of Texas Southwestern Medical Center, Dallas, Texas 75390-9148

CHUN-FANG HUANG (43), *Institute of Molecular Medicine College of Medicine, National Taiwan University, Taipei, Taiwan 100, Republic of China*

IRIT HUBER (33, 34), *Department of Biology, Technion-Israel Institute of Technology, Haifa 32000, Israel*

ROBERT TOD HUDSON (39), *Department of Molecular Biology and Microbiology, Case Western Reserve University School of Medicine, Cleveland, Ohio 44106-4960*

WALTER HUNZIKER (22), *Institute for Molecular and Cell Biology, Singapore 117609, Republic of Singapore*

CATHERINE L. JACKSON (31), *Service de Biochimie et Génétique Moléculaire, CEA/ Saclay, Gif-sur-Yvette, Cedex F-91191, France*

TREVOR R. JACKSON (37), *Department of Hematology, Royal Free and University College Medical School, Royal Free Campus, London NW3 2AF, United Kingdom*

GERALD C. JOHNSTON (34), *Department of Microbiology and Immunology, Dalhousie University, Halifax, Nova Scotia, Canada B3H 4H7*

MANDY JONGENEELEN (13), *Department of Cell Biology, Utrecht University School of Medicine, Utrecht 3584 CX, The Netherlands*

JES K. KLARLUND (30), *Ophthalmology and Visual Sciences Research Center, University of Pittsburgh School of Medicine, Pittsburgh, Pennsylvania 15213*

STUART KORNFELD (40), *Department of Internal Medicine, Washington University School of Medicine, St. Louis, Missouri 63110*

NICHOLAS T. KTISTAKIS (38), *Department of Signaling, Babraham Institute, Cambridge CB2 4AG, United Kingdom*

LYNNE A. LAPIERRE (23), *Institute for Molecular Medicine and Genetics, Departments of Medicine, Surgery, Cellular Biology and Anatomy, Medical College of Georgia and Augusta Veterans Affairs Medical Center, Augusta, Georgia 30912-3175*

ANTHONY LEE (48), *Department of Biochemistry and Biophysics, University of Pennsylvania School of Medicine, The Johnson Research Foundation, Philadelphia, Pennsylvania 19104-6059*

FANG-JEN S. LEE (43), *Institute of Molecular Medicine College of Medicine, National Taiwan University, Taipei, Taiwan 100, Republic of China*

MARK A. LEMMON (48), *Department of Biochemistry and Biophysics, University of Pennsylvania School of Medicine, The Johnson Research Foundation, Philadelphia, Pennsylvania 19104-6059*

ROGER LIPPÉ (15), *Max Planck Institute for Molecular Cell Biology and Genetics, European Molecular Biology Laboratory, Heidelberg 69117, Germany*

DANIEL LOUVARD (21), *Morphogenèse et Signalisation Cellulaires, URM 144, Institut Curie, Paris Cedex 05, France*

VARDIT MAKLER (33), *Department of Biology, Technion-Israel Institute of Technology, Haifa 32000, Israel*

WILLIAM A. MALTESE (4), *Department of Biochemistry and Molecular Biology, Medical College of Ohio, Toledo, Ohio 43614-5804*

ANNE-MARIE MARZESCO (21), *Morphogenèse et Signalisation Cellulaires, URM 144, Institut Curie, Paris Cedex 05, France*

JEANNE MATTESON (2), *Departments of Cell and Molecular Biology, The Scripps Research Institute, La Jolla, California 92037*

KOICHI MIURA (37), *Laboratory of Cellular Oncology, Division of Basic Sciences, National Cancer Institute, National Institutes of Health, Bethesda, Maryland 20892*

KARIN MOHRMANN (13), *Department of Cell Biology, Utrecht University School of Medicine, Utrecht 3584 CX, The Netherlands*

JON S. MORROW (42), *Departments of Pathology, and Molecular, Cellular, and Developmental Biology, Yale University, New Haven, Connecticut 06510*

JOEL MOSS (32, 35, 44), *Pulmonary-Critical Care Medicine Branch, National Heart, Lung, and Blood Institute, National Institutes of Health, Bethesda, Maryland 20892*

BRYAN D. MOYER (1, 2), *Departments of Cell and Molecular Biology, The Scripps Research Institute, La Jolla, California 92037*

AMY B. MUHLBERG (47), *Department of Cell Biology, The Scripps Research Institute, La Jolla, California 92037*

FUMIKO NAGANO (8), *Department of Molecular Biology and Biochemistry, Osaka University Graduate School of Medicine/ Faculty of Medicine, Osaka 565-0871, Japan*

HIROYUKI NAKANISHI (7), *Department of Molecular Biology and Biochemistry, Osaka University Graduate School of Medicine/ Faculty of Medicine, Osaka 565-0871, Japan*

JENNIFER NAVARRE (23), *Institute for Molecular Medicine and Genetics, Departments of Medicine, Surgery, Cellular Biology and Anatomy, Medical College of Georgia and Augusta Veterans Affairs Medical Center, Augusta, Georgia 30912-3175*

WALTER NICKEL (41), *Biochemie-Zentrum Heidelberg, Ruprecht-Karls Universität, Heidelberg D-69120, Germany*

PETER NOVICK (11, 12), *Department of Cell Biology, Yale University School of Medicine, New Haven, Connecticut 06520-8002*

SATOSHI ORITA (10), *Discovery Research Laboratories, Shionogi and Company, Limited, Osaka 565-0871, Japan*

JEAN H. OVERMEYER (4), *Department of Biochemistry and Molecular Biology, Medical College of Ohio, Toledo, Ohio 43614-5804*

GUSTAVO PACHECO-RODRIGUEZ (32, 44), *Pulmonary-Critical Care Medicine Branch, National Heart, Lung, and Blood Institute, National Institutes of Health, Bethesda, Maryland 20892*

XIAO-RONG PENG (52), *Programme in Cell Biology, Hospital for Sick Children, De-partment of Biochemistry, University of Toronto, Toronto, Ontario, Canada M5G 1X8*

JOHAN PERÄNEN (20), *Institute of Biotechnology, PB56, University of Helsinki, Helsinki FIN 00014, Finland*

PETER J. PETERS (22), *Dutch Cancer Research Institute, Amsterdam, The Netherlands*

ANNE PEYROCHE (31), *Service de Biochimie et Génétique Moléculaire, CEA/Saclay, Gif-sur-Yvette, Cedex F-91191, France*

ELAH PICK (33), *Department of Biology, Technion-Israel Institute of Technology, Haifa 32000, Israel*

PAK PHI POON (34), *Departments of Microbiology and Immunology, Biochemistry and Molecular Biology, Dalhousie University, Halifax, Nova Scotia, Canada B3H 4H7*

RICHARD T. PREMONT (36), *Department of Medicine, Duke University Medical Center, Durham, North Carolina 27710*

BARRY PRESS (19), *Dana Farber Cancer Institute, Harvard Medical School, Boston, Massachusetts 02115*

HARISH RADHAKRISHNA (26), *Department of Biology, Georgia Institute of Technology, Atlanta, Georgia 30332-0363*

PAUL A. RANDAZZO (37), *Laboratory of Cellular Oncology, Division of Basic Sciences, National Cancer Institute, National Institutes of Health, Bethesda, Maryland 20892*

RICHARD L. ROBERTS (16), *Department of Cell Biology and Physiology, Washington University School of Medicine, St. Louis, Missouri 63110*

SYLVIANE ROBINEAU (28), *Institut de Pharmacologie Moléculaire et Cellulaire, CNRS, Valbonne 06560, France*

MICHAEL G. ROTH (38), *Department of Biochemistry, University of Texas Southwestern Medical Center, Dallas, Texas 75235*

LILAH ROTHEM (33), *Department of Biology, Technion-Israel Institute of Technology, Haifa 32000, Israel*

MIRIAM ROTMAN (33), *Department of Biology, Technion-Israel Institute of Technology, Haifa 32000, Israel*

ANJA RUNGE (15), *Max Planck Institute for Molecular Cell Biology and Genetics, European Molecular Biology Laboratory, Heidelberg 69117, Germany*

MICHAEL SACHER (24), *Department of Cell Biology, Boyer Center for Molecular Medicine, Yale University School of Medicine, New Haven, Connecticut 06510*

LORRAINE C. SANTY (27), *Department of Cell Biology, University of Virginia, Health Sciences Center, Charlottesville, Virginia 22908*

TAKUYA SASAKI (8, 9, 10), *Department of Biochemistry, Tokushima University School of Medicine, Tokushima 770-8503, Japan*

AXEL J. SCHEIDIG (3), *Department of Physical Biochemistry, Max Planck Institute for Molecular Physiology, Dortmund 44202, Germany*

SANDRA L. SCHMID (47), *Department of Cell Biology, The Scripps Research Institute, La Jolla, California 92037*

SANJA SEVER (47), *Department of Cell Biology, The Scripps Research Institute, La Jolla, California 92037*

HIROMICHI SHIRATAKI (9), *Division of Molecular and Cell Biology, Institute for Medical Science, Dokkyo University School of Medicine, Mibu 321-0293, Japan*

ASSIA SHISHEVA (5), *Department of Physiology, Wayne State University School of Medicine, Detroit, Michigan 48201*

STEVEN SHOLLY (47), *Department of Cell Biology, The Scripps Research Institute, La Jolla, California 92037*

DIXIE-LEE SHURLAND (49), *Department of Biological Chemistry, University of California, School of Medicine, Los Angeles, California 90095-1737*

RICHARD A. SINGER (34), *Department of Biochemistry and Molecular Biology, Dalhousie University, Halifax, Nova Scotia, Canada B3H 4H7*

VLADIMIR I. SLEPNEV (50), *Department of Cell Biology, Yale University School of Medicine, New Haven, Connecticut 06510*

ELENA SMIRNOVA (49), *Department of Biological Chemistry, University of California, School of Medicine, Los Angeles, California 90095-1737*

PHILIP D. STAHL (16), *Department of Cell Biology and Physiology, Washington University School of Medicine, St. Louis, Missouri 63110*

YOSHIMI TAKAI (7, 8, 9, 10), *Department of Molecular Biology and Biochemistry, Osaka University Medical School, Osaka 700-8558, Japan*

KOHJI TAKEI (50), *Department of Biochemistry, Okayama University School of Medicine, Okayama-shi, Okayama 700-8558, Japan*

DANIEL R. TERBUSH (12), *Department of Biochemistry and Molecular Biology, Uniformed Services University of the Health Sciences, Bethesda, Maryland 20814*

WILLIAM S. TRIMBLE (52), *Programme in Cell Biology, Hospital for Sick Children, Department of Biochemistry, University of Toronto, Toronto, Ontario, Canada M5G 1X8*

ALEXANDER M. VAN DER BLIEK (49), *Department of Biological Chemistry, University of California, School of Medicine, Los Angeles, California 90095-1737*

PETER VAN DER SLUIJS (13), *Department of Cell Biology, Utrecht University School of Medicine, Utrecht 3584 CX, The Netherlands*

MARTHA VAUGHAN (32, 35, 44), *Pulmonary-Critical Care Medicine Branch, National Heart, Lung, and Blood Institute, National Institutes of Health, Bethesda, Maryland 20892*

NICOLAS VITALE (35, 36), *Center de Neurochimie, INSERM U-338, Strasbourg, Cedex 67084, France*

ANGELA WANDINGER-NESS (18, 19), *Department of Pathology, University of New Mexico Health Sciences Center, Albuquerque, New Mexico 87131*

DALE E. WARNOCK (47), *Department of Cell Biology, The Scripps Research Institute, La Jolla, California 92037*

JACQUES T. WEISSMAN (45), *Department of Cell Biology, The Scripps Research Institute, La Jolla, California 92037*

FELIX T. WIELAND (41), *Biochemie-Zentrum Heidelberg, Ruprecht-Karls Universität, Heidelberg D-69120, Germany*

ELKE WILL (6), *Department of Molecular Genetics, Max Planck Institute for Biophysical Chemistry, Göttingen D-37070, Germany*

AHMED ZAHRAOUI (21), *Morphogenèse et Signalisation Cellulaires, URM 144, Institut Curie, Paris Cedex 05, France*

MARINO ZERIAL (14, 15), *Max Planck Institute for Molecular Cell Biology and Genetics, European Molecular Biology Laboratory, Heidelberg 69117, Germany*

YUNXIANG ZHU (40), *Department of Internal Medicine, Washington University School of Medicine, St. Louis, Missouri 63110*

JAY ZIMMERMAN (19), *University of Chicago, Pritzker School of Medicine, Chicago, Illinois 60610*

Preface

GTPases are now recognized to regulate many different steps in membrane vesicular transport. They are involved in the assembly of vesicle coats (budding), movement along cytoskeletal elements, and in vesicle targeting and in fusion. They are clearly a key group of regulatory proteins that control transport through both the exocytic and endocytic pathways. GTPases involved in membrane transport include the Rab and ARF families, Sar1, and dynamin. Because these GTPases are switches, they function by either responding to or controlling the activity of a range of upstream and downstream effectors. These include posttranslational modifying enzymes (such as prenyltransferases and myristyltransferases), factors which effect guanine nucleotide binding [guanine nucleotide dissociation inhibitors (GDIs) and guanine nucleotide exchange factors (GEFs)], and factors which stimulate guanine nucleotide hydrolysis [GTPase-activating proteins (GAPs)]. Moreover, they may also interact with motors and structural elements dictating vesicle and organelle function.

The number of identified effectors directing or responding to transport GTPases is expanding rapidly. The purpose of this volume is to bring together the latest technologies that have developed over the past 5 years to study their function. Because each family contains a variety of isoforms, the techniques described for a particular GTPase family member are likely to be useful for other members of the same family. Moreover, the underlying conserved structural fold suggests that each of the various techniques are also applicable to other members of the larger superfamily of Ras-like GTPases. Given the abundance of both Rab and ARF GTPases and the intense interest of the cell biology community in their function, we have provided short editorial overviews for these two sections that describe the central features of their function and structural organization.

We are extremely grateful to the many investigators who have generously contributed their time and expertise to bring this wealth of technical experience into one volume. It should provide a valuable resource to address the many issues confronting our understanding of the role of these GTPases in cell biology.

WILLIAM E. BALCH
CHANNING J. DER
ALAN HALL

METHODS IN ENZYMOLOGY

VOLUME 282. Vitamins and Coenzymes (Part L)
Edited by DONALD B. MCCORMICK, JOHN W. SUTTIE, AND CONRAD WAGNER

VOLUME 283. Cell Cycle Control
Edited by WILLIAM G. DUNPHY

VOLUME 284. Lipases (Part A: Biotechnology)
Edited by BYRON RUBIN AND EDWARD A. DENNIS

VOLUME 285. Cumulative Subject Index Volumes 263, 264, 266–284, 286–289

VOLUME 286. Lipases (Part B: Enzyme Characterization and Utilization)
Edited by BYRON RUBIN AND EDWARD A. DENNIS

VOLUME 287. Chemokines
Edited by RICHARD HORUK

VOLUME 288. Chemokine Receptors
Edited by RICHARD HORUK

VOLUME 289. Solid Phase Peptide Synthesis
Edited by GREGG B. FIELDS

VOLUME 290. Molecular Chaperones
Edited by GEORGE H. LORIMER AND THOMAS BALDWIN

VOLUME 291. Caged Compounds
Edited by GERARD MARRIOTT

VOLUME 292. ABC Transporters: Biochemical, Cellular, and Molecular Aspects
Edited by SURESH V. AMBUDKAR AND MICHAEL M. GOTTESMAN

VOLUME 293. Ion Channels (Part B)
Edited by P. MICHAEL CONN

VOLUME 294. Ion Channels (Part C)
Edited by P. MICHAEL CONN

VOLUME 295. Energetics of Biological Macromolecules (Part B)
Edited by GARY K. ACKERS AND MICHAEL L. JOHNSON

VOLUME 296. Neurotransmitter Transporters
Edited by SUSAN G. AMARA

VOLUME 297. Photosynthesis: Molecular Biology of Energy Capture
Edited by LEE MCINTOSH

VOLUME 298. Molecular Motors and the Cytoskeleton (Part B)
Edited by RICHARD B. VALLEE

VOLUME 299. Oxidants and Antioxidants (Part A)
Edited by LESTER PACKER

VOLUME 300. Oxidants and Antioxidants (Part B)
Edited by LESTER PACKER

VOLUME 301. Nitric Oxide: Biological and Antioxidant Activities (Part C)
Edited by LESTER PACKER

Section I

Rab GTPases

[1] Structural Basis for Rab Function: An Overview

By BRYAN D. MOYER and WILLIAM E. BALCH

Rab proteins, members of the Ras superfamily of low molecular weight GTP-binding proteins (~20–25 kDa), modulate tubulovesicular trafficking between compartments of the biosynthetic and endocytic pathways.[1–3] Similar to Ras, Rab GTPases cycle between active, GTP-bound and inactive, GDP-bound states.[4] This brief introductory chapter summarizes Rab structure–function relationships in the context of membrane trafficking and serves as a prelude for the accompanying chapters, which describe specific methods for elucidating Rab function.

The unifying theme in research elucidating Rab structure–function relationships has been the Rab GTPase cycle model (Fig. 1) (reviewed in Refs. 1, 2, and 4). In the cytosol, Rab proteins are maintained in the GDP-bound state by interaction with a GDP dissociation inhibitor (GDI).[5] GDI delivers Rab–GDP to donor membranes where GDI may be displaced by a GDI displacement factor (GDF).[6] Subsequently, a guanine nucleotide exchange factor (GEF) is believed to stimulate exchange of GDP for GTP.[7,8] Transport intermediates containing activated Rab bud from donor membranes, where Rab–GTP recruits effector molecules required for trafficking to acceptor compartments.[1,3] Recent studies suggest that Rab effectors regulate the motility of transport intermediates along cytoskeletal elements and mediate the docking/fusion of transport intermediates with acceptor membranes.[9–12] Prior to or concomitant with membrane docking and fusion, a GTPase activating protein (GAP) is thought to stimulate Rab-mediated hydrolysis of GTP to GDP and recruited effector molecules dissociate from

[1] J. S. Rodman and A. Wandinger-Ness, *J. Cell Sci.* **113**, 183 (2000).
[2] O. Martinez and B. Goud, *Biochim. Biophys. Acta* **1404**, 101 (1998).
[3] F. Schimmöller, I. Simon, and S. R. Pfeffer, *J. Biol. Chem.* **273**(35), 22161 (1998).
[4] V. M. Olkkonen and H. Stenmark, *Int. Rev. Cytol.* **176**, 1 (1997).
[5] S.-K. Wu, K. Zeng, I. A. Wilson, and W. E. Balch, *Trends Biochem. Sci.* **21**, 472 (1996).
[6] A. B. Dirac-Svejstrup, T. Sumizawa, and S. R. Pfeffer, *EMBO J.* **16**(3), 465 (1997).
[7] T. Soldati, A. D. Shapiro, A. B. D. Svejstrup, and S. R. Pfeffer, *Nature* **369**, 76 (1994).
[8] O. Ullrich, H. Horiuchi, C. Bucci, and M. Zerial, *Nature* **368**, 157 (1994).
[9] A. Echard, F. Jollivet, O. Martinez, J.-J. Lacapère, A. Rousselet, I. Janoueix-Lerosey, and B. Goud, *Science* **279**, 580 (1998).
[10] S. Christoforids, H. M. McBride, R. D. Burgoyne, and M. Zerial, *Nature* **397**, 621 (1999).
[11] S. R. Pfeffer, *Nat. Cell Biol.* **1**, E17 (1999).
[12] M. G. Waters and S. R. Pfeffer, *Curr. Opin. Cell Biol.* **11**, 453 (1999).

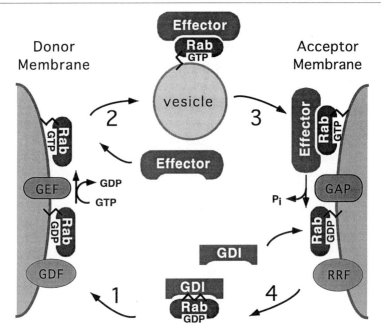

FIG. 1. Model of the Rab GTPase cycle. Rab–GDP/GDI complexes are targeted to donor membranes where GDF displaces GDI and a GEF stimulates Rab–GDP/GTP exchange (step 1). Transport intermediates containing Rab–GTP bud from donor membranes. Rab–GTP recruits effector molecules, which mediate the migration, docking, and fusion of transport intermediates to acceptor membranes (step 2). GTP is hydrolyzed to GDP by a GAP and effector molecules dissociate from Rab (step 3). GDI is recruited by RRF and extracts Rab–GDP from acceptor membranes for initiation of another round of the Rab GTPase cycle (step 4). See text for complete details. GDI, GDP dissociation inhibitor; GDF, GDI displacement factor; GEF, guanine nucleotide exchange factor; GAP, GTPase activating protein; RRF, Rab recycling factor.

Rab.[13] GDI, recruited to membranes by a putative Rab recycling factor (RRF),[14] then extracts Rab–GDP from acceptor membranes and the Rab–GDP/GDI complex recycles to donor membranes for initiation of another round of transport.

Rab GTPases contain conserved and unique sequence elements that mediate function, including GDP/GTP binding, subcellular targeting, and

[13] V. Rybin, O. Ullrich, M. Rubino, K. Alexandrov, I. Simon, M. C. Seabra, R. Goody, and M. Zerial, *Nature* **383,** 266 (1996).
[14] P. Luan, W. E. Balch, S. D. Emr, and C. G. Burd, *J. Biol. Chem.* **274**(21), 14806 (1999).

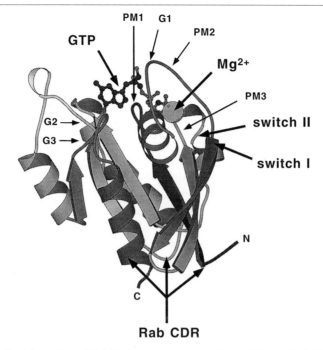

FIG. 2. Crystal structure of Rab3A. Pertinent structure elements that mediate Rab function are labeled and discussed in the text. N, N terminus; C, C terminus (site of geranylgeranyl lipid addition); Rab CDR, Rab complimentarity-determining region (site for specific effector binding); G1–G3, guanine base-binding motifs; PM1–3: phosphate/magnesium-binding motifs; GTP, guanosine triphosphate; Mg²⁺, divalent magnesium ion; switch I–II: regions undergoing large conformational changes during GTP binding and hydrolysis.

effector recognition (Fig. 2).[4] Rab proteins contain three highly conserved guanine base-binding motifs (termed G1 to G3) which mediate guanine nucleotide binding and three highly conserved phosphate/magnesium-binding motifs (termed PM1 to PM3), which bind and coordinate a divalent magnesium ion with the β- and γ-phosphates of GTP. On GAP-stimulated hydrolysis of GTP to GDP and loss of the terminal phosphate group, two regions in spatial proximity to the γ-phosphate, termed switch I (also called the effector domain) and switch II, undergo dramatic conformational changes, which result in reduced affinity for bound effector molecules and Rab inactivation. During Rab reactivation, GEF-stimulated conformational changes in the switch I, switch II, and P loop regions facilitate extrusion of GDP and incorporation of GTP.

The C-terminal regions of Rab proteins are highly divergent and contain two structural elements dictating function. First, the extreme C termini contain a conserved cysteine-based motif, which is posttranslationally modified by a geranylgeranyl lipid group and is required for interaction of Rab proteins with both GDI and membranes. Immediately upstream of the cysteine-based motif is a hypervariable region, which contains information directing Rab proteins to specific subcellular membranes.[4]

Determination of the crystal structure of Rab3A–GTP, alone or complexed with the effector molecule Rabphilin-3A, provides new insight into the structural basis for Rab-effector specificity.[15,16] Rab3A contacts Rabphilin-3A at two positions—first in the conserved switch I and switch II regions and second in nonconserved regions at the N terminus, central region, and C terminus. These later hypervariable regions coalesce into a deep pocket termed a Rab complementarity-determining region (Rab CDR). Because Rab CDRs are not conserved between family members, they are proposed to determine the specific interaction between individual GTP-bound Rab proteins and their effectors.[15] We are currently at a pivotal point in our understanding of the Rab GTPase family, which is now comprised of more than 40 members.[1,4] Genetic and biochemical methodologies are rapidly revealing the identity of novel Rab effector molecules. However the function of these effector molecules, in many instances, remains to be determined.

---

[15] C. Ostermeier and A. T. Brunger, *Cell* **96**, 363 (1999).
[16] J. J. Dumas, Z. Zhu, J. L. Connolly, and D. G. Lambright, *Structure* **7**, 413 (1999).

[2] Expression of Wild-Type and Mutant Green Fluorescent Protein–Rab1 for Fluorescence Microscopy Analysis

By Bryan D. Moyer, Jeanne Matteson, and William E. Balch

Introduction

The green fluorescent protein (GFP) has emerged as an important reporter molecule for studying complex biological processes such as organelle dynamics and protein trafficking.[1-4] GFP, a 238 amino acid ($\sim$27 kDa)

---

[1] M. Chalfie, Y. Tu, G. Euskirchen, W. W. Ward, and D. C. Prasher, *Science* **263**, 802 (1994).
[2] H.-H. Gerdes and C. Kaether, *FEBS Lett.* **389**, 44 (1996).
[3] J. Lippincott-Schwartz and C. L. Smith, *Curr. Biol.* **7**, 631 (1997).
[4] J. Lippincott-Schwartz, N. Cole, and J. Presley, *Trends Cell Biol.* **8**, 16 (1998).

protein from the jellyfish *Aequorea victoria,* generates a striking green fluorescence when viewed with conventional fluorescein isothiocyanate (FITC) optics, is visible in both living and fixed specimens, is resistant to photobleaching, does not require any exogenous cofactors or substrates (with the exception of molecular oxygen) to fluoresce, and, when ligated to other proteins, generally does not alter fusion protein function or localization.[5,6] Visualization of membrane dynamics in cells expressing GFP fusion proteins has recently revealed novel pathways and mechanisms of anterograde and retrograde endoplasmic reticulum (ER)–Golgi transport.[7–10] As part of our long-term goal to elucidate the molecular mechanism(s) by which Rab1 regulates ER to Golgi transport, we have generated fluorescent chimeric proteins in which GFP was ligated to wild-type (wt) or mutant forms of Rab1 and determined the subcellular distribution of GFP–Rab1 fusion proteins by fluorescence microscopy. By using GFP fluorescence as a marker for Rab1 localization, we avoid artifacts that might be introduced when studying Rab1 trafficking in cells stained with antibodies, such as the generation of false signals due to nonspecific antibody binding and the introduction of structural artifacts due to cell permeabilization.[11] This chapter describes the methodology we have found to work best in our laboratory for the expression and visualization of GFP–Rab1 fusion proteins in mammalian cells.

Preparation of Recombinant Expression Vectors

For transient GFP–Rab1 expression, we use the pET expression system.[12] Site-directed mutagenesis and restriction site introduction are performed using standard molecular biology techniques, and recombinant expression plasmids are purified using Qiagen plasmid kits (Qiagen, Valencia, CA). Humanized GFP, which has been codon optimized for expression

[5] A. B. Cubitt, R. Heim, S. R. Adams, A. E. Boyd, L. A. Gros, and R. Y. Tsien, *TIBS* **20,** 448 (1995).

[6] R. Y. Tsien, *Annu. Rev. Biochem.* **67,** 509 (1998).

[7] N. Nishimura, S. Bannykh, S. Slabough, J. Matteson, Y. Altschuler, K. Hahn, and W. E. Balch, *J. Biol. Chem.* **274**(22), 15937 (1999).

[8] J. White, L. Johannes, F. Mallard, A. Girod, S. Grill, S. Reinsch, P. Keller, B. Tzschaschel, A. Echard, B. Goud, and E. H. K. Stelzer, *J. Cell Biol.* **147**(4), 743 (1999).

[9] S. J. Scales, R. Pepperkok, and T. E. Kreis, *Cell* **90,** 137 (1997).

[10] J. F. Presley, N. B. Cole, T. A. Schroer, K. Hirschberg, K. J. M. Zaal, and J. Lippincott-Schwartz, *Nature* **389,** 81 (1997).

[11] G. Griffiths, R. G. Parton, J. Lucocq, B. Van Deurs, D. Brown, J. W. Slot, and H. J. Geuze, *Trends Cell Biol.* **3,** 214 (1993).

[12] F. W. Studier, A. H. Rosenberg, J. J. Dunn, and J. W. Dubendorff, *Methods Enzymol.* **185,** 60 (1990).

in mammalian cells and which contains the S65T mutation for increased fluorescence intensity,[13] was amplified by the polymerase chain reaction (PCR) using *Pfu* polymerase (Stratagene, La Jolla, CA), a proofreading enzyme that dramatically reduces base-misincorporation during PCR amplification. DNA sequence analysis confirmed that no errors were introduced during the PCR reaction. GFP primers contained *Nco*I (5′ sense primer) and *Nde*I (3′ antisense primer) sites to facilitate cloning. Rab1 cDNAs were digested from pET3c vector (Novagen, Madison, WI) using *Nde*I (5′ site) and *Bam*HI (3′ site). pET11d vector (Novagen) was digested with NcoI (5′ site) and *Bam*HI (3′ site) and combined with *Nco*I–*Nde*I GFP and *Nde*I–*Bam*HI Rab1 fragments in a three-piece ligation. Proceeding from the N to the C terminus, the resultant fusion protein consists of GFP followed by Rab1. GFP was deliberately positioned at the N terminus of Rab1 so as not to interfere with posttranslational addition of the C-terminal geranylgeranyl lipid group. Only a single extraneous amino acid (His), carried over from the *Nde*I restriction site, is positioned between GFP and Rab1 coding sequences.

Infection and Transfection

To transiently express GFP-Rab1 fusion proteins in mammalian cells, we use the recombinant T7 vaccinia virus system.[14,15] Expression is achieved by transfecting recombinant GFP-Rab1 plasmids containing the T7 RNA polymerase promoter into cells infected with recombinant vaccinia virus vTF7-3, which has been engineered to express the bacteriophage T7 RNA polymerase gene. This system is an effective tool for the rapid and transient expression of Rab1 proteins with altered guanine nucleotide binding properties, and has been used successfully by our laboratory to demonstrate a block in ER to Golgi transport by dominant negative Rab1 mutants.[16,17]

BHK-21 or HeLa cells may be used for transient expression studies. Cells are maintained in Dulbecco's modified Eagle's medium (DMEM) supplemented with 10% fetal bovine serum and 100 U/ml of penicillin and streptomycin at 37° in a humidified incubator (95% air/5% CO_2, v/v). Crude stocks of partially purified recombinant vTF7-3 vaccinia virus are generated

[13] R. Heim, A. B. Cubitt, and R. Y. Tsien, *Nature* **373,** 663 (1995).
[14] T. R. Fuerst, E. G. Niles, F. W. Studier, and B. Moss, *Proc. Natl. Acad. Sci. U.S.A.* **83,** 8122 (1986).
[15] C. Dascher, E. J. Tisdale, and W. E. Balch, *Methods Enzymol.* **257,** 165 (1995).
[16] E. J. Tisdale, J. R. Bourne, R. Khosravi-Far, C. J. Der, and W. E. Balch, *J. Cell Biol.* **119**(4), 749 (1992).
[17] C. Nuoffer, H. W. Davidson, J. Matteson, J. Meinkoth, and W. E. Balch, *J. Cell Biol.* **125**(2), 225 (1994).

as previously described.[18] Although vaccinia virus is relatively harmless unless it comes into direct contact with the eye, it is classified as a human pathogen and should be treated with caution. Vaccinations are available for laboratory personnel and biosafety level 2 guidelines must be followed. Our laboratory has dedicated one tissue culture hood specifically for vaccinia virus use. We always wear proper personal protective equipment, including goggles, a lab coat, and two pairs of gloves, and thoroughly rinse materials that have contacted vaccinia virus in 25% bleach. Whenever exiting the hood, we rinse the outer pair of gloves with 10% bleach and leave them in the hood. At the end of each experiment, the hood is exposed to UV light for 30 min to inactivate residual virus.

Materials

vTF7-3 virus stock
BHK-21 or HeLa cells (1- to 2-day-old culture at 60–80% confluency and growing on sterile No. 1 thickness glass coverslips in 35-mm dishes or 6-well plates)
1.5 μg Qiagen-purified GFP–Rab1 plasmid DNA
LipofectAMINE PLUS Reagent (Life Technologies, Rockville, MD)
Opti-MEM1 serum-free media (Life Technologies, Rockville, MD)
Bath sonicator
Hemacytometer
Falcon 2057 tubes (Becton Dickinson, Franklin Lakes, NJ)

Method

1. Determine the number of cells on a single coverslip using a hemacytometer.
2. Calculate the volume of virus required to achieve a desired multiplicity of infection (MOI), measured in plaque-forming units (pfu), by the following formula:

$$\frac{(\text{No. of cells/coverslip}) \times (\text{desired pfu/cell}) \times (10^3 \, \mu\text{l/ml})}{\text{viral titer (pfu/ml)}} = \mu\text{l virus/coverslip} \quad (1)$$

We work with vaccinia virus preparations with titers between 10^9 and 10^{10} pfu/ml and infect at an MOI of 15.

3. Thaw the virus stock on ice and sonicate at 4° for 10 sec to disrupt viral aggregates. Immediately place on ice for 30 sec to cool and repeat sonication once. Dilute the desired amount of virus (calculated

[18] C. Dascher, J. K. VanSlyke, L. Thomas, W. E. Balch, and G. Thomas, *Methods Enzymol.* **257**, 174 (1995).

in step 2) in 0.5 ml Opti-MEM1. Unused viral stock may be stored at 4° for 1–2 months or refrozen at −80° and should be resonicated prior to each use.

4. Wash cells twice with Opti-MEM1 (1 ml/wash) and overlay with 0.5 ml of diluted virus from step 3.

5. Let infection proceed for 30 min at room temperature while manually rocking the dish every 5–10 min to ensure that cells remain submerged in infection inoculum.

6. Prepare the DNA/lipid transfection mixture as outlined below:
   (a) Dilute 1.5 μg DNA and 7 μl LipofectAMINE PLUS reagent into a final volume of 100 μl Opti-MEM1 in a sterile Falcon 2057 tube. Vortex gently to mix and incubate for 15 min at room temperature.
   (b) Dilute 7 μl LipofectAMINE reagent into a final volume of 100 μl Opti-MEM1 in a second Falcon 2057 tube and vortex gently to mix (total volume is 200 μl now).
   (c) Combine DNA/LipofectAMINE PLUS and LipofectAMINE solutions, vortex briefly to mix, and incubate for 15 min at room temperature to allow DNA–liposome complexes to form.

7. Following 30 min of viral infection, aspirate the viral inoculum and wash the cells twice with Opti-MEM1.

8. Dilute the DNA–lipid mixture with 0.8 ml Opti-MEM1 and add to infected cells (total volume is 1.0 ml now).

9. Transfer cells to a 37° incubator and let transfection proceed for 4.5–6 hr.

Comment

It is essential to thoroughly vortex the LipofectAMINE transfection reagent before use to resuspend lipids that settle during storage. Using the transfection protocol outlined above, we routinely achieve transfection efficiencies of 50–60% and get 5- to 20-fold overexpression of protein compared to the endogenous pool, as determined by immunoblotting. Transfection efficiencies decrease with increasing plasmid size and DNA/ lipid ratios should be optimized for each cell line examined. Because cell morphology deteriorates over time as a result of vaccinia virus replication, infection and transfection solutions may be supplemented with hydroxyurea (Sigma, St. Louis, MO) (10 m*M*), an inhibitor of viral DNA replication, to preserve cell adherence and shape for subsequent microscopic examination.[19]

[19] C. Bucci, R. G. Parton, I. H. Mather, H. Stunnenberg, K. Simons, B. Hoflack, and M. Zerial, *Cell* **70**, 715 (1992).

Fixation and Mounting

GFP fluorescence is stable to fixation in formaldehyde, methanol, or acetone. However, because nonaldehyde fixatives allow cytosolic proteins to diffuse from cells and because Rab proteins cycle between membrane-bound and cytosolic pools, we use formaldehyde to fix cells expressing GFP–Rab1 fusion proteins. Formaldehyde fixation ensures that both membrane-bound and cytosolic pools of Rab1 proteins are cross-linked and retained for subsequent visualization.

Materials

PBS (phosphate-buffered saline: 137 mM NaCl, 2.7 mM KCl, 1.8 mM KH$_2$PO$_4$, and 8.1 mM Na$_2$HPO$_4$ at pH$_{final}$ 7.4)
10% formaldehyde (Polysciences, Inc., Warrington, PA, methanol free Ultrapure EM grade)
Aqua PolyMount mounting medium (Polysciences, Inc., Warrington, PA)
Nail Polish (Wet 'n' Wild 401 Clear Nail Protector; see "Comments" below for alternatives)
Forceps (very fine point, VWR Scientific Products, West Chester, PA)
Kimwipes or tissues
Glass slides

Method

1. Following 4.5–6 hr of transfection, wash cells three times with PBS. All steps may be performed at room temperature.
2. Fix cells with 2% formaldehyde in PBS for 10 min at room temperature.
3. Wash cells twice with PBS to remove formaldehyde. Following fixation, it is safe to work outside the biosafety hood and process the cells at a laboratory bench.
4. Carefully pick up coverslip with forceps and wick away excess fluid using a Kimwipe or tissue.
5. Place coverslip (cells facing down) onto a drop (30–40 μl) of Aqua PolyMount mounting medium on a glass slide.
6. Seal coverslip with nail polish and let harden at room temperature for 20–30 min before viewing.

Comment

Double immunofluorescence can be used to compare the distribution of a specific marker protein with GFP–Rab1 following standard methods. Since the excitation and emission spectra of GFP are similar to other green fluorophores such as FITC or Alexa 488 (Molecular Probes, Eugene, OR), it is necessary to use a secondary antibody coupled to Texas Red, or another fluorophore whose spectra do not overlap with GFP, when performing double-labeling studies. As an alternative to nail polish, which has been reported to quench GFP fluorescence,[1] Valap may be used to seal coverslips. Valap is a wax-based material made by mixing Vasoline, lanolin, and paraffin in a 1:1:1 ratio and liquefies when heated to 40–50°. On cooling, Valap solidifies and creates a hardened, air- and watertight seal around the specimen. As another alternative, ProLong antifade reagent (Molecular Probes) may be used instead of Aqua PolyMount to mount coverslips. ProLong-mounted cells do not require nail polish or Valap to seal coverslips; instead, ProLong requires a period of several hours to cure before cells can be visualized. Neither ProLong nor Valap have been reported to affect GFP fluorescence. If GFP fluorescence is dim, cells can be costained with commercially available GFP antibodies (Clontech, Palo Alto, CA) followed by secondary FITC or Alexa 488-coupled antibodies. By combining endogenous GFP fluorescence with exogenous GFP immunofluorescence, GFP signals can be artificially enhanced.

Fluorescence Microscopy

To visualize GFP–Rab1 fusion proteins, we use a Zeiss Axiovert 100 TV inverted microscope and either a 63X Plan-Apochromat/1.4 NA or 100X Plan-Neofluar/1.3 NA oil immersion objective (Carl Zeiss, Thornwood, NY). GFP fluorescence is excited using a 50-W mercury arc lamp and collected using the following filter set: excitation filter 485-nm bandpass, dichroic beamsplitter 510-nm longpass, and emission filter 540 ± 25-nm bandpass. Filter sets specifically matched to GFP excitation and emission spectra are also available (Chroma Optical Inc., Brattleboro, VT). Images are acquired using the Spot Digital Camera System and Spot 32 v2.1 software (Diagnostic Instruments, Sterling Heights, MI), imported into the hard drive of a computer (Datel Systems, Mansfield, MA) containing an Intel Pentium II processor (300 MHz) and a P6B40-A4X mainboard, and displayed on a Trinitron Multiscan 200 ES monitor (Sony Corp., Kansas City, MO). Images are saved in TIFF format and imported into Adobe Photoshop v5.0 for processing and printing.

The subcellular localization of various GFP–Rab1 fusion proteins expressed and visualized using the methods described above is shown in Fig. 1. In Fig. 1A wild-type (wt)-Rab1 is localized to the ER, peripheral punctate vesicular tubular clusters (VTCs), and the perinuclear Golgi apparatus. Double-labeling experiments using mannosidase II antibodies to label the Golgi apparatus and syntaxin 5 antibodies to label VTCs confirmed the localization of GFP–wt-Rab1 to these organelles. Similar results have

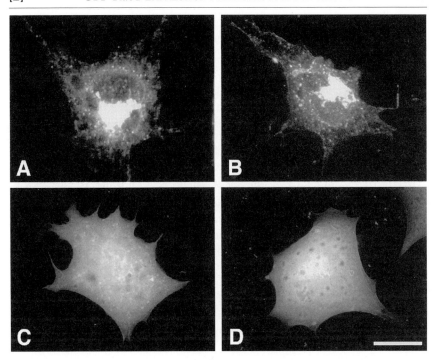

FIG. 1. Fluorescence micrographs of BHK-21 cells transiently expressing GFP–Rab1 fusion proteins. Cells were infected with vTF7-3 and transfected with recombinant GFP–Rab1 expression plasmids as described. After 5 hr of transfection, cells were fixed, mounted, and viewed by fluorescence microscopy. (A) GFP–wt-Rab1 (ER, punctate VTCs, and perinuclear Golgi region). (B) GFP–Rab1-Q67L (ER, punctate VTCs, and perinuclear Golgi region). (C) GFP–Rab1-S25N (VTCs/fragmented Golgi membranes, ER, and cytoplasm). (D) GFP–Rab1-N124I (cytoplasm). Bar: 10 μm.

been reported for wt-Rab1 without a GFP tag, demonstrating that GFP does not affect Rab1 localization.[20] In Fig. 1B Rab1-Q67L, a constitutively active mutant locked in the GTP-bound conformation and which does not affect ER to Golgi transport,[16] is localized to the ER, VTCs, and Golgi apparatus. In Fig. 1C Rab1-S25N, a constitutively inactive mutant locked in the GDP-bound conformation and which inhibits ER to Golgi transport at an early step,[17] is localized to punctate structures, which likely correspond to VTCs and fragmented Golgi membranes, ER membranes, and the cytoplasm. Figure 1D shows Rab1-N124I, a mutant which cannot stably bind guanine nucleotide and which inhibits ER to Golgi transport by blocking

[20] H. Plutner, A. D. Cox, S. Pind, R. Khosravi-Far, J. R. Bourne, R. Schwaninger, C. J. Der, and W. E. Balch, *J. Cell Biol.* **115**(1), 31 (1991).

fusion of ER-derived transport intermediates with Golgi membranes,[21] localized to the cytoplasm with minimal membrane localization.

Conclusion

The methods described in this chapter for the expression and visualization of GFP-Rab1 proteins by fluorescence microscopy are both rapid and straightforward. Cells can be infected with vaccinia virus, transfected with GFP–Rab1 plasmid, fixed, and viewed in a single day. The main advantage of using GFP as a reporter for Rab1 localization is the elimination of artifacts that might be introduced when examining Rab1 localization in permeabilized cells stained with antibodies. Similar to other Rab GTPases, fusion of GFP to Rab1 does not affect Rab1 subcellular localization.[8,22,23] Thus, GFP may be useful for determining the subcellular distributions of all members of the Rab GTPase family, which is now comprised of more than 40 proteins.[24] Looking to the future of Rab research, simultaneous visualization of two proteins in living cells in real time is now possible using different GFP color variants.[8] This methodology can now be used to elucidate the precise spatiotemporal relationships between Rab and Rab effectors during membrane trafficking in the living cell.

[21] S. N. Pind, C. Nuoffer, J. M. McCaffery, H. Plutner, H. W. Davidson, M. G. Farquhar, and W. E. Balch, *J. Cell Biol.* **125**(2), 239 (1994).

[22] R. L. Roberts, M. A. Barbieri, K. M. Pryse, M. Chua, J. H. Morisaki, and P. D. Stahl, *J. Cell Sci.* **112**, 3667 (1999).

[23] M. A. Barbieri, S. Hoffenberg, R. Roberts, A. Mukhopadhyay, A. Pomrehn, B. F. Dickey, and P. D. Stahl, *J. Biol. Chem.* **273**(40), 25850 (1998).

[24] V. M. Olkkonen and H. Stenmark, *Int. Rev. Cytol.* **176**, 1 (1997).

[3] Fluorescence Methods for Monitoring Interactions of Rab Proteins with Nucleotides, Rab Escort Protein, and Geranylgeranyltransferase

By KIRILL ALEXANDROV, AXEL J. SCHEIDIG, and ROGER S. GOODY

Introduction

To understand the manner in which biological macromolecules interact with each other, we need not only structural information, but also details of the kinetics and thermodynamics of the processes involved. This is particularly important for proteins acting in signal transduction or in the regulation of processes in, for example, protein transport and trafficking due to the dynamic nature of these interactions. Rab proteins are involved in several interactions with other protein components in the regulation of vesicular transport,[1] and as in other systems, signals must first be found before such interactions can be investigated. In this chapter, we describe several examples of fluorescence signals that can be used for such studies. As a model system we have chosen Rab7, a small GTPase involved in the biogenesis of late endosomes and lysosomes.[2,3] We have investigated its interaction with the Rab escort protein 1 (REP-1) and Rab geranylgeranyltransferase (Rab GGTase). REP proteins are required to present Rab proteins to the enzyme geranylgeranyltransferase to allow their C-terminal prenylation (double geranylgeranylation).[4] The details of the interactions involved in this process are poorly understood, but there was already an indication that REP has a higher affinity for the GDP than for the GTP form of Rab proteins. To quantitate this and other interactions more fully and to obtain information on the dynamics, fluorescence from different reporter groups associated with Rab7 was exploited, as described below.

Fluorescent Nucleotides

Principle

Like many other GTPases, the Rab proteins and their yeast homologs (Ypt proteins) tolerate modification of the ribose residue, and this has been taken advantage of to examine the interaction of nucleotides with Rab5 and Rab7.[5] In these studies, the methylanthraniloyl (mant) group is attached to either the 2'- or 3'-hydroxyl groups of GTP or GDP, and as shown in earlier work on the Ras protein, binding of such derivatives to the active site of the GTPase results in a large increase in fluorescence.[6] The fluorescence from the methylanthraniloyl group has been used extensively for examining the kinetics of interaction of GTPases with nucleotides, and also for mechanistic investigations of the effects of guanine nucleotide exchange

[1] O. Martinez and B. Goud, *Biochim. Biophys. Acta* **1404,** 101 (1998).
[2] A. Mukhopadhyay, K. Funato, and P. D. Stahl, *J. Biol. Chem.* **272,** 13055 (1997).
[3] R. Vitelli, M. Santillo, D. Lattero, M. Chiariello, M. Bifulco, C. B. Bruni, and C. Bucci, *J. Biol. Chem.* **272,** 4391 (1997).
[4] P. J. Casey and M. C. Seabra, *J. Biol. Chem.* **271,** 5289 (1996).
[5] I. Simon, M. Zerial, and R. S. Goody, *J. Biol. Chem.* **271,** 20470 (1996).
[6] J. John, R. Sohmen, J. Feuerstein, R. Linke, A. Wittinghofer, and R. S. Goody, *Biochemistry* **29,** 6058 (1990).

factors,[7,8] in both cases taking advantage of the fluorescence change occurring when nucleotides associate with or dissociate from the active site. However, there may also be changes in the nucleotide fluorescence when other proteins interact with the GTPases, as first seen in the case of Ran and its exchange factor RCC1[7] and subsequently on interaction of complexes between Ras and mant nucleotides with GTPase activating proteins (GAPs).[9,10] More recently, the fluorescence mant-nucleotide derivatives have been used for a detailed examination of the kinetics of the interaction of Ras with the Ras-binding domain of the Ras effector Raf.[11]

Methods

Preparation of Rab–MantGXP Complexes. To replace GDP bound to Rab proteins with the fluorescent analog mantGTP or mantGDP we took advantage of the fact that chelating of Mg^{2+} by EDTA dramatically increases the rate of nucleotide release. For Rab7, nucleotide exchange is typically performed with 50–200 nmol of protein. Mg^{2+} and GDP are removed from the protein storage buffer by passage over a NAP-5 gel filtration column (Pharmacia, Piscataway, NJ) equilibrated with buffer A (50 mM HEPES, pH 7.2, 3 mM DTE, 5 mM EDTA) according to the instructions of the manufacturer. A 10-fold excess of mantGXP over protein is then added and the sample incubated for 1 hr at room temperature. The unbound nucleotide is removed by passing the sample over a NAP-10 (Pharmacia) column equilibrated with buffer A. Another portion of mantGXP is added and the incubation is repeated. Finally, the excess mantGXP is removed by passing the sample over a PD-10 (Pharmacia) column equilibrated with buffer B (50 mM HEPES, pH 7.2, 3 mM DTE). The efficiency of nucleotide exchange is confirmed by HPLC on a Hypersil C_{18} reversed-phase column driven by a Beckman Gold HPLC system. The column is equilibrated with a buffer containing 100 mM K_2HPO_4/KH_2PO_4, pH 6.5, 10 mM tetrabutylammonium bromide, and 7.5% acetonitrile and developed with gradient of acetonitrile (7.5–20%) in the same buffer. The column is calibrated with standard solutions of mantGXP and GXP of known concentration under the conditions described above. Typically, exchange efficiency was more

[7] C. Klebe, H. Prinz, A. Wittinghofer, and R. S. Goody, *Biochemistry* **34,** 12543 (1995).

[8] C. Lenzen, R. H. Cool, H. Prinz, J. Kuhlmann, and A. Wittinghofer, *Biochemistry* **37,** 7420 (1998).

[9] R. Mittal, M. R. Ahmadian, R. S. Goody, and A. Wittinghofer, *Science* **273,** 115 (1996).

[10] M. R. Ahmadian, U. Hoffmann, R. S. Goody, and A. Wittinghofer, *Biochemistry* **36,** 4535 (1997).

[11] J. R. Sydor, M. Engelhard, A. Wittinghofer, R. S. Goody, and C. Herrmann, *Biochemistry* **37,** 14292 (1998).

than 90%. The protein is shock frozen in liquid nitrogen and stored at $-80°$ in small aliquots.

Fluorescence Measurements. Fluorescence spectra and long time-based fluorescence measurements are performed with an Aminco SLM 8100 spectrophotometer (Aminco, Silver Spring, MD). All reactions were followed at 25° in buffer C [25 mM HEPES, pH 7.2, 40 mM NaCl, 2 mM MgCl$_2$, 2 mM dithioerythritol (DTE), and 10 μM GDP]. The fluorescence of mantGDP and dansyl-labeled Rab7 is excited via tryptophan-FRET at 295 nm and measured at 440 nm. For the measurement of the direct fluorescence signal, dansyl-labeled Rab7 is excited at 333 nm and data collected at 440 nm. Stopped flow experiments are performed in a High-Tech Scientific SF61 apparatus (Salisbury, England). The fluorescence of mant or dansyl group is excited via FRET at 290 or directly at 333 nm and detected through a 389-nm cutoff filter. Data collection and primary analysis of rate constants are performed with the package from High-Tech Scientific; the secondary analysis is performed with the programs Grafit 3.0 (Erithacus Software) and Scientist 2.0 (MicroMath Scientific Software).

Results

Affinity of REP-1 for Rab7–Nucleotide Complexes. On interaction of Rab7–mantGDP with REP-1, there is a small change (increase) in the fluorescence of the mant group when this is excited directly at 365 nm. There is a much larger increase in the energy transfer signal seen when tryptophan fluorescence is excited at 290 nm while measuring the fluorescence intensity at the mant group emission wavelength. Both Rab7 and REP-1 contain tryptophan residues, so that it is not clear what the origin of this signal is, but recruitment of additional REP-1 tryptophans into the vicinity of the mant group on interaction with the Rab7–mantGDP complex is the most likely explanation.

Titrations are performed in a fluorescence cuvette containing 1 ml of buffer C. An appropriate concentration of Rab7–mantGDP is added to the cuvette and the fluorescence intensity at 440 nm is recorded on exciting at 290 nm. Addition of REP-1 in small aliquots results in an increase in intensity, which, if the starting concentration of Rab7–mantGDP is high enough, results in a plateau when excess REP-1 has been added. After correction for volume changes during the titration, the results can be fitted using the quadratic equation describing the fluorescence change expected under these conditions:

$$F = F_{min} + \frac{(F_{max} - F_{min})[(P + L + K) - \sqrt{(P + L + K)^2 - 4PL}]}{2P}$$

where F is the fluorescence intensity (in arbitrary units), F_{min} is the intensity at the beginning of the titration (i.e., the fluorescence intensity of Rab7–mantGDP), F_{max} is the fluorescence intensity of the REP-1–Rab7–mantGDP complex, P is the total Rab7–mantGDP concentration, L is the REP-1 concentration, and K is the dissociation constant for the interaction. In the fit to this equation, F_{min}, F_{max}, P, and K_d were allowed to vary in the program Grafit (Erithacus Software, Hovley, Surrey, UK).

The values obtained from the fit are 0.44 nM for K_d, and 222 nM for the effective concentration of Rab7–mantGDP. Because a nominal concentration of 250 nM was used, this suggests that the concentration of Rab7–mantGDP or REP-1, or possibly both is not well determined. For an optimal determination of the K_d value, it would be preferable to perform the titration at a lower concentration of Rab7–mantGDP. This is not conveniently achieved due to the less stable signal which then results. Because an independent determination of the K_d value can be derived from the transient kinetic experiments described below, this is not of great significance.

Equilibrium titrations are more difficult with the Rab7–mantGTP complex, since the fluorescence increase on interaction with REP-1 was much smaller than in the case of Rab7–mantGDP. However, as described below, this signal was adequate for transient kinetic experiments.

The interaction of Rab7 and REP-1 can also be monitored using the fluorescence of two dansyl labels covalently attached at the C terminus of Rab7 (method described below). There is a large increase of intensity of the fluorescence peak at 440 nm on exciting at 330 nm when REP-1 is added to the dansyl-labeled Rab7. The K_d value obtained from a titration curve was 0.6 nM, quite near to the value obtained for the Rab7–mantGDP complex, suggesting that the C-terminal modification does not interfere with the interaction between the two proteins.

Transient Kinetics of Rab7–REP-1 Interaction. The kinetics of the interaction of Rab7 and REP-1 can be examined using the signal from the fluorescent group of mantGDP and mantGTP. When the directly excited fluorescence of the mant group was used as a signal of binding, mixing of Rab7–mantGDP with REP-1 resulted in its rapid signal increase. Under the conditions used, there should be an exponential transient increase in fluorescence if the interaction with REP-1 is a simple one-step process. However, the curves obtained are biphasic, suggesting a more complex association mechanism. If energy transfer is used as a signal, the curves appear at first sight to be single exponentials, but on closer examination and with the knowledge obtained from the direct fluorescence signal, it becomes clear that they are also biphasic.

Information on the dissociation kinetics can be obtained by displacement of Rab7–mantGDP from its complex with REP-1 by an excess of Rab7–GDP. This leads to an estimate of the value for the dissociation rate constant of 0.012 sec^{-1}. The results in the two types of experiments performed can be explained in terms of the following mechanism[12]:

$$\text{REP} + \text{Rab–mantGDP} \underset{k_{-1}}{\overset{k_{+1}}{\rightleftharpoons}} \text{REP–Rab–mantGDP} \underset{k_{-2}}{\overset{k_{+2}}{\rightleftharpoons}} \text{REP–Rab*–mantGDP}$$

The values of the constants are $1.2 \times 10^7 \, M^{-1} \, \text{sec}^{-1}$ for k_{+1}, 0.012 sec^{-1} for k_{-1}, ca. 2.1 sec^{-1} for k_{+2}, and ca. 1.2 sec^{-1} for k_{-2}, although it should be noted that these are not very well specified by the data obtained. This leads to a calculated K_d value of ca. 0.6 nM, in good agreement with the value obtained in the equilibrium titration measurement. In the first step, there is a small increase in mant fluorescence (1.1%) followed by a slightly larger increase in the second step (1.4%). By contrast, most of the change in the FRET signal occurs in the first step (ca. 94%) with the remaining 6% being associated with the second step. These data suggest that there is a small change in the environment of the nucleotide in each of the two steps, but that after initial docking of REP-1 onto Rab7, which results in an relatively large increase of energy transfer from tryptophans of REP-1 to the nucleotide, there is not much change in the distance between the REP-1 tryptophans and the nucleotide in the second step.

Data on the interaction of Rab–mantGTP and REP-1 were more difficult to obtain, mainly because of much less efficient energy transfer from tryptophan to the nucleotide in the resulting complex. This could indicate a significantly different structure of the complex when compared with the corresponding mantGDP complex. However, the signal change was large enough to determine association and dissociation rate constants, leading to values of $1.25 \times 10^7 \, M^{-1} \, \text{sec}^{-1}$ and 0.2 sec^{-1}, respectively. The calculated K_d is therefore 16 nM, somewhat more than an order of magnitude larger than for the Rab7–mantGDP complex.

Generation of Fluorescent Proteins by Cysteine Labeling

Principle

The two C-terminal cysteines of Rab7, which are the site of geranylgeranylation, were chosen as convenient sites to introduce fluorescent groups (dansyl and rhodamine have been used, but we restrict the discussion to

[12] K. Alexandrov, I. Simon, A. Iakovenko, B. Holz, R. S. Goody, and A. J. Scheidig, *FEBS Lett.* **425**, 460 (1998).

the dansyl label here). This label allowed monitoring of the interaction between Rab7 and REP-1, leading to similar results to those described using the fluorescent nucleotides, and more importantly monitoring of the interaction of Rab7 with Rab GGTase.

Methods

Labeling of Rab7 with Dansyl Group. Typically 50 nmol of wild-type Rab7 is incubated with 1 μmol of 1,5-IAEDANS (Molecular Probes, Eugene, OR) in 300 μl of 100 mM Tris, pH 0.8, 1 mM MgCl$_2$, 100 μM GDP for 2 hr at 4°. After the indicated time period, the protein is passed over a PD-10 column (Pharmacia) preequilibrated with 20 mM HEPES, pH 7.2, 10 mM NaCl, 2 mM DTE, 1 mM MgCl$_2$, 100 μM GDP. Labeled protein is stored in multiple aliquots at −80°. The efficiency of labeling is determined by mass spectrometry and fluorescent yield measurements.

Results

Preparation of Dansyl-Labeled Rab7. The C terminus of Rab GTPases is known not to have a definite structure in solution. However, logic demands that the two C-terminal cysteines must be precisely positioned for the prenyl transfer reaction catalyzed by prenyltransferases. This obviously requires a relatively large structural rearrangement and hence must induce changes in the environment of the C-terminal cysteines. This makes C-terminal cysteines of Rab proteins a good target for modification with the fluorescent groups that could report on its interactions with subunits of Rab GGTase or other proteins. The labeling procedure yielded over 90% doubly dansylated Rab7 protein as confirmed by mass spectrometry and fluorescence yield measurements. Excitation and emission scans of the labeled protein revealed that fluorescence could be excited either directly at 333 nm or via fluorescence resonance energy transfer from tryptophan excited at 295 nm. This method has general applicability and has also been successfully used for labeling of Rab5, Ypt7, Ypt1, Ypt51, and H-Ras.

Interaction of Rab GGTase with Rab7–REP-1 Complexes: Equilibrium Titration Measurements. Titration of Rab GGTase to a 1 : 1 complex of the REP-1–dansylRab–GDP complex results in a large increase in fluorescence (Fig. 1A). Fitting to the data obtained gave a value of 13 nM for the K_d of the interaction. Since it seems likely that the dansyl groups, which are attached to the C-terminal cysteines at the sites of geranylgeranylation, will affect the affinity of Rab GGTase for the Rab–REP complex, information can be obtained on the affinity of unmodified Rab by using it as a competitor for the interaction with the fluorescent complex. For technical reasons, this is most conveniently done by titrating Rab GGTase to a mixture of the fluorescent REP-1–dansylRab7–GDP complex with the corresponding no-fluorescent complex (Fig. 1B). Evaluation of such data is difficult with classical approaches, since it involves solution of a cubic equation.[13] A much simpler and equally rigorous approach is to use the program Scientist, which allows the experimental system to be defined as a series of partial equations defining the equilibrium relationships between the species involved and defining the manner in which the final signal is produced. The model file for this program is given in the Appendix. The K_d value for the interaction of Rab7–GDP–REP-1 with Rab GGTase is 121 nM, or about an order of magnitude higher than for the dansyl-labeled complex. This suggests that the dansyl groups interact with the Rab GGTase, possibly with the geranylgeranyl pyrophosphate binding site.

The dansyl group signal can also be used for examining the kinetics of association of Rab GGTase with the Rab7–REP-1 complex.[15] As shown in Fig. 2A, there is a relatively rapid increase in the energy transfer signal on mixing in a stopped flow machine. The signal is biexponential, and the first phase is markedly dependent on the Rab GGTase concentration (Fig. 2A, inset). These results suggest that after an initial association reaction with a second-order rate constant of ca. $5 \times 10^7 \, M^{-1} \, sec^{-1}$, there is a slow step with a rate constant of ca. 4 sec^{-1}. However, the mechanism is likely to be more complex than this, since the dissociation kinetics, as shown in Fig. 2B, are also biphasic, with a second phase that is too slow to be the reversal of the second step implicated by the association measurements. Thus, there is probably at least one more step in an apparently complex binding mechanism that has not yet been completely characterized.

Use of Intein and *in Vitro* Ligation Technologies to Introduce Fluorescent Groups into Rab7

Principle

Labeling of the two C-terminal cysteines of Rab7 is a useful technique for examining the interaction with REP and Rab GGTase, as described above, but cannot be used in experiments in which these groups must be free for prenylation. In this case, a different approach is needed, and we have used a combination of intein biochemistry and the method of *in vitro*

---

[13] S. H. Thrall, J. Reinstein, B. M. Wohrl, and R. S. Goody, *Biochemistry* **35,** 4609 (1996).

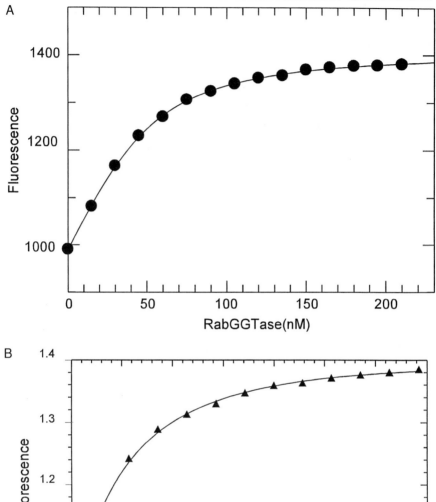

ligation to achieve this. Central to this method is the ability of certain protein domains (inteins) to excise themselves from the protein by combination of N → S(O)-acyl shift and transesterification reaction, thus leaving a thioester group attached to the C terminus of the remaining protein.[14] This thioester group can then be used to couple essentially any polypeptide to the thioester tagged protein by restoring the peptide bond.[14] The only requirement for the ligation reaction is the presence of the N-terminal cysteine to the target peptide.

Methods

Vector Construction, Protein Expression, Purification, and Ligation. We first generate an expression vector for C-terminal fusion of Rab7ΔC6 with intein by PCR amplifying the coding sequence of the former. The 3' oligonucleotide is designed in such a way that the resulting cDNA encoded a Rab7 protein truncated by 6 amino acids and fused to the N terminus of the intein. This product was subcloned into the pTYB1 expression vector (New England Biolabs, Beverly, MA).

To purify the CBD–intein–Rab7ΔC6 fusion protein, 1 liter of *Escherichia coli* BL21 cells transformed with pTYB2Rab7ΔC6 is grown to mid-log phase in Luria–Bertani medium and induced with 0.3 mM isopropyl-1-thio-D-galactopyranoside (IPTG) at 20° for 12 hr. After centrifugation, cells are resuspended in 60 ml of lysis buffer (25 mM Na$_2$HPO$_4$/NaH$_2$PO$_4$, pH 7.2, 300 mM NaCl, 1 mM MgCl$_2$, 10 μM GDP, 1.0 mM phenylmethylsulfonyl fluoride) and lysed using a fluidizer (Microfluidics Corporation, Newton, MA). After lysis, Triton X-100 is added to a final concentration of 1%. The lysate is clarified by ultracentrifugation and incubated with 9 ml of chitin beads (New England Biolabs) for 2 hr at 4°. The beads are washed extensively with the lysis buffer and incubated for 14 hr at room temperature with 40 ml of the cleavage buffer (25 mM Na$_2$HPO$_4$/NaH$_2$PO$_4$, pH 7.2, 300 mM NaCl, 1 mM MgCl$_2$, 10 μM GDP, 500 mM 2-mercaptoethanesul-

---

[14] G. J. Cotton, and T. W. Muir, *Chem. Biol.* **6**, R247 (1999).

FIG. 1. (A) Spectrofluorometric titration of dansRab7–REP-1 complex with Rab GGTase. Data were analyzed as described under "Methods" and led to K_d values of 13 nM. (B) Spectrofluorometric competition titration of dansRab7–REP-1 fluorescence by GGpp-free Rab GGTase in the presence of Rab7–REP-1 complex. The dansRab7–REP-1 complex concentration was 50 nM, wt-Rab7–REP-1 500 nM. Data were fitted using the program Scientist 2.0 and led to K_d values of 120 nM for the interaction of Rab GGTase with the unlabeled Rab7–REP-1 complex.

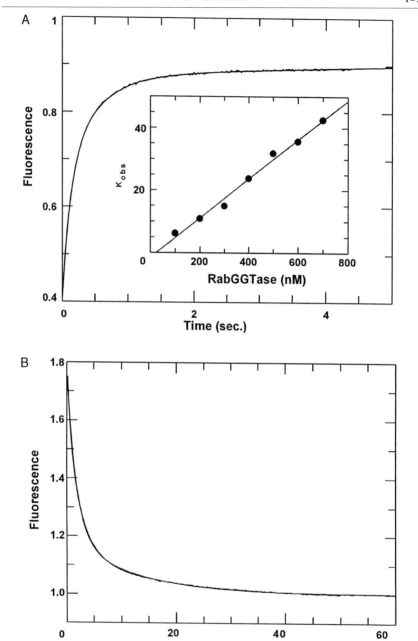

fonic acid). The Rab7ΔC6 thioester is concentrated using Centripreps 10 (Amicon, Danvers, MA) to a final concentration of 200 μM and stored frozen at $-80°$ until needed.

Peptide Synthesis and Protein Ligation. Peptide C-L(dans)-S-C-S-C is synthesized and HPLC purified to more than 90% purity by Interactiva (Ulm, Germany). The peptide is dissolved to a final concentration of 50 mM in 25 mM Tris, pH 7.2, and 5% CHAPS. In the ligation reaction the thioester-activated Rab7 is mixed with the peptide in a buffer containing 25 mM Na_2HPO_4/NaH_2PO_4, pH 7.2, 300 mM NaCl, 500 mM 2-mercapto-ethanesulfonic acid, 1 mM $MgCl_2$, 5% CHAPS, and 100 μM GDP and allowed to react overnight at room temperature. The final concentrations are 240 μM and 2 mM for Rab7 and peptide, respectively. Unreacted peptide and detergent are removed by passing the reaction mixture over a PD-10 desalting column (Pharmacia) equilibrated with 25 mM HEPES, pH 7.2, 40 mM NaCl, 2 mM $MgCl_2$, 100 μM GDP, and 2 mM DTE. The extent of ligation is determined by SDS–PAGE and mass spectrometry. For visualization of ligated fluorescent product, the reaction mixture is separated on a 15% SDS–PAGE gel and acetic acid-fixed gels are viewed in unfiltered UV light.

Results

Generation of Semisynthetic Rab7 Protein and Characterization of Its Interaction with REP-1 and Rab GGTase. We used a pTYB expression vector to generate a fusion protein containing Rab7ΔC6, intein, and chitin binding domain, fused in respective order. Addition of thiol reagent (mercaptoethanesulfonic acid in our case) to such protein promotes the rearrangement and disruption of the polypeptide bond between Rab7 and intein. The released Rab7 has a thioester group on its C terminus that can be used for the ligation reaction. We used a fluorescently labeled peptide

---

FIG. 2. (A) Time course of the fluorescent energy transfer signal change seen on mixing dansRab7–REP-1 complex (50 nM) with Rab GGTase (100 nM) in the stopped flow machine. The shown fit is to a double exponential equation with a rate constant (k_{ass1}) of 6.6 sec^{-1} and (k_{ass2}) of 2.4 sec^{-1}. *Inset:* Secondary plot of data from seven experiments of the above-described type. Circles represent k_{1obs} values plotted against the concentration of Rab GGTase. (B) Time course of the energy transfer signal change seen on mixing dansRab7–REP-1–Rab GGTase (0.1 μM) with Rab7–REP-1 (2.0 μM) in the stopped flow apparatus. Excitation was at 289 nm, and emission was detected through a 389-nm cutoff filter. The fit shown is to a double exponential equation with rate constants k_{diss1} and k_{diss2} for the displacement of dansRab7–REP-1 by Rab7–REP-1 of 0.39 and 0.03 sec^{-1}, respectively.

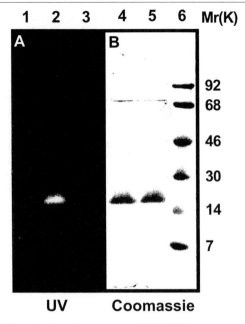

FIG. 3. SDS–PAGE gel of the thioester tagged Rab7Δ6C before (lane 2) and after ligation to a dansylated peptide (lane 5) photographed either in the UV light (A) or visible light after Coomassie blue staining (B). ESI-MS spectrum of thioester tagged Rab7ΔC6 (expected mass 22924 Da) (C) and Rab7A202CE203Ldans ligation product (D).

mimicking the missing six amino acids of Rab7 to restore a full-length protein. Figure 3 depicts a SDS–PAGE gel with thioester tagged Rab7ΔC6 and the ligated fluorescent product. The ligation reaction was highly efficient with yields over 90%. We designate the ligated protein Rab7A202CE203Ldans.

We assessed the fluorescent characteristics of the obtained protein by analyzing its excitation and emission spectra. Figure 4A shows that, consistent with the spectral characteristics of the incorporated group, the fluorescence had an excitation maximum at 340 nm and emission at 545 nm. Binding of Rab7 to its native substrate REP-1 was used to assess whether semisynthetic Rab7 was functionally active. REP-1 was added to a cuvette containing 380 nM of Rab7A202CE203Ldans to the final concentration of 500 nM. This addition resulted in a 6-fold increase of fluorescence (Fig. 4A). The emission maximum was strongly blue shifted and had a maximum at 493 nm. Further additions of REP-1 did not result in an in-

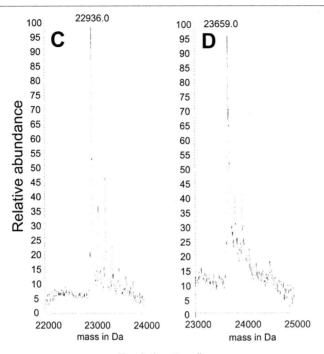

FIG. 3. (*continued*)

crease of fluorescence, indicating the formation of a stoichiometric Rab7A202CE203Ldans–REP-1 complex. We used this fluorescent signal change to determine the affinity of Rab7A202CE203Ldans for REP-1. We titrated 380 nM of Rab7A202CE203Ldans with REP-1. The obtained data were fit using a quadratic equation yielding a K_d of 4 nM and were consistent with a 1 : 1 stoichiometry (Fig. 4B). This is in reasonable agreement with the previously determined K_d of 1 nM for Rab7–REP-1 complex, thus indicating that incorporation of a dansyl group at position 202 of Rab7 docs not significantly perturb its interaction with REP-1.[12]

Next we examined whether introduction of the fluorescent group influenced interaction of the Rab7A202CE203Ldans–REP-1 complex with Rab GGTase. Addition of Rab GGTase to a cuvette containing 380 nM of the Rab7A202CE203Ldans–REP-1 complex resulted in a dose-dependent and

[15] K. Alexandrov, I. Simon, V. Yurchenko, A. Iakovenko, E. Rostkova, A. J. Scheidig, and R. S. Goody, *Eur. J. Biochem.* **265,** 160 (1999).

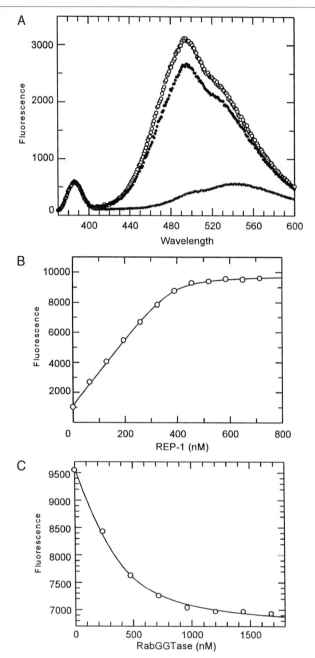

saturable fluorescence decrease by about 30% (Fig. 4A). This observation indicates that formation of the ternary Rab7–REP-1–Rab GGTase complex results in further environmental changes at the C terminus of Rab7. In the following experiment we titrated 380 nM of dansRab7–REP-1 complex with Rab GGTase and processed the data as in the previous case. The K_d was 111 nM (Fig. 4C), closely matching the value obtained by alternative methods. Therefore we concluded that introduction of a fluorescent group in position −5 of the Rab7 C terminus did not influence the interaction of Rab7 with the subunits of Rab GGTase.

Conclusion

Several lines of evidence suggest the physiological importance of Rab proteins in intercellular membrane transport. Nevertheless, the biochemistry of their function as well as the mechanism of their interaction with the other components of the docking and fusion machinery remain largely unknown. The elucidation of Rab function requires the dissection of such interactions at the molecular level. This stresses the need for development of sensitive biochemical assays for the study of such interactions. The fluorescent methods described in this chapter provide researchers with a number of tools for studying the intercommunications of Rab proteins with subunits of Rab GGTase and other molecules. It transpired in the course of this work that the different reactions were best resolved by more than one assay. Moreover, applying different methods to the same reaction allowed us to improve the reliability of the data and avoid misleading artifacts. We believe that the above-described methodology is generally applicable to study interaction of small GTPases with their interacting molecules.

FIG. 4. (A) Emission spectra of 380 nM of Rab7A202CE203Ldans alone (solid line), on addition of 500 nM REP-1 (open circles) or on further addition of 1 μM of Rab GGTase. Excitation was at 338 nm. (B) Titration of REP-1 to a nominal concentration of 380 nM Rab7A202CE203Ldans using direct fluorescence as a signal for binding (excitation wavelength, 338 nm; emission, 490 nm). The solid line shows the fit to a quadratic equation describing the binding curve and gives a value of 4 nM for the K_d and an effective REP-1 concentration of 391 nM. (C) Spectrofluorometric titration of Rab7A202CE203Ldans : REP-1 complex (380 nM) with Rab GGTase under the conditions described above. The K_d for the Rab7A202CE203Ldans–REP-1–Rab GGTase complex obtained from the fit of the data to a quadratic equation is 111 nM.

Appendix

Model File for the Program Scientist for the Titration Shown in Fig. 1B

Independent Variable:

C (concentration of RabGGTase)

Dependent Variables:

A (concentration of dansylRab7:REP complex),

B (concentration of Rab7:REP complex)

AC (concentration of dansylRab7:REP:RabGGTase complex)

BC (concentration of Rab7:REP:RabGGTase complex)

Cf (concentration of free RabGGTase)

F fluorescent yield

Parameters:

K1 (K_d for d_Rab7:REP:RabGGTase interaction)

K2 (K_d for Rab7:REP:RabGGTase interaction)

ATOT (total concentration of d_Rab7:REP)

BTOT (total concentration of Rab7:REP)

Ya (fluorescent yield of d_Rab7:REP complex)

Yac (fluorescence yield of d_Rab7:REP:RabGGTase complex)

BC= B*Cf/K1

AC=A*Cf/K2

ATOT=A+AC

BTOT=B+BC

C=Cf+AC+BC

0<A<ATOT

0<B<BTOT

0<Cf<C

F=A*Ya+AC*Yac

Initial conditions

C=0

Acknowledgments

We thank Heino Prinz for advice and help with mass spectrometry. This work was supported in part by a grant from DFG, number 545/1-2.

[4] Prenylation of Rab Proteins *in Vitro* by Geranylgeranyltransferases

By Jean H. Overmeyer and William A. Maltese

Introduction

Isoprenylation is a posttranslational modification of proteins that involves addition of a 15-carbon farnesyl or 20-carbon geranylgeranyl moiety in a thioether linkage to a C-terminal cysteine residue of the protein substrate.[1,2] In mammalian cells the majority of prenylated proteins belong to the Ras superfamily of small GTPases, which includes the large subgroup of Rab proteins.[3,4] Rab proteins must be modified by addition of 20-carbon geranylgeranyl moieties to one or two carboxyl-terminal cysteine residues in order to associate with membranes[5,6] and interact efficiently with GDP dissociation inhibitors (GDIs), accessory proteins that shuttle Rab proteins between the membrane and cytosolic compartments.[7,8,8a]

Ras and most other Ras-related GTPases outside of the Rab subgroup (Rho, Rac, etc.), are modified by a single prenyl moiety on a target cysteine embedded in a carboxyl-terminal C*aax* motif.[9] Within the latter motif the *a*'s generally are aliphatic amino acids, and the terminal *x* is a residue that specifies whether the protein will be modified by a farnesyl or geranylgeranyl moiety (discussed below). In contrast, the majority of Rab proteins end with *xx*CC, *x*C*x*C, or CC*xx* carboxyl-terminal amino acid motifs (*x* is any amino acid), and both cysteines are geranylgera-

[1] J. A. Glomset, M. H. Gelb, and C. C. Farnsworth, *Trends Biochem. Sci.* **15,** 139 (1990).
[2] W. A. Maltese, *FASEB J.* **4,** 3319 (1990).
[3] C. Nuoffer and W. E. Balch, *Annu. Rev. Biochem.* **63,** 949 (1994).
[4] F. Schimmoller, I. Simon, and S. R. Pfeffer, *J. Biol. Chem.* **273,** 22161 (1998).
[5] B. T. Kinsella and W. A. Maltese, *J. Biol. Chem.* **267,** 3940 (1992).
[6] J. H. Overmeyer and W. A. Maltese, *J. Biol. Chem.* **267,** 22686 (1992).
[7] O. Ullrich, H. Stenmark, K. Alexandrov, L. A. Hubar, K. Kaibuchi, T. Sasaki, Y. Takai, and M. Zerial, *J. Biol. Chem.* **268,** 18143 (1993).
[8] S. R. Pfeffer, A. B. Dirac-Svejstrup, and T. Soldati, *J. Biol. Chem.* **270,** 17057 (1995).
[8a] A. Shisheva, *Methods Enzymol.* **329,** 5 (2000) (this volume).
[9] P. J. Casey, *J. Lipid Res.* **33,** 1731 (1992).

nylated.[10] Exceptions to this rule are Rab8, Rab13, Rab18, and Rab23,[11] which contain C*aax* motifs similar to the Rho proteins and are therefore monoprenylated.

Progress in the enzymology of protein prenylation has revealed that different members of the Ras superfamily are modified by three distinct protein prenyltransferases. Farnesyltransferase (FTase) and geranylgeranyltransferase type I (GGTase I) are heterodimers that share a common α subunit, with unique β subunits.[12,13] These enzymes recognize protein substrates that contain carboxyl-terminal C*aax* motifs, with FTase showing a preference for substrates ending with M, S, A, or Q[14,15] and GGTase I exhibiting specificity for substrates ending with L or I.[16,17] Both enzymes can prenylate monomeric GTPase substrates irrespective of their nucleotide state[13] and can recognize small peptides that mimic the carboxyl-terminal C*aax* sequence.[14,18]

The properties of the enzyme complex that modify the Rab proteins are quite different. The catalytic component of this enzyme, termed Rab geranylgeranyltransferase or geranylgeranyltransferase type II (GGTase II), contains α and β subunits that are distinct from those found in FTase and GGTase I.[19] GGTase II cannot modify monomeric Rab substrates or carboxyl-terminal peptides.[20,21] Instead, the Rab protein must be associated with a chaperone termed Rab escort protein (REP)[22–24] to form a viable

[10] C. C. Farnsworth, M. Seabra, L. H. Ericsson, M. H. Gelb, and J. A. Glomset, *Proc. Natl. Acad. Sci. U.S.A.* **91,** 11963 (1994).

[11] M. Zerial and L. A. Huber, "Guidebook to the Small GTPases." Oxford University Press, New York, 1995.

[12] M. Seabra, Y. Reiss, P. J. Casey, M. S. Brown, and J. L. Goldstein, *Cell* **65,** 429 (1991).

[13] P. J. Casey and M. C. Seabra, *J. Biol. Chem.* **271,** 5289 (1996).

[14] Y. Reiss, S. J. Stradley, L. M. Gierasch, M. S. Brown, and J. L. Goldstein, *Proc. Natl. Acad. Sci. U.S.A.* **88,** 732 (1991).

[15] Y. Reiss, J. L. Goldstein, M. C. Seabra, P. J. Casey, and M. S. Brown, *Cell* **62,** 81 (1990).

[16] B. T. Kinsella, R. A. Erdman, and W. A. Maltese, *Proc. Natl. Acad. Sci. U.S.A.* **88,** 8934 (1991).

[17] P. J. Casey, J. A. Thissen, and J. F. Moomaw, *Proc. Natl. Acad. Sci. U.S.A.* **88,** 8631 (1991).

[18] K. Yokoyama, G. W. Goodwin, F. Ghomaschi, J. A. Glomset, and M. H. Gelb, *Proc. Natl. Acad. Sci. U.S.A.* **88,** 5302 (1991).

[19] S. A. Armstrong, M. C. Seabra, T. C. Sudhof, J. L. Goldstein, and M. S. Brown, *J. Biol. Chem.* **268,** 12221 (1993).

[20] S. L. Moores, M. D. Schaber, S. D. Mosser, E. Rands, M. B. O'Hara, V. M. Garsky, M. S. Marshall, D. L. Pompliano, and J. B. Gibbs, *J. Biol. Chem.* **266,** 14603 (1991).

[21] M. C. Seabra, M. S. Brown, C. A. Slaughter, T. C. Sudhof, and J. L. Goldstein, *Cell* **70,** 1049 (1992).

[22] D. A. Andres, M. C. Seabra, M. S. Brown, S. A. Armstrong, T. E. Smeland, F. P. M. Cremers, and J. L. Goldstein, *Cell* **73,** 1091 (1993).

substrate complex. REP is localized predominantly in the cytosol and binds preferentially to nascent Rab proteins when they are in the GDP-bound conformation.[25] There are two known isoforms of REP (REP1 and REP2), which differ slightly in their affinity for specific Rab proteins.[26] The details of the Rab digeranylgeranylation reaction are presently unclear. However, current evidence favors the concept of sequential addition of the two geranylgeranyl groups, with REP remaining tightly bound to the monoprenylated Rab intermediate and serving to recruit GGTase II to the substrate complex.[24] Upon completion of the reaction, REP can be released from the diprenylated Rab by incubation *in vitro* with detergents or phospholipid vesicles.[22,27] Moreover, REP can donate geranylgeranylated Rab proteins to intracellular membranes,[28,29] suggesting a potential role for REP in the subcellular targeting of Rabs.

Studies utilizing wild-type or functionally altered Rab mutants in reconstituted organelle preparations, permeabilized cells, or intact transfected cells provide valuable insights into the roles of different Rab proteins in vesicular trafficking. Because of the importance of the lipid modification for Rab membrane association and protein–protein interactions, rapid assays that indicate whether or not a particular amino acid substitution may affect the ability of a Rab protein to undergo prenylation can be critical to the interpretation of such studies. Moreover, when adding recombinant Rabs to reconstituted systems, it may be necessary to generate microgram quantities of recombinant Rab protein in the geranylgeranylated form. Because insect cells contain the necessary enzymes for protein prenylation, it is possible to obtain modified recombinant Rab proteins from Sf9 (*Spodoptera frugiperda* ovary) cells infected with appropriate baculovirus vectors.[30] However, the expense and time involved in preparation of hightiter virus stocks makes this method unsuitable for rapid initial screening of Rab mutants. In addition, to obtain large quantities of the prenylated

[23] K. Alexandrov, I. Simon, V. Yurchenko, A. Iakovenko, E. Rostkova, A. J. Scheidig, and R. S. Goody, *Eur. J. Biochem.* **265,** 160 (1999).

[24] J. S. Anant, L. Desnoyers, M. Machius, D. Borries, J. C. Hansen, K. D. Westover, J. Deisenhofer, and M. Scabra, *Biochemistry* **37,** 12559 (1998).

[25] M. C. Seabra, *J. Biol. Chem.* **271,** 14398 (1996).

[26] F. P. M. Cremers, S. A. Armstrong, M. C. Seabra, M. S. Brown, and J. L. Goldstein, *J. Biol. Chem.* **269,** 2111 (1994).

[27] F. Shen and M. C. Seabra, *J. Biol. Chem.* **271,** 3692 (1996).

[28] K. Alexandrov, H. Horiuchi, O. Steele-Mortimer, M. C. Seabra, and M. Zerial, *EMBO J.* **13,** 5262 (1994).

[29] J. H. Overmeyer, A. L. Wilson, R. A. Erdman, and W. A. Maltese, *Mol. Biol. Cell* **9,** 223 (1998).

[30] H. Horiuchi, O. Ullrich, C. Bucci, and M. Zerial, *Methods Enzymol.* **257,** 9 (1995).

Rab, it is typically necessary to extract the modified protein from cellular membranes and isolate it from nonprenylated Rab and other cellular proteins by ion-exchange chromatography.[30] A simpler approach is to express the Rab as a soluble protein in *Escherichia coli,* with addition of an epitope tag to facilitate rapid purification. However, because bacteria do not contain REP or GGTase II, the recombinant Rab must be converted to the geranylgeranylated form *in vitro.*

This chapter describes a simple method to geranylgeranylate recombinant Rab proteins directly in *E. coli* lysates, without the need to purify REP or GGTase II. The procedure can be used on an analytical scale to rapidly determine whether amino acid substitutions affect Rab prenylation. Alternatively, it can be scaled up to prepare modified recombinant proteins in microgram quantities sufficient for functional studies in reconstituted systems.

Methods

Preparation of Purified Recombinant Rab Substrates

cDNA encoding the Rab protein of interest is cloned into a prokaryotic expression vector (e.g., pET11a or pET17b from Novagen, Madison, WI) for protein production in *E. coli* BL21(DE3)pLysS. To facilitate purification and immunodetection of the substrate proteins, the *rab* cDNA sequence may be modified by polymerase chain reaction (PCR) to add an epitope tag, such as His$_6$ (hexahistidine) or myc (EQKLISEEDL) to the amino terminus of the protein.[31] Such modifications do not appear to affect the efficiency of Rab prenylation[29,32] or subcellular localization.[33-35] Expression of the recombinant Rab protein is induced by exposing the *E. coli* to 1.0 mM isopropyl-β-D-thiogalactopyranoside (IPTG) for 1–2 hr, followed by centrifugation of the cells. Detailed procedures for expression of recombinant Rab proteins in *E. coli* have been described.[36,37]

[31] J. H. Overmeyer, R. A. Erdman, and W. A. Maltese, *in* "Methods in Molecular Biology" (R. A. Clegg, ed.), p. 249. Humana Press Inc., Totowa, NJ, 1998.

[32] A. C. Schiedel, A. Barnekow, and T. Mayer, *FEBS Lett.* **376,** 113 (1995).

[33] W. H. Brondyk, C. J. McKiernan, E. S. Burstein, and I. G. Macara, *J. Biol. Chem.* **268,** 9410 (1993).

[34] Y. T. Chen, C. Holcomb, and H. P. H. Moore, *Proc. Natl. Acad. Sci. U.S.A.* **90,** 6508 (1993).

[35] F. Beranger, K. Cadwallader, E. Profiri, S. Powers, T. Evans, J. de Gunzberg, and J. F. Hancock, *J. Biol. Chem.* **269,** 13637 (1994).

[36] C. Nuoffer, F. Peter, and W. E. Balch, *Methods Enzymol.* **257,** 3 (1995).

[37] A. L. Wilson, K. M. Sheridan, R. A. Erdman, and W. A. Maltese, *Biochem. J.* **318,** 1007 (1996).

Prenylation of Recombinant Rabs in E. coli Lysates

Recombinant epitope-tagged Rab proteins can be geranylgeranylated directly in the *E. coli* cell lysate and then, if required, the modified protein can be isolated after the prenylation reaction. If this approach is used, it is first necessary to estimate the amount of expressed Rab protein present in the lysate before setting up the prenylation reaction. This can be done by preparing a small amount of lysate from a portion of the IPTG-induced *E. coli* culture and subjecting it to SDS–PAGE and immunoblot analysis, using any commercially available antibody against the specific Rab protein or the His_6 or myc epitope. The Rab concentration can then be estimated by relating the signal obtained from the lysate to a known amount of purified recombinant protein on the same blot.[37] The entire procedure can be completed in several hours if electrophoresis is performed on precast minigels and semidry protein transfer protocols are used. During this time, the remainder of the *E. coli* culture can be held at 4°, so that once the Rab expression level is known, the remaining cells can be disrupted in an appropriate amount of lysis buffer and the prenylation reaction can be carried out without the need to freeze the lysate.

Because the guanine nucleotide state of the Rab protein has a major effect on its ability to bind to REP and serve as a substrate for GGTase II,[25,32,33,37,38] it is critical that the lysate is prepared under conditions that promote GDP loading. In our hands, optimal results have been obtained when *E. coli* are disrupted by freeze–thawing (−80°) the cell pellet in Buffer B [50 mM HEPES, pH 7.4, 1.0 mM dithiothreitol (DTT), 1.0 mM GDP, 10 μM $MgCl_2$]. The lysate is cleared by centrifugation at 9000g, 4°, 5 min and then preincubated for 15 min at 37° before adding it to the prenylation reaction. Millimolar concentrations of guanine nucleotide are known to inhibit GGTase II activity[35] but this is not a problem because the GDP concentration is reduced 10-fold by dilution in the subsequent prenylation reaction.

Enzyme Sources

Although REP and GGTase II can be purified from rat brain[39] or from Sf9 cells expressing the recombinant proteins,[40] these procedures can be expensive and laborious. An acceptable alternative is to use a

[38] K. Alexandrov, I. Simon, A. Iakovenko, B. Holz, R. S. Goody, and A. J. Scheidig, *FEBS Lett.* **425**, 460 (1998).

[39] M. C. Seabra, J. L. Goldstein, T. C. Sudhof, and M. S. Brown, *J. Biol. Chem.* **267**, 14497 (1992).

[40] S. A. Armstrong, M. S. Brown, J. L. Goldstein, and M. Seabra, *Methods Enzymol.* **257**, 30 (1995).

relatively crude enzyme-enriched fraction prepared according to the early part of the purification scheme described by Seabra et al.[39] Briefly, one or two rat brains are homogenized in cold 50 mM Tris-HCl, pH 7.5, 1.0 mM EDTA, 1.0 mM EGTA, 0.2 mM PMSF, and 0.1 mM leupeptin (1.0–1.5 g tissue/ml). The homogenate is centrifuged at 60,000g for 70 min and the supernatant is subjected to a stepwise ammonium sulfate fractionation. The protein precipitating between 30 and 50% ammonium sulfate is dissolved in and dialyzed overnight against a buffer containing 20 mM Tris, pH 7.5, 1.0 mM DTT, and 20 μM ZnCl$_2$ (Buffer C). This fraction (AS$_{30-50}$) is diluted to 10–15 mg/ml in Buffer C and used as the enzyme source in the prenylation reaction. We have stored this material in aliquots at $-80°$ for at least 6 months with no loss of Rab GGTase activity. Although REP and GGTase II can be further purified and segregated from most of the FTase and GGTase I activity by ion-exchange chromatography on a Mono Q column as described,[39] this step is not essential because FTase and GGTase I do not interfere with Rab prenylation by GGTase II.

Prenylation Reactions

For prenylation of Rab proteins on an analytical scale, reactions are carried out in a total volume of 50 μl containing: 5 μl 10× reaction buffer [500 mM HEPES, pH 7.4, 10.0 mM DTT, 50.0 mM MgCl$_2$, 100 μM GDP, and 3 mM Nonidet P-40 (NP-40)], 10 μl AS$_{30-50}$ as a source of REP and GGTase II, and an aliquot of cleared and preincubated E. coli lysate (see above) containing approximately 0.3 μg Rab protein (approximately 0.25 μM final concentration in reaction mixture). Alternatively, if purified enzymes are available, 20 ng of REP and 20 ng of GGTase II may be substituted for the AS$_{30-50}$. NP-40 is added to the prenylation reaction to allow REP to be recycled.[22] In the absence of added detergent, Andres et al.[22] have demonstrated that the reaction will reach completion following one round of Rab prenylation. The detergent is required to dissociate the final complex between REP and digeranylgeranyl-Rab, so that REP can be used to support prenylation of additional molecules of Rab protein within the same reaction. The reaction is started with the addition of 1.0 μl [^{3}H]GGPP (1.0 μCi/μl, 10–30 Ci/mmol), which may be purchased from American Radiochemical Co., St. Louis, MO, or Dupont/NEN, Boston, MA. Maximal prenylation of the substrate protein is obtained after 2–3 hr at 37°.

To modify larger amounts of recombinant Rab, the volume of the reaction may be scaled up (e.g., to 500 μl). In addition, after initial confirmation that sufficient incorporation of [^{3}H]GG into the substrate protein has occurred (see below), unlabeled GGPP (American Radiolabeled Chemical Inc; 0.6–2 μM final concentration) can be used instead of the more expensive [³H]GGPP in the larger scale reactions.

The proportion of total Rab substrate converted to the geranylgera-nylated form can be estimated on the basis of electrophoretic mobility shift assays as described.[41] However, because the introduction of amino acid substitutions into some Rab proteins may cause anomalous electrophoretic mobility in the absence of prenylation, we favor measurements based on the incorporation of [³H]GGPP. An aliquot of the reaction mixture (20% or more) is mixed with an equal volume of 2× Laemmli sample buffer[42] and subjected to SDS–PAGE and fluorography.[37] The amount of [³H]GG transferred to the Rab protein is then determined by excising the region of the dried gel (22–30 kDa) containing the Rab protein, solubilizing it in 0.75% H_2O_2 at 65° for 16 hr, and counting the sample in a liquid scintillation spectrometer. Based on the specific radioactivity of the [³H]GGPP and the approximate amount of Rab protein in the reaction, Wilson *et al.*[37] estimated that 24% of recombinant Rab1B can be converted to the digeranylgera-nylated form in reactions utilizing *E. coli* lysate.

Recovery of Geranylgeranylated Rab Protein

The Rab protein that has been prenylated *in vitro* can be segregated from excess nonprenylated Rab by phase separation in Triton X-114 detergent.[6,43] The hydrophobic prenyl groups will cause the geranylgeranylated Rab to partition into the detergent phase while the nonprenylated protein remains in the aqueous phase. Triton X-114 is added to the sample at a final concentration of 1.0% and set on ice for 10 min. Following a 2-min incubation at 37°, the phases are separated by centrifugation at top speed in a microfuge. If the prenylated Rab protein contains the His_6 tag, it may be purified to near homogeneity by diluting the detergent phase with imidazole buffer (20 mM imidazole, 50 mM NaH_2PO_4, pH 8.0, 300 mM NaCl, 10 μM GDP) and subjecting the protein to affinity chromatography on a nickel-agarose spin column (Qiagen).[37] The purified geranylgeranylated Rab protein eluted from the resin should be concentrated and stored at −80° in Buffer B (50 mM HEPES, pH 7.4, 1.0 mM DTT, 1.0 mM GDP, 10 μM $MgCl_2$, and protease inhibitors) supplemented with 20% (v/v) glyccrol.

If the ultimate goal is to generate a complex between the prenylated Rab protein and GDI, an alternative approach to isolating the prenylated

[41] J. C. Sanford, L. Foster, Z. Kapadia, and M. Wessling-Resnick, *Anal. Biochem.* **224,** 547 (1995).
[42] U. K. Laemmli, *Nature* **227,** 680 (1970).
[43] J. F. Hancock, A. I. Magee, J. E. Childs, and C. J. Marshall, *Cell* **57,** 1167 (1989).

Rab may be used. This entails the addition of recombinant GDI or a cytosolic preparation from mammalian cells to the prenylation reaction, followed by gel filtration chromatography to separate the 80-kDa Rab–GDI complex (which contains the prenylated Rab) from the monomeric, nonprenylated protein that elutes around 20–30 kDa. Details of this method have been described by Riederer *et al.*[44]

Comments

The members of the Rab protein family that contain C*aax* motifs deserve special mention in connection with the issue of prenyltransferase specificity. In a recent study of Rab8 (which terminates with CVLL), we determined that unlike other Rabs with double-cysteine motifs, this protein can serve as a substrate *in vitro* for GGTase I.[45] However, Rab8 also is a good substrate for GGTase II, provided that REP is present. Interestingly, in intact cells it appears that Rab8 is modified predominantly through the pathway involving REP/GGTase II. Nevertheless, as a practical matter, if it becomes necessary to prenylate a Rab mutant where conformational issues may impair association with REP (e.g., amino acid substitutions in Rab effector domain[46] or switch-2[29] regions), it is possible to modify the protein with a single geranylgeranyl moiety using GGTase I (which is present in the AS_{30-50} enzyme preparation) provided that the protein has a naturally occurring C*aa*L motif, or is engineered to contain such a motif. If this approach is used, it is important to note that it is not entirely clear whether the functional properties of a Rab protein that is normally digeranylgeranylated might be altered if it contains only a single prenyl group. However, previous studies of Rab9 have shown that a monogeranylgeranylated version of the protein can associate efficiently with GDI.[47]

Another potential application of the assay described in this chapter is to distinguish which enzyme may be responsible for the prenylation of a particular GTPase substrate. As noted earlier, the AS_{30-50} contains REP and all three of the mammalian prenyltransferases. Because specific inhibitors of

[44] M. A. Riederer, T. Soldati, A. B. Dirac-Svejestrup, and S. R. Pfeffer, *Methods Enzymol.* **257**, 15 (1995).

[45] A. L. Wilson, R. A. Erdman, F. Castellano, and W. A. Maltese, *Biochem. J.* **333**, 497 (1998).

[46] A. L. Wilson and W. A. Maltese, *J. Biol. Chem.* **268**, 14561 (1993).

[47] T. Soldati, M. A. Riederer, and S. R. Pfeffer, *Mol. Biol. Chem* **4**, 425 (1993).

FTase[48–50] and GGTase I[51] have been described, it is possible to use these compounds to eliminate one or both of these activities in the AS_{30-50} fraction while GGTase II remains active, thereby allowing one to deduce which enzyme is required for prenylation of a given substrate protein. Furthermore, competition experiments can be performed for GGTase I and FTase activities, by addition of increasing concentrations of tetrapeptides corresponding to their respective substrate *Caax* recognition sequences.[18,20,52] Due to the complex nature of substrate recognition by REP/GGTase II (discussed above), this activity cannot be inhibited by analogous addition of peptides.[20,21]

[48] M. Nigam, C.-M. Seong, Y. Qian, A. D. Hamilton, and S. M. Sebti, *J. Biol. Chem.* **268,** 20695 (1993).
[49] E. C. Lerner, Y. Qian, M. A. Blaskovich, R. D. Fossum, A. Vogt, J. Sun, A. D. Cox, C. J. Der, A. D. Hamilton, and S. M. Sebti, *J. Biol. Chem.* **270,** 26802 (1995).
[50] A. Vogt, Y. Qian, M. A. Blaskovich, R. D. Fossum, A. D. Hamilton, and S. M. Sebti, *J. Biol. Chem.* **270,** 660 (1995).
[51] A. Vogt, Y. Qian, T. McGuire, A. D. Hamilton, and S. M. Sebti, *Oncogene* **13,** 1991 (1996).
[52] J. L. Goldstein, M. S. Brown, S. Stradley, Y. Reiss, and L. Gierasch, *J. Biol. Chem.* **266,** 15575 (1991).

[5] Antibody and Oligonucleotide Probes to Distinguish Intracellular Expression and Localization Patterns of Rab GDP-Dissociation Inhibitor Isoforms

By ASSIA SHISHEVA

Introduction

Among the number of accessory proteins required for Rab function, the GDP dissociation inhibitor (GDI) proteins have emerged as crucial for Rab progression through the membrane/cytosol localization cycle and recycling.[1] GDI's abilities to form stable cytosolic complexes with GDP-loaded prenylated Rabs, to present the active complexes to membranes, and to retrieve "spent" Rab proteins for returning them back to their membrane of origin, allow multiple rounds of membrane traffic and fusion.[1] Consistent with this important role, a depletion of yeast GDI *in vivo* results in multiple defects in intracellular transport.[2] The need for escort services

[1] S. Pfeffer, B. A. Dirac-Svejstrup, and T. Soldati, *J. Biol. Chem.* **270,** 17057 (1995).
[2] M. D. Garrett, J. E. Zahner, C. M. Cheney, and P. J. Novick, *EMBO J.* **13,** 1718 (1994).

is conferred by the fact that, due to the addition of geranylgeranyl lipids to their carboxyl termini, Rabs are hydrophobic peripheral membrane proteins. Rab escort protein-1, which is structurally related to GDIs, has also been assigned an escort role but only in the initial delivery of newly synthesized Rabs.[3,4] Thus, GDIs likely represent the only proteins fully competent to chaperone Rabs to their correct destination. Consistent with this requirement are several reports documenting additional important functions of GDI on membranes. Hence, GDI-1 contributes to the specific recognition of the Rab membrane targets,[5] stimulates GTP/GDP exchange,[6] and dissociates thereafter back into the cytosol, a step likely promoted by GDP dissociation factor(s)[7] and cellular redox state.[8]

Three mammalian GDI genes, GDI-1 (also known as GDIα), GDI-2, and GDIβ, have been cloned to date.[9-13] They encode ubiquitously expressed, highly homologous protein isoforms, tightly conserved in mammals. The three GDIs are found equally capable of operating with various Rabs in cell-free systems, indicating that, at least in resting cells, no specificity toward an individual Rab or subset of Rab proteins is associated with GDI function. Thus, although the possibility that one or more of the Rab proteins are uniquely regulated by only one of the three GDI isoforms is challenging, it is not supported (but not excluded) by our data and by that of others.[10,12,14] However, given the differential intracellular localization of GDI-1 vs. GDI-2 and the subtle regulation of Rab–GDI complex formation in the context of living cells,[13,15–18] functional differences among GDIs can be anticipated.

[3] K. Alexandrov, H. Horiuchi, O. Steele-Mortimer, M. C. Seabra, and M. Zerial, *EMBO J.* **13**, 5262 (1994).

[4] A. L. Wilson, R. A. Erdman, and W. A. Maltese, *J. Biol. Chem.* **271**, 10932 (1996).

[5] A. B. Dirac-Svejstrup, T. Soldati, A. Shapiro, and S. R. Pfeffer, *J. Biol. Chem.* **269**, 15427 (1994).

[6] H. Horiuchi, A. Giner, B. Hoflack, and M. Zerial, *J. Biol. Chem.* **270**, 11257 (1995).

[7] A. B. Dirac-Svejstrup, T. Sumizawa, and S. R. Pfeffer, *EMBO J.* **16**, 465 (1997).

[8] S. R. Chinni, M. Brenz, and A. Shisheva, *Expt. Cell Res.* **242**, 373 (1998).

[9] Y. Matsui, A. Kikuchi, S. Araki, Y. Hata, J. Kondo, Y. Teranishi, and Y. Takai, *Mol. Cell. Biol.* **10**, 4116 (1990).

[10] A. Shisheva, T. C. Südhof, and M. P. Czech, *Mol. Cell. Biol.* **14**, 3459 (1994).

[11] N. Nishimura, H. Nakamura, Y. Takai, and K. Sano, *J. Biol. Chem.* **13**, 14191 (1994).

[12] I. Janoueix-Lerosey, F. Jollivet, J. Camonis, P. N. Marche, and B. Goud, *J. Biol. Chem.* **270**, 14801 (1995).

[13] W. Chen, Y. Feng, D. Chen, and A. Wandinger-Ness, *Mol. Biol. Cell.* **9**, 3241 (1998).

[14] C. Yang, V. L. Stepnev, and B. Goud, *J. Biol. Chem.* **269**, 31891 (1994).

[15] A. Shisheva, J. Buxton, and M. P. Czech, *J. Biol. Chem.* **269**, 23865 (1994).

[16] A. Shisheva, S. Doxsey, J. Buxton, and M. P. Czech, *Eur. J. Cell Biol.* **68**, 143 (1995).

[17] A. Shisheva and M. P. Czech, *Biochemistry* **36**, 6564 (1997).

[18] A. Shisheva, S. R. Chinni, and C. DeMarco, *Biochemistry* **38**, 11711 (1999).

Because of the high homology among the three isotypes, it is typically difficult and sometimes impossible to distinguish their expression profiles in cells and tissues. In this article we describe a strategy that has been used in our laboratory to distinguish the endogenous expression and localization patterns of the GDI family members. Methods we use to express and purify GDI-2 and GDIβ in bacterial hosts are also described.

Distinguishing between the Expression Pattern of GDI-1 and GDI-2 at the Protein Level

General Considerations

The overall degree of the deduced amino acid sequence identity between mouse GDI-2 and rat or mouse GDI-1 is 86%.[10] Together with the similar molecular size of these two isotypes predicted by the deduced molecular weight (M_r of mGDI-2 and rGDI-1 are 50484 and 50687, respectively[10]), these results suggest that a selective immunodetection of intracellular GDI-1 vs. GDI-2 by fluorescence microscopy and Western blotting analyses will be difficult. However, pairwise comparison of the GDI-1 and GD-2 deduced amino acid sequences[10] reveals a couple of hydrophilic polypeptide stretches along the sequence, for which the level of homology is significantly lower. Thus, the peptides encompassing amino acids from 389 to 404 and from 431 to 447 of GDI-1 exhibit only 56 and 45% identity to the corresponding peptides in GDI-2, respectively. This result implies that antibodies directed against these most divergent peptide regions may specifically recognize GDI-1 and GDI-2, allowing their selective intracellular detection.

Procedures

Antibody Production and Purification. Antibodies to GDI-1 and GDI-2 are raised in rabbits against an mGDI-1 peptide (amino acids 431–447) and mGDI-2 peptide (amino acids 387–404).[10] The services of East Acres (Southbridge, MA) are used for immunization and bleeding of the animals. The peptides are coupled via an NH_2-terminal tyrosine to keyhole limpet hemocyanin (Peptide Core facility, UMass Medical Center, Worcester, MA). For Western blotting, the anisera are used in a dilution 1 : 20,000 (R3357; anti-GDI-1) or 1 : 10,000 (R3361; anti-GDI-2). While crude anti-GDI-2 antiserum selectively recognizes GDI-2, the blot is typically dirty and multiple dots are observed on Western blot analysis. Therefore we routinely perform the immunoblotting with the immunoglobulin G (IgG) fraction ($\sim$0.2 μg/ml) purified from the R3361 antiserum on a protein

A-Sepharose 4 Fast Flow column (Pharmacia Biotech, Piscataway, NJ), following common protocols.[19]

For immunofluorescence microscopy analyses the crude anti-GDI antisera are affinity purified on the corresponding glutathione S-transferase (GST) fusion proteins.[16] GST–GDI-2 and GST–GDI-1 are produced in bacterial hosts as previously described.[10] Following purification on glutathione-agarose beads (Sigma, St. Louis, MO) the fusion proteins are further purified by SDS–PAGE. Briefly, aliquots of the proteins (20–50 μg/lane, solubilized in Laemmli sample buffer[20]) are loaded on a 10.5% SDS–polyacrylamide gel and resolved by electrophoresis. The proteins from the gel are then electrotransferred onto a nitrocellulose membrane (0.45 mm, BA85; Schleicher & Schuell, Keene, NH) under standard conditions,[21] and visualized by Ponceau S (Sigma, St. Louis, MO) staining. The bands, corresponding to the electrophoretic mobility of the full-length fusion proteins (80 kDa for GST–GDI-1 or 76 kDa for GST–GDI-2)[10] are cut as rectangular strips (1 $\times$ 3 cm). The strips, containing typically ~200 μg of the fusion protein, are rinsed two times in phosphate-buffered saline (PBS) and blocked for 1 hr at room temperature with bovine serum albumin (BSA) (1%). The strips are then rinsed in PBS and incubated with a twofold diluted (50 mM HEPES, pH 7.6, 100 mM NaCl) crude antiserum (3 ml), which has been precleared on a microcentrifuge. The strips are washed four times (15 min each) with PBS. To elute the antibodies, the strip is overlaid with 150 μl of 100 mM glycine, pH 3.0, and incubated for exactly 60 sec at room temperature.[19] The eluate is quickly collected with a pipette and transferred to a tube containing 15 μl of 1 M Tris, pH 8.0, to bring the pH back to normal. The elution step is repeated twice. The two eluates are then combined and the affinity-purified antibodies are stored at $-20°$ in aliquots.

Cell Cultures. We have used various types of cultured cells for the analyses of the intracellular expression and localization of GDI-1 and GDI-2, including Chinese hamster ovary (CHO) T and 3T3-L1 adipocytes. 3T3-L1 mouse fibroblasts are grown to confluence on 100-mm plates (for biochemical analyses) or on 18-mm round glass coverslips (for immunofluorescence microscopy studies) in Dulbecco's modified Eagle's medium (DMEM) containing 10% calf serum, 50 units/ml penicillin, and 50 μg/ml streptomycin sulfate. The cells are differentiated into adipocytes as described previously.[10] CHO-T cells are grown to confluence on 100-mm plates for biochemical studies, or to 75% confluence on 18-mm glass coverslips

[19] E. Harlow and D. Lane, "Antibodies: A Laboratory Manual," p. 311. Cold Spring Harbor, New York, 1988.
[20] U. K. Laemmli, *Nature* **227,** 680 (1970).
[21] H. Towbin, T. Staehlin, and J. Gordon, *Proc. Natl. Acad. Sci. U.S.A.* **76,** 4350 (1979).

for immunofluorescence microscopy studies, respectively, in Ham's F12 medium, containing 10% fetal bovine serum (FBS) and the above antibiotics.

Preparation of Total Cellular Membrane Fraction and Cytosol for Western Blotting. 3T3-L1 adipocytes and CHO cells are fractionated to total membranes and cytosol. Briefly, the cells are scraped in homogenization buffer (20 mM Tris-HCl, pH 7.4, 255 mM sucrose, 1 mM EDTA, containing 1× protease inhibitor mixture [1 mM phenylmethylsulfonyl fluoride (PMSF), 1 mM benzamidine, 5 μg/ml leupeptin, 5 μg/ml aprotinin, and 1 μg/ml pepstatin]) and homogenized in a motor-driven Teflon/glass homogenizer. Total membrane and cytosol fractions are obtained by two sequential centrifugations: 3 min at 800g 4° (Micromax, IEC, Needham Heights, MA) and 15 min at 200,000g 4° (TL-100 Ultracentrifuge, Beckman Instruments, Inc., Palo Alto, CA). Pellets from the second spin are resuspended in 10 mM Tris-HCl, pH 7.5, containing 1 mM EDTA and 1× protease inhibitor mixture. Equal amounts of membrane and cytosolic proteins are solubilized in Laemmli sample buffer,[20] resolved by SDS–PAGE, and transferred to a nitrocellulose membrane under standard conditions.[21] The expression and differential localization of GDI-1 vs. GDI-2 is determined by immunoblotting using anti-GDI-1 antiserum or anti-GDI-2 IgG fraction.

The nitrocellulose membrane is blocked for 2 hr in TBS-T buffer (10 mM Tris-HCl, pH 7.6, 150 mM NaCl, 0.05% Tween 20), containing 3% (w/v) nonfat dry milk, 2% (w/v) BSA and then incubated (16 hr at 4°) with anti-GDI-1 or anti-GDI-2 antibodies, appropriately diluted in TBS-T buffer. The blots are rinsed twice and washed three times (15 min each) in TBS-T buffer. After incubation with horseradish peroxidase-labeled goat anti-rabbit IgG (Boehringer Mannheim Corp., Indianapolis, IN) bound antibodies are visualized by the chemiluminescence detection system (Renaissance, NEN, Boston, MA).

Immunofluorescence Microscopy. To determine differences in the intracellular localization of GDI-1 vs. GDI-2, immunofluorescence microscopy analysis is performed in CHO cells and 3T3-L1 adipocytes.[15,16] Subconfluent CHO-T cells grown on glass coverslips are washed three times in PBS, and fixed in 4% formaldehyde in PBS (10 min at 25°). Fixation and extraction with methanol (6 min at −20°) should be applied in the experiments with 3T3-L1 adipocytes for better visualization of the antigens. Cells are then washed three times in PBS, and permeabilized with 0.5% Triton X-100 and 1% FBS in PBS for 15 min. Cells are then incubated with affinity purified rabbit polyclonal anti-GDI-1 or anti-GDI-2 antibodies (2–3 μg IgG/ml in PBS, containing 1% FBS and 0.5% Triton X-100) for 90 min at 25°. We typically spot 30 μl of the appropriately diluted solution of the affinity-purified antibody on a piece of Parafilm (American Company, Greenwich,

CT), placed in a box over a pre-wet 3MM chromatography paper (Whatman, Clifton, NJ). We place the coverslip over the antibody drop with its face (cell) side down. At the end of the incubation, the coverslips are transferred in a 6-well culture plate with its face side up and washed four times (5 min each) in PBS, containing 1% FBS and 0.5% Triton X-100. The coverslips are incubated (30 min at 25°) with fluorescein isothiocyanate (FITC)-coupled goat anti-rabbit IgG (Tago, Inc., Burlingame, CA) as a secondary antibody. After the incubation the cells are thoroughly washed, first in PBS + 1% FBS + 0.5% Triton X-100 (6 times, 5 min each), then only in PBS (3 times), fixed in 4% formaldehyde (10 min, 25°), and finally washed in PBS (5 times). The coverslips are mounted on slides with a Slow Fade Kit (Molecular Probes, Eugene, OR). Negative controls are performed utilizing rabbit IgG and the secondary conjugate to assess the level of general background. Immunofluorescence analysis is performed with the digital imaging microscope (Nikon Diaphot 200), using a Nikon Apo 60/1.4 or 100/1.4 immersion lens.

Applications

We have used this approach successfully to distinguish the expression and localization patterns of GDI-1 vs. GDI-2 in 3T3-L1 adipocytes, CHO-T cells (Figs. 1 and 2), and other cultured cell lines by Western blotting and immunofluorescence microscopy.[15,16] The fact that both proteins exhibit different electrophoretic mobility (55- and 46-kDa protein bands for GDI-1 and GDI-2, respectively; Fig. 2) makes it easy to distinguish those two proteins by SDS–PAGE.[10] However, the contribution of GDIβ to GDI-2 detection should be determined separately as discussed below.

Distinguishing Expression Pattern between GDI-2 and GDIβ at mRNA Level

General Considerations

We and others have isolated a clone that is highly homologous to but distinct from GDI-2, named GDIβ (since GDI-1 is often called GDIα).[12,22] Our full-length GDIβ cDNA is obtained by screening a mouse 3T3-422A adipocyte cDNA library (a gift from Dr. B. Spiegelman) with an oligonucleotide probe, generated by the RT–PCR technique using mouse 3T3-L1 adipocyte mRNA and primers designed from the conserved regions of rGDI-1 and mGDI-2 cDNAs.[10,22] The degree of the deduced amino acid

[22] A. Shisheva, upublished data (1996).

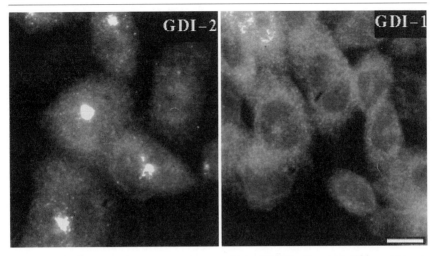

Fig. 1. Immunofluorescence microscopy analysis in CHO-T cells reveals differential localization of GDI-1 vs. GDI-2 (or GDIβ). Cells were fixed, permeabilized, and stained with affinity-purified specific anti-GDI-1 or anti-GDI-2 antibodies. Primary antibodies were then detected with FITC-conjugated goat anti-rabbit IgG. A strong focus of fluorescence in the juxtanuclear region for GDI-2, as opposed to a completely diffuse pattern for GDI-1 is apparent. Bar: 10 μm. Since the GDI-2 antibodies used in this study cross-react with the GDIβ isoform, codetection of GDIβ should be expected. (From Shisheva et al.[16] and reproduced by permission of Urban & Fischer Verlag.)

sequence identity between mouse GDI-2 and GDIβ is quite high: 96%. Random and few substitutions (most of them being conservative; 11 out of the 18 substitutions) along the deduced amino acid sequence of GDIβ vs. GDI-2 limit the design and production of specific antibodies to distinguish between the endogenous GDI-2 and GDIβ isotype expression and localization pattern by Western blotting and immunofluorescence microscopy analyses. The 89% identity between GDI-2 and GDIβ at the nucleotide level and their similar mRNA transcript sizes (2.4 kb)[10,22] predict cross-hybridization of these two isotypes, restricting their selective detection by Northern blot analysis. Also unfavorable is the fact that GDI-2 and GDIβ proteins, unlike GDI-2 and GDI-1, display the same electrophoretic mobility as judged by the appearance of His$_6$ or GST fusion proteins on Coomassie stained gels. Thus, both His$_6$–GDI-2 and His$_6$–GDIβ exhibit an apparent molecular mass of 49 kDa and both GST–GDI-2 and GST–GDIβ, an apparent molecular mass of 76 kDa (Fig. 3). Furthermore, on immunoblots equal amounts of the recombinant proteins appear as bands with identical intensity on probing with several antibodies directed against a GDI-2 peptide, the GST–GDI-2 fusion protein, or the GDIβ protein. These results

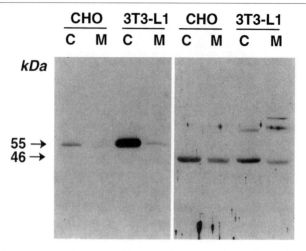

Fig. 2. Endogenous GDI-1 and GDI-2 (or GDIβ) display different electrophoretic mobility, allowing their selective detection by Western blotting analysis. Total membrane (30 μg; M) and cytosolic (20 μg; C) proteins of indicated cell types were analyzed by SDS–PAGE (duplicate gels) and immunoblotting with specific anti-GDI-1 or anti-GDI-2 antisera. GDI-1 migrates as a 55-kDa protein, while GDI-2 migrates as a 46-kDa protein (shown by arrows), facilitating their selective detection even with antisera that recognize both isoforms. The GDI-1 isoform appears to be less abundant in CHO cells, compared to 3T3-L1 adipocytes, whereas GDI-2 (and perhaps GDIβ) is equally expressed in those two cell types. It is also apparent that the relative proportion of GDI-2 on membranes markedly exceeds that of GDI-1. Because the GDI-2 antibodies cross-react with the GDIβ isoform, which exhibits the GDI-2 electrophoretic mobility (see Fig. 3), codetection of GDIβ should be expected. (Adapted from Shisheva et al.[15])

demonstrate that no specific detection of GDI-2 vs. GDIβ protein is presently achievable and indicate that we and others may have detected additively the GDI-2 and GDIβ isotypes by Northern and Western blotting or by immunofluorescence microscopy analysis.[10,12,14–16]

The length of the isolated mouse GDI-2 (from skeletal muscle cDNA library)[10] and GDIβ (from mouse 3T3-F422 adipocyte cDNA library)[22] cDNAs is 2.4 kbp. This exactly corresponds to the reported mRNA transcript size of GDIs (2.4 kb),[10] implying that our clones probably represent their complete cDNA sequences. Analysis of the sequencing information reveals that, while both clones are highly homologous in their coding regions, the corresponding 3′ UTRs are different in size (627 and 923 bp of mGDI-2 [accession number AF095729] and mGDIβ [accession number AF095728] 3′ UTRs, respectively) and display limited sequence homology (Fig. 4). These results indicate that oligonucleotide probes, generated from

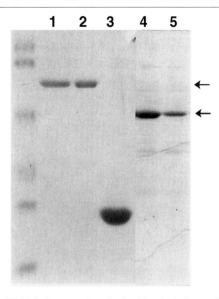

Fig. 3. GDI-2 and GDIβ fusion proteins display identical electrophoretic mobility. GST–GDI-2 fusion protein was generated as we previously described.[10] An expression vector of GDI-2 tagged with His$_6$ epitope (His$_6$–GDI-2) was engineered by ligating the EcoRI fragment of pGEX-1 GDI-2,[10] comprising the coding region, into the EcoRI site of pRSET B (Invitrogen, San Diego, CA). GST–GDIβ was generated by first eliminating the flanking nucleotide sequence (the 5' noncoding region) of the full-length mGDIβ cDNA,[22] and a subsequent subcloning of the coding region in pGEX4T-1 (Pharmacia Biotech, Piscataway, NJ) in frame with the GST gene. For this purpose, an oligonucleotide fragment was synthesized by PCR using our mouse adipocyte GDIβ clone and a primer pair that carry appropriate restriction sites (EcoRI for the sense primer; 5'-CCGG*AATTC*ATGAGGAG-3', and HindIII for the anisense primer; 5'-ACA*AAGCTT*CCCTCAATCACT-3'). The PCR product (a 321-bp oligonucleotide fragment) was digested with HindIII and EcoRI, and together with the HindIII/EcoRI fragment (1.7 kbp) of pBluescript SK (−) GDIβ cDNA was ligated into the EcoRI site of pGEX4T-1. To generate the His$_6$–GDIβ construct, the EcoRI fragment of pGEX4T-1 GDIβ was inserted into the EcoRI site from the polylinker of pRSET B (Invitrogen, San Diego, CA). GST–GDI-2 and GST–GDIβ fusion proteins were produced in E. coli strain XA-90 and purified on glutathione-agarose beads (Sigma, St. Louis, MO) as we previously described.[10] GDI-2 and GDIβ, both tagged with His$_6$ epitope at their amino terminus, were produced in E. coli strain BL21(DE3). His$_6$–GDI-2 and His$_6$–GDIβ proteins were purified on a Ni^{2+}-NTA-agarose (Qiagen, Chatsworth, CA) column utilizing 0–200 imidazole gradient following the manufacturer protocols. Purified GST–GDI-2 (lane 1), GST–GDIβ (lane 2), GST alone (lane 3), His$_6$–GDI-2 (lane 4), and His$_6$–GDIβ (lane 5) were analyzed by SDS–PAGE. Coomassie stained gels are shown, which demonstrate the identical electrophoretic mobility of recombinant GDI-2 and GDIβ fusion proteins (arrows). Molecular mass markers indicated on the left are 108, 80, 49, 34, 27.3, and 16.9 kDa (Bio-Rad, Hercules, CA).

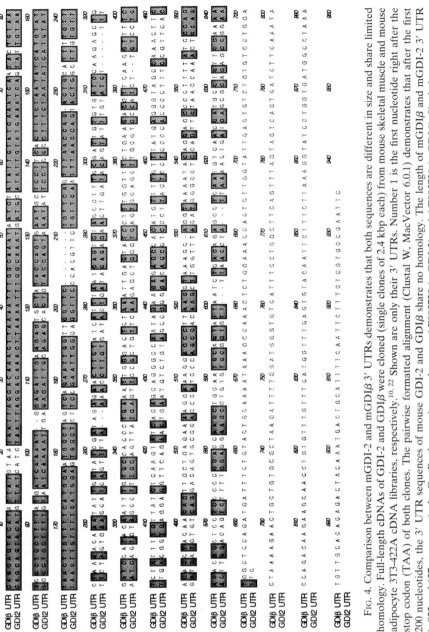

Fig. 4. Comparison between mGDI-2 and mGDIβ 3′ UTRs demonstrates that both sequences are different in size and share limited homology. Full-length cDNAs of GDI-2 and GDIβ were cloned (single clones of 2.4 kbp each) from mouse skeletal muscle and mouse adipocyte 3T3-422A cDNA libraries, respectively.[10,22] Shown are only their 3′ UTRs. Number 1 is the first nucleotide right after the stop codon (TAA) of both clones. The pairwise formatted alignment (Clustal W, MacVector 6.0.1) demonstrates that after the first 200 nucleotides, the 3′ UTR sequences of mouse GDI-2 and GDIβ share no homology. The length of mGDIβ and mGDI-2 3′ UTR is 923 and 627 nt, respectively. GenBank accession numbers are AF095729 (mGDI-2) and AF095728 (mGDIβ). Shaded areas represent nucleotide identity.

the highly divergent regions of GDI-2 and GDIβ 3' UTRs may provide a tool to distinguish the expression of GDI-2 vs. GDIIβ in cells and tissues.

Procedures

Design of Oligonucleotide Probes. Figure 4 demonstrates that after the first 180 bp, GDI-2 and GDIβ 3' UTRs share no homology. Therefore, in the probe generation we have sought to eliminate the initial fragments of their 3' UTRs, which display ~80% homology. GDIβ 3' UTR probe is generated by using the convenient and unique *Bam*HI site that cuts GDIβ 3' UTR at nucleotide 309. For this purpose, mGDIβ cDNA cloned in pBluescript SK (−)[22] is digested with *Eco*RI/*Bam*HI. The reaction is subjected to electrophoresis on a 1.2% agarose gel. The cDNA fragment corresponding to 616 bp is cut from the gel and saved for further use. The remaining two larger fragments of ~2.9 and ~1.8 kbp, representing the plasmid and the rest of GDIβ clone, respectively, are discarded.

Because convenient restriction sites at the GDI-2 3' UTR sequence are lacking, we generate the GDI-2 3' UTR probe by PCR. A 19-mer sense primer is designed to encompass nucleotides from 184 to 202 (5'-CTTCCGAATTGGCAGCTTC-3'), and is paired with a KS primer (5' CGAGGTCGACGGTATCG-3') from the pBluescript II SK (+), positioned conveniently right after the *Eco*RI cloning site of the GDI-2 insert.[10] A 480-bp cDNA fragment is amplified using pBluescript II (+) GDI-2 cDNA (2.4 bp) as a template under standard PCR conditions.[23] The PCR reaction is subjected to electrophoresis on a 2% agarose gel. The piece containing the amplified 480-bp oligonucleotide is cut from the gel and stored at 4° for further use.

ClustalW formatted alignment (MacVector, Oxford Molecular Group, Inc; Campbell, CA) reveals that the selected cDNA fragments from GDI-2 and GDIβ 3' UTRs display <40% homology in a pairwise comparison with the 3' UTR complete sequence of GDIβ and GDI-2, respectively. This low level of homology together with the multiple gaps observed in the alignment makes the cross-hybridization under stringent conditions unlikely, allowing the selective detection of GDI-2 vs. GDIβ transcript.

Probe Labeling and Northern Blot Analysis. The two oligonucleotide probes are purified from the agarose gel with a Compass DNA purification kit (American Bioanalytical, Natick, MA). The concentrations of the purified oligonucleotides are estimated from the agarose gel relative to the DNA standards (New England BioLabs, Beverly, MA). Twenty-five nano-

[23] R. K. Saiki, D. H. Gelfand, S. Stoffel, S. J. Scharf, R. Higuchi, G. T. Horn, K. B. Mullis, and H. A. Erlich, *Science* **239,** 487 (1988).

grams of the oligonucleotide is ^{32}P-labeled by random priming.[24] Procedures for RNA isolation and Northern blotting have been described in detail elsewhere.[10]

Applications

We have used this approach to distinguish the expression of GDI-2 vs. GDIβ in a variety of cultured cells and tissues. The results thus far demonstrate that the predominant form expressed in most cells and tissues is GDIβ. However, in cultured L6 myocytes, the principal form is GDI-2, the GDIβ being beyond the detection limit of the Northern blot analysis.

[24] A. P. Feinberg and B. Vogelstein, *Anal. Biochem.* **132,** 6 (1983).

[6] Expression, Purification, and Biochemical Properties of Ypt/Rab GTPase-Activating Proteins of Gyp Family

By ELKE WILL, ŠTEFAN ALBERT, and DIETER GALLWITZ

Introduction

Ypt/Rab GTPases are regulators of vesicular protein transport in exo- and endocytosis.[1,2] They are highly conserved from yeast to humans. In the budding yeast *Saccharomyces cerevisiae,* this group of the Ras superfamily has 11 members (Ypt1p, Ypt31/32p, Sec4p, Ypt51/52/53p, Ypt6p, Ypt7p, Ypt10p, and Ypt11p).[1] Only those involved in the biosynthetic pathway are essential for cell viability.[1] Like Ras and other Ras-like proteins, Ypt/Rab GTPases have a slow intrinsic GTP hydrolyzing activity (<0.0015 sec^{-1} at 30°).[3] This can be enhanced more than 10^5-fold by specific GTPase-activating proteins (GAPs).[4] Hydrolysis to GDP of the Ypt/Rab-bound GTP is thought to terminate the functional cycle, dissociate the GTPase from effector proteins,[5] and make the regulator available for a new round of action.[6]

[1] T. Lazar, M. Götte, and D. Gallwitz, *TIBS* **22,** 468 (1997).
[2] P. Novick and M. Zerial, *Curr. Opin. Cell Biol.* **9,** 496 (1997).
[3] S. Albert and D. Gallwitz, *J. Biol. Chem.* **274,** 33186 (1999).
[4] S. Albert, E. Will, and D. Gallwitz, *EMBO J.* **18,** 5216 (1999).
[5] C. Ostermeier and A. T. Brunger, *Cell* **96,** 363 (1999).
[6] V. Rybin, O. Ullrich, M. Rubino, K. Alexandrov, I. Simon, C. Seabra, R. Goody, and M. Zerial, *Nature* (*London*) **383,** 266 (1996).

Ypt/Rab-specific GAPs, termed Gyp (*GAP* for *Ypt*), were originally identified from yeast by high expression cloning.[7-9] A sophisticated computer search predicted[10] and biochemical analyses proved[3,4,11] that in yeast, Ypt/Rab GAPs form a family with several structurally related members. Related proteins were also found in other species,[10] and one of human origin was recently shown to be a GAP for Rab6.[11] Although *in vitro* each Gyp has a preferred substrate, none of them is specific for only one Ypt/Rab GTPase.[3,4,7,9,12] The catalytic domain of Gyp1p and of Gyp7p was identified and both were shown to contain a conserved arginine residue essential for catalytic activity.[4] This makes likely the same basic mechanism for the acceleration of the slow intrinsic GTPase activity of Ras, Rho, and Ypt/Rab proteins by their cognate GAPs, which provide a catalytic arginine ("arginine finger").[4,13,14]

We here describe procedures for the isolation and purification of full-length or catalytically active fragments of Gyp1p and Gyp7p from yeast and bacteria, and for measuring the GAP-accelerated GTPase activity of their preferred substrates.

Cloning and Expression of GYP Genes and of Fragments Encoding Their Catalytic Domains

General Considerations

The sequence of the entire *S. cerevisiae* genome is known. Therefore, the genes encoding Gyp1p (preferred substrates Sec4p, Ypt51p, and Ypt1p),[4,12] Gyp6p (preferred substrate Ypt6p),[7] and Gyp7p (preferred substrate Ypt7p),[8,9] which were originally identified by high expression cloning,[7,8] and those that code for related proteins[3,10] can be easily isolated by polymerase chain reaction (PCR) amplification. Appropriate primers (with restriction enzyme cleavage sites and six histidine codons for C-terminal tagging of the proteins) are designed to allow insertion into yeast or bacterial

[7] M. Strom, P. Vollmer, T. J. Tan, and D. Gallwitz, *Nature* (*London*) **361,** 736 (1993).
[8] P. Vollmer and D. Gallwitz, *Methods Enzymol.* **257,** 118 (1995).
[9] P. Vollmer, E. Will, D. Scheglmann, M. Strom, and D. Gallwitz, *Eur. J. Biochem.* **260,** 284 (1999).
[10] A. F. Neuwald, *TIBS* **22,** 243 (1997).
[11] M. Cuif, F. Possmayer, H. Zander, N. Bordes, F. Jollivet, A. Couedel-Courteille, I. Janoueix-Lerosey, G. Langsley, M. Bornens, and B. Goud, *EMBO J.* **18,** 1772 (1999).
[12] L. L. Du, R. N. Collins, and P. J. Novick, *J. Biol. Chem.* **273,** 3253 (1998).
[13] M. R. Ahmadian, P. Stege, K. Scheffzek, and A. Wittinghofer, *Nature Struct. Biol.* **4,** 686 (1997).
[14] K. Rittinger, P. A. Walker, J. F. Eccleston, S. J. Smerdon, and S. J. Gamblin, *Nature* (*London*) **389,** 758 (1997).

expression vectors for the production of full-length GAPs or fragments thereof. It was previously observed that full-length Gyp proteins could not be produced in *Escherichia coli* in appreciable amounts to allow purification and biochemical characterization.[4,7,9] Expression in yeast is therefore an alternative, especially in studies of other yeast proteins suspected to have GAP activity. Whereas all Ypt/Rab GAPs tested could be produced as highly active, C-terminally hexahistidine (His$_6$)-tagged proteins in reasonable or substantial quantities,[3,4] some of them in *E. coli*, at least one GAP, Gyp7p and its active C-terminal fragment, could not be obtained as catalytically active GST fusion.[4] Therefore, the production of C-terminally His$_6$-tagged versions is the method of choice for large-scale preparations of Ypt/Rab GAPs.

PCR Amplification

To minimize errors during DNA amplification, DNA polymerase with proofreading activity (DeepVent, New England Biolabs, Frankfurt/Main, Germany) and high concentration of DNA template are used. Each gene (or fragment) is amplified in triplicate, and PCR products are cloned and tested separately for the production of active GAP. For PCR, 100-μl reactions contain 10 μl of 10× concentrated buffer (Thermopol, New England Biolabs, Frankfurt/Main, Germany), 2–4 μg of yeast genomic DNA, 100 pmol each of the appropriate primers, 200 μM of each of the four dNTPs, and 50 μM MgCl$_2$. The reaction mixtures are preheated for 2 min at 94° before adding 2.5 units of DNA polymerase. Thirty reaction cycles are performed as follows: 94° for 30 sec, 56° for 30 sec (depends on T_m of primers), and 72° (2 min for each 1000 bp, because polymerases with proofreading activity possess slower processivity). The PCR products are cleaved with the appropriate restriction enzymes to allow insertion into the cut plasmid pYES2 (Invitrogen) for expression in yeast or into a pET vector[15] for expression in *E. coli*.

Ypt/Rab GAP Expression in Yeast

The Gyp proteins are produced in the protease-deficient yeast strains BJ5459 (Mat**a** *ura3-52 trp1 lys2-801 leu2Δ1 his3Δ200 pep4::HIS3 prb1Δ1.6R can1 GAL*) (Yeast Genetic Stock Center, University of California, Berkeley) or cl3-ABYS-86 (Matα *ura3-Δ5 leu2-3, 112 his⁻ pra1-1 prb1-1 prc1-1 cps1-3 can^R*).[16] In a first screen, we transform yeast recombinant vectors

[15] F. W. Studier, A. H. Rosenberg, J. J. Dunn, and J. W. Dubendorff, *Methods Enzymol.* **185,** 60 (1990).
[16] M. Tsukada and D. Gallwitz, *J. Cell Sci.* **109,** 2471 (1996).

from separate PCR clonings, pick one transformant each, grow the cells in 10 ml liquid medium, and induce the GAL1 promoter-controlled *GYP* genes with galactose. Total cellular protein obtained by alkali lysis of cell pellets is searched for GAP production by immunoblot analysis with a monoclonal anti-His$_6$ antibody (Invitrogen). In parallel, GAP activity is determined in the same cells disrupted with glass beads.[8] In addition, plasmid inserts of positive transformants can be subjected to sequence verification.

Milligram quantities of the 73-kDa Gyp1p, its C-terminal active domain Gyp1–46p (amino acids 249–637), of the 87-kDa Gyp7p and its catalytically active fragment Gyp7–47p (amino acids 359–745) were successfully obtained as C-terminal His$_6$-tagged versions. In the following, we describe the purification procedure for Gyp7–47–His$_6$. Cells of strain cl3-ABYS-86 newly transformed with the recombinant vector pYES2-GYP7-46 which harbors the *URA3* selective marker are grown at 30° in 100 ml PM selective medium (peptone 140-containing, Life Technologies, Karlsruhe, Germany) containing 2% (w/v) glucose. They are pelleted and grown in 1-liter PM medium containing 2% (w/v) α-D-raffinose (to avoid glucose repression of the *GAL1* promoter). At an OD$_{600}$ of 1–2, cells are pelleted for 5 min at 4000 rpm (Sorvall RC3B centrifuge) and resuspended for induction in 2 liters of prewarmed PM medium containing 2% galactose. After vigorous shaking in a 3-liter Erlenmeyer flask for 15–20 hr at 30°, cells have reached an OD$_{600}$ of >5 and are harvested by centrifugation.

Affinity Chromatography on Ni^{2+}-Agarose

Up to 20 g (wet weight) of pelleted cells are resuspended in 40 ml of cold lysis buffer A [50 mM NaH$_2$PO$_4$, pH 8.0, 0.3 M NaCl, 10 mM imidazole, 1 mM Pefabloc and protease inhibitor cocktail, complete (Roche Diagnostics, Mannheim, Germany)]. Cells are disintegrated to more than 95% by three successive passages through a French Press (Gaulin, Lübeck, Germany) at 1300 bar. The lysate is centrifuged for 10 min at 8000g and 4° to remove cell debris, and cytosolic extracts are obtained by subsequent ultracentrifugation for 1 hr at 70,000g. The supernatant is mixed with a slurry of 0.5 ml Ni^{2+}-NTA (Qiagen, Hilden, Germany) per liter of culture and incubated for 1–3 hr at 4° in a rotating 50-ml Falcon plastic tube. The Ni^{2+}-agarose is placed into a small column and washed successively with 50 ml lysis buffer A and 25 ml wash buffer (50 mM NaH$_2$PO$_4$, pH 8.0, 0.3 M NaCl, 20 mM imidazole). Bound proteins are eluted with 5 ml wash buffer containing 250 mM imidazole. Fractions of 200 μl are collected, protein concentrations of the fractions are determined using Bio-Rad Protein Assay (Bio-Rad, Richmond, CA), and protein-containing fractions are analyzed by SDS–PAGE. In general, this procedure yielded 0.5–2 mg of Gyp7–47–His$_6$ protein with a purity of 50–80%.

Ion-Exchange Chromatography and Gel Filtration

For further purification, pooled fractions with about 10 mg of affinity-purified Gyp7–47–His$_6$ from several 2-liter preparations are diluted with 3 volumes of Mono Q buffer [20 mM Tris-Cl, pH 8.0, 2 mM MgCl$_2$, 1 mM EGTA, 1 mM dithiothreitol (DTT), 10% glycerol] and loaded onto a FPLC (fast protein liquid chromatography) Mono Q HR10/10 column (Pharmacia, Piscataway, NJ). The column is washed with 30 ml Mono Q buffer (flow rate 1 ml/min), and bound proteins are eluted with a linear gradient of 0–0.6 M NaCl in Mono Q buffer. Gyp7–47–His$_6$ eluted at 0.4 M NaCl with a purity higher than 85%.

Five ml of the Gyp7–47–His$_6$ protein-containing Mono Q eluate is then subjected to gel filtration using a Sephacryl S-200 column (HiPrep 16/60, Pharmacia) equilibrated with Mono Q buffer without glycerol. Proteins are separated in the same buffer at a flow rate of 0.5 ml/min and 2-ml fractions are collected. About 4–5 mg of Gyp7–47–His$_6$ protein with >95% purity is obtained.

This procedure is also used for the preparation of full-length Gyp proteins.[3]

Ypt/Rab GAP Expression in E. coli

A C-terminally His$_6$-tagged version of the Gyp1p catalytic domain, Gyp1–46–His$_6$, can be expressed in substantial amounts in *E. coli*. The gene fragment amplified by PCR is integrated into the bacterial expression vector pET22 (Novagen, Madison, WI) to obtain pET22-GYP1-46. The recombinant plasmid was transformed into *E. coli* strain BL21(DE3) (Novagen). For protein production, 100 ml of LB + ampicillin (100 μg/ml) is inoculated with a single colony of the transformed strain and incubated at 37° overnight. One liter of prewarmed medium is inoculated with the overnight culture to an OD$_{600}$ of 0.1 and incubated at 30° until an OD$_{600}$ of 0.5 is reached. Recombinant protein expression is induced by adding isopropyl-thiogalactoside (IPTG) to a final concentration of 1 mM. Cells are shaken for 4–5 hr at 30°, pelleted, washed once with cold lysis buffer A (see above), and resuspended in 30 ml of this buffer. Cells are disrupted by sonication (Cell Disrupter B15, Branson) at an output level 4 under cooling. A cleared lysate is obtained by centrifugation for 15 min at 16,000g (HB4 rotor, Sorvall RC5B centrifuge) and mixed with 1 ml of Ni^{2+}-NTA agarose (Qiagen) per liter of culture.

Isolation of Gyp1–46–His$_6$ protein by Ni^{2+} affinity and Mono Q ion-exchange chromatography is performed as described above for proteins expressed in yeast. Per liter of *E. coli* culture, about 15 mg of Gyp1–46–His$_6$ protein can be obtained with a purity higher than 95%.

Analysis of Ypt/Rab GAP Activity

Although it has been reported that GAPs with specificity for mammalian Rab3 and Rab6 proteins require the substrate GTPase to be C-terminally prenylated,[11,17] the structurally related Ypt/Rab GAPs from yeast,[3,4,7,9,12] which contain the Gyp domain,[4] accept nonprenylated substrates. For all assays, Ypt/Rab GTPases were produced in milligram quantities in *E. coli* as previously described.[18] The preparation of α- or γ-^{32}P-loaded Ypt proteins and filter assays for GAP activity has been described in detail previously.[8] For quantitative GAP assays and kinetic analyses, GTPases loaded with unlabeled GTP were used and GTP–GDP ratios were followed by HPLC analysis.

HPLC-Based GAP Assay

Mg^{2+} is known to stabilize the binding of guanine nucleotide to the GTPase by mediating the interactions between conserved amino acid residues and β and γ phosphates of GTP. Removal of Mg^{2+} cations by chelating agents, such as EDTA, weakens this interaction significantly, which then allows the release of the bound nucleotide and its eventual replacement by a nucleotide present in the solution.

GTP-loaded GTPase can be prepared for several assays and stored at $-75°$ until use. For GTP loading, we use the following protocol. To 200 μl of purified Ypt/Rab protein at a concentration of at least 2 mg/ml ($\sim$80 μM) in reaction buffer (50 mM Tris-Cl, pH 8.0, 5 mM MgCl$_2$, 1 mM DTT), a 50-fold molar excess of GTP (4 mM final concentration) and 4 μl of 0.5 M EDTA (10 mM final concentration) are added. During incubation at room temperature for 20 min, two NAP5 columns (Pharmacia) are equilibrated with ice-cold reaction buffer, and the incubation mixture is immediately passed over one of the columns in the cold to separate protein and free nucleotide. Drop fractions are collected and protein-containing fractions (identified with Bio-Rad Protein Assay) are pooled and passed over the second column. The protein concentration is determined in the pool of peak fractions. The protein-bound GTP is assessed by high-performance

[17] K. Fukui, T. Sasaki, K. Imazumi, Y. Matsuura, H. Nakanishi, and Y. Takai, *J. Biol. Chem.* **272**, 4655 (1997).

[18] P. Wagner, L. Hengst, and D. Gallwitz, *Methods Enzymol.* **219**, 369 (1992).

liquid chromatography (HPLC) analysis of an aliquot on a calibrated 5-μm Hypersil column (250 × 4.6 mm, Bischoff, Germany) run under iso-cratic conditions.[19] Aliquots of the GTP-loaded GTPase are shock frozen in liquid nitrogen and stored at −75°.

Because the HPLC-based method does not distinguish between protein-bound and -free nucleotide, attention must be paid to separate the protein from free nucleotide. We have tested the completeness of GTP separation by incubating a small amount of freshly prepared GTP-loaded protein with an equimolar amount of appropriate GAP protein for 5 min. Such an amount of GAP causes complete hydrolysis of protein-bound but not of -free GTP. Free GTP remains unhydrolyzed. Ypt/Rab proteins also possess an intrinsic ability to exchange the bound nucleotide with one present in the solution, but this activity is very slow ($\sim$10^{-5} per min).[20] For the reaction time used (5 min), the effect of spontaneous exchange can be neglected.

For the determination of the intrinsic and GAP-accelerated GTPase activity of different Ypt/Rab proteins, 4 nmol of GTP-loaded protein in 200 μl reaction buffer is incubated at 30°. Depending on their specific activity, 1–50 pmol of purified GAPs is added to the reaction mix that is prewarmed to 30° for 2 min. Aliquots of 15 μl are taken at different time intervals (1–20 min) and pipetted into cooled microtubes that are immediately transferred to liquid nitrogen. To determine the GTP–GDP ratio, the frozen aliquots are shortly thawed in a boiling water bath and immediately subjected to the HPLC column (see above) that was run at 2 ml/min with buffer F (100 mM KH$_2$PO$_4$, pH 6.5, 10 mM tetrabutylammo-nium bromide, 0.2 mM sodium azide, 4% acetonitrile) on the HPLC system Gold (Beckman) with the pump module 126 and the detector module 166. Calibration of the column was done with GDP solutions (more stable than GTP) of known concentrations. Guanine nucleotides were detected by absorbance at 254 nm.

From the GTP and GDP peak areas, the relative amount of GTP is calculated according to

$$\text{Relative amount of GTP} = \frac{\text{GTP}}{\text{GTP} + \text{GDP}} \qquad (1)$$

and plotted as a function of time. This plot is fitted with the simple exponen-tial decay function

$$y = Y_0 + e^{-kt} \qquad (2)$$

[19] J. Tucker, G. Sczakiel, J. Feuerstein, J. John, R. S. Goody, and A. Wittinghofer, *EMBO J.* **5**, 1351 (1986).
[20] I. Simon, M. Zerial, and R. S. Goody, *J. Biol. Chem.* **271**, 20470 (1996).

TABLE I
INTRINSIC GTP HYDROLYSIS RATES OF *S. cerevisiae*
YPT GTPASES

GTPase	Rate $(\text{min}^{-1})^{a}$
Ypt1p	0.0025
Sec 4p	0.0016
Ypt31p	0.0064
Ypt32p	0.0083
Ypt51p	0.0052
Ypt52p	0.0862
Ypt53p	0.0102
Ypt6p	0.0002
Ypt7p	0.0023

[a] Measured at 30°.

[where Y_0 is the GTP/(GTP + GDP) ratio at the start of the reaction]. The resulting curve allows for the reading of the rate constant k of the reaction.

The intrinsic GTPase activities for different *S. cerevisiae* Ypt proteins determined by this procedure are given in Table I.

Kinetic Analysis of GTPase–GAP Interaction

Initially, K_m and k_{cat} values of the Gyp7p–Ypt7p interaction have been obtained by classical Michaelis–Menten kinetics. Under single turnover conditions, Ypt7p–GTP can be considered the substrate and Ypt7p–GDP the product of the reaction. Because the intrinsic rate of GTP hydrolysis is negligible compared to the GAP-activated rate, Gyp7p is regarded as an enzyme despite the fact that the catalytic center of the reaction is present on Ypt7p. To determine initial rates (V) of the GAP-catalyzed GTP hydrolysis at different Ypt7p–GTP concentrations (S_0), quantitative GAP assays were performed with 2.5, 5, 10, 15, 20, or 200 μM Ypt7p–GTP and 10 nM Gyp7–47p (E_0) as described above. By nonlinear fitting to the Michaelis–Menten equation

$$V = \frac{k_{cat}[E_0][S_0]}{[S_0] + K_m} \tag{3}$$

we obtained a K_m of 44 μM and a k_{cat} of 33.2 sec^{-1}.

As an alternative method of determining K_m and k_{cat} from a single reaction, the integrated Michaelis–Menten equation[21] can be used as has been described for the interaction of Ras and Ras–GAP.[22] The single reaction was started at a high substrate concentration (100 μM Ypt–GTP or more), and the remaining concentration of Ypt–GTP was determined by HPLC at different time points after GAP addition. The fitting procedure involves numerical integration and simulation, and leads to a representation of the concentration of Ypt/Rab–GTP as a function of time. Fitting was performed using a model file and the software Scientist (kindly provided by R. Goody, Max Planck Institute for Molecular Physiology, Dortmund, Germany).

To investigate whether the integrated Michaelis–Menten equation was applicable for Gyp–Ypt interactions, 100 μM Ypt7p–GTP (C_0) was incubated with 20 nM Gyp7–47p (E_0) in reaction buffer (see above). The resulting time curve was evaluated as described above. The calculated K_m of 42 μM and k_{cat} of 25.8 sec^{-1} are comparable to the values obtained using the classical equation. The agreement between both methods indicates that the enzyme (Gyp protein) is stable during the reaction and that the reaction product (Ypt–GDP) has no significant inhibitory effect. The integrated Michaelis–Menten equation was also used to analyze the interaction of Gyp1p and Gyp3p with their preferred substrates.[3,4]

Acknowledgments

We are indebted to Roger Goody (Dortmund) for help with the kinetic analyses of Gyp–Ypt interactions, Ursula Welscher-Altschäffel for expert technical assistance, and Ingrid Balshüsemann for secretarial assistance. This work was supported by the Max Planck Society, the Deutsche Forschungsgemeinschaft, and Fonds der Chemischen Industrie.

[21] R. G. Duggleby and R. B. Clarke, *Biochim. Biophys. Acta* **1080**, 231 (1991).
[22] T. Schweins, M. Geyer, H. R. Kalbitzer, A. Wittinghofer, and A. Warshel, *Biochemistry* **35**, 14225 (1996).

[7] Purification and Properties of Rab3 GDP/GTP Exchange Protein

By HIROYUKI NAKANISHI and YOSHIMI TAKAI

Introduction

The Rab small G protein family consists of nearly 30 members and is implicated in intracellular vesicle trafficking, such as exocytosis, endocytosis, and transcytosis.[1–6] All the Rab family members have unique C-terminal structures, which undergo posttranslational modifications with geranylgeranyl moieties in most cases. The Rab family members cycle between the GDP-bound inactive and GTP-bound active forms and between the cytosol and membranes. The Rab3 subfamily belongs to the Rab family and consists of four members, Rab3A, Rab3B, Rab3C, and Rab3D.[2] Rab3A and Rab3C are present in cells with Ca^{2+}-dependent exocytosis, particularly in neurons, whereas Rab3B and Rab3D are expressed in epithelial cells and adipocytes, respectively.[7,8] Of these subfamily members, Rab3A has been most extensively studied. Evidence is accumulating that Rab3A is involved in Ca^{2+}-dependent exocytosis, particularly neurotransmitter release.[2] The Rab3A knockout mouse analysis has revealed that Rab3A plays two roles: one is to efficiently dock synaptic vesicles to the presynaptic plasma membrane; and the other is to regulate the efficiency of the fusion process.[9,10] It has also been reported that Rab3A is involved in the formation of long-term potentiation in hippocampus.[11]

The precise mechanisms of Rab3A in the regulation of these docking and fusion processes remain to be clarified, but Rab3A is regulated by at

[1] C. Nuoffer and W. E. Balch, *Annu. Rev. Biochem.* **63,** 949 (1994).

[2] Y. Takai, T. Sasaki, H. Shirataki, and H. Nakanishi, *Genes Cells* **1,** 615 (1996).

[3] P. Novick and M. Zerial, *Curr. Opin. Cell Biol.* **9,** 496 (1997).

[4] V. M. Olkkonen and H. Stenmark, *Int. Rev. Cytol.* **176,** 1 (1997).

[5] F. Schimmöller, I. Simon, and S. R. Pfeffer, *J. Biol. Chem.* **273,** 22161 (1998).

[6] O. Martinez and B. Goud, *Biochim. Biophys. Acta* **1404,** 101 (1998).

[7] E. Weber, G. Berta, A. Tousson, P. St. John, M. W. Green, U. Gopalokrishnan, T. Jilling, E. J. Sorscher, T. S. Elton, D. R. Abrahamson, and K. L. Kirk, *J. Cell Biol.* **125,** 583 (1994).

[8] G. Baldini, T. Hohl, H. Y. Lin, and H. L. Lodish, *Proc. Natl. Acad. Sci. U.S.A.* **89,** 5049 (1992).

[9] M. Geppert, V. Y. Bolshakov, S. A. Siegelbaum, K. Takei, P. Decamill, R. E. Hammer, and T. C. Südhof, *Nature* **369,** 494 (1994).

[10] M. Geppert, Y. Goda, C. F. Stevens, and T. C. Südhof, *Nature* **387,** 810 (1997).

[11] P. E. Castillo, R. Janz, T. C. Südhof, T. Tzounopoulos, R. C. Malenka, and R. A. Nicoll, *Nature* **388,** 590 (1997).

0076-6879/00 $35.00

least three types of regulators: Rab3 GDP/GTP exchange protein (GEP),[12] Rab3 GTPase-activating protein (GAP) (see Rab3 GAP in this volume[12a]),[13] and Rab GDP dissociation inhibitor (GDI).[14,15] Of these regulators, Rab GDI is a general regulator of all the Rab family members, whereas Rab3 GEP and Rab3 GAP are specific for the Rab3 subfamily.[2] Rab3 GEP stimulates the GDP/GTP exchange reaction and thereby the conversion from the GDP-bound form to the GTP-bound form. Rab3 GAP stimulates the GTPase activity and thereby the conversion from the GTP-bound form to the GDP-bound form. Rab GDI has three activities: (1) Rab GDI preferentially interacts with the GDP-bound form and keeps it in the GDP-bound form in the cytosol by preventing it from converting to the GTP-bound form by the action of each Rab GEP and from the association with each target membrane; (2) Rab GDI transports its complexed Rab family member to each target membrane where the GDP-bound form dissociates from Rab GDI by the action of each putative Rab GDI displacement factor, followed by the conversion to the GTP-bound form by the action of each Rab GEP; and (3) after the GTP-bound form accomplishes its function and is converted to the GDP-bound form by the action of each Rab GAP, Rab GDI forms a complex with it and translocates it to the cytosol. All three regulators are abundant in brain, but they are ubiquitously expressed.

This chapter describes the assays for Rab3 GEP activity, the procedures for the purification of native Rab3 GEP from rat brain, the procedures for the purification of recombinant Rab3 GEP from COS7 cells, and the properties of Rab3 GEP.

Materials

(p-Amidinophenyl)methanesulfonyl fluoride (APMSF) is purchased from Wako Pure Chemicals (Osaka, Japan). Leupeptin, chloroquine, and bovine serum albumin (BSA) (fraction V) are from Sigma Chemical Co. (St. Louis, MO). 3-[(3-Cholamidopropyl)dimethylammonio]-1-propanesulfonic acid (CHAPS) and sodium cholate are from Dojindo Laboratories (Kumamoto, Japan). Guanosine 5'-(3-O-thio)triphosphate (GTPγS) is from Boehringer Mannheim (Indianapolis, IN). [³H]GDP (518 GBq/mmol) and

[12] M. Wada, H. Nakanishi, A. Satoh, H. Hirano, H. Obaishi, Y. Matsuura, and Y. Takai, *J. Biol. Chem.* **272,** 3875 (1997).

[12a] F. Nagano, T. Sasaki, and Y. Takai, *Methods Enzymol.* **329,** [8], (2001) (this volume).

[13] K. Fukui, T. Sasaki, K. Imazumi, Y. Matsuura, H. Nakanishi, and Y. Takai, *J. Biol. Chem.* **272,** 4655 (1997).

[14] T. Sasaki, A. Kikuchi, S. Araki, Y. Hata, M. Isomura, S. Kuroda, and Y. Takai, *J. Biol. Chem.* **265,** 2333 (1990).

[15] Y. Matsui, A. Kikuchi, S. Araki, Y. Hata, J. Kondo, Y. Teranishi, and Y. Takai, *Mol. Cell. Biol.* **10,** 4116 (1990).

[^{35}S]GTPγS (40.7 TBq/mmol) are obtained from Amersham-Pharmacia Biotech. Inc. (Piscataway, NJ) and Du Pont–New England Nuclear (Boston, MA), respectively. BA-85 nitrocellulose filters (pore size, 0.45 μm) are purchased from Schleicher & Schuell (Dassel, Germany). Q-Sepharose FF, phenyl-Sepharose, Mono Q HR10/10, Mono Q PC1.6/5, HiLoad 16/60 Superdex 200, and DEAE-dextran are from Amersham-Pharmacia Biotech. Inc. Hydroxyapatite is from Calbiochem-Novabiochem Co. (La Jolla, CA). A high-performance liquid chromatography (HPLC) hydroxyapatite column (0.78 × 10 cm) is from Koken Co. Ltd. (Tokyo, Japan). Centriprep 30 is from Millipore Co. (Bedford, MA). Dulbecco's modified Eagle's medium (DMEM) and fetal calf serum (FCS) are from Life Technologies Inc. (Rockville, MD). All other chemicals are of reagent grade.

Lipid-modified Rab3A, Rab3B, Rab3C, Rab3D, Rab2, Rab5A, Rab10, and Rab11 are purified from the membrane fraction of *Spodoptera frugiperda* cells (Sf9 cells) infected with the baculovirus carrying each cDNA.[16,17] Lipid-unmodified Rab3A is purified from Rab3A-overexpressing *Escherichia coli* as a fusion protein with N-terminal glutathione S-transferase (GST), of which the glutathione S-transferase carrier is cleaved off from Rab3A by digestion with thrombin.[18] These Rab family members are dissolved in a buffer containing 20 mM Tris-HCl at pH 7.5, 5 mM MgCl$_2$, 1 mM EDTA, 1 mM dithiothreitol (DTT), and 0.6% CHAPS. Rab GDI is purified from bovine brain cytosol.[14]

The various buffers used in the purification of Rab3 GEP are as follows:

Buffers for Purification of Rab3 GEP from Rat Brain

Buffer A: 0.32 M sucrose, 1 mM NaHCO$_3$, 1 mM MgCl$_2$, 0.5 mM CaCl$_2$, and 1 μM AMPSF

Buffer B: 0.32 M sucrose, 1 mM NaHCO$_3$, and 10 μg/ml leupeptin

Buffer C: 6 mM Tris-HCl at pH 8.0 and 10 μg/ml leupeptin

Buffer D: 20 mM Tris-HCl at pH 7.5 and 1 mM DTT

Buffer E: 20 mM potassium phosphate at pH 7.8, 1 mM DTT, 0.6% CHAPS, and 10% glycerol (v/v)

Buffer F: 20 mM bis-Tris-Cl at pH 5.5, 0.5 mM EDTA, 1 mM DTT, 0.6% CHAPS, and 10% glycerol (v/v)

Buffer G: 20 mM Tris-HCl at pH 7.5, 0.5 mM EDTA, 1 mM DTT, 0.6% CHAPS, 0.45% sodium cholate, 10% glycerol (v/v), and 0.15 M NaCl

---

[16] H. Horiuchi, O. Ullrich, C. Bucci, and M. Zerial, *Methods Enzymol.* **257,** 9 (1995).

[17] A. Kikuchi, H. Nakanishi, and Y. Takai, *Methods Enzymol.* **257,** 57 (1995).

[18] A. Miyazaki, T. Sasaki, K. Araki, N. Ueno, K. Imazumi, F. Nagano, K. Takahashi, and Y. Takai, *FEBS Lett.* **350,** 333 (1994).

Buffers for Purification of Recombinant Rab3 GEP from COS7 Cells

Buffer A: 20 mM Tris-Cl at pH 7.9, 140 mM NaCl, 3 mM KCl, 1 mM CaCl$_2$, 0.5 mM MgCl$_2$, and 0.9 mM Na$_2$PO$_4$
Buffer B: 20 mM Tris-Cl at pH 7.5, 1 mM DTT, and 0.6% CHAPS

Methods

Assays for Rab3 GEP Activity

The Rab3 GEP activity to stimulate the GDP/GTP exchange reaction of Rab3A is assayed by measuring either the dissociation of [^{3}H]GDP from or binding of [^{35}S]GTPγS to lipid-modified Rab3A.

Dissociation Assay. [^{3}H]GDP-bound Rab3A is made by incubating Rab3A (3 pmol) at 30° for 20 min with 3 μM [^{3}H]GDP (7–9 × 10^3 cpm/pmol) in a reaction mixture (5 μl) containing 50 mM Tris-HCl at pH 8.0, 5 mM MgCl$_2$, 10 mM EDTA, 0.5 mM DTT, and 0.12% CHAPS. The reaction is stopped by adding 2 μl of 100 mM MgCl$_2$ and 5 μl of a solution containing 50 mM Tris-HCl at pH 8.0, 5 mM MgCl$_2$, 0.5 mM EDTA, and 1 mM DTT, and the mixture is immediately cooled on ice. The sample to be assayed is incubated at 30° for 10 min with [^{3}H]GDP-bound Rab3A (3 pmol) in a reaction mixture (50 μl) containing 50 mM Tris-HCl at pH 8.0, 12 mM MgCl$_2$, 2 mM EDTA, 0.2 mg/ml BSA, 12 μM GTPγS, and 0.06% CHAPS. The reaction is stopped by adding 2 ml of an ice-cold solution containing 20 mM Tris-HCl at pH 7.5, 25 mM MgCl$_2$, and 100 mM NaCl to the reaction mixture, followed by rapid filtration on BA-85 nitrocellulose filters and washing with 2 ml of the same solution four times. The radioactivity trapped on the filters is determined by liquid scintillation counting.

Binding Assay. The sample to be assayed is incubated at 30° for 10 min with GDP-bound Rab3A (3 pmol) in a reaction mixture (50 μl) containing 50 mM Tris-HCl at pH 8.0, 12 mM MgCl$_2$, 2 mM EDTA, 0.2 mg/ml BSA, 12 μM [^{35}S]GTPγS (6–8 × 10^3 cpm/pmol), and 0.06% CHAPS. GDP-bound Rab3A used is the sample purified from Sf9 cells because it is purified as the GDP-bound form. The reaction is stopped and the radioactivity trapped on the filters is counted as described above.

Purification of Rab3 GEP from Rat Brain

The steps used in the purification of Rab3 GEP from rat brain are as follows: (1) preparation of the synaptic soluble fraction from rat brain; (2) Q-Sepharose FF column chromatography; (3) phenyl-Sepharose column chromatography; (4) hydroxyapatite column chromatography; (5) Mono Q HR10/10 column chromatography; (6) HiLoad 16/60 Superdex 200 column chromatography; and (7) HPLC hydroxyapatite column chromatography. All the purification procedures are carried out at 0–4°.

Preparation of Synaptic Soluble Fraction from Rat Brain. The synaptic soluble fraction is prepared as follows[19]: Cerebra are rapidly removed from 80 rats after decapitation. Homogenization is performed with 12 up-and-down strokes of a Potter–Elvehjem Teflon–glass homogenizer in buffer A (cerebral tissues 10 g wet weight per 40 ml of buffer A). The homogenates are combined, diluted to 800 ml with buffer A, and filtrated through four layers of gauze. The homogenate is centrifuged at 1400g for 10 min. The pellet is resuspended with three strokes of the homogenizer in 700 ml of the same buffer and centrifuged at 710g for 10 min. The resultant pellet, designated as P1 fraction, contains nuclei and cell debris. The supernatants from the two steps of centrifugation are combined and centrifuged at 13,800g for 10 min. The pellet is resuspended and rehomogenized in 700 ml of the same buffer and then centrifuged again at 13,800g for 10 min. The resultant pellet, designated as P2 fraction, contains myelin, mitochondria, and synaptosomes. This pellet is resuspended with three strokes of the homogenizer in 200 ml of buffer B. The suspension is diluted with 900 ml of buffer C, stirred for 45 min, and centrifuged at 32,800g for 20 min. The supernatant is further centrifuged at 78,000g for 120 min. The final supernatant is pooled as the synaptic soluble fraction. This synaptic soluble fraction can be stored at −80° for at least 3 months.

Q-Sepharose FF Column Chromatography. A half of the synaptic soluble fraction (550 ml, 0.9 g of protein) is adjusted to 0.2 M NaCl and applied to a Q-Sepharose FF column (2.6 × 10 cm) equilibrated with buffer D containing 0.2 M NaCl. Elution is performed with 350 ml of buffer D containing 0.5 M NaCl at a flow rate of 5 ml/min. Fractions of 10 ml each are collected. The Rab3 GEP activity appears in fractions 5–19.

Phenyl-Sepharose Column Chromatography. The active fractions of the Q-Sepharose column chromatography (150 ml, 159 mg of protein) are collected, and NaCl is added to give a final concentration of 2 M. The sample is applied to a phenyl-Sepharose column (2.6 × 10 cm) equilibrated with buffer D containing 2 M NaCl. Elution is performed with a 360-ml linear gradient of NaCl (2–0 M) in buffer D, followed by 180 ml of buffer D, at a flow rate of 3 ml/min. Fractions of 6 ml each are collected. The Rab3 GEP activity appears in fractions 52–63.

Hydroxyapatite Column Chromatography. The active fractions of the phenyl-Sepharose column chromatography (72 ml, 8.6 mg of protein) are collected and applied to a hydroxyapatite column (1.0 × 30 cm) equilibrated

[19] A. Mizoguchi, T. Ueda, K. Ikeda, H. Shiku, H. Mizoguchi, and Y. Takai, *Mol. Brain Res.* **5**, 31 (1089).

with buffer E. Elution is performed with a 75-ml linear gradient of potassium phosphate (20–100 mM) in buffer E and a subsequent 75-ml linear gradient (100–300 mM) in buffer E, followed by a 50-ml linear gradient (300–500 mM) in buffer E, at a flow rate of 1 ml/min. Fractions of 2.5 ml each are collected. The Rab3 GEP activity appears in fractions 46–54.

Mono Q HR10/10 Column Chromatography. The active fractions of the hydroxyapatite column chromatography (22.5 ml, 2.2 mg of protein) are collected, diluted with an equal volume of buffer F, and applied to a Mono Q HR10/10 column equilibrated with buffer F. Elution is performed with a 60-ml linear gradient of NaCl (0.2–0.5 M) in buffer F at a flow rate of 1 ml/min. Fractions of 1 ml each are collected. The Rab3 GEP activity appears in fractions 24–33.

HiLoad 16/60 Superdex 200 Column Chromatography. The active fractions of the Mono Q HR10/10 column chromatography (10 ml, 0.44 mg of protein) are collected, concentrated to about 2 ml by a centrifugal ultrafiltration concentrator (Centriprep 30), and applied to a HiLoad 16/60 Superdex 200 column (1.6 × 60 cm) equilibrated with buffer G. Elution is performed with the same buffer at a flow rate of 0.25 ml/min. Fractions of 2 ml each are collected. The Rab3 GEP activity appears in fractions 26–30. The position of this peak corresponds to a molecular mass of about 270 kDa. When an aliquot of each fraction (20 μl) is subjected to sodium dodecyl sulfate–polyacrylamide gel electrophoresis (SDS–PAGE) (6.5% polyacrylamide gel), followed by protein staining with silver, one protein band with a molecular mass of about 200 kDa coincides well with the GEP activity.

HPLC Hydroxyapatite Column Chromatography. The active fractions of the Superdex 200 column chromatography (10 ml, 45 μg of protein) are collected. These collected fractions can be stored at −80° for at least 3 months. The other half of the synaptic soluble fraction is also subjected to the successive column chromatographies in a manner similar to that described above. The active fractions of the two Superdex 200 column chromatographies are combined and applied to the HPLC hydroxyapatite column (0.78 × 10 cm) equilibrated with buffer E. Elution is performed with a 12.5-ml linear gradient of potassium phosphate (20–100 mM) in buffer E, followed by a 50-ml linear gradient of potassium phosphate (100–500 mM) in buffer B, at a flow rate of 0.5 ml/min. Fractions of 1 ml each are collected. The Rab3 GEP activity appears in two peaks in fractions 29–33 and 34–38. When an aliquot of each fraction (20 μl) is subjected to SDS–PAGE (6.5% polyacrylamide gel), followed by protein staining with silver, protein bands with a molecular mass of about 200 kDa coincide well with the GEP activity of both peaks. These bands are Rab3 GEPs. The first (5 ml, 15.5 μg of protein) and second (5 ml, 7.5 μg of protein) peaks are separately collected as Rab3 GEPI and GEPII, respectively. Rab3 GEPI and GEPII can be stored at −80° for at least 6 months.

Purification of Recombinant Rab3 GEP from COS7 Cells

The steps used in the purification of recombinant Rab3 GEP from COS7 cells are as follows: (1) expression of recombinant Rab3 GEP in COS7 cells and (2) Mono Q PC1.6/5 column chromatography.

Expression of Recombinant Rab3 GEP in COS7 Cells. An about 5.7-kbp Rab3 GEP cDNA containing the complete coding region is inserted into the pCMV5 vector. This construct, pCMV5-Rab3 GEP, is transfected to COS7 cells with the DEAE-dextran method as follows[20]: COS7 cells (1×10^6 cells) are plated on a 10-cm dish and cultured for 1 day in DMEM supplemented with 10% FCS. The cells are washed with phosphate-buffered saline (PBS) and incubated at 37° for 30 min with 44 μg of pCMV5-Rab3 GEP in 2 ml of buffer A containing 0.5 mg/ml DEAE-dextran. After the solution is aspirated, the cells are incubated at 37° for 3 hr with 100 mM chloroquine in DMEM supplemented with 10% FCS. The cells are washed with PBS, followed by incubation at room temperature for 2 min in DMEM containing 20% glycerol. The cells are gently washed with PBS and incubated at 37° for 48 hr in DMEM supplemented with 10% FCS.

Mono Q PC 1.6/5 Column Chromatography. The COS7 cells expressing Rab3 GEP are washed with PBS and scrapped. The cells are homogenized with a Teflon–glass homogenizer in 2 ml of buffer B containing 1 μM APMSF and 10 μg/ml leupeptin, and centrifuged at 100,000g for 60 min. The supernatant (2 ml, 4.2 mg of protein) is applied to a Mono Q PC1.6/5 column equilibrated with buffer B containing 0.1 M NaCl. Elution is performed with a 1.5-ml linear gradient of NaCl (0.1–0.5 M) in buffer B at a flow rate of 50 μl/min. Fractions of 50 μl each are collected. The Rab3 GEP activity appears in fractions 19–21. The active fractions (150 μl, 0.1 mg of protein) are collected and used as recombinant Rab3 GEP. This recombinant Rab3 GEP can be stored at −80° for at least 3 months.

Properties of Rab3 GEPI, GEPII, and Recombinant Rab3 GEP

Rab3 GEPI, GEPII, and recombinant Rab3 GEP show similar properties, including the substrate specificity, the requirement of the lipid modifications of Rab3A, and the ineffectiveness to Rab3A complexed with Rab GDI.

---

[20] Y. Hata and T. C. Südohof, *J. Biol. Chem.* **270**, 13022 (1995).

Substrate Specificity. The Rab3 GEP activity is assayed in a manner similar to that described above except that various Rab family members are used instead of Rab3A as substrates. The Rab family members are purified from the membrane fraction of Sf9 cells. These Rab family members are the lipid-modified forms because they are sensitive to Rab GDI. Rab3 GEPI, GEPII, and recombinant Rab3 GEP stimulate the dissociation of [^{3}H]GDP from and the binding of [^{35}S]GTPγS to Rab3A and Rab3C in dose-dependent and time-dependent manners. These Rab3 GEPs partially stimulate the GDP/GTP exchange reaction of Rab3D, but hardly Rab3B. The Rab3 GEPs do not catalyze the reaction of other Rab subfamily members, including Rab2, Rab5A, Rab10, and Rab11.

Requirement of Lipid Modifications of Rab3A. The Rab3 GEP activity is assayed in a manner similar to that described above except that lipid-unmodified Rab3A is used instead of lipid-modified Rab3A as a substrate. Lipid-unmodified Rab3A is purified from *E. coli.* Rab3 GEPI, GEPII, and recombinant Rab3 GEP do not stimulate the dissociation of [^{3}H]GDP from and the binding of [^{35}S]GTPγS to lipid-unmodified Rab3A.

Ineffectiveness to Rab3A Complexed with Rab GDI. In the dissociation assay, [^{3}H]GDP-bound Rab3A is premixed with various doses of Rab GDI (0–1.0 μM) and then incubated with Rab3 GEPI, GEPII, or recombinant Rab3 GEP. In the binding assay, GDP-bound Rab3A is similarly premixed with Rab GDI and incubated with Rab3 GEPI, GEPII, or recombinant Rab3 GEP. Rab3A used is the lipid-modified form and forms a complex with Rab GDI. The Rab3 GEP activities of Rab3 GEPI, GEPII, and recombinant Rab3 GEP are inhibited by Rab GDI in a dose-dependent manner.

Comments

We have described the purification of Rab3 GEPI and GEPII from rat brain. On the basis of the amino acid sequences of the peptides of Rab3 GEPII, the cDNA of recominant Rab3 GEP is cloned. However, when the peptide maps of Rab3 GEPI and GEPII are determined, they are apparently identical. Moreover, an antibody raised against recombinant Rab3 GEP recognizes both Rab3 GEPI and GEPII. The exact relationship between Rab3 GEPI and GEPII is not known, but it is likely that they are splicing variants.

Because Rab3 GEPI and GEPII are separated by the last column chromatography, it is difficult to calculate the exact purification folds and yields of Rab3 GEPI and GEPII through the purification procedures. However, on the assumption that they are purified with the same yields, the purification folds of Rab3 GEPI and GEPII are calculated to be about 2000 and

6000 folds, respectively, of the synaptic soluble fraction. The purification yields of Rab3 GEPI and GEPII are calculated to be both about 4%. It can be estimated that the amounts of Rab3 GEPI and GEPII are both about 0.01% of the total proteins in the synaptic soluble fraction. Rab3 GEPI and GEPII are about 20 and 60% pure, respectively, as estimated by SDS–PAGE followed by protein staining with Coomassie brilliant blue. On the basis of these observations, it can be estimated that the k_{cat} values of Rab3 GEPI and GEPII are both about 40 nmol/min/nmol.

Northern and Western blot analyses indicate that Rab3 GEPI and GEPII are expressed in all the rat tissues examined with the highest expression in brain.[12] Western blot analysis of the subcellular fractions of rat brain indicates that Rab3 GEPI and GEPII are highly enriched in the synaptic soluble fraction.[21] Consistently, immunofluorescence microscopic analysis of primary culture hippocampal neurons from rat embryo indicates that Rab3 GEPI and GEPII are localized at the synaptic release sites.[21] Therefore, the preparation of the synaptic soluble fraction is a crical step in the purification of Rab3 GEPI and GEPII.

[21] H. Oishi, T. Sasaki, F. Nagano, W. Ikeda, T. Ohya, M. Wada, N. Ide, H. Nakanishi, and Y. Takai, *J. Biol. Chem.* **273,** 34580 (1998).

[8] Purification and Properties of Rab3 GTPase-Activating Protein

By FUMIKO NAGANO, TAKUYA SASAKI, and YOSHIMI TAKAI

Introduction

The Rab small G protein family consists of nearly 40 members in mammal and 11 members in yeast, and is implicated in intracellular vesicle trafficking.[1–6] Like other family members, the Rab family members (Rab proteins) cycle between the GDP-bound inactive and GTP-bound active forms, and the GTP-bound form interacts with their specific effector proteins. The GTP-bound form is converted by the action of the intrinsic

[1] C. Nuoffer and W. E. Balch, *Annu. Rev. Biochem.* **63,** 949 (1994).
[2] Y. Takai, T. Sasaki, H. Shirataki, and H. Nakanishi, *Genes Cells* **1,** 615 (1996).
[3] P. Novick and M. Zerial, *Curr. Opin. Cell Biol.* **9,** 496 (1997).
[4] V. M. Olkkonen and H. Stenmark, *Int. Rev. Cytol.* **176,** 1 (1997).
[5] F. Schimmöller, I. Simon, and S. R. Pfeffer, *J. Biol. Chem.* **273,** 22161 (1998).
[6] O. Martinez and B. Goud, *Biochim. Biophys. Acta* **1404,** 101 (1998).

GTPase activity to the GDP-bound form, which then releases the bound effector proteins. Because GTPase-activating proteins (GAPs) stimulate this reaction, they are assumed to play a crucial role in terminating the functions of the substrate small G proteins. However, it is not yet clear whether GTP hydrolysis is important for Rab proteins to accomplish their functions.[2,7] Rab3 GAP, which is the first GAP specific for Rab proteins in mammal, has originally been purified with Rab3A as a substrate from rat brain synaptic soluble fraction.[8] This GAP is specifically active on the Rab3 subfamily members (Rab3A, Rab3B, Rab3C, and Rab3D), and prefers the lipid-modified form to the lipid-unmodified form. Of the Rab3 subfamily members, Rab3A and Rab3C are implicated in Ca^{2+}-dependent exocytosis, particularly in neurotransmitter release.[2] Evidence is accumulating that Rab3A is involved in the docking and/or fusion processes. Although the precise mechanism of Rab3A in the regulation of these processes remains to be clarified, the GTP-bound form might interact with a prefusion complex, thereby preventing fusion.[2] The GTPase-deficient mutant of Rab3A inhibits Ca^{2+}-dependent exocytosis from PC12 cells and chromaffin cells,[9,10] suggesting that Rab3 GAP plays a crucial role in the function of Rab3A.

Rab3 GAP shows two bands with molecular weights of about 130,000 (p130) and 150,000 (p150) on sodium dodecyl sulfate–polyacrylamide gel electrophoresis (SDS–PAGE).[8] The cDNAs of p130 and p150 have been cloned from a human brain cDNA library, and the encoded proteins show no homology to any known protein.[8,11] By Northern blot analysis, both p130 and p150 are shown to be ubiquitously expressed. The subcellular fractionation analysis in rat brain indicates that both p130 and p150 are enriched in the synaptic soluble fraction.[11,12] p150 is coimmunoprecipitated with p130 from this fraction.[11] Recombinant p150 forms a heterodimer with recombinant p130 as estimated by sucrose density gradient ultracentrifugation. Recombinant p130 exhibits the GAP activity toward the Rab3 subfamily members and the catalytic domain is located at the C-terminal region.

[7] T. C. Südhof, *Neuron* **18,** 519 (1997).

[8] K. Fukui, T. Sasaki, K. Imazumi, Y. Matsuura, H. Nakanishi, and Y. Takai, *J. Biol. Chem.* **272,** 4655 (1997).

[9] R. W. Holz, W. H. Brondyk, R. A. Senter, L. Kuizon, and I. G. Macara, *J. Biol. Chem.* **269,** 10229 (1994).

[10] L. Johannes, P. M. Lledo, M. Roa, J. D. Vincent, J. P. Henry, and F. Darchen, *EMBO J.* **13,** 2029 (1994).

[11] F. Nagano, T. Sasaki, K. Fukui, T. Asakura, K. Imazumi, and Y. Takai, *J. Biol. Chem.* **273,** 24781 (1998).

[12] H. Oishi, T. Sasaki, F. Nagano, W. Ikeda, T. Ohya, M. Wada, N. Ide, H. Nakanishi, and Y. Takai, *J. Biol. Chem.* **273,** 34580 (1998).

In contrast, recombinant p150 neither shows the Rab3 GAP activity nor affects the activity of recombinant p130. These results indicate that Rab3 GAP consists of the catalytic (p130) and noncatalytic (p150) subunits. The role of the noncatalytic subunit of Rab3 GAP is unknown, but Sar1 GAP also consists of the catalytic (Sec23) and noncatalytic (Sec24) subunits, both of which are required for the function of Sar1 in the vesicle budding from the endoplasmic reticulum.[13]

This chapter describes the assays for the Rab3 GAP activity, the procedures for the purification of native Rab3 GAP from rat brain synaptic soluble fraction, the procedures for the purification of recombinant hexahistidine (His$_6$)-tagged Rab3 GAP from *Escherichia coli*, and the properties of Rab3 GAP.

Materials

Dithiothreitol (DTT), EGTA, Nonidet P-40 (NP-40), and Triton X-100 are purchased from Nacalai Tesque (Kyoto, Japan). EDTA, HEPES, and 3-[(3-cholamidopropyl)dimethylammonio]-1-propanesulfonic acid (CHAPS) are purchased from Dojindo Laboratories (Kumamoto, Japan). (*p*-Amidinophenyl)methanesulfonyl fluoride (APMSF) and isopropyl-β-D-thiogalactopyranoside (IPTG) are purchased from Wako Pure Chemicals (Osaka, Japan). [γ-^{32}P]GTP (185 TBq/mmol) and [α-^{32}P]GTP (110 TBq/mmol) are obtained from Amersham Pharmacia Biotech (Milwaukee, WI). BA-85 nitrocellulose filters (pore size, 0.45 μm) are purchased from Schleicher & Schuell (Dassel, Germany). All other chemicals are of reagent grade.

Lipid-modified and lipid-unmodified Rab3As are purified from the membrane and soluble fractions, respectively, of *Spodoptera frugiperda* cells (Sf9 cells) infected with the baculovirus carrying the Rab3A cDNA.[14] The lipid-modified form of Rab2, Rab3B, Rab3C, Rab3D, Rab5A, and Rab11 are purified from the membrane fraction of Sf9 cells infected with the baculovirus carrying each cDNA in a similar manner. All Rab proteins are dissolved in a buffer containing 20 m*M* Tris-HCl at pH 8.0, 5 m*M* MgCl$_2$, 1 m*M* EDTA, 1 m*M* DTT, and 0.6% CHAPS.

Expression plasmids, pRSET-p130 and pRSET-p150, are constructed by the following procedures. The 2946-bp fragment containing the complete p130 cDNA coding region with the *Kpn*I sites upstream of the initiator methionine codon and downstream of the termination codon is synthesized by polymerase chain reaction (PCR). This fragment is digested by *Kpn*I and inserted into the *Kpn*I site of pRSETB (Invitrogen BV, Groningen,

---

[13] L. Hicke, T. Yoshihisa, and R. Schekman, *Mol. Biol. Cell* **3**, 667 (1992).
[14] A. Kikuchi, H. Nakanishi, and Y. Takai, *Methods Enzymol.* **257**, 57 (1995).

Netherlands). The 4182-bp fragment containing the complete p150 cDNA coding region with the *Eco*RI sites upstream of the initiator methionine codon and downstream of the termination codon is obtained as follows: The N-terminal fragment (base pairs 1–1151) with the *Eco*RI site upstream of the initiator methionine codon and with the *Xba*I site in the coding region is synthesized by PCR. This fragment is digested by *Eco*RI and *Xba*I and inserted into the *Eco*RI and *Xba*I sites of pBluescript (Stratagene, La Jolla, CA). The C-terminal fragment (base pairs 290–4182) is obtained from screening a human brain cDNA library by hybridization and inserted into the *Eco*RI site of pBluescript. The fragment (base pairs 1152–4182) is digested by *Xba*I from the latter plasmid and inserted into the *Xba*I site of the former plasmid. The ligated fragment is digested by *Eco*RI and inserted into the *Eco*RI site of pRSETB. An *E. coli* strain DE3 is transformed with pRSET-p130 and pRSET-p150.

Methods

Assay for Rab3 GAP Activity

The Rab3 GAP activity to stimulate the intrinsic GTPase activity of Rab3A is assayed by three methods as follows: (1) standard assay (filter assay), (2) overlay assay, and (3) thin-layer chromatography assay.

Standard Assay (Filter Assay). Lipid-modified Rab3A (3 pmol) is incubated at 30° for 10 min in a reaction mixture (10 μl) containing 25 mM Tris-HCl at pH 8.0, 10 mM EDTA, 5 mM MgCl$_2$, 0.5 mM DTT, 0.3% CHAPS, and 1.5 μM [γ-^{32}P]GTP (1 $\times$ 10^4 cpm/pmol). The reaction is stopped by adding 2.5 μl of 80 mM MgCl$_2$. To this mixture (12.5 μl), the sample to be assayed is added in a total volume of 50 μl and further incubated at 30° for 5 min. The reaction is stopped by adding 2 ml of an ice-cold solution containing 20 mM Tris-HCl at pH 7.5, 25 mM MgCl$_2$, and 100 mM NaCl to the reaction mixture, followed by rapid filtration on BA-85 nitrocellulose filters and washing with the same solution three times. The radioactivity retained on the filter is determined by Cerenkov counting.

Overlay Assay. The sample to be assayed is subjected to SDS–PAGE. After semidry Western blotting, the nitrocellulose filter-bound proteins are renatured in phosphate-buffered saline (PBS) containing 1% bovine serum albumin, 0.5 mM MgCl$_2$, 0.1% Triton X-100, and 5 mM DTT. The filter is incubated at 25° for 10 min with [γ-^{32}P]GTP–Rab3A (3 pmol), which is prepared with the same method as described above, in a buffer containing 25 mM HEPES/NaOH at pH 7.0, 1.25 mM MgCl$_2$, 0.05% Triton X-100, and 2.5 mM DTT. After the filter is washed with PBS containing 25 mM HEPES/NaOH at pH 7.0, 5 mM MgCl$_2$, and 0.05% Triton X-100, the hydrolysis of [γ-^{32}P]GTP bound to Rab3A is analyzed with Fujix BAS 2000 Imaging Analyzer.

Thin-Layer Chromatography Assay. [α-^{32}P]GTP–Rab3A (3 pmol) is made as described above except that [α-^{32}P]GTP is used instead of [γ-^{32}P]GTP. The sample to be assayed is mixed with [α-^{32}P]GTP–Rab3A in a reaction mixture (50 μl) containing 35 mM Tris-HCl at pH 8.0, 12 mM MgCl$_2$, 2 mM EDTA, 0.2 mM EGTA, 1 mM DTT, and 0.06% CHAPS at 30° for 5 min. The mixture is applied to a nitrocellulose filter and then rinsed three times with an ice-cold solution containing 20 mM Tris-HCl at pH 7.5, 25 mM MgCl$_2$, and 100 mM NaCl. Guanine nucleotides bound to Rab3A are eluted by immersing the filter in a buffer containing 20 mM Tris-HCl at pH 8.0, 20 mM EDTA, 2% SDS, 1 mM GDP, and 1 mM GTP at 65° for 5 min. The released nucleotides are separated on a polyethylene-imine-cellulose thin-layer chromatography plate (Macherey-Nagel, Düren, Germany) with 1 M KH$_2$PO$_4$ at pH 3.4. After developing the thin-layer chromatography plate, the plate is dried, and the GDP and GTP spots are visualized with a short-wave ultraviolet lamp and analyzed with Fujix BAS 2000 Imaging Analyzer.

Purification of Native Rab3 GAP

The various buffers used in the isolation of native Rab3 GAP are as follows:

    Buffer A: 20 mM Tris-HCl at pH 7.5, 0.5 mM EGTA, 0.5 mM EDTA, and 1 mM DTT

    Buffer B: 20 mM potassium phosphate at pH 7.5, 0.5 mM EDTA, and 1 mM DTT

The steps used in the purification of native Rab3 GAP are as follows: (1) preparation of the synaptic soluble fraction from rat brain, (2) Q-Sepharose FF column chromatography, (3) hydroxyapatite column chromatography, (4) heparin-Sepharose CL-6B column chromatography, and (5) Mono Q PC 1.6/5 column chromatography. All the purification procedures are performed at 0–4°.

Preparation of Synaptic Soluble Fraction from Rat Brain. The synaptic soluble fraction is prepared from 200 rat brains as described (see the chapter on Rab3 GEP by H. Nakanishi and Y. Takai in this volume[15]).

Q-Sepharose FF Column Chromatography. One-fifth of the synaptic soluble fraction (450 ml, 315 mg of protein) is directly applied to a Q-Sepharose FF column (2.6 × 23 cm) equilibrated with buffer A. After the column is washed with 600 ml of buffer A, elution is performed with

---

[15] H. Nakanishi and Y. Takai, *Methods Enzymol.* **329**, [7], (2001) (this volume).

a 600-ml linear gradient of NaCl (0–0.5 M) in buffer A, followed by 120 ml of 0.5 M NaCl in buffer A at a flow rate of 5 ml/min. Fractions of 8 ml each are collected. One peak of the Rab3 GAP activity appears in fractions 62–70. These fractions (72 ml, 36 mg of protein) are collected. The rest of the synaptic soluble fraction is subjected to the same Q-Sepharose column chromatography four times in a similar manner.

Hydroxyapatite Column Chromatography. The samples of the five Q-Sepharose column chromatographies are pooled and diluted with 720 ml of buffer B. The sample is applied to a hydroxyapatite column (2.6 × 6.6 cm) equilibrated with buffer B. After the column is washed with 350 ml of the same buffer, elution is performed with a 500-ml linear gradient of potassium phosphate (20–212 mM) in buffer B, followed by a 150-ml linear gradient (212–500 mM) and 150 ml of 500 mM potassium phosphate in buffer B at a flow rate of 1.25 ml/min. Fractions of 10 ml each are collected. One peak of the Rab3 GAP activity appears in fractions 29–40. These fractions (120 ml, 18 mg of protein) are collected.

Heparin-Sepharose CL-6B Column Chromatography. The sample is diluted with 240 ml of buffer A and applied to a heparin-Sepharose CL-6B column (0.5 × 5 cm) equilibrated with buffer A. After the column is washed with 20 ml of the same buffer, elution is performed with 0.5 M NaCl in buffer A at a flow rate of 0.5 ml/min. Fractions of 1 ml each are collected. One peak of the Rab3 GAP activity appears in fractions 2–6. These fractions (5 ml, 4 mg of protein) are collected.

Mono Q PC 1.6/5 Column Chromatography. One-fifth of the sample is diluted with 2 ml of buffer A and applied to a Mono Q PC 1.6/5 column equilibrated with 280 mM NaCl in buffer A. After the column is washed with 2 ml of the same buffer, elution is performed with a 3-ml linear gradient of NaCl (280–500 mM) in buffer A, followed by a 0.5-ml linear gradient of NaCl (0.5–1 M) and 0.5 ml of 1 M NaCl in buffer A at a flow rate of 0.1 ml/min. Fractions of 0.1 ml each are collected. One peak of the Rab3 GAP activity appears in fractions 10 and 11. This GAP activity coincides well with two protein bands with molecular weights of about 130,000 and 150,000 as estimated by SDS–PAGE. These fractions (0.2 ml, 14 μg of protein) are collected. The rest of the heparin-Sepharose sample is subjected to the same Mono Q column chromatography four times in a similar manner. The samples of the five Mono Q column chromatographies are pooled and stored at −80° until use.

Expression and Purification of Recombinant Rab3 GAP

The buffer used in the isolation of recombinant Rab3 GAP is buffer C: 50 mM sodium phosphate at pH 7.5 and 50 mM NaCl.

The steps used in the purification of recombinant Rab3 GAP are as follows: (1) cultivation of *E. coli* and induction of His$_6$-tagged Rab3 GAP, (2) preparation of crude supernatant, and (3) affinity purification of His$_6$-tagged Rab3 GAP.

Cultivation of E. coli and Induction of His$_6$-Tagged Rab3 GAP. *Escherichia coli* DE3 transformed with pRSET-p130 or pRSET-p150 is cultured at 37° in 1 liter of LB medium containing 50 μg per ml ampicillin to an OD$_{595}$ of 0.2. After the addition of IPTG at a final concentration of 1 m*M* for p130 (0.1 m*M* for p150), cells are further cultured at 30° for p130 (25° for p150) for 4 hr. All procedures after this step are performed at 0–4°. Cells are harvested, suspended in 20 ml of PBS, and washed with 20 ml of PBS. The cell pellet is frozen at −80°.

Preparation of Crude Supernatant. The cell pellet is quickly thawed at 37° and suspended in 10 ml of buffer C containing 1 mg per ml lysozyme and 400 μ*M* APMSF, and the cell suspension is sonicated at a setting of 60 by an ultrasonic processor (Taitec, Tokyo, Japan) on ice for 10 sec six times at 1-min intervals. The homogenate is centrifuged at 100,000*g* for 1 hr. The supernatant is used for the affinity purification.

Affinity Purification of His$_6$-Tagged Rab3 GAP. One ml of Ni^{2+}-NTA-agarose beads (Qiagen K.K., Tokyo, Japan) is washed in a batch twice with 4 ml of distilled water; once with 4 ml of a buffer containing 50 m*M* sodium phosphate at pH 6.3, 300 m*M* NaCl, and 200 m*M* imidazole; three times with 4 ml of a buffer containing 20 m*M* sodium phosphate at pH 7.8 and 0.5 *M* NaCl. Forty milliliters of the crude supernatant prepared as described above is then incubated with the beads on a rotating wheel for 2 hr. After the incubation, the beads are spun down at 800*g* for 5 min and washed in a batch once with 10 ml of buffer C and once with 10 ml of buffer C containing 40 m*M* imidazole. Then, the beads are packed onto a 5-ml disposable syringe, washed with 10 ml of buffer C containing 40 m*M* imidazole, and eluted with 5 ml of buffer C containing 500 m*M* imidazole. The eluate is dialyzed with 1 liter of buffer A three times. The purity and protein concentrations are analyzed by SDS–PAGE, followed by protein staining with Coomassie Brilliant Blue.

Properties of Rab3 GAP

Activity of Recombinant Rab3 GAP. The Rab3 GAP activity is assayed in a manner similar to that described above except that recombinant Rab3 GAP is used instead of native Rab3 GAP. Native Rab3 GAP and recombinant p130 stimulate the GTPase activity of Rab3A in dose-dependent and time-dependent manners, but the specific activity of recombinant p130 is weaker than that of native Rab3 GAP. Recombinant p150 does not show

Rab3 GAP activity under the conditions where native Rab3 GAP and recombinant p130 show activity. In addition, recombinant p150 does not affect the Rab3 GAP activity of recombinant p130.

Substrate Specificity. The Rab3 GAP activity is assayed in a manner similar to that described above except that various Rab proteins are used instead of Rab3A as substrates. Rab proteins are purified from the membrane fraction of Sf9 cells. These Rab proteins are the lipid-modified forms because they are sensitive to Rab GDI. Native Rab3 GAP and recombinant p130 stimulate the GTPase activity of Rab3B, Rab3C, and Rab3D as well as that of Rab3A, but the GAP activity against Rab3B is slightly weaker than that against Rab3A, Rab3C, and Rab3D. These Rab3 GAPs do not catalyze the reaction of other Rab proteins, including Rab2, Rab5A, and Rab11.

Requirement of Lipid Modifications of Rab3A. The Rab3 GAP activity is assayed in a manner similar to that described above except that lipid-unmodified Rab3A is used instead of lipid-modified Rab3A as a substrate. Native Rab3 GAP or recombinant p130 does not stimulate the GTPase activity of lipid-unmodified Rab3A.

Coimmunoprecipitation of p130 and p150 from Rat Brain Synaptic Soluble Fraction. Rat brain synaptic soluble fraction (2.7 mg of protein) is incubated with 26 μg of the anti-p130 or anti-p150 polyclonal antibody bound to 40 μl of protein A-Sepharose in 2 ml of a solution containing 10 mM Tris-HCl at pH 8.0, 1 mM EDTA, 150 mM NaCl, and 1% NP-40. Each immunoprecipitate is subjected to SDS–PAGE, followed by protein staining with Coomassie Brilliant Blue or by immunoblotting with the anti-p130 and anti-p150 polyclonal antibodies. In these experiments, protein bands corresponding to p130 and p150 are immunoprecipitated with either antibody at a molar ratio of about 1:1. These results indicate that p130 and p150 form a complex in the synaptic soluble fraction.

Complex Formation of Recombinant p130 and p150. Recombinant p150 (50 pmol) is incubated with recombinant p130 (50 pmol) at 4° for 20 min. Recombinant p130 alone (50 pmol), recombinant p150 alone (50 pmol), or the mixture of recombinant p130 and p150 is subjected to ultracentrifugation using 4.8 ml of a continuous sucrose density gradient (5–40% sucrose in buffer A). Centrifugation is performed at 220,000g for 14 hr. Fractions of 150 μl each are collected. A 20-μl aliquot of each fraction is subjected to SDS–PAGE, followd by protein staining with silver. When purified Rab3 GAP is subjected to the same continuous sucrose density gradient ultracentrifugation, p130 and p150 appear in a single peak with a molecular weight of 300,000. In these experiments, recombinant p130 and p150 appear in a single peak with a molecular weight of about 110,000 and 170,000, respectively. As to the mixture of recombinant p130 and p150, both the proteins mostly shift to the position with a molecular weight of 300,000. The molar ratio of p130 and p150 in the peak fraction is estimated to be about 1 : 1 by SDS–PAGE. These results indicate that p130 interacts directly with p150 and forms a heterodimer.

Comments

We have shown the three assay methods for the Rab3 GAP activity. By use of the standard assay, many samples can be assayed for a short period. Therefore, this assay is selected for the purification procedures of native Rab3 GAP. In contrast, the overlay assay is useful for the detection of the protein band showing the Rab3 GAP activity. However, these two assay methods are not accurate and do not distinguish the GAP activity with the nonspecific phosphatase activity or the GTP dissociation activity. Therefore, the GAP activity should be confirmed by thin-layer chromatography assay.

Of the purification steps for native Rab3 GAP from rat brain synaptic soluble fraction, Mono Q column chromatography by use of the SMART system (Amersham Pharmacia Biotech) is the most important step to obtain a large amount of purified Rab3 GAP. Small total gel volume and dead volume in this system contribute to low nonspecific adsorption, resulting in the superior recovery. In addition, because there is less dilution in this system, it is possible to achieve sample concentration, which also increases the recovery.

[9] Rabphilin-3: A Target Molecule for Rab3 Small G Proteins

By Hiromichi Shirataki, Takuya Sasaki, and Yoshimi Takai

Introduction

Rab3A, a member of the Rab small G protein family, is implicated in Ca^{2+}-dependent exocytosis, particularly in neurotransmitter release.[1,2] Recent analyses of Rab3A-deficient mice have revealed an important insight into Rab3A function. In the hippocampal CA1 region of the mice, synaptic depletion is much faster, although two forms of short-term synaptic

[1] Y. Takai, T. Sasaki, H. Shirataki, and H. Nakanishi, *Genes Cells* **1,** 615 (1996).
[2] L. Jr. Gonzalez and R. H. Scheller, *Cell* **96,** 755 (1999).

plasticity, paired-pulse facilitation and posttetanic potentiation, are unaffected.[3] These findings suggest that Rab3A plays a role in the recruitment of synaptic vesicles for exocytosis. In the Rab3A-deficient hippocampal cultured cells, Rab3A is found to be involved in the late step in the vesicle fusion.[4] Moreover, Rab3A-deficient mice show reduced postsynaptic long-term potentiation formation in the CA3 region.[5] Thus, Rab3A is not essential for basal synaptic transmission, but modulates the synaptic vesicle trafficking, thereby contributing to synaptic plasticity.

Rab3A is converted between the GDP-bound and GTP-bound forms. GTP-Rab3A interacts with a target molecule and functions.[1] Rabphilin-3 has been identified as a target molecule for Rab3A by the use of a cross-link technique.[6,7] Rabphilin-3, which is highly expressed in brain, is mainly localized on synaptic vesicles.[7-9] Rabphilin-3, a protein with a calculated molecular weight of 77,976, is composed of at least three different domains: the Rab3 domain (amino acids 40–170), which interacts with Rab3 and is located at the N-terminal region; the M (middle) domain (amino acids 171–402), which is phosphorylated by protein kinase A and calmodulin-dependent protein kinase II and located at the middle region; and the C2 domain (amino acids 403–704), which interacts with Ca^{2+} and phospholipid and is located at the C-terminal region where there are two repeats of C2-like domains as described for synaptotagmin.[7,10-13] Rabphilin-3 has a weak activity to activate the Rab3A GTPase activity (GAP activity), a strong activity to inhibit the Rab3A GAP activity (GIP activity), and a weak activity to stimulate the GDP/GTP exchange reaction of Rab3A (GEP

[3] M. Geppert, V. Y. Bolshakov, S. A. Siegelbaum, K. Takei, P. De Camilli, R. E. Hammer, and T. C. Südhof, *Nature* **369,** 493 (1994).

[4] M. Geppert, Y. Goda, C. F. Stevens, and T. C. Südhof, *Nature* **387,** 810 (1997).

[5] P. E. Castillo, R. Janz, T. C. Südhof, T. Tzounopoulos, R. C. Malenka, and R. A. Nicoll, *Nature* **388,** 590 (1997).

[6] H. Shirataki, K. Kaibuchi, T. Yamaguchi, K. Wada, H. Horiuchi, and Y. Takai, *J. Biol. Chem.* **267,** 10946 (1992).

[7] H. Shirataki, K. Kaibuchi, T. Sakoda, S. Kishida, T. Yamaguchi, K. Wada, M. Miyazaki, and Y. Takai, *Mol. Cell. Biol.* **13,** 2061 (1993).

[8] A. Mizoguchi, Y. Yano, H. Hamaguchi, H. Yanagida, C. Ide, A. Zahraoui, H. Shirataki, T. Sasaki, and Y. Takai, *Biochem. Biophys. Res. Commun.* **202,** 1235 (1994).

[9] C. Li, K. Takei, M. Geppert, L. Daniell, K. Stenius, E. R. Chapman, R. Jahn, P. De Camilli, and T. C. Südhof, *Neuron* **13,** 885 (1994).

[10] T. Yamaguchi, H. Shirataki, S. Kishida, M. Miyazaki, J. Nishikawa, K. Wada, S. Numata, K. Kaibuchi, and Y. Takai, *J. Biol. Chem.* **268,** 27164 (1993).

[11] B. Stahl, J. H. Chou, C. Li, T. C. Südhof, and R. Jahn, *EMBO J.* **15,** 1799 (1996).

[12] S. Numata, H. Shirataki, S. Hagi, T. Yamamoto, and Y. Takai, *Biochem. Biophys. Res. Commun.* **203,** 1927 (1994).

[13] M. Kato, T. Sasaki, K. Imazumi, K. Takahashi, K. Araki, H. Shirataki, Y. Matsuura, A. Ishida, H. Fujisawa, and Y. Takai, *Biochem. Biophys. Res. Commun.* **205,** 1776 (1994).

activity).[14,15] Although the precise function of Rabphilin-3 is not known, it has been revealed from overexpression and injection experiments of various truncated forms of Rabphilin-3 that it is implicated in neurotransmitter release.[16–18] Recently, another target molecule for Rab3A, Rim, has been identified.[19] In contrast to Rab3A and Rabphilin-3, Rim is clearly absent from synaptic vesicles but enriched on the presynaptic plasma membrane, especially at the active zone. Although the precise function of Rim is still obscure, it has been revealed from overexpression experiments of its N-terminal fragment that it is also implicated in neurotransmitter release.

A mode of action of Rabphilin-3 in neurotransmitter release has not yet been clarified. However, since various truncated forms of Rabphilin-3 affect Ca^{2+}-dependent exocytosis as described above, it is thought that Rabphilin-3 functions through interactions with other molecules. β-Adducin,[20] GTP cyclohydrolase I,[21] α-actinin,[22] and rabaptin5,[23] a target molecule for Rab5, have been identified as Rabphilin-3-interacting molecules.

This chapter first describes the purification method for HA-tagged Rabphilin-3 from the membrane fraction of overexpressing *Spodoptera frugiperda* cells (Sf9 cells). The chapter then describes the methods for detecting the interactions of Rabphilin-3 with α-actinin and rabaptin5, and the effects of Rabphilin-3 on the α-actinin-induced actin filament bundling and the receptor-mediated endocytosis.

Materials

A plasmid for expression of HA-tagged Rabphilin-3 in Sf9 cells is constructed as follows. A DNA fragment encoding the HA (YPYDVPDYA)

[14] S. Kishida, H. Shirataki, T. Sasaki, M. Kato, K. Kaibuchi, and Y. Takai, *J. Biol. Chem.* **268,** 22259 (1993).

[15] Y. Fujita, T. Sasaki, K. Araki, K. Takahashi, K. Imazumi, M. Kato, Y. Matsuura, and Y. Takai, *FEBS Lett.* **353,** 67 (1994).

[16] S. H. Chung, Y. Takai, and R. W. Holz, *J. Biol. Chem.* **270,** 16714 (1995).

[17] R. Komuro, T. Sasaki, S. Orita, M. Maeda, and Y. Takai, *Biochem. Biophys. Res. Commun.* **219,** 435 (1996).

[18] M. E. Burns, T. Sasaki, Y. Takai, and G. J. Augustine, *J. Gen. Physiol.* **111,** 243 (1998).

[19] Y. Wang, M. Okamoto, F. Schmitz, K. Hofmann, and T. C. Südhof, *Nature* **388,** 593 (1997).

[20] M. Miyazaki, H. Shirataki, H. Kohno, K. Kaibuchi, A. Tsugita, and Y. Takai, *Biochem. Biophys. Res. Commun.* **205,** 460 (1994).

[21] K. Imazumi, T. Sasaki, K. Takahashi, and Y. Takai, *Biochem. Biophys. Res. Commun.* **205,** 1409 (1994).

[22] M. Kato, T. Sasaki, T. Ohya, H. Nakanishi, H. Nishioka, M. Imamura, and Y. Takai, *J. Biol. Chem.* **271,** 31775 (1996).

[23] T. Ohya, T. Sasaki, M. Kato, and Y. Takai, *J. Biol. Chem.* **273,** 613 (1998).

epitope with the methionine codon is inserted into the *Bam*HI site of pAcYM1 *Autographa californica* baculovirus transfer vector to express the HA-tagged fusion proteins (pAcYM1-HA). The 2.1-kb fragment, containing the complete Rabphilin-3 coding region with *Kpn*I sites upstream of the initiator methionine codon and downstream of the termination codon, is synthesized by the polymerase chain reaction (PCR).[24] This fragment is digested with *Kpn*I and inserted into the *Kpn*I-cut pAcYM1-HA vector to express the HA-tagged Rabphilin-3 cDNA under the control of the polyhedrin promoter in Sf9 cells.[25] Plasmids for expression of the HA-tagged N-terminal fragment (amino acids 1–280) and HA-tagged C-terminal fragment (amino acids 396–704) of Rabphilin-3 are constructed by the same procedures. Mammalian expression plasmids, pEFBOS-HA and pEFBOS-*myc,* are generated to express fusion proteins with the N-terminal HA and *myc* epitopes, respectively.[26] The same cDNA fragment of Rabphilin-3 as described above is inserted into *Kpn*I-cut pEFBOS-HA. The cDNA fragment encoding human transferrin receptor is inserted into *Bam*HI-cut pEFBOS-*myc.*

Anti-Rabphilin-3 polyclonal and anti-α-actinin monoclonal antibodies are prepared as described.[8,27] Native α-actinin is purified from chicken gizzard as described.[28] Recombinant chicken lung type α-actinin is purified from overexpressing *Escherichia coli.*[22] Glutathione *S*-transferase (GST)–rabaptin5, a GST-fusion protein containing the rat rabaptin5 fragment, which corresponds to an amino acid position from 385 to 753 of human rabaptin5, is purified from overexpressing *E. coli* as described.[23] An anti-HA antibody is from Berkeley Antibody Co. (Richmond, CA). Fluorescein isothiocyanate (FITC)–transferrin is from Molecular Probes (Eugene, OR). Rhodamine-conjugated anti-mouse donkey antibody is from Chemicon International (Temecula, CA). Lipofectin is from Life Technologies, Inc. (Rockville, MD). (*p*-Amidinophenyl)methanesulfonyl fluoride (APMSF), Triton X-100, formaldehyde, and uranyl acetate are from Wako Pure Chemicals (Osaka, Japan). HEPES, EGTA, sodium cholate, and dithio-threitol (DTT) are from Nacalai Tesque (Kyoto, Japan). EDTA and 3-[(3-cholamidopropyl)dimethylammonio]-1-propanesulfonate (CHAPS) are from Dojindo Laboratories (Kumamoto, Japan). Leupeptin, antipain, benzamidine, ATP, and Nonidet P-40 (NP-40) are from Sigma (St. Louis,

[24] J. Sambrook, E. F. Fritsch, and T. Maniatis, *Molecular cloning: A laboratory manual,* 2nd Ed. Cold Spring Harbor Laboratory, Cold Spring Harbor, NY, 1989.

[25] Y. Matsuura, R. D. Possee, H. A. Overton, and D. H. L. Bishop, *J. Gen. Virol.* **68,** 1233 (1987).

[26] S. Orita, T. Sasaki, R. Komuro, G. Sakaguchi, M. Maeda, H. Igarashi, and Y. Takai, *J. Biol. Chem.* **271,** 7257 (1996).

[27] M. Imamura and T. Masaki, *J. Biol. Chem.* **267,** 25927 (1980).

[28] J. R. Feramisco and K. Burridge, *J. Biol. Chem.* **255,** 1194 (1980).

MO). A Centricon-30 microconcentrator is from Amicon Inc. (Beverly, MA). Protein A-Sepharose CL-4B and glutathione-Sepharose 4B are from Amersham Pharmacia Biotech (Buckinghamshire, England).

Methods

The various buffers used in the purification procedures and the assays are as follows:

Buffer A: 20 mM HEPES/NaOH at pH 7.4, 1 mM dithiothreitol (DTT), 1 mM EDTA, 1 mM EGTA, 30 μM APMSF, 25 μM leupeptin, 25 μM antipain, and 0.5 mM benzamidine

Buffer B: 20 mM HEPES/NaOH at pH 7.4, 1 mM DTT, and 1% sodium cholate

Buffer C: 20 mM HEPES/NaOH at pH 7.4 and 1 mM DTT

Purification of HA-Tagged Rabphilin-3

The steps used in the purification of HA-tagged Rabphilin-3 are as follows: (1) preparation of the membrane fraction from Sf9 cells, (2) heparin-Sepharose CL-6B column chromatography, and (3) Superose 12 HR10/30 column chromatography.

Preparation of Membrane Fraction from Sf9 Cells. The following procedures are carried out at 0–4°. The Sf9 cells expressing HA-tagged Rabphilin-3 (1 × 10^8 cells) are collected, washed three times with 30 ml of phosphate-buffered saline (PBS), and suspended with 30 ml of buffer A. The suspension is sonicated at a setting of 60 by an ultrasonic processor (Taitec, Tokyo, Japan) on ice for 30 sec four times at 1-min intervals, followed by centrifugation at 100,000g for 1 hr. The pellet is suspended with 30 ml of buffer A containing 2.5% sodium cholate with gentle stirring for 30 min. After the centrifugation at 100,000g for 1 hr, the supernatant is divided into three tubes and stored at −80°. The supernatant can be stored for at least 3 months at −80°.

Heparin-Sepharose CL-6B Column Chromatography. One-third of the supernatant (9 ml, 67.5 mg of protein) is applied to a heparin-Sepharose column (0.5 × 5 cm) equilibrated with buffer B. After the column is washed with 10 ml of the same buffer, elution is performed with a 15-ml linear gradient of NaCl (0–1.0 M). Fractions of 0.5 ml each are collected. HA-tagged Rabphilin-3 appears as a single peak in fractions 65–76. The active fractions are pooled and concentrated to 1.5 ml by Centricon-30. The concentrate is divided into three tubes and can be stored for at least 3 months at −80°.

Superose 12 HR10/30 Column Chromatography. One-third of the concentrate (0.5 ml, 3.2 mg of protein) is applied to a Superose 12 column (1.0 × 30 cm) equilibrated with buffer B containing 1 *M* NaCl. Elution is performed with 20 ml of the same buffer. Fractions of 0.5 ml each are collected. HA-tagged Rabphilin-3 appears as a single peak in fractions 23–25. These procedures are repeated three times in the same way. The collected active fractions are dialyzed three times against buffer C containing 0.5 *M* NaCl and concentrated to 1 ml by Centricon-30. The concentrate (1 ml, 0.6 mg of protein) is used as HA-tagged Rabphilin-3. HA-tagged Rabphilin-3 can be stored for at least 3 months at −80°. The freezing and thawing of HA-tagged Rabphilin-3 should not be repeated more than three times.

HA-tagged Rabphilin-3 is detected by immunoblotting with the anti-Rabphilin-3 antibody.[29] However, unless this antibody is available, HA-tagged Rabphilin-3 can be detected by immunoblotting with the anti-HA antibody. In analysis of HA-tagged Rabphilin-3 on SDS–PAGE, SDS–PAGE is performed on 10% gels. Protein concentrations are determined with bovine serum albumin as a reference protein.[30]

The HA-tagged N-terminal fragment (amino acids 1–280) and HA-tagged C-terminal fragment (amino acids 396–704) of Rabphilin-3 are purified by the same procedures as described above except that the Sf9 cells expressing each fragment are used.

Properties of Rabphilin-3

Assay for Interaction of Rabphilin-3 with α-Actinin. Native α-actinin (100 pmol) is incubated at 4° for 90 min with HA-tagged Rabphilin-3 (20 pmol) bound to 25 μl of protein A-Sepharose through the anti-HA antibody in a buffer containing 20 m*M* Tris-HCl at pH 8.0, 1 m*M* DTT, 1 m*M* EGTA, 0.26% CHAPS, and 150 m*M* NaCl. After washing three times with the same buffer, α-actinin associated with the beads is detected by SDS–PAGE, followed by immunoblotting with the anti-α-actinin monoclonal antibody.

α-Actinin interacts with both HA-tagged Rabphilin-3 and its N-terminal fragment but to little extent with its C-terminal fragment. The stoichiometry of binding between α-actinin and HA-tagged Rabphilin-3 is at least 0.1 under the assay conditions. GTPγS–Rab3A inhibits the interactions of α-actinin with HA-tagged Rabphilin-3 and its N-terminal fragment. These results indicate that α-actinin interacts with Rabphilin-3 free of Rab3A.

Assay for Effect of Rabphilin-3 on Actin Filament Bundling Activity of α-Actinin. Recombinant chicken lung type α-actinin (120 n*M*) is incubated

[29] H. Towbin, T. Staehelin, and J. Gordon, *Proc. Natl. Acad. Sci. U.S.A.* **76,** 4350 (1979).
[30] M. M. Bradford, *Anal. Biochem.* **72,** 248 (1976).

with HA-tagged Rabphilin-3 (0.2 μM) at 4° for 1 hr in a buffer containing 20 mM Tris-HCl at pH 7.2, 0.6 mM DTT, 4 mM EDTA, 8 mM MgCl$_2$, 0.12% CHAPS, and 100 mM KCl. After the incubation, the sample is mixed with F-actin (1.0 mg/ml) in a buffer containing 20 mM Tris-HCl at pH 7.2, 2 mM EDTA, 0.65 mM EGTA, 6 mM MgCl$_2$, 0.06% CHAPS, 100 mM KCl, and 180 μM ATP. In one set of experiments, after the sample is sucked into a 0.1-ml micropipette and incubated at 25° for 1 hr, the time for a stainless steel ball to fall a fixed distance in the pipette is measured and converted into viscosity in centipoise using various concentrations of sucrose solution at 20° as a standard. In another set of experiments, after incubation at 25° for 1 hr, the sample is negatively stained with 2% uranyl acetate and viewed with a Hitachi electron microscope (model H-7100).[22]

Both HA-tagged Rabphilin-3 and its N-terminal fragment increase the viscosity, whereas its C-terminal fragment does not affect the viscosity. None of them affect the viscosity in the absence of α-actinin. In transmission electron microscopic analysis, both HA-tagged Rabphilin-3 and its N-terminal fragment stimulate the α-actinin-induced actin filament bundling, whereas its C-terminal fragment does not. Neither HA-tagged Rabphilin-3 nor its N-terminal fragment alone causes actin filaments to associate into a bundle. These results suggest that Rabphilin-3 serves as a linker for Rab3A and cytoskeleton.

Assay for Interaction of Rabphilin-3 with Rabaptin5. HA-tagged Rabphilin-3 (50 pmol) is incubated with GST-rabaptin5 (100 pmol) bound to glutathione-Sepharose 4B beads in a buffer containing 20 mM Tris-HCl at pH 7.5, 1 mM DTT, 5 mM MgCl$_2$, 0.5% NP-40, and 20 mM NaCl at 4° for 90 min. After washing four times with the same buffer, the bound proteins are eluted by addition of 100 μl of a buffer containing 20 mM Tris-HCl at pH 8.0, 1 mM DTT, 5 mM MgCl$_2$, 0.5% NP-40, 20 mM NaCl, and 10 mM reduced glutathione. The eluates are subjected to SDS–PAGE, followed by protein staining with Coomassie Brilliant Blue.

The stoichiometry of binding between HA-tagged Rabphilin-3 and GST-rabaptin5 is about 0.1 under the assay conditions. HA-tagged Rabphilin-3 binds to GST-rabaptin5 through its N-terminal Rab3 domain. GTPγS-Rab3A inhibits the interaction of HA-tagged Rabphilin-3 with GST-rabaptin5. GDP-Rab3A slightly inhibits it, because GDP-Rab3A also weakly binds to Rabphilin-3. These results indicate that rabaptin5 interacts with Rabphilin-3 free of Rab3A.

Assay for Effect of Rabphilin-3 on Receptor-Mediated Endocytosis. PC12 cells are transfected with the plasmid encoding *myc*-tagged human transferrin receptor and plasmid encoding HA-tagged N-terminal fragment of Rabphilin-3 using lipofectin reagent. Forty-eight hours after the transfection, the cells are incubated at 37° for 1 hr with serum-free Dulbecco's modified Eagle's medium (DMEM) containing 0.1 mg/ml FITC-transferrin. The cells are then fixed with 4% formaldehyde and permeabilized with 0.2% Triton X-100. The HA epitope is detected with the anti-HA antibody as the first antibody and rhodamine-conjugated anti-mouse donkey antibody as the second antibody. The cells are analyzed with an LSM 410 confocal laser scanning microscope (Carl Zeiss, Oberkochen, Germany).

HA-tagged N-terminal fragments of Rabphilin-3 inhibit the incorporation of FITC–transferrin into the cells. This inhibition is overcome by a dominant active mutant of Rab3A, Rab3AQ81L, or rabaptin5. These results suggest that Rabphilin-3 is implicated in endocytosis after Rabphilin-3 complexed with GTP-Rab3A regulates exocytosis.

Comments

HA-tagged Rabphilin-3 is very sticky under low ionic conditions. Therefore, HA-tagged Rabphilin-3 should be stored at $-80°$ under high ionic conditions (over 0.5 M NaCl) or conditions where ionic detergents (for example, 0.2% sodium cholate) are present.

It has been shown that abnormalities of synaptic transmission and synaptic plasticity, which are observed in Rab3A-deficient mice, are not observed in Rabphilin-3-deficient mice.[31] However, this does not necessarily indicate that Rabphilin-3 is not involved in neurotransmitter release. It is possible that Rim, another target molecule for Rab3A, compensates for the loss of function of Rabphilin-3 in the mice, that abnormalities of synaptic transmission and synaptic plasticity which are still not detected by the methods performed so far are present, and that in addition to the loss of Rabphilin-3, loss of other molecules is required for abnormalities of synaptic transmission and synaptic plasticity.

---

[31] O. M. Schlüter, E. Schnell, M. Verhage, T. Tzonopoulos, R. A. Nicoll, R. Janz, R. C. Malenka, M. Geppert, and T. C. Südhof, *J. Neurosci.* **19,** 5834 (1999).

[10] Doc2α as Modulator of Ca²⁺-Dependent Exocytosis

By SATOSHI ORITA, TAKUYA SASAKI, and YOSHIMI TAKAI

Introduction

The C2 domain, first found in protein kinase C, is known to interact with Ca²⁺ and phospholipid.[1–3] The C2-like domain was subsequently found in many other important intracellular signaling elements, including phospholipase Cγ,[4] Ras GAP,[5] and phospholipase A2.[6] All of these proteins have one C2-like domain. In contrast to these proteins, synaptotagmin and Rabphilin-3 have two C2-like domains.[7,8] Synaptotagmin was originally found to be specifically located on synaptic vesicles and shown to regulate neurotransmitter release as a Ca²⁺ sensor.[9–11] Rabphilin-3 has a Rab3A-binding domain upstream of the two C2-like domains, suggesting that Rabphilin-3A may serve as both a downstream target molecule of Rab3A and a Ca²⁺ sensor in neurotransmitter release.[8,12]

Doc2 (Double C2) was originally isolated as a novel Rabphilin-3-like molecule having two C2-like domains.[13] Doc2 consists of two isoforms, Doc2α and Doc2β.[13,14] Doc2α is specifically expressed in neuronal cells

[1] Y. Takai, A. Kishimoto, Y. Iwasa, Y. Kawahara, T. Mori, and Y. Nishizuka, *J. Biol. Chem.* **254,** 5049 (1979).

[2] Y. Takai, A. Kishimoto, U. Kikkawa, T. Mori, and Y. Nishizuka, *Biochem. Biophys. Res. Commun.* **91,** 1218 (1979).

[3] Y. Nishizuka, *Nature* **334,** 661 (1988).

[4] M. L. Stahl, C. R. Ferenz, K. L. Kelleher, R. W. Kriz, and J. L. Knopf, *Nature* **332,** 269 (1988).

[5] U. S. Vogel, R. A. F. Dixon, M. D. Schaber, R. E. Diehl, M. S. Marshall, E. M. Scolnick, I. S. Sigal, and J. B. Gibbs, *Nature* **335,** 90 (1988).

[6] J. D. Clark, L. L. Lin, R. W. Kriz, C. S. Ramesha, L. A. Sultzman, A. Y. Lin, N. Milona, and J. L. Knopf, *Cell* **65,** 1043 (1991).

[7] M. S. Perin, V. A. Fried, G. A. Mignery, R. Jahn, and T. C. Südhof, *Nature* **345,** 260 (1990).

[8] H. Shirataki, K. Kaibuchi, T. Sakoda, S. Kishida, T. Yamaguchi, K. Wada, M. Miyazaki, and Y. Takai, *Mol. Cell. Biol.* **13,** 2061 (1993).

[9] W. D. Matthew, L. Tsavaler, and L. F. Richardt, *J. Cell. Biol.* **91,** 257 (1981).

[10] N. Borse, A. G. Petrenko, T. C. Südhof, and R. Jahn, *Science* **256,** 1021 (1992).

[11] M. Geppert, Y. Goda, R. E. Hammer, C. Li, T. W. Rosahl, C. F. Stevens, and T. C. Südhof, *Cell* **79,** 717 (1994).

[12] T. Yamaguchi, H. Sirataki, S. Kishida, M. Miyazaki, J. Nishikawa, K. Wada, S. Numata, K. Kaibuchi, and Y. Takai, *J. Biol. Chem.* **268,** 27164 (1993).

[13] S. Orita, T. Sasaki, A. Naito, R. Komuro, T. Ohtsuka, M. Maeda, H. Suzuki, H. Igarashi, and Y. Takai, *Biochem. Biophys. Res. Commun.* **206,** 439 (1995).

[14] G. Sakaguchi, S. Orita, M. Maeda, H. Igarashi, and Y. Takai, *Biochem. Biophys. Res. Commun.* **217,** 1053 (1995).

0076-6879/00 $35.00

and localized on synaptic vesicles in nerve terminal, whereas Doc2β is ubiquitously expressed.[13–16] Doc2α and Doc2β have no Rab3A-binding domain but have a Doc2-specific region (DSR) upstream of the two C2-like domains. Both isoforms interact with Munc13-1 through the region within DSR.[17]

Munc13-1 was isolated as a mammalian counterpart of *Caenorhabditis elegans* unc-13, which is implicated in Ca^{2+}-dependent neurotransmitter release.[18,19] Munc13-1 is specifically expressed in neuronal cells and localized on the presynaptic plasma membrane. Munc13-1 has three C2-like domains and one C1-like domain that interacts with diacylglycerol or phorbol ester.[19,20] The interaction of Doc2α with Munc13-1 is induced by the binding of diacylglycerol produced during Ca^{2+}-dependent exocytosis to the C1-like domain of Munc13-1.[17,21] The growth hormone (GH) coexpression assay system of PC12 cells shows evidence that the Doc2α–Munc13-1 interactions are indeed involved in Ca^{2+}-dependent exocytosis.[17,22] Electrophysiological analysis of superior cervical ganglion neurons in culture indicates that the Doc2α–Munc13-1 interactions play a role in a step before the final fusion step of synaptic vesicles with the presynaptic plasma membrane.[23] Analysis of Doc2α-deficient mice suggests that Doc2α is implicated in the efficient recruitment of synaptic vesicles from the reserve vesicle pool to the readily releasable vesicle pool (mature and fusion-competent vesicle pool).[24] Doc2α-deficient mice show impairment in long-term potentiation and passive avoidance task, suggesting that Doc2α modulates neurotransmitter release, thereby contributing to memory formation. Moreover, analyses of Munc13-1-deficient mice, *C. elegans* unc-13 null mutants, and *Dro-*

[15] A. Naito, S. Orita, A. Wanaka, T. Sasaki, G. Sakaguchi, M. Maeda, H. Igarashi, M. Tohyama, and Y. Takai, *Mol. Brain Res.* **44,** 198 (1997).

[16] M. Verhage, K. J. Vries, H. Röshol, J. P. Burbach, W. H. Gispen, and T. C. Südhof, *Neuron* **18,** 453 (1997).

[17] S. Orita, A. Naito, G. Sakaguchi, M. Maeda, H. Igarashi, T. Sasaki, and Y. Takai, *J. Biol. Chem.* **272,** 16081 (1997).

[18] I. N. Maruyama, and S. Brenner, *Proc. Natl. Acad. Sci. U.S.A.* **88,** 5729 (1991).

[19] N. Brose, K. Hofmann, Y. Hata, and T. C. Südhof, *J. Biol. Chem.* **270,** 25273 (1995).

[20] A. Betz, U. Ashery, M. Rickmann, I. Augustin, E. Neher, T. C. Südhof, J. Rettig, and N. Brose, *Neuron* **21,** 123 (1998).

[21] R. R. Duncan, A. Betz, M. J. Shipston, N. Brose, and R. H. Chow, *J. Biol. Chem.* **274,** 27347 (1999).

[22] S. Orita, T. Sasaki, R. Komuro, G. Sakaguchi, M. Maeda, H. Igarashi, and Y. Takai, *J. Biol. Chem.* **271,** 7257 (1996).

[23] S. Mochida, S. Orita, G. Sakaguchi, T. Sasaki, and Y. Takai, *Proc. Natl. Acad. Sci. U.S.A.* **95,** 11418 (1998).

[24] G. Sakaguchi, T. Manabe, K. Kobayashi, S. Orita, T. Sasaki, A. Naito, M. Maeda, H. Igarashi, G. Katsuura, H. Nishioka, A. Mizoguchi, S. Itohara, T. Takahashi, and Y. Takai, *Eur. J. Neurosci.* **11,** 4262 (1999).

sophila unc-13 (Dunc-13) null mutants indicate that Munc13-1 and its counterparts are essential for maturation of synaptic vesicles to the readily releasable vesicles.[25-27] Thus, evidence is accumulating that the Doc2α–Munc13-1 interactions modulate Ca^{2+}-dependent neurotransmitter release.

This chapter focuses on the biochemical analysis of the Doc2α–Munc13-1 interactions. The chapter first describes the procedures for the purification of glutathione *S*-transferase (GST)–Doc2α and the analysis of the Doc2α–Munc13-1 interactions in a cell-free system. The chapter then describes the procedures for the analysis of the Doc2α–Munc13-1 interactions in an intact cell system.

Materials

Dithiothreitol (DTT), EGTA, HEPES, Nonidet P-40, and Triton X-100 are purchased from Nacalai Tesque (Kyoto, Japan). EDTA is from Dojindo Laboratories (Kumamoto, Japan). (*p*-Amidinophenyl)methanesulfonyl fluoride (APMSF) and isopropyl-β-D-thiogalactopyranoside (IPTG) are from Wako Pure Chemicals (Osaka, Japan). A GST expression vector, pGEX-2T, glutathione-Sepharose 4B, protein A-Sepharose CL-4B, and enhanced chemiluminescence (ECL) detection kit are purchased from Amersham-Pharmacia Biotech (Milwaukee, WI). [^{35}S]Methionine (37 TBq/mmol) is from NEN Life Science Products (Boston, MA). TNT T7 quick coupled transcription/translation system is from Promega (Madison, WI). T7 RNA polymerase recombinant vaccinia virus (LO-T7) is kindly supplied by M. Kohara (Tokyo Metropolitan Institute of Medical Science, Tokyo, Japan). OPTI-MEM and LipofectAMINE reagent are from Life Technologies (Rockville, MD). A human GH radioimmunoassay kit is from Nichols Institute (San Juan Capistrano, CA). All other chemicals are of reagent grade.

A plasmid for expression of GST–Doc2α is constructed as follows. The cDNA fragment encoding the N-terminal fragment (amino acids 1–90) of human Doc2α is inserted into pGEX-2T to construct pGEX-2T-Doc2N. An *Escherichia coli* strain JM109 is transformed with this plasmid and the resulting strain is used as a source for GST–Doc2α. *In vitro* and *in vivo* expression plasmids pGEM-HA and pBluescript-*myc* are generated to express fusion proteins with the N-terminal HA and myc epitopes, respectively. The cDNA fragment encoding human Doc2α is inserted into pGEM-

[25] I. Augustin, C. Rosenmund, T. C. Südhof, and N. Brose, *Nature* **400,** 457 (1999).

[26] J. E. Richmond, W. S. Davis, and E. M. Jorgensen, *Nat. Neurosci.* **2,** 959 (1999).

[27] B. Aravamudan, T. Fergestad, W. S. Davis, C. K. Rodesch, and K. Broadie, *Nat. Neurosci.* **2,** 965 (1999).

HA to construct pGEM-HA-Doc2. The cDNA fragment encoding rat Munc13-1 is inserted into pBluescript-*myc* to construct pBluescript-*myc*-Munc13-1. Plasmids for the GH release assay are constructed as follows. pEFBOS-HA and pEFBOS-*myc* are generated to express fusion proteins with the N-terminal HA and *myc* epitopes, respectively. The cDNA fragment encoding human Doc2α is inserted into pEFBOS-HA to construct pEFBOS-HA-Doc2. The cDNA fragment encoding the Doc2-interacting domain of Munc13-1 (amino acids 851–1461) is inserted into pEFBOS-*myc* to construct pEFBOS-*myc*-Munc13.

Methods

Purification of GST–Doc2α and Assay for Doc2α–Munc13-1 Interactions in Cell-free System

The various buffers used in the purification of GST–Doc2α and the assay for Doc2α–Munc13-1 interactions in a cell-free system are as follows:

Buffer A: 25 mM Tris-HCl at pH 7.5, 1 mM DTT, 0.5 mM EDTA, 320 mM sucrose, and 10 μM APMSF

Buffer B: 25 mM Tris-HCl at pH 7.5, 1 mM DTT, and 0.5 mM EDTA

Buffer C: 50 mM Tris-HCl at pH 8.0 and 5 mM reduced glutathione

Buffer D: 150 mM NaCl, 50 mM HEPES/NaOH at pH 7.4, and 1 mM EGTA

The steps used in the purification of GST–Doc2α and the assay for Doc2α–Munc13-1 interactions in a cell-free system are as follows: (1) cultivation of *E. coli* and induction of GST–Doc2α, (2) preparation of crude supernatant, (3) affinity purification of GST–Doc2α, and (4) assay for Doc2α–Munc13-1 interactions in a cell-free system.

Cultivation of E. coli and Induction of GST–Doc2α. JM109 transformed with pGEX-2T-Doc2N is cultured at 25° in 1 liter of LB medium containing 50 μg of ampicillin per milliliter to an OD$_{595}$ of 0.2. After addition of IPTG at a final concentration of 0.1 mM, cells are further cultured for 4 hr. All procedures after this step are performed at 0–4°. Cells are harvested, suspended in 20 ml of phosphate-buffered saline (PBS), and washed with 20 ml of PBS. The cell pellet is frozen at −80°.

Preparation of Crude Supernatant. The cell pellet is quickly thawed at 37° and suspended in 20 ml of buffer A and the cell suspension is sonicated at a setting of 60 by Ultrasonic Processor (Taitec, Tokyo, Japan) on ice for 30 sec four times at 30-sec intervals. The homogenate is centrifuged at 100,000g for 1 hr. The supernatant is used for the affinity purification.

Affinity Purification of GST–Doc2α. Glutathione-Sepharose 4B beads are packed onto a 5-ml disposable syringe (bed volume, 2 ml). The beads are equilibrated with 20 ml of buffer B. Twenty milliliters of the crude supernatant prepared as described above is applied to the column and the pass fraction is reapplied to the column. After the column is washed with 20 ml of buffer B, GST–Doc2α is eluted with 5 ml of buffer C. This eluate is dialyzed against buffer B used as purified GST–Doc2α. Purified GST–Doc2α can be kept at $-80°$ for at least 6 months without loss of activity.

Assay for Doc2α–Munc13-1 Interactions in a Cell-Free System. One microgram of pBluescript-*myc*-Munc13 is transcribed and translated *in vitro* using 50 μl of TNT T7 quick master mix from TNT T7 quick coupled transcription/translation system (Promega) and 4 μl of [^{35}S] methionine (37 TBq/nmol) at 30° for 1 hr. Two micrograms of GST–Doc2α or GST is immobilized onto 20 μl of gulutathione-Sepharose 4B beads. The immobilized beads are added to 500 μl of buffer D containing 15 μl of *in vitro* translated Munc13-1 and gently mixed at 4° for 4 hr in the presence or absence of 100 nM 12-*O*-tetradecanoylphorbol 13-acetate (TPA). The beads are washed four times with buffer D and the bound proteins are eluted by addition of 100 μl of buffer D containing 20 mM glutathione. Each eluate is subjected to SDS–PAGE followed by autoradiography.

In these experiments, Munc13-1 binds to GST–Doc2α and does not bind to GST.[17] Moreover, the Doc2α–Munc13-1 interactions are stimulated by TPA. These results, together with the finding that phorbol ester directly interacts with the C1 domain of Munc13-1,[20] indicate that the binding of phorbol ester to the C1 domain of Munc13-1 enhances the the Doc2α–Munc13-1 interactions.

Assay for Doc2α–Munc13-1 Interactions in Intact Cell System

The various buffers used in the assays for Doc2α–Munc13-1 interactions in an intact cell system are as follows:

Low K$^+$ solution: 140 mM NaCl, 4.7 mM KCl, 2.5 mM CaCl$_2$, 1.2 mM MgSO$_4$, 1.2 mM KH$_2$PO$_4$, 20 mM HEPES/NaOH at pH 7.4, and 11 mM glucose

High K$^+$ solution: 85 mM NaCl, 60 mM KCl, 2.5 mM CaCl$_2$, 1.2 mM MgSO$_4$, 1.2 mM KH$_2$PO$_4$, 20 mM HEPES/NaOH at pH 7.4, and 11 mM glucose

Cell Culture. Stock cultures of PC12 cells are maintained at 37° in a humidified atmosphere of 10% CO$_2$ and 90% air (v/v) in Dulbecco's modified Eagle's medium (DMEM) containing 10% fetal calf serum (FCS), 5% horse serum (HS), penicillin (100 U/ml), and streptomycin (100 μg/ml).

Assay for High K⁺- or TPA-Induced Doc2α–Munc13-1 Interactions in PC12 Cells.

INFECTION AND TRANSFECTION. PC12 cells that are to be infected with the vaccinia virus (LO-T7) are plated at a density of 5×10^5 cells per 35-mm dish. Prior to the addition of virus, the cells are incubated for 18–24 hr and washed once with DMEM. Two hundred microliters of the virus stock, diluted in PBS containing 1% FCS, is then added to the cells at a multiplicity of infection (MOI) of 5 plaque-forming units (pfu)/cell. The infected cells are incubated at room temperature for 30 min with intermittent agitation.

While the cells are incubated with the virus, a solution containing plasmid DNAs and LipofectAMINE reagent is prepared as follows: 2 μg of pGEM-HA-Doc2 and 2 μg of pBluescript-*myc*-Munc13 are mixed with 100 μl of OPTI-MEM. In a separate tube, 15 μl of LipofectAMINE reagent is mixed with 100 μl of the same medium. The contents of two tubes are then mixed, allowed to be incubated at room temperature for 15 min, and diluted with 0.8 ml of OPTI-MEM.

After the cells are infected, the solution containing virus is aspirated, and the solution containing plasmid DNAs and LipofectAMINE reagent is added. The cells are then incubated at 37° for 1 hr. The cells are washed with DMEM and incubated in 2 ml of DMEM containing 10% FCS and 5% HS.

IMMUNOPRECIPITATION. Immunoprecipitation was performed 6 hr after the transfection. PC12 cells are washed with low K⁺ solution and incubated at 37° for 10 min with 1 ml of high K⁺ solution, low K⁺ solution, or low K⁺ solution containing 100 nM TPA. The cells are washed with PBS twice and lysed in 500 μl of a lysis buffer containing 20 mM Tris-HCl at pH 7.5 and 150 mM NaCl. HA-Doc2α expressed in the cells is immunoprecipitated by gently mixing the cell lysate with 3 μg of the anti-HA monoclonal antibody bound to 20 μl of protein A-Sepharose CL-4B beads at 4° for 2 hr. The beads are washed four times with the lysis buffer and the bound proteins are eluted by 100 μl of Laemmli sample buffer. Twenty microliters of each eluate is subjected to SDS–PAGE and the separated proteins are electrophoretically transferred to a polyvinylidene difluoride membrane sheet. The sheet is processed using the ECL detection kit to detect *myc*-Munc13-1 with the biotinated mouse anti-*myc* antibody.

In these experiments, Munc13-1 is coimmunoprecipitated with Doc2α.[17] The coimmunoprecipitation of Munc13-1 with Doc2α is markedly enhanced when the cells are stimulated by high K⁺ or TPA, which induces Ca^{2+}-dependent exocytosis. However, when the C1 domain deletion mutant of Munc13-1 is used instead of full-length Munc13-1, the Doc2α–Munc13-1 interactions are observed even without the stimulation of PC12 cells by

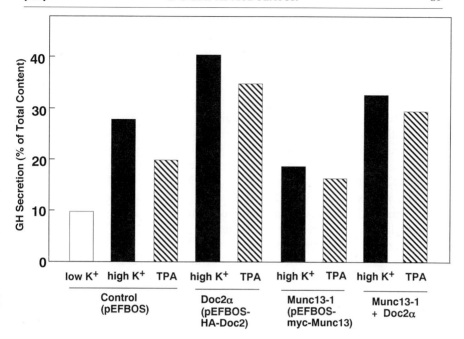

FIG. 1. Involvement of the Doc2α–Munc13-1 interactions in Ca^{2+}-dependent exocytosis from PC12 cells. PC12 cells are cotransfected with pXGH5 encoding human GH and pEFBOS-HA-Doc2 encoding Doc2α and/or pEFBOS-*myc*-Munc13 encoding the Doc2-interacting domain of Munc13-1. Data are expressed as the average percentage released of the total GH stores.

high K^+ or TPA. These results, together with the earlier findings that high K^+ induces diacylglycerol formation,[28] suggesting that the binding of diacylglycerol to the C1 domain of Munc13-1 induces a conformational change of the Doc2-interacting domain of Munc13-1, there by enhancing its interaction with Doc2α during Ca^{2+}-dependent exocytosis.

Assay for GH Release from PC12 Cells. The activity of Doc2α to regulate Ca^{2+}-dependent exocytosis is assayed by measuring GH release from PC12 cells cotransfected with pXGH5 encoding human GH and pEFBOS-HA-Doc2. In this assay system, expressed GH is stored in dense core vesicles of PC12 cells and released in response to various agonists in an extracellular Ca^{2+}-dependent manner.[29,30]

[28] T. D. Wakade, S. V. Bhave, A. S. Bhave, R. K. Malhotra, and A. R. Wakade, *J. Biol. Chem.* **266,** 6424 (1991).

[29] E. S. Schweitzer and R. B. Kelly, *J. Cell Biol.* **101,** 667 (1985).

[30] P. F. Wick, R. A. Senter, L. A. Parsels, M. D. Uhler, and R. W. Holz, *J. Biol. Chem.* **268,** 10983 (1993).

TRANSFECTION. PC12 cells are plated at a density of 5×10^5 cells per 35-mm dish and are incubated for 18–24 hr. The cells are then cotransfected with 2 μg of pXGH5 and 2 μg of the pEFBOS-HA-Doc2 and/or pEFBOS-*myc*-Munc13 using 15 μl of LipofectAMINE reagent and 1 ml of OPTI-MEM. Six hours after the transfection, the cells are washed with DMEM and incubated in 2 ml of DMEM containing 10% FCS and 5% HS.

GH RELEASE EXPERIMENTS. GH release experiments are performed 48 hr after the transfection. PC12 cells are washed with low K^+ solution and incubated at 37° for 10 min with 2 ml of high K^+ solution or low K^+ solution. After high K^+ or low K^+ solution is removed, the cells are lysed in 2 ml of high K^+ solution containing 0.5% Triton X-100 on ice. The amounts of the GH released into high K^+ or low K^+ solution and retained in the cell lysate are measured using a radioimmunoassay kit according to the manufacturer's instruction (Nichols Institute).

In these experiments, overexpression of Doc2α enhances not only the high K^+-induced GH release but also TPA-induced GH release (Fig. 1).[17,22] Overexpression of the Doc2-interacting domain of Munc13-1 reduces both high K^+- and TPA-induced GH release. Coexpression with Doc2α suppresses this reduction.

Comments

Generally, protein–protein interactions *in vitro* are assayed by affinity column chromatography, gel filtration, or sucrose density gradient ultracentrifugation by use of pure samples, and the stoichiometry of the interactions and a K_d value are estimated. However, it is practically difficult to prepare a pure recombinant sample of Munc13-1 because of its low expression level in *E. coli* or Sf9 cells. Therefore, Munc13-1 is prepared by the *in vitro* translation/transcription method, and the *in vitro* Doc2α–Munc13-1 interactions are assayed by use of GST–Doc2α and *in vitro* translated Munc13-1. Protein–protein interactions *in vivo* are generally assayed by immunoprecipitation of endogenous proteins. However, it is practically difficult to carry out this assay since the anti-Doc2α and anti-Munc13-1 antibodies for immunoprecipitation are not available. Therefore, the epitope tagging technique and the vaccinia virus expression method are used.

[11] Purification and Properties of a GTPase-Activating Protein for Yeast Rab GTPases

By LI-LIN DU and PETER NOVICK

Introduction

Rab proteins constitute the largest family of small GTP binding proteins.[1] A total of 11 genes encoding Rab proteins is found in the yeast genome. Among them, Sec4p was the first to be implicated in vesicle traffic.[2] When the function of Sec4p is blocked, cells accumulate Golgi-derived secretory vesicles.

GTP binding proteins cycle between the GTP-bound active form and the GDP-bound inactive form. The conversion between these two forms is catalyzed by guanine nucleotide exchange factors (GEFs) and GTPase activating proteins (GAPs). Yeast proteins Gyp6p and Gyp7p were the first Rab GAPs to be cloned,[3,4] but they are not active on Sec4p. GAP activity for Sec4p was first observed in yeast cell lysate.[5] To clone a Sec4p-specific GAP, we searched the sequence database and found an uncharacterized yeast open reading frame that shares homology with Gyp6p and Gyp7p. We named this protein Gyp1p and demonstrated that it is a GAP for Sec4p.[6] Independently, a more systematic analysis of the homology shared by yeast Rab GAPs has been published.[7] The homologous domain defined in that paper has been shown to be the GAP catalytic domain in Gyp1p and Gyp7p.[8] This domain is not limited to yeast proteins or to Rab GAPs. A human protein called CAPCenA is a Rab6 GAP and shares this domain.[9] Interestingly, one *Schizosaccharomyces pombe* protein sharing this homology, Cdc16, has been shown to be one sub-

[1] P. Novick and M. Zerial, *Curr. Opin. Cell Biol.* **9,** 496 (1997).
[2] A. Salminen and P. J. Novick, *Cell* **49,** 527 (1987).
[3] M. Strom, P. Vollmer, T. J. Tan, and D. Gallwitz, *Nature* **361,** 736 (1993).
[4] P. Vollmer and D. Gallwitz, *Methods Enzymol.* **257,** 118 (1995).
[5] N. C. Walworth, P. Brennwald, A. K. Kabcenell, M. Garrett, and P. Novick, *Mol. Cell Biol.* **12,** 2017 (1992).
[6] L. L. Du, R. N. Collins, and P. J. Novick, *J. Biol. Chem.* **273,** 3253 (1998).
[7] A. F. Neuwald, *Trends Biochem. Sci.* **22,** 243 (1997).
[8] S. Albert, E. Will, and D. Gallwitz, *EMBO. J.* **18,** 5216 (1999).
[9] M. H. Cuif, F. Possmayer, H. Zander, N. Bordes, F. Jollivet, A. Couedel-Courteille, I. Janoueix-Lerosey, G. Langsley, M. Bornens, and B. Goud, *EMBO J.* **18,** 1772 (1999).

METHODS IN ENZYMOLOGY, VOL. 329
0076-6879/00 $35.00

unit of a heterodimeric GAP for a GTPase involved in cell cycle regulation.[10]

In this chapter, we present the methods that we used in the study of Gyp1p.

Construction of Expression Plasmids

Polymerase chain reaction (PCR) was used to amplify the open reading frames of Gyp1p, Ypt1p, Ypt6p, Ypt7p, Ypt32p, and Ypt51p (Research Genetics, Huntsville, AL) with the following primer pair:

Forward primer:
5'-CGGGATCCACTCGAGCATATGGAATTCCAGCTGACCACC-3'
Back primer:
5'-GGAATTCCATATGCTCGAGGATCCCCGGGAATTGCCATG-3'

The PCR products were digested with *Nde*I and *Bam*HI, and then cloned into the expression vector pET15b (Novagen, Madison, WI) so that each open reading frame is fused in frame to the hexahistidine (His$_6$) tag sequence in the vector. The restriction map of the Gyp1p expression plasmid, pNB1032, is shown in Fig. 1. The plasmids for the expression of Ypt1p, Ypt6p, Ypt7p, Ypt32p, and Ypt51p were designated pNB842, pNB843, pNB844, pNB845, and pNB846, respectively. The construct for the expression of His$_6$-tagged Sec4p, pNB639, has been described.[11] These plasmids were then transformed into the *Escherichia coli* strain BL21(DE3). Several transformants were screened by small-scale induction. The transformant giving the highest level of expression was frozen at $-80°$ in 15% glycerol.

Expression and Purification of His$_6$-Tagged Proteins from Bacteria

Solutions

Buffer stock: 10× sonication and wash buffer stock, 0.5 M sodium phosphate, pH 8.0 or 6.0, 3 M NaCl; 5 M imidazole, adjust to pH 8.0. Make the following buffers just before use:

Sonication buffer: 50 mM sodium phosphate, pH 8.0, 0.3 M NaCl, 5 mM 2-mercaptoethanol, 1 mM phenylmethylsulfonyl fluoride (PMSF), 0.1% Tween 20

Wash buffer: 50 mM sodium phosphate, pH 6.0, 0.3 M NaCl, 10% glycerol, 5 mM 2-mercaptoethanol

[10] K. A. Furge, K. Wong, J. Armstrong, M. Balasubramanian, and C. F. Albright, *Curr. Biol.* **8,** 947 (1998).

[11] R. N. Collins, P. Brennwald, M. Garrett, A. Lauring, and P. Novick, *J. Biol. Chem.* **272,** 18281 (1997).

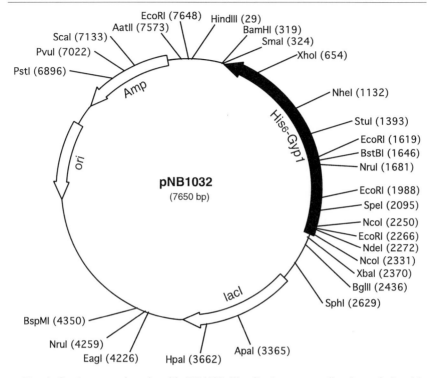

FIG. 1. Gyp1p expression plasmid pNB1032. The Gyp1p open reading frame is fused in frame to a His$_6$ tag.

2× column buffer: 40 mM Tris-HCl, pH 8.0, 2 mM MgCl$_2$, 10 mM 2-mercaptoethanol, 20% glycerol

2× storage buffer: 40 mM Tris-HCl, pH 8.0, 2 mM MgCl$_2$, 0.2 M NaCl, 2 mM dithiothreitol (DTT)

Method

1. Inoculate 2 ml Luria–Bertani (LB) medium containing 0.1 mg/ml ampicillin with frozen stock. Grow overnight. For better retention of plasmids, grow to A_{600} = 0.2–1 during the day, then store at 4° overnight.

2. Centrifuge 1 ml of culture in a 1.7-ml microcentrifuge tube at top speed for 20 sec. Remove the supernatant, resuspend the cells, and add them to 50 ml LB containing ampicillin. Grow to A_{600} = 0.2–1. Spin down the cells in a 50-ml tube at 3600 rpm for 15 min in a

Beckman (Palo Alto, CA) GPR tabletop centrifuge. Resuspend the cells, add them to 1 liter of medium.

3. Grow to $A_{600} \approx 0.6$. Save 1 ml of culture as uninduced control and for plasmid stability test. Add isopropyl-β-D-thiogalactopyranoside (IPTG) to 0.4 mM. Grow at 37° for 2 hr (or cool down the culture on ice then induce at 25° for 4 hr to increase solubility).

4. Harvest the cells by centrifuging at 7000g for 5 min. Resuspend in 40 ml of sonication buffer. Freeze the cells at −20°.

5. Thaw the cells under flowing tap water. Sonicate on ice for 3 × 1 min, with at least 1 min of cooling in between. Centrifuge the lysate at 20,000g for 15 min. Transfer the supernatant with a 25-ml pipette to a 60-ml syringe, and filter through a 0.45-μm filter to a 50-ml tube.

6. Add 0.5 ml Ni^{2+}-NTA resin (Qiagen, Chatsworth, CA) preequilibrated with sonication buffer to the tube. Rotate for >1 hr at 4°. Spin down the resin at 3600 rpm for 2 min in Beckman tabletop centrifuge. Wash the resin twice with 40 ml of sonication buffer.

7. Wash the resin twice with 40 ml of wash buffer for 10 min each time at 4°.

8. Transfer the resin to a 1-ml Bio-Spin column (Bio-Rad, Richmond, CA). Wash with 5 ml of column buffer containing 1 M NaCl and 40 mM imidazole.

9. Wash with 1 ml of column buffer. Elute with 1.5 ml of column buffer containing 0.1 M NaCl and 250 mM imidazole.

10. Check the purity with sodium dodecyl sulfate–polyacrylamide gel electrophoresis (SDS–PAGE). Do not boil the sample in the presence of imidazole. Heat at 37° for 10 min.

11. Use Centricon (Millipore, Bedford, MA) to change the buffer to 2× storage buffer. Then add an equal volume of 100% glycerol for long-term storage at −20°.

All of the His$_6$-tagged Rab proteins were induced for 2 hr at 37°, and purified to more than 90% purity with the above method, as shown in Fig. 2A. Gyp1p was induced for 4 hr at 25°. Most of the induced Gyp1p is insoluble. The soluble Gyp1p was purified with the above method, some truncated forms copurified with the 70-kDa full-length protein, as shown in Fig. 2B.

Measuring GTPγS Binding Capacity of Rab Proteins

Reagents

^{35}S-labeled GTPγS (NEN, Boston, MA), 12.5 mCi/ml in 10 mM Tricine, pH 7.6, 10 mM DTT: The specificity is 1000 Ci/mmol, so the

A

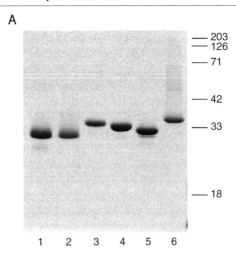

— 203
— 126

— 71

— 42

— 33

— 18

1   2   3   4   5   6

B

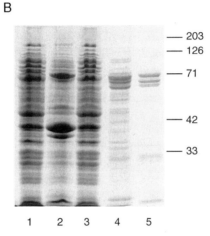

— 203
— 126

— 71

— 42

— 33

1   2   3   4   5

Fig. 2. (A) Ni^{2+}-NTA resin purified His$_6$-tagged yeast Rab proteins were electrophoresed on a 13.5% SDS–PAGE gel, and visualized by Coomassie blue staining. Lanes 1–6, Sec4p, Ypt1p, Ypt6p, Ypt7p, Ypt32p, and Ypt51p, respectively. (B) Purification of His$_6$-tagged Gyp1p. Lane 1, postsonication supernatant; lane 2, postsonication pellet; lane 3, proteins not bound to Ni^{2+}-NTA resin; lane 4, 40 mM imidazole wash; lane 5, 250 mM imidazole elution.

actual concentration of GTPγS is 12.5 μM. The size that we order is 250 μCi, i.e., 20 μl volume. It can be diluted fivefold with 50 mM Tris-HCl, pH 8.0, 10 mM DTT, and aliquoted to 5 tubes and frozen at −80°. The scintillation counting efficiency of ^{35}S is almost 100%, so 1 μCi corresponds to 2 × 10^6 cpm. The activity of the original material would be 2.5×10^7 cpm/μl, and the activity of the diluted stock would be 5×10^6 cpm/μl. In our experiment, the fresh [^{35}S]GTPγS gave counts 20% higher than this calculation.

GTPγS (Boehringer Mannheim, Indianapolis, IN): To make 100 mM of stock, dissolve it in water, then adjust to pH 8.0 with Tris, calibrate the concentration by absorbance at 253 nm. Its $\varepsilon_{253} = 1.37 \times 10^4$, i.e., the A_{253} of 20 μM GTPγS, is 0.274. Store at $-80°$. A more convenient way is to buy the 100 mM GTPγS solution from Boehringer Mannheim.

Assay buffer: 50 mM Tris-HCl, pH 8.0, 1 mg/ml bovine serum albumin (BSA), 1 mM DTT, 1 mM ATP, 5 mM MgCl$_2$. Make it fresh each time. (*Note:* This is the assay buffer for Sec4p and Ypt6p; use 0.5 mM MgCl$_2$, 1 mM EDTA instead of 5 mM MgCl$_2$ for Ypt1p, Ypt7p, Ypt32p, and Ypt51p.)

Stop buffer: 20 mM Tris-HCl, pH 8.0, 25 mM MgCl$_2$.

Binding Assay

Dilute GTPγS to 1 μM in assay buffer and spike with [^{35}S]GTPγS so the final activity is about 4×10^4 cpm/μl (1/125 volume of stock if it is fresh). At this dilution, the concentration of GTPγS from the radioactive material is less than 3% of the total GTPγS, so it is negligible.

Dilute Rab proteins to 0.8 μM with assay buffer.

For a typical experiment with three time points (10, 30, and 60 min), mix 15 μl of assay buffer and 20 μl of 1 μM spiked GTPγS in a 500-μl microfuge tube, add 5 μl of 0.8 μM Rab, and vortex to start the reaction. The final concentration is 0.1 μM Rab, 0.5 μM GTPγS. The total volume is 40 μl. Do a negative control tube without Rab.

Take out 10 μl of the reaction mix at 10, 30, and 60 min, mix with 0.95 ml ice-cold stop buffer. Do the filter binding as below to determine the radioactivity associated with 1 pmol Rab protein.

Drop 10 μl of the reaction mix on a 25-mm HA filter (Millipore). Dry it and count directly with a scintillation counter to determine the total radioactivity of 5 pmol GTPγS.

At room temperature, 60 min is sufficient for the GTPγS binding to go to completion. The binding capacity can be calculated as: (radioactivity associated with 1 pmol Rab at 60 min $-$ control)/radioactivity of 1 pmol GTPγS.

Measuring Protein-Associated Radioactivity with Filter Binding

We use a Millipore 1225 sampling manifold for this method. A repetitive pipette is useful for washing.

Keep the samples of early time points on ice until all the samples have been collected. Have stop buffer ready on ice. For the first 12 samples, wet the wells with stop buffer, place in each well a 25-mm HA filter. Put in 12 filters even if you have fewer than 12 samples to keep it airtight. Put on the sample cap and screw the nut tight. Prewash the filters by adding 3 ml of stop buffer to each well, connecting the vacuum to aspirate the buffer through the filter. Disconnect the vacuum. Add the reaction samples to the wells, then connect the vacuum. Wash the wells three times with 3 ml of stop buffer each time. Keep the vacuum connected after the last wash. Unscrew the nut and remove the cap. Use tweezers to remove the filters. Dry the filters under an infrared lamp. Follow the same procedure for the rest of the samples except that the wells do not need to be wet anymore.

Put the dry filters into scintillation tubes, add 4 ml of scintillation liquid and count.

Guanine Nucleotide Analysis of GTPase Reaction

Reagents

[α-^{32}P]GTP (NEN): 3000 Ci/mmol, 10 mCi/ml in 50 mM Tricine (pH 7.6). The actual GTP concentration is about 3.33 μM. Make a stock solution: Dilute the [α-^{32}P]GTP 10 times in 10 μM GTP, 50 mM Tris-HCl, pH 8.0, 1 mM DTT, and 1 mM ATP.

Polyethyleneimine-cellulose thin-layer chromatography (TLC) plates with fluorescent indicator (Fisher Scientific, Springfield, NJ): Predevelop in distilled water and dry before use.

10× buffer A: 500 mM Tris-HCl, pH 8.0, 10 mg/ml BSA, 10 mM DTT, 10 mM ATP.

Preloading GTPase

Incubate 2 μM Sec4p (as defined by its GTPγS binding capacity) with 0.4 μM spiked [α-^{32}P]GTP in 1× buffer A plus 5 mM MgCl$_2$. Incubate at room temperature for 30–60 min.

Reaction

For a 50-μl reaction, mix 5 μl of 10× buffer A, 5 μl of 50 mM MgCl$_2$, add water and other addition (e.g., GAPs) to 40 μl, add 10 μl preloaded Sec4p, and vortex to start reaction.

Take 10 μl of the reaction mix at different time points, add it into 10 μl 2% SDS, 40 mM EDTA, mix, and incubate at 65° for 10 min. Spot a 2-μl sample on a TLC plate, and allow it to dry. 4 μl nonradioactive

1 mM GTP, GDP, and GMP are spotted on the plate alongside the samples to determine the migration of these nucleotides. Develop the TLC plate in 1 M LiCl until the front reaches the top. Dry the plate, then expose to X-ray film. The positions of nucleotide standards can be viewed under a handheld UV light.

GAP Assay

Reagents

[γ-^{32}P]GTP (NEN): 6000 Ci/mmol, 10 mCi/ml in 50 mM Tricine (pH 7.6), stored at 4°. The actual GTP concentration is about 1.67 μM. Make a stock solution: 10 μM GTP in 50 mM Tris-HCl, pH 8.0, 1 mM DTT, and 1 mM ATP. Add [γ-^{32}P]GTP so the final activity is about 4 × 10^5 cpm/μl (1/50 volume of [γ-^{32}P]GTP, if it is fresh).

10× buffer B: 500 mM sodium N-2 hydroxyethylpiperazine-N'-2-ethanesulfonic acid (HEPES)–NaOH, pH 6.8, 10 mg/ml BSA, 10 mM DTT, and 10 mM ATP. (The optimal pH for Gyp1p catalyzed GTP hydrolysis by Sec4p is between 6.5 and 7.0. At pH 8.0, the GTP hydrolysis rate is about 70% of that at optimal pH.)

Stop buffer: 20 mM Tris-HCl, pH 8.0, 5 mM MgCl$_2$, 100 mM NaCl.

Preloading GTPase

Incubate 1 μM of Rab (as defined by its GTPγS binding capacity) with 5 μM of spiked GTP in 1× buffer B plus 5 mM MgCl$_2$. (*Note:* This is the buffer for Sec4p and Ypt6p; use 0.5 mM MgCl$_2$ and 1 mM EDTA instead of 5 mM MgCl$_2$ for other Rabs.) Incubate at room temperature for 30–60 min.

Reaction

For a 50-μl reaction, mix 5 μl 10× buffer B, 5 μl 50 mM MgCl$_2$, 5 μl 10 mM GTP, add water and other addition (e.g., GAPs) to 45 μl, add 5 μl preloaded 1 μM Rab, and vortex to start reaction. The final concentration of Rab–GTP is 0.1 μM. (Because of the presence of a large excess of nonradioactive GTP in the reaction mixture, Rab proteins cannot bind radioactive GTP again after hydrolysis, so the activity measured is the single round turnover rate. In this assay condition, the exchange of Rab-associated radioactive GTP by free GTP in the buffer is negligible.)

The GTPase rate can be measured by either the release of γ-phosphate into solution or the decrease of protein associated GTP. Both methods are detailed below. In our hands, the charcoal binding method is easier to conduct, but the filter binding method may be a better option for some applications, for example, the measurement of GAP activity in a crude lysate.

Using Charcoal Binding to Measure Rate of γ-Phosphate Release

Take 10 μl of the reaction mix at different time points, add it to 790 μl ice-cold 5% (w/v) activated charcoal (Sigma, St. Louis, MO) in 50 mM NaH_2PO_4. Vortex, then spin down the charcoal at top speed in a microcentrifuge for 5 min, transfer 400 μl of the supernatant into scintillation tubes, add 4 ml scintillation liquid, and count. Plot radioactivity in the supernatant versus time. Use the slope of the linear range of the curve as the initial GTPase rate. The first-order rate can be obtained by dividing the initial rate by the concentration of Rab–GTP.

Using Filter Binding to Determine Rate of Loss of Protein-Associated GTP

Take 10 μl of the reaction mix at different time points, add it to 0.95 ml ice-cold stop buffer. Do the filter binding as described in GTPγS binding protocol to get the radioactivity associated with 1 pmol Rab protein. Data can be processed as described above.

Summary of Results

The methods presented here have been used to characterize Gyp1p.[6] We found that Gyp1p stimulates the hydrolysis of Sec4p-bound GTP to GDP, but Gyp1p itself does not have GTPase activity. Gyp1p can act together with the GEF of Sec4p, Sec2p, to enhance the steady-state rate of GTP hydrolysis by Sec4p. Under single turnover conditions, the accelerated rate of GTP hydrolysis is directly proportional to the concentration of Gyp1p, and the maximal enhancement is more than 200-fold, all indicating the catalytic nature of the interaction. Gyp1p has a broad specificity. Among the six Rab proteins that we tested, Gyp1p can act on Sec4p, Ypt1p, Ypt7p, and Ypt51p, but not on Ypt6p and Ypt32p.

Acknowledgment

This work was supported by grants from the National Institutes of Health to P.N.

[12] Purification and Characterization of Yeast Exocyst Complex

By Daniel R. TerBush, Wei Guo, Steven Dunkelbarger, and Peter Novick

Introduction

The targeting of transport vesicles to their acceptor membranes requires the participation of complex assemblies of tethering and docking proteins that act in concert with Rab proteins and SNAREs to ensure specificity.[1] One of the first docking complexes purified and identified was the yeast exocyst complex.[2,3] The yeast exocyst complex consists of an assembly of eight proteins (Sec3p, Sec5p, Sec6p, Sec8p, Sec10p, Sec15p, Exo70p, and Exo84p; Fig. 1) that is recruited from the cytoplasm to sites of active exocytosis on the plasma membrane.[2,4,5] Sec3p is localized to exocytic sites even in the absence of membrane traffic, suggesting that it functions as a spatial landmark for exocytosis. Another subunit of the exocyst, Sec15p, was recently shown to directly interact with the Rab GTPase Sec4p specifically in its GTP bound form, identifying it as a downstream effector of Sec4p.[6] The current working model is that the exocyst constitutes the first bridge between the surface of the secretory vesicle, where Sec15 interacts with Sec4, and the plasma membrane, where Sec3p defines the site of exocytosis.

Seven of the eight yeast exocyst complex proteins were identified through a combination of genetic and biochemical approaches[2,3,7–9] and the eighth by its homology to the rexo84p component of the rat brain exocyst complex.[5,10] The protein–protein interactions of the yeast exocyst complex have recently been determined as well.[5,6] The biochemical mechanism used by the exocyst complex to dock a secretory vesicle should serve as a model for the mammalian exocyst complex and possibly other functionally, al-

[1] S. R. Pfeffer, *Nat. Cell Biol.* **1,** E17 (1999).

[2] D. R. TerBush and P. Novick, *J. Cell Biol.* **130,** 299 (1995).

[3] D. R. TerBush, T. M. Maurice, D. Roth, and P. Novick, *EMBO J.* **15,** 6483 (1996).

[4] F. P. Finger, T. E. Hughes, and P. Novick, *Cell* **92,** 559 (1998).

[5] W. Guo, A. Grant, and P. Novick, *J. Biol. Chem.* **274,** 23558 (1999).

[6] W. Guo, D. Roth, C. Walch-Solimena, and P. Novick, *EMBO J.* **18,** 1071 (1999).

[7] P. Novick, C. Field, and R. Schekman, *Cell* **21,** 205 (1980).

[8] R. Bowser and P. Novick, *J. Cell Biol.* **112,** 1117 (1991).

[9] R. Bowser, H. Muller, B. Govindan, and P. Novick, *J. Cell Biol.* **118,** 1041 (1992).

[10] S.-C. Hsu, A. E. Ting, C. D. Hazuka, S. Davanger, J. W. Kenny, Y. Kee, and R. H. Scheller, *Neuron* **17,** 1209 (1996).

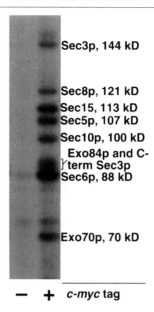

— + c-myc tag

FIG. 1. Size and identification of the eight exocyst complex proteins that coimmunoprecipitate. Yeast strains lacking (NY13) or containing (DTY166) a c-myc epitope-tagged Sec8p were labeled with [^{35}S]cysteine/methionine. Lysates were prepared and immunoprecipitated with MAB9E10. The immunoprecipitates were washed with buffer A containing 300 mM NaCl at 4° for 3 hr prior to harvesting and separating on a 7% SDS–PAGE slab gel. Autoradiography was then performed on the dried gel. The C terminus of Sec3p and Exo84p both migrate at about 91 kDa.

though not structurally, related docking complexes.[1] To facilitate the further analysis of these other docking complexes, the methods used to purify and characterize the yeast exocyst complex are described below.

Preliminary Characterization of Exocyst Complex

The exocyst complex is purified by immunoprecipitation from yeast strains expressing *SEC8* tagged with single and triple c-myc epitopes (Table I). Preliminary studies are done to define the buffer conditions that stabilize the exocyst complex and give an optimal yield of the complex in the immunoprecipitate. The three most important considerations are the presence of reducing agents, the pH, and the type of salt. The presence of 1 mM dithiothreitol (DTT) must be maintained during all steps of the purification to stabilize the complex during salt washes of the immunoprecipitate and thereby increase the immunoprecipitation efficiency. At pH 7.2 or above, Sec15p begins to dissociate from the complex and therefore the pH is

TABLE I

YEAST STRAINS

Strain number	Genotype
NY1008	MATa, ura3-52, sec8Δ::URA3, leu2-3,112::(LEU2, SEC8-c-myc),L-A-+[a]
DTY166	MATa, ura3-52, sec8Δ::URA3, leu2-3,112::(LEU2, SEC8-c-myc), L-A-0[a]
NY1427	MATα, leu2-3, 112::(LEU2, SEC8-3X-c-myc), ura3-52, sec8Δ::URA3, L-A-0[a]

[a] L-A-+ and L-A-0 refer to the presence or absence, respectively, of the yeast dsRNA virus, L-A. L-A is also called ScV-L1 in the literature.[14]

usually maintained between pH 6.5 and 6.8. The choice of salt is less important. Sodium acetate, potassium acetate, and NaCl are all suitable, but NaPO₄ should be avoided because it is disruptive to complex integrity.

Greater immunoprecipitation efficiency of the complex is obtained by preincubating the yeast lysate with the antibody rather than by preabsorbing the antibody to the protein A-Sepharose CL-4B beads (Sigma, St. Louis, MO) prior to addition of the lysate. Approximately threefold more antibody is required to immunoprecipitate an equivalent amount of the exocyst complex, if the antibody is preabsorbed to the protein A-Sepharose beads prior to addition to the lysate. This suggests that the exocyst complex is so large that it cannot penetrate into the interior regions of the bead and may only be bound to the surface. Rather than chemically cross-linking preabsorbed antibody to the beads to reduce the amount of antibody found in the subsequently solubilized immunoisolates, as is commonly done, we developed a method of sequential protein concentration and separation to remove immunoglobulin G (IgG) and other contaminants from each specific band of the immunoprecipitate.[11]

Exocyst Complex Purification

Immunoprecipitation Protocol

The exocyst can be purified by immunoprecipitation and sequential slab gel/funnel tube gel SDS–PAGE electrophoresis.[11] Yeast strains are constructed that expressed either Sec8p triple (NY1427)[3] or single (NY1008)[2] c-*myc* epitope tags as the sole copy of Sec8 protein (Table I). A major contaminant of the precipitate is a protein expressed by the L-A virus-like particle of yeast and therefore strain DTY1427 was "cured" of the double-stranded virus by sequentially streaking the strain on YPD [1%

[11] D. R. TerBush and P. Novick, *J. Biomol. Tech.* **10,** 149 (1999).

yeast extract (Difco, Detroit, MI), 2% Bacto-peptone (Difco), 2% dextrose] plates incubated at 39°. This eliminates the virus as determined by Western blot for the viruslike particle coat protein (data not shown). Seven unique exocyst complex proteins coimmunoprecipitate with the c-*myc*-tagged Sec8 protein[3,5] and their identities are shown in Fig. 1. A C-terminal proteolytic fragment of Sec3p and full-length Exo84p almost exactly comigrate at 91 kDa. The complex is purified once from 6.4 g of protein obtained from strain NY1008 and once from 4.8 g of protein obtained from strain NY1427. The data shown in Figs. 2 and 3 and the description below are for the purification from the NY1008 strain.

Overnight yeast cultures (6 liters) are grown to a final A_{600} of between 1 and 2 in YPD. The yeast were harvested by centrifugation and washed by resuspension in 10 mM Tris (pH 7.5) containing 10 mM sodium azide. The yeast are then converted to spheroplasts and lysed in buffer A [20 mM PIPES (pH 6.8), 100 mM NaCl, 0.5% Tween 20, 1 mM dithiothreitol (DTT), 1 mM EDTA, 1 mM phenylmethylsulfonyl fluoride (PMSF), 5 μg/ml pepstatin A, and 2 μg/ml each of aprotinin, antipain, leupeptin, and chymostatin (protein inhibitors, Boehringer-Mannheim, Indianapolis, IN)] as previously reported.[3] The lysate is spun for 30 min at 30,000g. The supernatant is removed and the concentration of protein in this fraction is determined using bovine IgG as a standard.[12] The lysate is diluted to 4 mg/ml protein concentration, divided into four equal aliquots, and each precleared by incubation with 10 ml of a 50% slurry of Sepharose CL-4B beads for 1.5 hr at 4°. Each aliquot is then spun at 10,000g for 10 min, the supernatant removed to fresh tubes, and 0.6 ml of anti-c-*myc* antibody (MAB9E10, ascites) is added to each aliquot.[13] After incubation overnight at 4°, 1.25 g of hydrated protein A-Sepharose CL-4B beads are added to each aliquot and incubated for 1.5 hr at 4°. The immunoprecipitates are cleared by spinning for 10 min at 3700 rpm in a general-purpose tabletop centrifuge at 4°. The immunoprecipitates are washed 4 times with buffer A (all washes lacked the protease inhibitors), and the immunoisolated proteins are recovered by boiling the protein A beads for 5 min in an equal volume of 2× sample buffer (about 6 ml per aliquot). The beads are pelleted and the sample buffer removed. The beads are then mixed a second time with an equal amount of 1× sample buffer, reboiled, spun, and the supernatants pooled. The material from this immunoprecipitation is used to purify the proteins corresponding to bands D (Sec5p), E (Sec10p), and H (Exo70p)

[12] M. M. Bradford, *Anal. Biochem.* **72**, 248 (1976).

[13] G. I. Evan, G. K. Lewis, G. Ramsay, and J. M. Bishop, *Mol. Cell. Biol.* **5**, 3610 (1985).

[14] J. A. Bruenn, *in* "RNA Genetics" (R. D. J. Holland and P. Ahlquist, eds.), pp. 195–209. CDC Press, Boca Raton, FL, 1988.

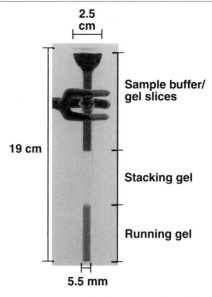

FIG. 2. Concentration of exocyst proteins by funnel tube gel electrophoresis. The bands corresponding to a given unknown were excised from the SDS–PAGE slab gel, equilibrated with water, and then an equal volume of 2× sample buffer. The sample buffer/gel slice mixture (about 5 ml) was loaded into a FT gel for concentration. The running and stacking gels of the funnel tube were each 5 cm tall (1.4 ml of unpolymerized gel mixture; dye was added to the running gel to facilitate photography) and the running gel was 8% acrylamide. The electrophoresis tank used was improvised from a 25-year-old Bio-Rad (Hercules, CA) tube gel apparatus capable of handling 11-mm outside diameter tubes. After the running and stacking gels had polymerized, the FT was glued into the electrophoresis tank using RTV Multi-Purpose Sealant Model 732 (Dow Corning, Midland, MI). After the sealant had set, the tank was filled top and bottom with running gel buffer and electrophoresed by limiting the voltage to 150 V and the current to 7 mA per tube to avoid heating. When the trailing edge of dye is about 4 or 5 mm into the running gel, the running gel is extruded from the tube and the top 4 or 5 mm of the running gel is sliced off. The protein enters the running gel just behind the dye. The running gel can be easily extruded from the FT by placing the bottom 10 cm of the tube under running hot water until the glass is quite warm to the touch. The gel can then be pushed out of the bottom of the FT with a glass rod.[11]

in Fig. 3. Bands A (Sec3p) and F (mixture of Exo84p and a C-terminal proteolytic fragment of Sec3p) are purified identically, except that lysates are made from yeast strain NY1427 and 4.8 g of starting protein are used.

First Separation by SDS–PAGE

Sample buffer corresponding to one-quarter of the total volume (about 12 ml) is multiply loaded into a one-well 16-cm × 14-cm × 1.5-mm, 7%

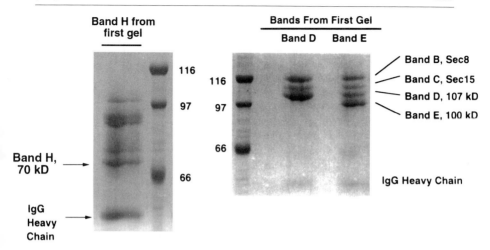

Fig. 3. Second separating SDS–PAGE slab gel showing selected proteins. The gel plugs from the FT gels for bands D, E, and H were chopped up and equilibrated with 2× sample buffer. The gel pieces were then packed into separate wells of SDS–PAGE slab gels and electrophoresed. Note that the concentrated protein from bands D, E, and H was contaminated with higher molecular weight exocyst proteins, which smeared down the first separating gel, and by IgG heavy chain, which smeared all the way to the top of the first separating gel. These bands are cleanly resolved on the second separating gel. After the second separating gel, the purified proteins were concentrated into a single gel slice with one more round of FT gel electrophoresis before in-gel digestion.

acrylamide SDS–PAGE slab gel. Four gels total are used for the initial slab gel electrophoresis. After separation, each gel is stained and destained and the bands corresponding to each unknown protein are excised. Because separation is poor due to the large amount of IgG released from the beads (2.4 mg per gel), the bands corresponding to the unknown exocyst components had to be concentrated and separated again on an SDS–PAGE slab gel to remove IgG and other unresolved proteins. The gel slices from the four separating gels corresponding to each band are pooled, washed with water until the pH is between 5 and 6, and then equilibrated with an equal volume of 2× sample buffer for 2 hr. The equilibrated gel slices can be frozen for storage before further purification at this point.

Concentration by Funnel Tube Gel

The sample buffer/gel slice mixture equivalent to two SDS–PAGE separating gels is loaded into a funnel tube (FT) gel for protein concentration (about 5 ml per tube)[11] (see Fig. 2 legend for details of the electrophoresis protocol). Thus, two FT gels are used to concentrate a single unknown after the initial SDS–PAGE separation. The top of the running gel that contains the concentrated protein was then excised and chopped into small cubes (about 1 mm) and incubated for 2 hr in an equal volume of 2× sample buffer. The buffer and the cubes are then loaded onto a second separating SDS–PAGE slab gel (7% acrylamide, 2 to 4 wells, 8 mm by 1.5 mm, per unknown). The second separating gel is stained and destained as above, and the bands corresponding to the unknown proteins are excised and pooled. One lane each of the second separating gel for bands D (Sec5p), E (Sec 10p) and H (Exo 70p) is shown in Fig. 3. The second separating gel gives a clean separation of all the concentrated proteins. Gel slices from the second separating gel corresponding to a given unknown are then processed exactly as above and concentrated in a single FT gel. The resulting 5.5-mm by 3-mm gel slice is fixed overnight with at least three solution changes in 50% (v/v) methanol and 10% (v/v) glacial acetic acid. The gel slices containing the polypeptides of bands A, D, E, F, and H (Fig. 3) were then submitted to the W. M. Keck Laboratory for amino acid analysis, tryptic and/or endoproteinase Lys-C in-gel digestion, HPLC separation, mass spectrometry, and microsequencing of the proteolytic peptides.

Yields of Exocyst Complex Proteins

The technique of sequential SDS–PAGE separating gel/FT concentrating gel electrophoresis used to purify and concentrate the unknown exocyst complex proteins is very efficient. We estimate that 640 μg of the exocyst proteins corresponding to bands D, E, and H (Fig. 3) are present in the 6.4 g of starting lysate (NY1008) used for their purification (each exocyst protein represents about 0.01% of total cellular protein).[3] In the large-scale immunoprecipitation, 5% of the total c-*myc*-Sec8p is immunoprecipitated (data not shown). Because the exocyst complex proteins are present in a 1 : 1 ratio to Sec8p,[3] the sample buffer (48 ml) loaded onto the first separating gels contains about 32 μg of the band D, E, and H proteins (Fig. 3).

After two rounds of SDS–PAGE separating gel and FT concentrating gel electrophoresis, we recovered 20 μg of band D protein (62%), 9.7 μg of band E protein (30%), and 8 μg of band H (25%) (Fig. 3). The results for the other exocyst components were comparable (data not shown). We were able to obtain peptide sequence from bands D, E, F, and H (Fig. 3) and determine the identity of band A by mass pattern matching.[3] The results for band F (a mixture of Sec3p and Exo84p) deserve special mention. The original sequencing and mass data identified the C terminus of Sec3p as band F, but subsequent work has shown that this band also contains Exo84p.[5] The identity of proteins corresponding to these bands is shown in Fig. 1 and elsewhere.[3]

Structural Characterization of Exocyst Complex

Protein–Protein Interactions of Exocyst Complex

It has been proposed that Sec3p functions as a spatial landmark for polarized secretion on the plasma membrane.[4] It has also been demonstrated that Sec15p interacts with the GTP-bound form of Sec4p and is able to associate with secretory vesicles.[6] To effectively direct secretory vesicles to the spatial landmark, there must be a trail of protein–protein interactions that connects Sec15p on the secretory vesicles to Sec3p marking the plasma membrane. The assembly of the exocyst complex may provide this link. We therefore decided to investigate the subunit interactions of the exocyst complex.

To efficiently achieve this goal, we use the strategy outlined below (Fig. 4A). The cDNAs for individual components are subcloned into expression vectors (such as pcDNA3), and the proteins are *in vitro* translated, both individually and in all pairwise combinations, in rabbit reticulocyte lysates. *In vitro* transcription/translation reactions are performed using the TnT system (Promega). Transcription is under the control of the T7 promoter, enhanced by the CMV sequence of the pcDNA3 vector (Invitrogen). Protease inhibitors and RNAase inhibitors are added to all the reactions. ^{35}S-labeled Promix (Amersham) is added to a final concentration of 1 μCi/μl for radioactive labeling of the synthesized proteins. The lysates are then diluted 1:100 in binding buffer (20 mM PIPES, pH 6.8, 0.5% Tween 20, 150 mM NaCl, 1 mM EDTA, 1 mM DTT). To identify the interactions between the subunits, coimmunoprecipitation experiments are carried out on all combinations of cosynthesized proteins. Because we do not have antibodies directed against all the exocyst proteins, we tagged the individual cDNAs with specific epitope sequences, including HA, c-*myc*, and FLAG in the expression constructs. The tagged proteins synthesized *in vitro* are then immunoprecipitated using commercially available, specific, monoclonal antibodies against the epitope sequences (BabCO), and the precipitates are analyzed for the presence of the untagged subunit. Interactions are detected by coimmunoprecipitation of the untagged protein with the tagged proteins. As a control, we evaluate the precipitation of the untagged protein by the epitope antibody in the absence of the tagged binding partner.

As one example, we show data supporting a direct interaction between Sec10p and Sec15p (Fig. 4B). The anti-c-*myc* antibody is able to precipitate c-*myc*-Sec10p from the *in vitro* translation mixture. When Sec15p and c-*myc*-Sec10p are cosynthesized, the antibody is able to precipitate Sec15p in addition to c-*myc*-Sec10p. However, Sec15p is not precipitated by anti-c-*myc* antibody in the absence of c-*myc*-Sec10p. This experiment indicates

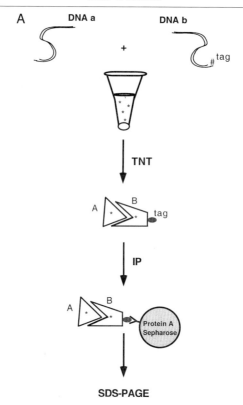

B In vitro Translation   Anti-Sec10p-myc IP

—Sec15p—
—Sec10p—
-myc

10   15   10+15   10   15   10+15

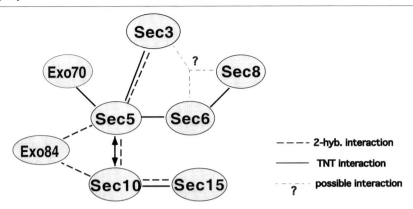

FIG. 5. Protein–protein interactions of the exocyst components. Interactions identified by TnT and coimmunoprecipitation are indicated by solid lines. The interactions identified by yeast two-hybrid assay are indicated by dashed lines. Possible interactions between Sec3p and Sec8p and/or Sec6p that are not identified using the above two methods are indicated with a question mark.

that Sec15p specifically binds Sec10p. The advantage of this approach is that unlike many other methods commonly used for studying *in vitro* protein–protein interactions, researchers do not have to biochemically purify the proteins of interest. In addition to the strategy mentioned above, we have also utilized the yeast two-hybrid assay to probe the pairwise interactions between exocyst components.

Summary of Interactions

All possible pairwise interactions have been examined using these strategies. In total, we have found that Sec15p binds directly to Sec10p. The

FIG. 4. (A) The coprecipitation assay for protein–protein interactions. To study the interaction of two proteins (A and B), the cDNA for each protein was subcloned into an expression vector and the proteins were *in vitro* translated in rabbit reticulocyte lysates in the presence of [^{35}S]methionione. We tagged one of the proteins with a specific epitope sequence, such as HA, c-*myc*, or FLAG in the expression construct. The proteins synthesized *in vitro* were then immunoprecipitated using commercially available monoclonal antibodies against the epitope sequence, and the precipitates were analyzed for the presence of the nontagged subunit. (B) Sec10p binds Sec15p. An example of a positive interaction between individual subunits of the exocyst, Sec10p and Sec15p. *Left: in vitro* translated proteins. *Right:* immunoprecipitation experiment: left-hand lane shows the immunoprecipitation of Sec10p tagged by c-*myc* sequence; center lane, as a negative control, shows that the antibody does not immunoprecipitate the untagged Sec15p. When Sec15p and c-*myc*-Sec10p were cosynthesized, the antibody was able to precipitate Sec15p in addition to c-*myc*-Sec10p (right-hand lane).

binding of the Sec10p/Sec15p subcomplex to the rest of the exocyst is mediated by the interaction of Sec10p with Sec5p. Sec5p also makes key links to Exo70p, Sec6p, and Sec3p. Thus Sec5p appears to be at the core of the complex. In addition, Sec6p directly binds Sec8p. Exo84p interacts with both Sec5p and Sec10. A map summarizing the identified interactions is shown in Fig. 5.

These results are fully consistent with earlier studies in which partially assembled exocyst structures were immunoprecipitated from lysates of various mutant strains.[2] For example, in *sec10-2* mutant cells, the mutant Sec10-2p is absent from the complex immunopurified using 9E10 antibody against the c-*myc* epitope fused to Sec8p. In addition to Sec10p, Sec15p is also absent from the complex although Sec15p is normally expressed. This result indicates that Sec15p requires Sec10p to associate with the exocyst complex. Our finding that Sec15p physically interacts with Sec10p is consistent with these data. In *sec5-24* mutant cells, Sec10p, Sec15p, and Exo70p are all absent as is the mutant Sec5-24p. This is consistent with our finding that Sec5p binds to Sec10p, Exo70p, and Sec6p, and through its binding to Sec6p, Sec5p is connected to Sec8p.

Except for the positive interactions shown in the map, all other pairwise combinations were found to be negative by these methods. The identified interactions link all the individual components of the exocyst. However, we speculate that additional interactions may exist that are not detected by the strategies mentioned above. For example, Sec3p may bind Sec6p and/or Sec8p since in a *sec5* mutant, only Sec3p, Sec6p, and Sec8p are in the immunoisolated complex, although the amount of Sec3p is reduced.[2]

Acknowledgments

We would like to thank Ralph Stevens and Alan Brown for making the funnel tubes. We would also like to thank all the members of the W. M. Keck Sequencing Facility at Yale for their help and patience, in particular, Kathryn Stone, Myron Crawford, Edward Papacoda, Mary Lo Presti, and Kenneth Williams. We would like to thank James Jamieson for the use and gift of the Bio-Rad tube gel tank. Finally, we would like to thank David Toft for the L-A viral coat protein monoclonal antibody and Reed Wickner for his advice on how to "cure" yeast strains of the L-A virus. This research was supported by NIH grant GM35370, by an NIH postdoctoral fellowship GM14884-03 for DRT, and by USUHS grant RO73EK. The opinions and assertions contained herein are private ones of the authors and are not to be construed as official or reflecting the views of the Department of Defense or the Uniformed Services University of the Health Sciences.

[13] Expression and Properties of Rab4 and Its Effector Rabaptin-4 in Endocytic Recycling

By Peter van der Sluijs, Karin Mohrmann, Magda Deneka, and Mandy Jongeneelen

Introduction

Rab4 is associated with early endocytic compartments and regulates receptor recycling from early endosomes back to the plasma membrane.[1–5] To understand how Rab4 mediates its function, we searched for effector proteins interacting with the GTP-bound form of Rab4 using the two-hybrid system. One of the clones we identified in this screen encoded a Rabaptin-5 homolog,[6] which we named Rabaptin-4.[7] Rabaptin-4 interacts preferentially with Rab4 and Rab5 in the GTP-bound form, a property shared by Rabaptin-5.[8] Rab4 binds to an N-terminal domain of Rabaptin-4, whereas two binding sites are present for Rab5, one in the N-terminal and one in the C-terminal half of Rabaptin-4. Rabaptin-4 is recruited by Rab4 to the perinuclear region where it colocalizes with endocytosed transferrin (Tf) and cellubrevin. Because Rabaptin-4 binds Rab4–GTP and reduces the intrinsic GTP hydrolysis rate of Rab4, we think that Rab4 and Rabaptin-4 coordinately regulate transport from early endosomes. We describe here some of the assays that we used to study the function of Rabaptin-4.

Methods

Expression of His$_6$-Tagged Rab4 and His$_6$-Tagged Rabaptin-4ΔC in Escherichia coli

Expression Constructs. To investigate binding of Rab4 to the N terminus of Rabaptin-4 *in vitro*, we first generated the expression plasmid pRSET/

[1] P. van der Sluijs, *et al., Proc. Natl. Acad. Sci. USA.* **88**, 6313 (1991).

[2] P. van der Sluijs, M. Hull, P. Webster, B. Goud, and I. Mellman, *Cell* **70**, 729 (1992).

[3] P. van der Sluijs, *et al., EMBO J.* **11**, 4379 (1992).

[4] G. Bottger, B. Nagelkerken, and P. van der Sluijs, *J. Biol. Chem.* **271**, 29191 (1996).

[5] E. Daro, P. van der Sluijs, T. Galli, and I. Mellman, *Proc. Natl. Acad. Sci. U.S.A.* **93**, 9559 (1996).

[6] H. Stenmark, G. Vitale, O. Ullrich, and M. Zerial, *Cell* **83**, 423 (1995).

[7] B. Nagelkerken, E. van Anken, M. van Raak, L. Gerez, K. Mohrmann, N. van Uden, J. Holthuizen, L. Pelkmans, and P. van der Sluÿs, *Biochem. J.* **346**, 593 (2000).

[8] G. Vitale, V. Rybin, S. Christoforidis, P. O. Thornqvist, M. McCaffrey, H. Stenmark, and M. Zerial, *EMBO J.* **17**, 1941 (1998).

C–Rabaptin-4ΔC. This construct encodes amino acids 11–388 of Rabaptin-4 and is prepared by releasing an *Eco*RI–*Hin*dIII fragment from pGADGH-#16B[7] and ligating it in the same sites of pRSET/C (Invitrogen, Leek, The Netherlands).

Buffers

Buffer A: 300 mM NaCl, 50 mM Tris-HCl, pH 8.0, 20 mM imidazole, 2 mg/ml lysozyme, 3 μg/ml aprotinin, 1 mM phenylmethylsulfonyl fluoride (PMSF), 10 μg/ml leupeptin, 1 μM pepstatin

Buffer B: 300 mM NaCl, 50 mM KOH–MES, pH 6.5, 10% (v/v) glycerol, 20 mM imidazole, 3 μg/ml aprotinin, 1 mM PMSF, 10 μg/ml leupeptin, 1 μM pepstatin

Buffer C: 50 mM KOH–PIPES, pH 8.0, 150 mM KCl, 1 mM EGTA, 5 mM MgCl$_2$, 1 mM dithiothreitol (DTT) 0.1% CHAPS

Expression and Purification Protocol. Hexahistidine-tagged (His$_6$) Rab4 and His$_6$-tagged Rabaptin-4ΔC are produced as fusion proteins in *E. coli*. Constructs are transformed into competent *E. coli* BL21(DE3)pLysS. Transformants are grown at 37° and selected on Luria–Bertani (LB) agar plates containing 100 μg/ml ampicillin + 50 μg/ml chloramphenicol. The next day a colony is transferred to 2 ml LB + 100 μg/ml ampicillin + 50 μg/ml chloramphenicol. Precultures are grown for 6 hr at 37°, diluted into 50 ml, and grown overnight. Next morning the culture is diluted 20 times with LB + 100 μg/ml ampicillin + 50 μg/ml chloramphenicol. After the culture reaches OD$_{600}$ ~ 0.6, isopropylthiogalactoside (IPTG) (Boehringer Mannheim, Almere, The Netherlands) is added to a final concentration of 0.1 mM and the cells are grown for 3 hr at 37°. Bacteria are harvested after centrifugation of the culture for 10 min at 5000 rpm in a Sorvall GSA rotor. Pellets are resuspended with cold phosphate-buffered saline (PBS), combined, and centrifuged for 10 min at 5000 rpm in a Sorvall SS34 rotor. At this stage, pellets are snap-frozen in liquid nitrogen and stored at −80°.

The cell pellet is resuspended in 50 ml cold buffer A and sonicated 3 times for 45 sec on ice. Cell debris is pelleted by centrifugation at 4° for 15 min at 16,000 rpm in a Sorvall SS34 rotor. The supernatant containing soluble protein is retrieved and loaded on a 5-ml HiTrap affinity column (Pharmacia Biotech, Roosendaal, The Netherlands) and washed with buffer B until OD$_{280}$ is ~0. Proteins are eluted at a flow rate of 0.4 ml/min using a continuous 0.02–0.5 M imidazole gradient in buffer B and collected in fractions of 0.25 ml. Column fractions (5 μl) are analyzed by 12.5% SDS–PAGE and stained with Coomassie Brilliant Blue. Peak fractions are pooled and desalted on two PD10 columns (Pharmacia), equilibrated with buffer C. Fractions of 0.5 ml are collected, and protein content is measured with the BCA protein assay (Pierce, Rockford, IL). Peak fractions are pooled, snap-frozen in liquid nitrogen, and stored at -80°.

Comments. For the expression of His_6-tagged Rab4 and His_6-tagged Rabaptin-4ΔC, we recommend starting with fresh transformants because they provide the most reproducible results in terms of protein yield. We have tried to increase the yield of His_6-tagged Rabaptin-4ΔC by growing cultures at room temperature; however, this provided little improvement. For purification of His_6-tagged Rab4, all buffers are supplemented with 5 μM GDP to stabilize the small GTPase. Typical yields are 1.5 mg of His_6-tagged Rab4 and 0.3 mg of His_6-tagged Rabaptin-4ΔC per liter of culture. The purity of recombinant proteins is ~90% as assessed by silver staining.

In Vitro Binding Assay

A simple and rapid procedure to assay guanine nucleotide-dependent binding of Rab4 to Rabaptin was developed using an enzyme-linked immunosorbent assay (ELISA) format in which His_6-tagged Rabaptin-4 is coated in microtiter plates. Subsequently wells are incubated with increasing amounts of His_6-tagged Rab4 in the presence of GDP or GTP and bound Rab4 is determined with Rab4 antibody.

Reagents

Binding buffer: 100 mM $NaHCO_3/Na_2CO_3$, pH 9.5
TBST: 10 mM Tris-HCl, pH 8.0, 150 mM NaCl, 0.05% Tween 20
Blotto: 5% skimmed milk powder in TBST
Stock solutions of GDP (0.25 M) and GTP (0.25 M)

Procedure. Wells are coated for 1 hr at room temperature with 300 ng His_6-tagged Rabaptin-4ΔC in binding buffer and blocked for 1 hr with Blotto. Increasing amounts of His_6-tagged Rab4 are added in the presence of 5 mM $MgCl_2$, and 1 mM GDP or GTP (final concentrations) in TBST and incubated for 1 hr in a cold room. Wells are washed 5 times with 200 μl Blotto and incubated for 45 min with a rabbit antibody against Rab4 (1 : 2000). Wells are again washed 5 times with Blotto, and incubated 45 min with horseradish peroxidase (HRP)-conjugated goat anti-rabbit antibody (Jackson ImmunoResearch Laboratories, Westgrove, PA). After 5 washes, HRP enzyme activity is detected colorimetrically with 0.1 mg/ml 3,3′,5,5′-tetramethylbenzidine (Sigma, Zwijndrecht, The Netherlands) in 0.03‰ H_2O_2, 0.111 M sodium acetate, pH 5.5. Reactions are stopped after 5–10 min by addition of 100 μl 2 M H_2SO_4 and OD_{450} is assayed in a microtiter plate reader. Results of a representative binding assay are shown in Fig. 1.

Comments. As controls we use Blotto, or His_6-tagged Rabaptin-4ΔN (amino acids 469–830). When wells are coated with these proteins, a con-

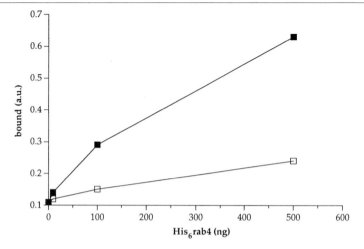

FIG. 1. *In vitro* binding of His$_6$-tagged Rab4 to His$_6$-tagged Rabaptin-4ΔC. His$_6$-tagged Rabaptin-4ΔC is immobilized in a microtiter plate and incubated with His$_6$-tagged Rab4 in the presence of GDP (open squares) or GTP (closed squares). Binding is detected using anti-Rab4 antibodies and a HRP-conjugated secondary antibody. An approximately fourfold increase in binding is observed in the presence of GTP as compared to GDP, showing that the interaction prefers the GTP-bound conformation of His$_6$-tagged Rab4. Reprinted with permission from *Biochemical Journal* **346,** 593 (2000).

stant background signal of ~0.3 OD$_{450}$ units is obtained, irrespective of the amount of His$_6$-tagged Rab4 that is added. This background is subtracted from experimental data points. Three to four times more Rab4 is bound to rabaptin in the presence of GTP as compared to GDP. Replacement of GTP by the nonhydrolyzable analog GTPγS should further increase this ratio.

Intrinsic GTPase Assay for His$_6$-Tagged Rab4 in Presence of His$_6$-Tagged Rabaptin-4ΔC

To investigate the effect of the bound effector molecule on the intrinsic GTPase activity of Rab4 we use a slightly modified charcoal precipitation assay[9] that measures release of hydrolyzed [^{32}P]P from [γ-^{32}P]GTP prebound to His$_6$-tagged Rab4.

Procedure. Reactions are done at 37° in a volume of 350 μl and contain 200 n*M* active His$_6$-tagged Rab4 (determined as described in Ref. 10), 500

[9] A. K. Kabcenell, B. Goud, J. K. Northup, and P. J. Novick, *J. Biol. Chem.* **265,** 9366 (1990).
[10] A. D. Shapiro, M. A. Riederer, and S. R. Pfeffer, *J. Biol. Chem.* **268,** 6925 (1993).

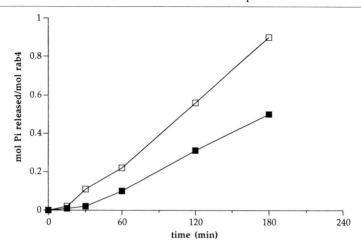

FIG. 2. Effect of His$_6$-tagged Rabaptin-4ΔC on intrinisic GTP hydrolysis of His$_6$-tagged Rab4. [γ-^{32}P]GTP loaded His$_6$-tagged Rab4 is incubated with His$_6$-tagged Rabaptin-4ΔC (closed square) or bovine serum albumin (open squares) for different periods of time. In the presence of His$_6$-tagged Rabaptin-4ΔC the intrinsic rate of GTP hydrolysis is reduced 45% compared to the incubations with BSA. Reprinted with permission from *Biochemical Journal* **346,** 593 (2000).

nM [γ-^{32}P]GTP (7.4 $\times$ 10^{11} Bq/mmol), and 2 μM His$_6$-tagged Rabaptin-4ΔC or BSA (control) in 50 mM KOH–PIPES, pH 8.0, 150 mM KCl, 1 mM EDTA, 10 mM MgCl$_2$, 1 mM DTT, 0.1% CHAPS. After different periods of time, 50-μl aliquots are retrieved and added to 750 μl 5% (w/v) activated charcoal (Serva, Heidelberg, Germany) in 50 mM NaH$_2$PO$_4$ on ice. The mixture is vortexed and centrifuged in a microcentrifuge at 3000g for 10 min at 4°. The supernatant is saved and 500 μl is mixed with 5 ml Ultima Gold scintillation fluid (Packard Biosystems, Groningen, The Netherlands) to quantitate hydrolyzed [^{32}P]P$_i$. Background signals are determined in parallel reactions lacking His$_6$-tagged Rab4 and subtracted from experimental values.

Comments. In the presence of a 10-fold molar excess of His$_6$-tagged Rabaptin-4ΔC, the intrinsic GTP hydrolysis rate of His$_6$-tagged Rab4 is reduced from 0.046 to 0.025 min^{-1}, as shown in Fig. 2. An equimolar amount of His$_6$-tagged Rabaptin-4ΔC does not effect GTP hydrolysis. To formally show that Rabaptin-4 is a true effector of Rab4 requires knowledge of whether it can effect the exchange of Rab4-bound guanine nucleotide. We have tested this in an *in vitro* guanine nucleotide exchange assay and found

[11] B. Nagelkerken, *et al., Electrophoresis* **18,** 2694 (1997).

that a 10-fold molar excess of His_6-tagged Rabaptin-4ΔC does not effect guanine nucleotide exchange on His_6-tagged Rab4.

Expression in Mammalian Cells

Expression Constructs. For the analysis of Rab4 binding to Rabaptin-4, and recruitment assays of Rabaptin-4 to early endosomes, we generated constructs from which Rabaptin-4, the N-terminal region (aa 11–388), and C-terminal region (aa 469–830) are expressed as hexahistidine fusion proteins in mammalian cells. An *EcoRI–XbaI* Rabaptin-4 fragment encoding aa 11–830 is excised from pGADGH-#16B to generate pcDNA3.1His/B-Rabaptin (Invitrogen). The N-terminal Rabaptin-4 construct pcDNA3.1His/B-Rabaptin-4ΔC is made by inserting a *BamHI–DraI* fragment between the *BamHI–EcoRV* sites of pcDNA3.1His/B. The C-terminal construct pcDNA3.1His/A-Rabaptin-4ΔN is made by inserting a Rabaptin-4 *HindII–XbaI* fragment between the *EcoRV–XbaI* sites of pcDNA3.1His/A. For NHRab5 expression (canine Rab5a with an N-terminal X31 influenza hemagglutinin epitope tag, we use the previously described Rab5 expression plasmid pcDNA3-NHRab5.[11]

Transfection Protocol. 20 μg sterile DNA is mixed with 0.5 ml 137 mM NaCl, 5 mM KCl, 0.7 mM Na_2HPO_4, 6 mM dextrose, 21 mM HEPES, pH 7.1 (HBS), in a polycarbonate tube. A calcium phosphate precipitate is prepared by adding 31 μl 2 M $CaCl_2$ while flicking the tube 2–3 times. After 45 min, the cloudy suspension is added to an aspirated 10-cm plate of Chinese hamster ovary (CHO) cells that is 25% confluent. Cells are incubated 30 min at room temperature on a rocking platform. At the end of this period, cells receive 10 ml of minimal essential medium-α (αMEM), 10% fetal calf serum (FCS), and are grown in a 37° incubator. After 4 hr, medium is aspirated and 2.5 ml 15% glycerol in HBS is added for precisely 3.5 min at 37°. The cells are washed twice with culture medium and grown in αMEM, 10% FCS. After 24 hr, cells are trypsinized and grown for a week in αMEM, 10% FCS, 0.6 mg/ml G418 (Gibco-BRL, Breda, The Netherlands). For immunofluorescence microscopy, cells are seeded on round (11-mm diameter) #1 coverslips. Transfectants receive 5 mM sodium butyrate (Fluka, Buchs, Germany) 17 hr prior to the experiments to enhance expression of cytomegalovirus promotor (CMV)-driven expression plasmids.

Comments. The glycerol shock is used to enhance transfection efficiency and is optional. When this step is omitted, the medium containing calcium phosphate and DNA should be replenished with fresh culture medium the next morning. Sodium butyrate affects the ultrastructure of secretory and endocytic compartments at concentrations above 5 mM. We grow trans-

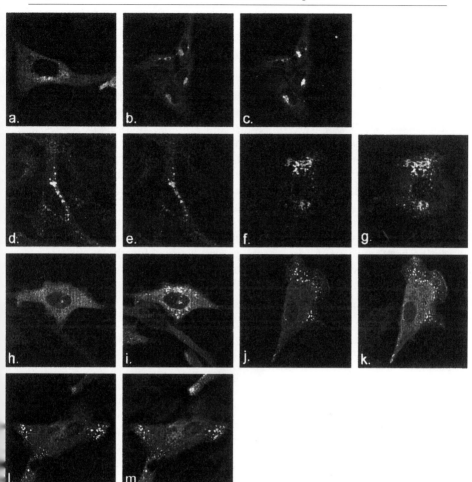

FIG. 3. Recruitment of His$_6$-tagged Rabaptin-4 to early endosomes. In CHO cells transfected with His$_6$-tagged Rabaptin-4, His$_6$-tagged Rabaptin-4 predominantly shows a diffuse cytosolic labeling (a). In His$_6$-tagged Rabaptin-4–Rab4 transfectants, His$_6$-tagged Rabaptin-4 (b) and Rab4 (c) colocalize on aciniform organelles in the perinuclear region. This is an endocytic organelle since it contains endocytosed Cy3-labeled transferrin (d) that also is decorated with His$_6$-tagged Rabaptin-4 (e). The region required for targeting His$_6$-tagged Rabaptin-4 to Rab4 containing endosomes is localized in the N terminus of Rabaptin-4, as the His$_6$-tagged Rabaptin-4ΔC (f) colocalizes with Rab4 (g) in a His$_6$-tagged Rabaptin-4ΔC–Rab4 transfectant, while His$_6$-tagged Rabaptin-4ΔN (h) does not colocalize with Rab4 (i) in His$_6$-tagged Rabaptin-4ΔN–Rab4 CHO cells. Both His$_6$-tagged Rabaptin-4ΔN (j) and His$_6$-tagged Rabaptin-4ΔC (l) are are recruited to Rab5 containing endosomes (k, m) in His$_6$-tagged Rabaptin-4ΔN–NHRab5 and His$_6$-tagged Rabaptin-4ΔC–NHRab5 CHO cells, showing that Rabaptin-4 contains two domains that bind Rab5. Reprinted with permission from *Biochemical Journal* **346,** 593 (2000).

fected cells for a week on selection to increase the number of transfectants and because it improves morphology of the cells.

Immunofluorescence Microscopy

To detect compartments of the transferrin pathway, Cy3-labeled transferrin is internalized in cells grown on coverslips as described.[5] Coverslips are fixed in 1 ml 3% paraformaldehyde for 1 hr and subsequently washed with PBS. Excess fixative is quenched in 1 ml 50 mM NH$_4$Cl in PBS for 5 min. The cells are permeabilized and blocked for 30 min in 1 ml 0.1% saponin, 0.5% bovine serum albumin in PBS (blocking buffer) and incubated with primary antibodies in a volume of 25 μl. Rab4 and NHRab5 are detected with affinity purified antibodies raised against glutathione S-transferase(GST)–Rab4 and the NH epitope tag.[4,11] The Rabaptin constructs are detected with the XPRESS monoclonal mouse antibody (Invitrogen). The coverslips are washed for three periods of 15 min with 1 ml blocking buffer and then incubated in 25 μl for 30

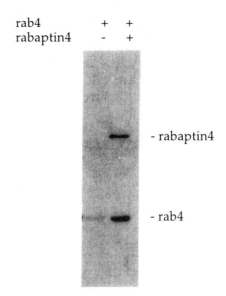

FIG. 4. *In vivo* binding assay of Rab4 to rabaptin4. A Rab4 CHO cell line was either mock-transfected or transfected with pcDNA3.1His/B-Rabaptin. Postnuclear supernatants are incubated with Ni^{2+}-NTA resin to bind His$_6$-tagged Rabaptin-4. Resin fractions are washed with RIPA buffer and resolved by SDS–PAGE on 12.5% gels under reducing conditions followed by Western blotting using antibodies against Rab4 and His$_6$-tagged Rabaptin-4. In the cells expressing His$_6$-tagged Rabaptin-4, ~6 times more Rab4 is isolated on the beads than in mock transfected Rab4 CHO cells.

min with indocarbocyanine Cy3-conjugated goat anti-mouse IgG and 5-[(4,6-dichlorotriazin-2-yl)amino] (DTAF) fluorescein-conjugated goat anti-rabbit IgG (Jackson ImmunoResearch Laboratories), diluted 200 times and 50 times, respectively, in blocking buffer. The coverslips are washed 3 times with blocking buffer, once with PBS, and finally mounted in Moviol 4-88 (Calbiochem, La Jolla, CA) containing 2.5% 1,4-diazabi-cyclo[2.2.2]octane (Sigma, Zwijndrecht, The Netherlands) and examined with a 63× planapo objective on a Leitz DMIRB fluorescence microscope (Fig. 3) (Leica, Voorburg, The Netherlands).

Coprecipitation of Rab4 with His_6-Tagged Rabaptin-4 on Ni^{2+}-NTA Resin

To demonstrate the association of Rab4 and Rabaptin-4 *in vivo,* we use Rab4 CHO transfected with pcDNA3.1His/B-Rabaptin and control (mock transfected) Rab4 CHO cells.

Procedure. Three 10-cm plate subconfluent cells are homogenized in 1 ml 100 mM KOH–HEPES, pH 7.4, 0.1 mM GTP, 1 mM MgCl$_2$, 50 mM NaF, and 1 mM PMSF by passing them 10 times through a 25-gauge needle. Homogenates are spun 10 min at 1500 rpm to prepare a postnuclear supernatant. The postnuclear supernatant is next incubated with 50 μl Ni^{2+}-NTA beads (Qiagen-Westburg, Leusden, The Netherlands) by end-over-end rotation for 1 hr at 4° to collect His_6-tagged Rabaptin-4–Rab4 complexes. Resin fractions are washed four times with 1 ml 150 mM NaCl, 100 mM Tris-HCl, pH 8.3, 0.1% SDS, 0.5% deoxycholate (DOC), 0.5% Nonidet P-40. After the third wash, the resuspended beads are transferred to a clean tube, pelleted, and boiled for 5 min in reducing Laemmli sample buffer. Samples are resolved by SDS–PAGE on 12.5% minigels, transferred to PVDF membrane, and analyzed by Western blotting using the anti-Xpress antibody (at 1 : 4000 dilution) to detect His_6-tagged Rabaptin-4 and a rabbit antibody against Rab4 (at 1 : 2000 dilution) (Fig. 4).

Acknowledgment

This work was supported by grants from NWO-MW, NWO-ALW and the Dutch Cancer Society (to PvdS).

[14] Purification of EEA1 from Bovine Brain Cytosol Using Rab5 Affinity Chromatography and Activity Assays

By Savvas Christoforidis and Marino Zerial

Introduction

Vesicle docking and fusion as well as organelle dynamics are processes that are regulated by Rab proteins[1-8] via interactions with effector molecules. Therefore, understanding the molecular mechanisms underlying intracellular transport requires the identification and characterization of Rab effectors.

Transport through the early endocytic pathway is regulated by the small GTPase Rab5.[6] Work over many years has established multiple functions for Rab5 within the early endocytic route, including distinct roles in vesicle budding from the plasma membrane, motility along microtubules, and vesicle docking and fusion.[2,6,9,10] Consistent with this idea, we have shown that at least 22 cytosolic proteins can bind to the active form of Rab5.[11] Among these proteins we found, the previously characterized Rabaptin-5/Rabex-5 complex (see related chapter in this issue[11a]) and identified EEA1 as a new Rab5 effector.[12] EEA1 was initially described as an autoantigen in

[1] A. Salminen and P. J. Novick, Cell 49, 527 (1987).

[2] C. Bucci, R. G. Parton, I. H. Mather, H. Stunnenberg, K. Simons, B. Hoflack, and M. Zerial, Cell 70, 715 (1992).

[3] S. Ferro-Novick and P. Novick, Annu. Rev. Cell Biol. 9, 575 (1993).

[4] M. Sogaard, K. Tani, R. R. Ye, S. Geromanos, P. Tempst, T. Kirchhausen, J. E. Rothman, and T. Sollner, Cell 78, 937 (1994).

[5] V. Rybin, O. Ullrich, M. Rubino, K. Alexandrov, I. Simon, M. C. Seabra, R. Goody, and M. Zerial, Nature 383, 266 (1996).

[6] P. Novick and M. Zerial, Curr. Opin. Cell Biol. 9, 496 (1997).

[7] A. Mayer and W. Wickner, J. Cell Biol. 136, 307 (1997).

[8] X. Cao, N. Ballew, and C. Barlowe, EMBO J. 17, 2156 (1998).

[9] H. McLauchlan, J. Newell, N. Morrice, A. Osborne, M. West, and E. Smythe, Curr. Biol. 8, 34 (1998).

[10] E. Nielsen, F. Severin, J. M. Backer, A. A. Hyman, and M. Zerial, Nat. Cell Biol. 1, 376 (1999).

[11] S. Christoforidis, H. M. McBride, R. D. Burgoyne, and M. Zerial, Nature 397, 621 (1999).

[11a] R. Lippe, H. Horiuchi, A. Runge, and M. Zerial, Methods Enzymol. 329, [15], (2001) (this volume).

[12] A. Simonsen, R. Lippé, S. Christoforidis, J.-M. Gaullier, A. Brech, J. Callaghan, B.-H. Toh, C. Murphy, M. Zerial, and H. Stenmark, Nature 394, 494 (1998).

lupus erythematosus associated with early endosomes.[13] This protein was found to be required for fusion between early endosomes,[12,14] and shown to be an essential component of the endosome membrane docking machinery.[11] Besides interacting with Rab5, EEA1 also binds to phosphatidylinositol 3-phosphate (PI3P) via a conserved FYVE finger domain.[15–19] This finding provided a molecular explanation for the previously observed requirement for phosphatidylinositol-3-OH kinase (PI3K) in endosome fusion.[20–22]

Recently, we identified two PI3 kinases among the Rab5 interacting proteins, p85α/p110β and hVPS34, as new Rab5 effectors.[23] This finding implies that Rab5 not only interacts directly with EEA1 but also regulates the local production of PI3P, thereby creating a second binding site for EEA1 on the endosomal membrane. Therefore, on activation of Rab5, the combinatorial interactions between EEA1, Rab5, and PI3P would ensure the specific recruitment of the effector to the early endosome.

Here we describe the experimental system we developed to biochemically purify and characterize native EEA1 from bovine brain cytosol. This method is based on a combination of conventional biochemical techniques, and primarily on the use of affinity chromatography. Affinity chromatography has been used in the past and led to the identification of three RhoA effectors.[24,25] We have improved this method by optimizing a number of parameters such as increasing the density of Rab5 on the beads, the yield of nucleotide exchange, and the specificity of the elution step. This approach

[13] F. Mu, J. M. Callaghan, O. Steele-Mortimer, H. Stenmark, R. G. Parton, P. L. Campbell, J. McCluskey, J. P. Yeo, E. P. C. Tock, and B. H. Toh, *J. Biol. Chem.* **270,** 13503 (1995).

[14] I. G. Mills, A. T. Jones, and M. J. Clague, *Curr. Biol.* **8,** 881 (1998).

[15] V. J. Patki, W. S. Lane, B. H. Toh, H. S. Shpetner, and S. Corvera, *Proc. Natl. Acad. Sci. U.S.A.* **94,** 7326 (1997).

[16] J.-M. Gaullier, A. Simonsen, A. D'Arrigo, B. Bremnes, H. Stenmark, and R. Aasland, *Nature* **394,** 432 (1998).

[17] V. Patki, D. C. Lawe, S. Corvera, J. V. Virbasius, and A. Chawla, *Nature* **394,** 433 (1998).

[18] C. G. Burd and S. D. Emr, *Mol. Cell* **2,** 157 (1998).

[19] C. Wiedemann and S. Cockcroft, *Nature* **394,** 426 (1998).

[20] G. Li, C. D'Souza-Schorey, M. A. Barbieri, R. L. Roberts, A. Klippel, L. T. Williams, and P. D. Stahl, *Proc. Natl. Acad. Sci. U.S.A.* **92,** 10207 (1995).

[21] A. T. Jones and M. Clague, *Biochem. J.* **311,** 31 (1995).

[22] D. J. Spiro, W. Boll, T. Kirchhausen, and M. Wessling-Resnick, *Mol. Biol. Cell* **7,** 355 (1996).

[23] S. Christoforidis, M. Miaczynska, K. Ashman, M. Wilm, L. Zhao, S. C. Yip, M. D. Waterfield, J. M. Backer, and M. Zerial, *Nat. Cell Biol.* **1,** 249 (1999).

[24] K. Amano, H. Mukai, Y. Ono, K. Chihara, T. Matsui, Y. Hamajima, K. Okawa, A. Iwamatsu, and K. Kaibuchi, *Science* **271,** 648 (1996).

[25] T. Matsui, M. Amano, T. Yamamoto, K. Chichara, M. Nakafuku, M. Ito, T. Nakano, K. Okawa, A. Iwamatsu, and K. Kaibuchi, *EMBO J.* **15,** 2208 (1996).

has the advantage of allowing a comprehensive analysis of the Rab5 effectors in addition to EEA1. Therefore, we have developed a strategy that will allow us to better understand the complexity and dynamics of Rab5-regulated transport within the endocytic pathway.

Purification of Native EEA1 from Bovine Brain

EEA1 was purified from cytosol derived from bovine brain, and the entire procedure involves four principal steps: (1) preparation of bovine brain cytosol, (2) preparation of a Rab5 affinity chromatography column, (3) isolation of Rab5 effectors via affinity chromatography, and (4) purification of EEA1 using ion-exchange and size-exclusion chromatography. A flowchart of the entire purification procedure is shown in Fig. 1.

Preparation of Bovine Brain Cytosol

Solutions

Wash buffer: Phosphate-buffered saline (PBS) consisting of 2.7 mM KCl, 1.8 mM KH$_2$PO$_4$, 8.1 mM Na$_2$HPO$_4$, 137 mM NaCl, pH 7.2

Homogenization buffer (HB): 20 mM HEPES, 100 mM NaCl, 5 mM MgCl$_2$, 1 mM dithiothreitol (DTT), pH 7.5

Procedure. Fourteen fresh bovine brains are put on ice and processed immediately. All steps of the purification are performed in the cold room, unless otherwise stated. First, meninges and cerebellum are removed and discarded and the remaining tissue washed twice with ice-cold PBS. Protease inhibitors are added freshly to ice-cold homogenization buffer (2.05 liters) in the following concentrations: 6 μg/ml chymostatin, 0.5 μg/ml leupeptin, 10 μg/ml antipain hydrochloride, 2 μg/ml aprotinin, 0.7 μg/ml pepstatin, 10 μg/ml 4-Amidinophenylmethane sulfonyl fluoride (APMSF) (Sigma, St. Louis, MO). 220 ml of homogenization buffer (containing the protease inhibitors) is added to each 1.5 brain and homogenized in a Waring blender (Bender + Hobein, Zurich, Switzerland) on the highest speed for 50 sec. The total homogenate is centrifuged in a GS3 rotor (Sorvall) at 4200g at 4° for 50 min. The resulting postnuclear supernatant (PNS) is further centrifuged in a Ti 45 rotor (Beckman ultracentrifuge) at 100,000g, 4°, for 1 hr. The 100,000g supernatant (cytosol, 1.5 liter) is dialyzed twice against 50 liters (each time) of homogenization buffer (without protease inhibitors) to remove endogenous nucleotides that would interfere with the desired nucleotide state of Rab5 at later steps. The dialyzed cytosol is aliquoted in 50-ml Falcon tubes and snap-frozen in liquid nitrogen for storage at −80°. Potential aggregates resulting from dialysis and snap-freezing are removed by preclearing the thawed cytosol at 100,000g for 60 min at 4°

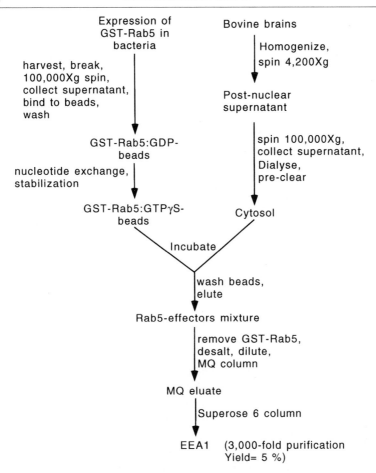

FIG. 1. Schematic representation of EEA1 isolation from bovine brain cytosol. Yield and purification fold are indicated in the final preparation of EEA1.

prior to use. This method typically results in a preparation of 1.5 liter cytosol at a protein concentration of 14 mg/ml.

Preparation of Rab5 Affinity Chromatography Column

Solutions

Lysis buffer: PBS, 5 mM 2-mercaptoethanol, 5 mM MgCl$_2$, 200 μM GDP, 5 μg/ml DNase, 5 μg/ml RNase, 6 μg/ml chymostatin, 0.5 μg/ml leupeptin, 10 μg/ml antipain hydrochloride, 2 μg/ml aprotinin, 0.7 μg/ml pepstatin, 10 μg/ml APMSF

Nucleotide exchange buffer (NE): 20 mM HEPES, 100 mM NaCl, 10 mM EDTA, 5 mM MgCl$_2$, 1 mM DTT, pH 7.5

Nucleotide stabilization buffer (NS): 20 mM HEPES, 100 mM NaCl, 5 mM MgCl$_2$, 1 mM DTT, pH 7.5

Procedure. Rab5 cDNA, tagged at the N terminus with GST in the pGEX-5X3 vector (Pharmacia, Piscataway, NJ), is transformed into DH5α bacteria cells. An overnight culture is grown in Luria–Bertani (LB) medium containing 100 μg/ml ampicillin. 150 flasks, each containing 0.8 liter LB + ampicillin (100 μg/ml), are inoculated with 20 ml of overnight culture and grown to an OD$_{600}$ of 0.7–0.9 at 37°. The cultures are induced with 1 mM isopropylthiogalactoside (IPTG) (final) at 37° for 3 hr. Cells are harvested by centrifugation at 6000 rpm, for 15 min, 4°, in a Sorvall Avanti J-20 centrifuge, in a JLA 8100 rotor. The supernatant is discarded and the bacterial pellet resuspended in cold lysis buffer in a total volume of 2.5 liters, snap-frozen in liquid nitrogen, and kept at −20° until use.

Frozen bacteria are thawed and lysed using a French press at 800–1000 psi. The lysate is centrifuged at 33,000 rpm in a Ti 45 rotor, for 1 hr at 4°, and the supernatant incubated with 20 ml glutathione Sepharose beads (Pharmacia) (prewashed with PBS, 5 mM 2-mercaptoethanol) for 2 hr under rotation at 4°. Then the beads are pelleted at 500g for 15 min, loaded on an empty column, and washed with 200 ml PBS, 5 mM 2-mercaptoethanol, 5 mM MgCl$_2$, and 200 $\mu$$M$ GDP. This preparation yields 800 mg of glutathione *S*-transferase (GST)–Rab5 (Fig. 2C).

The procedure described above results in the production of GST–Rab5 : GDP form (inactive) due to the high intrinsic rate of GTP hydrolysis by Rab5.[5] To obtain the active (GTP) from of Rab5 on the beads, we developed a two-step reaction, first a nucleotide exchange (NE) reaction to release the bound nucleotide, followed by the stable loading of empty Rab5 with the desired nucleotide, such as GTPγS, a slowly hydrolyzble analog of GTP. The nucleotide exchange reaction is based on the use of EDTA to strip Mg^{2+} ions and exchange GDP for GTPγS on Rab5. Following the nucleotide exchange, Rab5 is stabilized in the active form in the presence of excess GTPγS and magnesium in the absence of EDTA. Nucleotide exchange is performed by first dividing the beads (20 ml) containing GST–Rab5 in 20 small empty columns (Bio-Rad, Hercules, CA). All volumes for incubations and washes described below refer to each column. The columns are washed with 5 ml NE buffer containing 10 $\mu$$M$ GTPγS and then closed and incubated with 2 ml NE buffer containing 1.5 mM GTPγS, for 30 min, at room temperature under very slow rotation. Then the columns are set up on racks, the solution drained out, and the beads

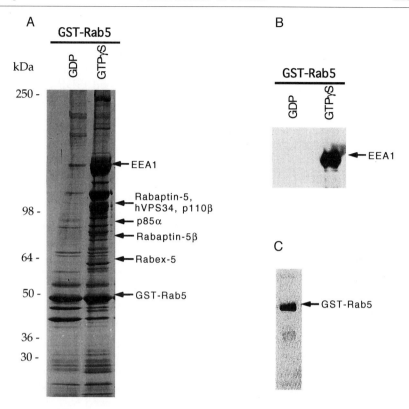

Fig. 2. Rab5 affinity chromatography reveals 22 Rab5 interacting proteins, and in this mixture of proteins EEA1 is the most abundant. (A) GST–Rab5 was expressed in bacteria, bound to glutathione Sepharose beads (see part C) and the corresponding nucleotide form was prepared. After incubation with cytosol, proteins bound to either form of GST–Rab5 were eluted. 50 μl of the eluate was separated on SDS–PAGE and the gel was silver stained. The positions of the known (so far) Rab5 effectors are shown on the right-hand side and molecular weight standards on the left-hand side of the gel. (B) Western blotting analysis of the eluate (10 μl) obtained from the Rab5 affinity chromatography using anti-EEA1 rabbit antibodies raised against a synthetic peptide corresponding to the N terminus of EEA1. (C) GST–Rab5 was produced in bacteria and bound to glutathione Sepharose beads. 10 μl of the beads were used to elute the protein in order to test its purity by SDS–PAGE analysis (loaded 5 μg) followed by staining with Coomassie blue.

washed with 5 ml NE buffer containing 10 μM GTPγS. Subsequently, the columns are closed again and incubated as above. The whole procedure is repeated a total of 3 times, which allows for complete removal of all the GDP released from Rab5. Then, the columns are washed with 5 ml NS buffer containing 100 μM GTPγS and incubated for 20 min with 2 ml NS buffer, 2 mM GTPγS, at room temperature and slow rotation (excess of Mg^{2+} at this step serves to stabilize Rab5 in the GTPγS active form). To test for the binding specificity of Rab5:GTP interacting proteins, an identically treated GST–Rab5:GDP column is prepared except that the NE and NS buffers contained GDP instead of GTPγS at the same concentrations.

Isolation of Rab5 Effectors via Affinity Chromatography

Solutions

Wash buffer A: 20 mM HEPES, 100 mM NaCl, 5 mM MgCl$_2$, 1 mM DTT, pH 7.5

Wash buffer B: 20 mM HEPES, 250 mM NaCl, 5 mM MgCl$_2$, 1 mM DTT, pH 7.5

Wash buffer C: 20 mM HEPES, 250 mM NaCl, 1 mM DTT, pH 7.5

Elution buffer: 20 mM HEPES, 1.5 M NaCl, 20 mM EDTA, 1 mM DTT, pH 7.5

Procedure. Beads containing GST–Rab5:GTPγS are incubated with cytosol prepared as described above (50 ml of cytosol per 0.7 ml of beads) for 2 hr at 4° under rotation in the presence of 100 μM GTPγS. Then the beads are pelleted at 500g, for 15 min, at 4°, and loaded back on the empty plastic columns used before. The supernatant (unbound cytosol) is divided in the 20 columns and passed through again. The columns are washed first with 10 ml NS buffer containing 10 μM GTPγS, followed by a wash with 10 ml buffer B containing 10 μM GTPγS, and a third wash with 2 ml buffer C. Then columns are taken at room temperature and loaded with 0.3 ml elution buffer containing 5 mM GDP and allowed to drain (void volume). Subsequently, the beads are incubated in the columns with 1 ml elution buffer containing 5 mM GDP for 20 min with occasional mixing at room temperature. Then the eluate is collected and 0.7 ml new elution buffer containing 5 mM GDP passed through the column and pooled with the previous eluate, to give a final volume of 1.7 ml eluate per column. EDTA used at this step in the elution buffer serves to remove Mg^{2+} nucleotide from Rab5, thereby releasing the effectors from the column. The eluate (34 ml total volume) has a protein concentration of 0.3 mg/ml, and the pattern of molecules is shown in Fig. 2A. The presence of EEA1 in this preparation is shown by Western blotting using anti-EEA1 specific antibodies (Fig. 2B). EEA1 is the most abundant protein in the mixture of Rab5 effectors (Fig. 2A).

The GST–Rab5:GDP column is prepared like the corresponding GTPγS column with the following change: all buffers contain GDP instead of GTPγS except for the elution buffer, which contains GTPγS (1 mM) instead of GDP. The inactive form of Rab5 results in a different protein pattern than that of the active form (Fig. 2A), indicating the specificity of the method.

Purification of EEA1 Using Ion-Exchange and Size-Exclusion Chromatography

Solutions

Buffer D: 20 mM HEPES, 150 mM NaCl, 1 mM DTT, pH 7.5
Buffer E: 20 mM HEPES, 1 mM DTT, pH 7.5
Buffer F: 20 mM HEPES, 1 M NaCl, 1 mM DTT, pH 7.5

Procedure. The eluate from the affinity column prepared as described above contains an excess of GDP, EDTA, and NaCl as well as GST–Rab5, which leaks from the beads. These components must be removed prior to further purification of EEA1. Glutathione Sepharose beads (1.5 ml) are washed with 15 ml PBS, and blocked with 10 mg/ml BSA in PBS for 1 hr under rotation. Then the beads are washed with 15 ml PBS to remove excess BSA, and incubated with the eluate at 4° for 45 min. Beads are removed and the procedure is repeated again with 1.5 ml fresh beads to obtain the complete removal of GST–Rab5. These incubations remove leaked GST–Rab5 without significant loss of effectors (Fig. 3A), due to the presence of EDTA and GDP, which prevent the interaction between Rab5 and effectors from occurring.

To remove NaCl, EDTA, and GDP, the eluate is passed through PD10 desalting columns (Pharmacia) previously equilibrated with buffer D, at 4°. The desalted eluate (50 ml) is diluted with buffer E to a final volume of 150 ml to reduce the salt concentration enough for subsequent ion-exchange chromatography. The diluted sample is loaded on a 1-ml Mono Q fast protein liquid chromatography column (FPLC) (Pharmacia) at a flow rate of 0.5 ml/min and the column washed with 10 ml buffer E. Bound proteins are step-eluted with buffer F at a flow rate of 0.1 ml/min, 0.15-ml fractions are collected, and the protein concentration is determined. Fractions with the highest protein concentrations are pooled (fractions 9–15, Fig. 3B) to give a final volume of up to 1 ml. This sample is then injected into a size-exclusion column (Superose 6, 25-ml, FPLC column from Pharmacia) preequilibrated with buffer D. The column is run at a flow rate of 0.12 ml/min (Fig. 4A), and 0.4-ml fractions are collected and analyzed by SDS–PAGE (Fig. 4B). Fractions containing the effectors are finally snap-frozen in liquid nitrogen and stored at −80°. EEA1 is obtained in the first peak [fractions 17–20 (Fig. 4A

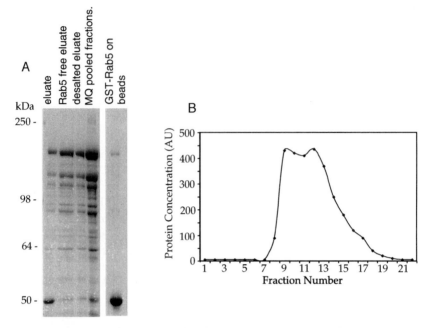

FIG. 3. Anion-exchange Mono Q (MQ) column chromatography. (A) Comparison of protein pattern by SDS–PAGE analysis and Coomassie blue staining of samples obtained from the purification procedure of EEA1 prior and after the MQ chromatography step. The following samples were used: eluate obtained from the affinity column (10 μl), eluate after the removal of GST–Rab5 (30 μl), desalted eluate (40 μl), and pooled fractions after the MQ column (10 μl). The last lane shows GST–Rab5 retained on glutathione beads after its removal from the eluate. (B) Elution profile of proteins from the MQ column chromatography. Fractions 9 to 15 contained the highest protein concentration (expressed here as arbitrary units) and were pooled.

and B)] and is well separated from the bulk of the other proteins. The peak fraction of EEA1 (fraction 18) shows a protein concentration of 0.3 mg/ml. Using Western blotting analysis as a quantitation method, and the purified EEA1 as a standard, we found that the endogenous concentration of EEA1 in cytosol is approximately 4 μg/ml. Therefore, we estimate that the above method results in approximately 3000-fold purification of EEA1.

Functional Assays of Purified EEA1

The activity of purified EEA1 was tested by two means. First we tested whether the whole purification procedure affects the ability of EEA1 to

A

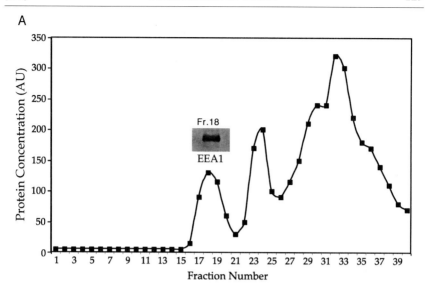

B

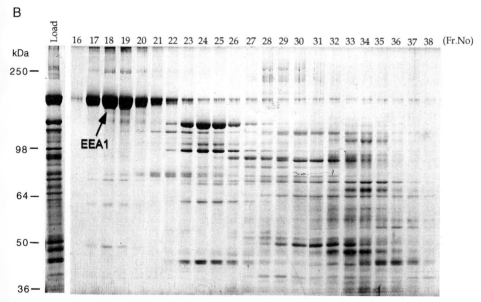

FIG. 4. Final purification step of EEA1 using size-exclusion Superose 6 column chromatography. (A) Elution profile of Rab5 effectors on a Superose 6 column. Protein concentration is expressed in arbitrary units. The peak of EEA1 (fraction 18) was identified by Western blotting (*Inset*). (B) SDS–PAGE analysis and Coomassie blue staining of the Superose 6 fractions. Fraction number is shown on the top of the gel, and molecular weight standards are depicted on the left-hand side. For comparison, 5 μl of the sample loaded on the column was run in parallel in the left-hand side of the gel. The position of EEA1 (peak in fraction 18) is shown.

interact with Rab5 using a ligand overlay assay, and second whether pure EEA1 is able to stimulate fusion between early endosomes in the absence of cytosol.

Solutions

Glutathione elution buffer: 100 mM Tris-HCl, 5 mM 2-mercaptoethanol, 15 mM glutathione, 5 mM MgCl$_2$

Blocking buffer: 50 mM HEPES, 5 mM magnesium acetate, 100 mM potassium acetate, 1 mM DTT, 1% (w/v) bovine serum albumin (BSA), 0.1% (w/v) Triton X-100, 0.1% (w/v) Tween 20, 1 μM ZnCl$_2$, pH 7.4

Binding buffer: 12.5 mM HEPES, 1.5 mM magnesium acetate, 75 mM potassium acetate, 1 mM DTT, 0.2% (w/v) BSA, 0.005% (w/v) Triton X-100, 4 mM n-octylglucoside, pH 7.4

Procedure. The ligand overlay assay is based on the renaturation of proteins immobilized on a nitrocellulose filter in order to test for potential protein–protein interactions with a soluble ligand, which can be easily detected. In this case, purified EEA1 is renatured on nitrocellulose and incubated with soluble GST–Rab5 : GTPγS or GST–Rab5 : GDP, followed by the detection of bound GST–Rab5 using anti-GST antibodies. Active (GTPγS) and inactive (GDP) forms of GST–Rab5 are made as described above for the preparation of the affinity column. The stable nucleotide-bound Rab5 is eluted from the beads with glutathione elution buffer with 1 mM of the corresponding nucleotide. Glutathione is removed by desalting on a PD-10 column preequilibrated with NS buffer containing 100 μM of the corresponding nucleotide. This procedure resulted typically in 3 mg/ml of GST–Rab5 of either form.

Purified EEA1 (Superose 6 column, fraction 18) is loaded on a SDS–PAGE gel (without prior boiling) and transferred to nitrocellulose membrane. The blot is then washed 2× for 10 min with PBS containing 0.1% Tween 20, and blocked overnight by incubation with blocking buffer (all incubations are done at 4° unless otherwise stated). Then the blots are incubated 2× for 1 hr with binding buffer, followed by incubation with 1 μg/ml GST–Rab5 : GTPγS or GST–Rab5 : GDP in binding buffer containing 100 μM GTPγS or GDP, and 1% BSA, for 1 hr. Subsequently, the blots are washed 5× for 5 min with binding buffer containing 10 μM of the appropriate nucleotide (all subsequent washes contained this concentration of nucleotide), and incubated with anti-GST antibody (Pharmacia) at a dilution of 1 : 2000 in binding buffer containing 1% BSA. The blots are washed 5× for 5 min with binding buffer and incubated with rabbit anti-goat IgG conjugated to HRP at a dilution of 1 : 1500 in binding buffer containing 1% BSA, followed by five 5-min washes with binding buffer,

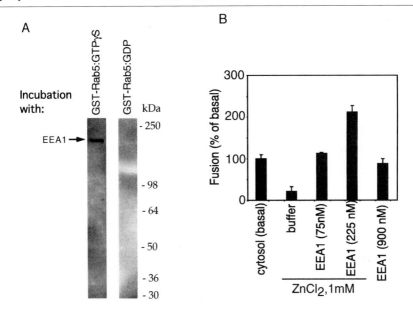

FIG. 5. Assays of purified EEA1. (A) Purified EEA1 (fraction 18, 400 ng) was loaded on a 7–17% gradient SDS–PAGE followed by transferring on a nitrocellulose filter. Then the blots were renatured, incubated with the indicated form of Rab5 (1 μg/ml), and the position of bound GST–Rab5 was identified using goat anti-GST antibodies, followed by incubation with rabbit anti-goat IgG coupled to HRP and detection by ECL. The expected molecular weight of EEA1 (on the left-hand side) and molecular weight standards (on the right-hand side) are shown in the figure. (B) Fusion between early endosomes in the presence of cytosol (basal), or buffer, or different concentrations of EEA1 in the presence or absence of $ZnCl_2$. The extent of fusion is expressed as percent of basal.

and 2 quick washes with PBS. Finally, the blots are treated with the ECL (enhanced chemiluminescence) detection system (Amersham) and exposed to autoradiography. This method shows that purified EEA1 is able to bind GST–Rab5 : GTPγS but not GST–Rab5 : GDP (Fig. 5A).

The activity of EEA1 is also tested in an *in vitro* early endosome fusion assay that has been described before.[26] In this case, isolated early endosomes are mixed with purified EEA1[11] and the percent of fusion is measured using an Origen analyzer (IGEN, Gaithersburg, MD). The fraction from the size-exclusion chromatography containing EEA1 (fraction 18) is able to support fusion between early endosomes in the absence of cytosol (Fig. 5B). When EEA1 is used in the absence of $ZnCl_2$, fusion is significantly reduced and requires much higher concentration of the fraction containing

[26] H. Horiuchi, R. Lippé, H. M. McBride, M. Rubino, P. Woodman, H. Stenmark, V. Rybin, M. Wilm, K. Ashman, M. Mann, and M. Zerial, *Cell* **90**, 1149 (1997).

EEA1 in order to compensate for the reduction of $ZnCl_2$. It is likely that the purification procedure (treatment with 20 mM EDTA during elution from the affinity column) has stripped the Zn^{2+} ions from the FYVE finger of EEA1, consistent with previous reports.[27,28] Our efforts to elute the effectors from the affinity column in the absence of EDTA using high salt concentrations and detergent proved unsuccessful.

The above two criteria used for testing the activity of EEA1 (binding to active Rab5 and stimulation of endosome fusion) show that the described purification method results in an active preparation of this Rab5 effector.

Acknowledgments

We are grateful to Harald Stenmark for the cDNA of GST–Rab5, and Heidi McBride and Marta Miaczynska for critical reading of the manuscript. S.C. is a recipient of a Max Planck fellowship. This work was supported by the Max Planck Gesellschaft, grants from the Human Frontier Science Program (RG-432/96), EU TMR (ERB-CT96-0020), and Biomed (BMH4-97-2410) (M.Z.).

[27] H. Stenmark, R. Aasland, B. H. Toh, and A. D'Arringo, *J. Biol. Chem.* **271,** 24048 (1996).

[28] S. Misra and J. H. Hurley, *Cell* **97,** 657 (1999).

[15] Expression, Purification, and Characterization of Rab5 Effector Complex, Rabaptin-5/Rabex-5

By ROGER LIPPÉ, HISANORI HORIUCHI, ANJA RUNGE, and MARINO ZERIAL

Introduction

The small GTPase Rab5 is a molecule regulating the endocytic pathway and is an important component of the docking and fusion apparatus.[1–3]

[1] J.-P. Gorvel, P. Chavrier, M. Zerial, and J. Gruenberg, *Cell* **64,** 915 (1991).

[2] C. Bucci, R. G. Parton, I. H. Mather, H. Stunnenberg, K. Simons, B. Hoflack, and M. Zerial, *Cell* **70,** 715 (1992).

[3] G. Li, M. A. Barbieri, M. I. Colombo, and P. D. Stahl, *J. Biol. Chem.* **269,** 14631 (1994).

Hence, in concert with the SNARE machinery,[4–6] it regulates heterotypic fusion between clathrin-coated vesicles and early endosomes, as well as homotypic fusion between early endosomes. In addition, Rab5 has also been shown to regulate the motility of early endosomes along microtubules[7] and the formation of clathrin-coated vesicles at the plasma membrane.[8] So far, several Rab5 effectors have been documented, including EEA1,[9–11] the PI3 kinases hVPS34 and p85–p110,[12] Rabaptin-5,[13,14] Rabaptin-5β,[15] and recently a large number of other potential effectors.[11] Given the multiple roles of Rab5, the characterization of its effectors is essential.

Rabaptin-5 was the first identified Rab5 effector[13] and is essential for the Rab5 regulated docking and fusion machinery.[13,14] It reduces hydrolysis of the Rab5-bound GTP into GDP, possibly by preventing a GAP from binding to Rab5 and hence preserving Rab5 in its active conformation.[16] Interestingly, Rabaptin-5 does not support fusion of early endosomes by itself and must be bound to yet another Rab5 interacting molecule called Rabex-5 to be functional.[14] Rabex-5 is a guanine nucleotide exchange factor (GEF) that mediates the GDP-to-GTP nucleotide exchange on Rab5. Surprisingly, the effector Rabaptin-5 and the exchange factor Rabex-5 are stably associated in cytosol, which is the first identification of an effector–GEF complex for the Rab family.[14] Although the combined function of this complex is to drive the equilibrium of Rab5 toward its GTP active

[4] S. H. Low, S. J. Chapin, C. Wimmer, S. W. Whiteheart, L. G. Komuves, K. E. Mostov, and T. Weimbs, *J. Cell Biol.* **141,** 1503 (1998).

[5] K. Seron, V. Tieaho, C. Prescianotto-Baschong, T. Aust, M. O. Blondel, P. Guillaud, G. Devilliers, O. W. Rossanese, B. S. Glick, H. Riezman, S. Keranen, and R. Haguenauer-Tsapis, *Mol. Biol. Cell* **9,** 2873 (1998).

[6] H. M. McBride, V. Rybin, C. Murphy, A. Giner, R. Teasdale, and M. Zerial, *Cell* **98,** 377 (1999).

[7] E. Nielsen, F. Severin, J. M. Backer, A. A. Hyman, and M. Zerial, *Nat. Cell Biol.* **1,** 376 (1999).

[8] H. McLauchlan, J. Newell, N. Morrice, A. Osborne, M. West, and E. Smythe, *Curr. Biol.* **8,** 34 (1998).

[9] A. Simonsen, R. Lippé, S. Christoforidis, J. M. Gaullier, A. Brech, J. Callaghan, B. H. Toh, C. Murphy, M. Zerial, and H. Stenmark, *Nature* **394,** 494 (1998).

[10] I. G. Mills, A. T. Jones, and M. J. Clague, *Curr. Biol.* **8,** 881 (1998).

[11] S. Christoforidis, H. M. McBride, R. D. Burgoyne, and M. Zerial, *Nature* **397,** 621 (1999).

[12] S. Christoforidis, M. Miaczynska, K. Ashman, M. Wilm, L. Zhao, S. C. Yip, M. D. Waterfield, J. M. Backer, and M. Zerial, *Nat. Cell Biol.* **1,** 249 (1999).

[13] H. Stenmark, G. Vitale, O. Ullrich, and M. Zerial, *Cell* **83,** 423 (1995).

[14] H. Horiuchi, R. Lippé, H. M. McBride, M. Rubino, P. Woodman, H. Stenmark, V. Rybin, M. Wilm, K. Ashman, M. Mann, and M. Zerial, *Cell* **90,** 1149 (1997).

[15] H. Gournier, H. Stenmark, V. Rybin, R. Lippé, and M. Zerial, *EMBO J.* **17,** 1930 (1998).

[16] V. Rybin, O. Ullrich, M. Rubino, K. Alexandrov, I. Simon, C. Seabra, R. Goody, and M. Zerial, *Nature* **383,** 266 (1996).

form, the exact role and mechanism of action of the complex await a more detailed characterization. It has been suggested, however, that the association of Rabaptin-5 and Rabex-5 may be functionally important to generate local clusters of active Rab5 at the docking and fusion site.[14] In addition, Rabaptin-5 specifically interacts not only with Rab5 through its carboxyl terminus, but also with Rab4 through a distinct binding domain located at the amino end of Rabaptin-5, suggesting the complex may also be important to link endocytosis to recycling from early endosomes to the plasma membrane.[17]

Following the discovery of Rabaptin-5, a related molecule, Rabaptin-5β, was subsequently identified. It shares significant sequence homology (40% homology and 60% similarity) with Rabaptin-5 and, as a result, also binds to Rabex-5.[15] Importantly, Rabaptin-5β, which is also recognized by the Rabaptin-5α antibody we use, can readily be distinguished from Rabaptin-5α on the basis of its different molecular mass (70 and 100 kDa, respectively). Thus, Rabaptin-5 and Rabaptin-5β are not found in the same complex because Rabex-5 associates exclusively with either Rabaptin protein.[14,15] Although the Rabaptin-5β complex is also involved in the homotypic fusion between early endosomes, it appears to work cooperatively with the functionally predominant Rabaptin-5 complex,[15] hence our focus on the latter complex.

To further understand the significance and function of the Rabaptin-5/Rabex-5 complex and to molecularly dissect the Rab5-mediated pathway, purified molecules are needed. Two main approaches exist, namely, the purification of native complex from cytosol and the expression and purification of complex as recombinant proteins. This chapter first details the purification of the native Rabaptin-5/Rabex-5 complex from bovine brain cytosol. It subsequently describes the recent purification of the recombinant complex from insect cells using baculovirus.

Description of Method

Purification of Native Complex

Preparation of Bovine Brain Cytosol. The purification of the native complex is performed from bovine brain cytosol (Fig. 1). For this purpose, obtain four fresh brains from the slaughterhouse and quickly put on ice to avoid protein degradation. All steps must be performed at 4°. Wash them in cold phosphate-buffered saline (PBS) to remove the blood and prepare

[17] G. Vitale, V. Rybin, S. Christoforidis, P. Thornqvist, M. McCaffrey, H. Stenmark, and M. Zerial, *EMBO J.* **17,** 1941 (1998).

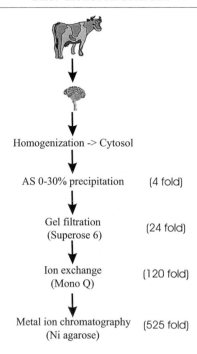

Homogenization -> Cytosol

AS 0-30% precipitation   (4 fold)

Gel filtration   (24 fold)
(Superose 6)

Ion exchange   (120 fold)
(Mono Q)

Metal ion chromatography   (525 fold)
(Ni agarose)

FIG. 1. Purification scheme of the native Rabaptin-5/Rabex-5 complex from bovine brain. Schematic representation of the purification protocol to obtain native Rabaptin-5 complex from bovine brain cytosol. The fold purification is indicated for each step. It is determined by Western analysis against Rabaptin-5, using the L1-46 polyclonal antibody, and by determination of the total protein content with the Bio-Rad protein assay.

the cytosol as follows. First remove the meninges, cerebellum, and remains of the spinal cord and homogenize the brains in a blender with 600 ml of ice-cold homogenization buffer [20 mM HEPES, pH 7.4, 5 mM MgCl$_2$, 1 mM dithiothreitol (DTT), 100 mM NaCl, and a cocktail of protease inhibitors, including 6 μg/ml chymostatin, 0.5 μg/ml leupeptin, 10 μg/ml antipain, 2 μg/ml aprotinin, 0.7 μg/ml pepstatin, and 10 μg/ml APMSF]. Spin the homogenate at 4200g for 50 min at 4°. Keep the postnuclear supernatant and centrifuge at 100,000g for 1 hr at 4°. Collect the cytosol and proceed immediately to purification, as this is important for the activity of the complex (see Comments).

Ammonium Sulfate Precipitation. The Rabaptin-5 complex mostly precipitates from cytosol at 30% ammonium sulfate but requires 40% ammonium sulfate for a quantitative recovery. To achieve this, place the cytosol in a large beaker at 4° atop a stirrer and slowly add 40% of ammonium sulfate (243 g per liter of cytosol). Make sure there is extensive stirring to avoid high local concentrations of ammonium sulfate. Once all the ammonium sulfate is added, stir for another 30 min. Spin the suspension at 100,000g for 10 min and solubilize the pellet with a Dounce homogenizer in 50 ml of HMD (20 mM HEPES, pH 7.4, 5 mM MgCl$_2$, 1 mM DTT). Dialyze the sample against HMD/100 mM NaCl. At this point, samples can be snap-frozen in 20-ml aliquots. Two other Rab5 effectors, EEA1 and Rabaptin-5β, which are also precipitated during this step, are separated from the Rabaptin-5 complex at later stages.

Fractionation by Gel Filtration. The second purification step consists of the separation of the 0–40% ammonium sulfate fraction on a 125-ml (bed volume) preparative grade Superose 6 column (Pharmacia, Stockholm). Preequilibrate the column with filtered and degassed HMD/100 mM NaCl buffer. If the sample was frozen, spin one of the 20-ml aliquots at 150,000g for 30 min (4°) and filter the supernatant first through a 0.45 μm then a 0.22 μm filter to remove aggregates. This is enough material for two runs on the column. Load half of the sample, perform the fractionation at 0.5 ml/min in HMD/100 mM NaCl, and collect 2-ml fractions. Wash the column with HMD/100 mM NaCl until the UV reading has returned to its baseline. Repeat immediately with the second half of the sample. Meanwhile take a 15 μl aliquot of each fraction, run a 10% SDS–PAGE gel, and identify the Rabaptin-5 positive fractions by Western analysis using an anti-Rabaptin-5 antibody such as L1-46.[13] It is critical at this stage to work as fast as possible to preserve the activity of the complex. If possible, use minigels and short incubation times for the Western analysis. Alternatively, the complex can be monitored at this stage by an anti-Rabex-5 antibody, if available. As a reference point, the Rabaptin-5 complex should elute approximately 10–14 fractions before the hemoglobin peak, easily visible by eye during the fractionation. Pool the positive fractions (around six fractions of 2 ml per run). Note that this step efficiently purifies the Rabaptin-5 complex away from the Rab5 effector EEA1, but not yet from the related Rabaptin-5β.

Ion-Exchange Chromatography. In addition to further purifying the complex, the ion-exchange column serves the purpose of a concentration step as well. Pool the positive fractions from two gel filtration runs (usually 20–22 ml in total) and load all onto a 1-ml prepacked Mono Q column (Pharmacia, Stockholm) preequilibrated with HMD. Fractionate at 0.5 ml/min by first washing the column with 10 column volumes of HMD, eluting with 15–20 column volumes of a 0–500 mM NaCl gradient (starting buffer, HMD; final buffer, HMD/500 mM NaCl), and a final wash step of five column volumes of HMD/1 M NaCl. Collect 2-ml fractions during the initial wash step, and 0.5-ml fractions during the subsequent elution step. Identify the positive fractions by Western analysis as above. The complex elutes at 200–300 mM salt in a peak of about five fractions (2.5 ml).

Final Purification Step on Metal Ion Column. Purification on nickel columns is typically used for recombinant proteins tagged with a His$_{6-9}$ tail, tags which are obviously absent in native proteins. Although Rabaptin-5 contains 16 histidine residues (1.9% of total amino acids), it has no single contiguous stretch of histidine residues. The same is similarly true for its partner protein Rabex-5 (six histidine residues scattered throughout the whole protein). Despite this, Rabaptin-5 binds to nickel-coated beads (Horiuchi and Zerial, unpublished observation, 1999). Therefore, it is possible to use such an unconventional approach to purify the native Rabaptin-5 complex.

To purify the complex, proceed as follows. Wash 300–400 μl (packed volume) of nickel NTA agarose beads (Qiagen, Hilden) twice with HMD. Incubate in batch the Mono Q Rabaptin-5 containing fractions (2–2.5 ml) with the nickel beads for 1–2 hr in two 2-ml Eppendorf tubes. Transfer the beads into a single empty 10-ml chromatographic column (Bio-Rad, Munich) and drain the excess liquid by gravity flow. Wash the beads with 10–15 ml of wash solution (HMD/10 mM imidazole). Do not use a higher concentration of imidazole because the native protein is eluted starting at 25 mM imidazole (Lippé, Runge, Horiuchi, and Zerial, unpublished observation, 1999). Note that the Rabaptin-5β is quantitatively eluted by the 10 mM wash. Elute the Rabaptin-5 complex with 1 ml of elution buffer (HMD/100 mM imidazole), omitting the first two drops to get more concentrated material. Aliquot the nickel agarose eluate in small volumes and snap-freeze in liquid nitrogen before storing at $-80°$.

Comments. Rabex-5 will elute in both the 10 mM imidazole wash and the 100 mM imidazole elution because it forms distinct complexes with both Rabaptin-5 and Rabaptin-5β. Therefore, it is not advisable to monitor the purification of the Rabaptin-5 complex with an antibody directed against Rabex-5. The expected yield from four bovine brains is low (1 ml of roughly 200 nM complex), but the material is active and in sufficient amounts for functional assays. The association of Rabaptin-5 and Rabex-5 is maintained during all of these steps, indicating that they interact with high affinity. It is therefore not necessary to follow Rabex-5 during the purification. By this procedure, the complex is not purified to homogeneity, but it is of reasonable purity (Fig. 2). In our hands, further purification of the complex is deleterious for its activity and is not recommended. The complex is highly sensitive to freeze-thawing and care should be taken to avoid such processing. In addition, the successful purification of the complex is dependent on avoiding freezing during the purification and working as quickly as possible.

Analysis of Activity of Complex. The Rabaptin-5/Rabex-5 complex is an essential component of the Rab5-mediated endocytic pathway. It has

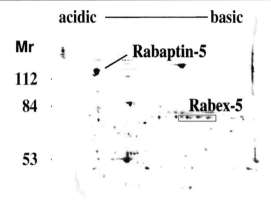

FIG. 2. Silver staining of the purified native complex. Analysis of the purified native complex by isoelectrofocusing electrophoresis (first dimension) and SDS–PAGE (second dimension). The gel was then silver stained. The positions of Rabaptin-5 and Rabex-5 are indicated, as confirmed by Western blot analysis. The molecular mass markers in the second dimension are indicated in kilodaltons.

been shown to be necessary in the *in vitro* early endosome homotypic fusion assay and to stimulate fusion when added exogenously to the fusion reaction.[14] In addition, depletion of the complex from the cytosol used in the assay significantly impairs the fusion of the endosomes. Rescue of such depleted cytosol can be achieved by the addition of exogenous Rabaptin-5 complex.[14] These observations therefore provide means to assay for the functionality of the purified complex. Finally, the nucleotide exchange activity of Rabex-5 provides an additional and complementary method for evaluating the quality of the preparation.[14]

In Vitro Fusion Assay. To evaluate the purified Rabaptin-5 complex, one can use an *in vitro* homotypic endosome fusion assay. Although the detailed description of such assay is beyond the scope of this chapter and has also been reported elsewhere,[14] the general guidelines are here provided. Briefly, the assay consists of the incubation of biotinylated transferrin-labeled early endosomes with anti-transferrin antibody-labeled early endosomes in the presence of HeLa cytosol, an ATP-regenerating system, an unlabeled holotransferrin quencher, buffer, and water in a total volume of 20 μl. Measurements of fusion scores are taken for the content mixing between the endosomes containing antibody and its antigen, using streptavidin-coated magnetic beads and an Origen analyzer. To test the Rabaptin-5 complex, simply replace 10 μl of water by 10 μl of complex (Fig. 3A).

Alternatively, the complex can be used in the fusion assay to rescue cytosol depleted of the Rabaptin-5 complex. To deplete the complex from cytosol, wash 10 μl of protein A agarose beads twice with phosphate-buffered saline (PBS) and incubate with 10 μl of crude anti-Rabaptin-5 antibody for 2 hr on a rotator at 4° in a total volume of 1 ml in PBS. Spin the beads at low g force for 30–60 sec and wash them twice with cytosol buffer (50 mM HEPES, pH 7.4, 50 mM KCl, 10 mM EGTA, 2 mM MgCl$_2$). In this case, a wash means a 20-min incubation of the beads with cytosol buffer at 4° on the rotator to allow desorption of the nonspecific material from the beads. Add the beads to 100 μl of the cytosol used in the fusion assay and incubate 30 min on the rotator (4°). Sediment the beads and collect the depleted cytosol. Verify the depletion by SDS–PAGE and Western blot analysis using a Rabaptin-5 antibody. Use this cytosol in the fusion assay and rescue the inhibitory effect of the depletion with 10 μl of purified native complex (Fig. 3A).

Guanine Nucleotide Exchange Assay. Dilute recombinant GTPase (e.g., Rab5) produced in *Escherichia coli* or purified from the membrane fraction of overexpressing Sf9 insect cells[18,19] to 1 mg/ml in exchange buffer (20 mM Hepes/NaOH, pH 7.2, 5 mM MgCl$_2$, 1 mM DTT). Incubate 10 pM of the GTPase with different amounts of Rabaptin-5 complex or Rabex-5 (0–50 nM), together with 2 μM [^{35}S]GTPγS (20,000 cpm/pmol) for a final reaction volume of 20–25 μl in exchange buffer. Incubate the samples at 30° for different times (0–25 min). Stop the exchange reaction by adding 2 ml of ice-cold buffer (20 mM Tris, pH 7.4, 20 mM MgCl$_2$, 100 mM NaCl) to each sample, filter through nitrocellulose filters (2-cm-diameter BA85, Schleicher & Schuell, Keene, NH), wash the filters twice with 4 ml of the above ice-cold buffer, and dry them. Count the filter-bound radioactivity with a beta scintillation counter. Express the results as folds of the counts obtained with complex or Rabex-5 compared with those obtained without complex or Rabex-5. See Fig. 3B for an example.

Purification of Recombinant Complex

Although the purification of the Rabaptin-5 complex from bovine cyto-sol has been instrumental in identifing Rabex-5, further characterization of the role and significance of the complex requires the use of recombinant proteins. Furthermore, this is needed to improve the complex yields and to have it free of the contaminating proteins present in the native material. Finally, a recombinant complex would be essential to ultimately reconstitute the entire Rab5 docking and fusion apparatus *in vitro*. Although its expres-

---

[18] K. Alexandrov, H. Horiuchi, O. Steele-Mortimer, M. C. Seabra, and M. Zerial, *EMBO J.* **13,** 5262 (1994).

[19] F. P. Cremers, S. A. Armstrong, M. C. Seabra, M. S. Brown, and J. L. Goldstein, *J. Biol. Chem.* **269** (1994).

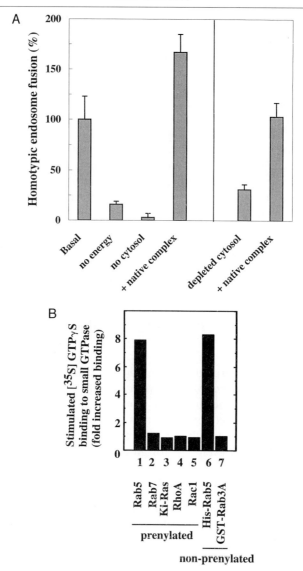

sion in bacteria is possible, expression in an insect cell system is preferable due to potential posttranslational modifications of the complex and the ease of expression of multiple subunits in insect cells. For these reasons, we have chosen to express the Rabaptin-5 complex in insect cells using baculoviruses.

Construction of Vectors. The Bac-to-Bac baculovirus expression system (Gibco, Karlsruhe) is used for these experiments. This system is particularly useful, because it is versatile and relies solely on bacteria to perform all the cloning steps and the clonal expansion of recombinant viral genomes, thereby avoiding tedious and time-consuming plaque purification steps. In addition, the blue/white selection procedure of the Bac-to-Bac system allows a quick and easy distinction between wild-type and recombinant viral genomes. The recombination efficiency, in fact a transposition event, is also much higher than in insect cells (10–25%).[20] Finally, the presence of extensive multiple cloning sites (MCS) and the strong *Autographa californica* polyhedrin promoter are an asset. Cloning of Rabaptin-5 and Rabex-5 is designed to permit their coexpression to allow the formation of complexes *in vivo* and their subsequent copurification. Although a complex of Rabaptin-5 and Rabex-5 can be formed *in vitro* from individually produced recombinant proteins, complex formation *in vivo* is a key aspect since it is significantly more efficient (Lippé, Horiuchi, and Zerial, unpublished observation, 1999). Subclone the full-length Rabaptin-5 cDNA into the MCS of the proper pFAST BAC HT donor vector (a, b, or c depending on the frame), yielding a histidine-tagged recombinant protein. In addition, subclone Rabex-5 into the MCS of the pFAST BAC1 donor vector, yielding an untagged recombinant protein.

[20] D. Anderson, R. Harris, D. Polayes, V. Ciccaron, R. Donahue, G. Gerard, J. Jessee, and V. Luckow, *Focus* **17,** 53 (1995).

FIG. 3. Functional analysis of the purified native complex. (A) Analysis of the purified complex with the use of the *in vitro* homotypic early endosome assay. Basal fusion represents the fusion in presence of donor and acceptor early endosomes, cytosol, and energy regenerating system. The depleted cytosol lane consists of the above but using cytosol that has been depleted of the Rabaptin-5 complex using the L1-46 anti Rabaptin-5 antibody. For the native complex lanes, 100 nM of exogenous purified bovine Rabaptin-5/Rabex-5 was added to the reaction. (B) Analysis of the complex in the nucleotide exchange assay. The ability of the complex to load [^{35}S]GTPγS nucleotide onto various recombinant small GTPases, including Rab5, is shown. Prenylated GTPases (Rab5, Rab7, Ki-Ras, RhoA, and Rac1) were purified from overexpressing insect cells.[18] Nonprenylated GTPases (Rab5 and Rab3A) were produced in and purified from *Escherichia coli*. In this experiment, the data indicate the fold of stimulation of [^{35}S]GTPγS binding in 20 min at 30° in the presence of 20 nM Rabex-5 over the [^{35}S]GTPγS binding in the absence of Rabex-5. Although thus far specific for Rab5, a nucleotide exchange activity of Rabex-5 on GTPases other than the ones tested here cannot formally be excluded.

Generation of Recombinant Viruses and Viral Stocks. Generate the recombinant viruses separately for each construct by transforming 1 ng of the recombinant donor plasmids into the *A. californica* baculoviral genome present in DH10Bac *E. coli* cells (100 μl of bacteria) according to the manufacturer's instructions. It is critical to incubate the transformed cells 4 hr at 37° prior to plating them on selection plates [50 μg/ml kanamycin, 7 μg/ml gentamycin, 10 μg/ml tetracycline, 100 μg/ml Bluo-Gal (Sigma, Taufkirchen), and 40 μg/ml isopropylthiogalactoside (IPTG)]. Three hundred micrograms of X-Gal per ml of culture can alternatively be use with equivalent results instead of the Bluo-Gal. Develop the blue color for 2 days, pick the resulting white recombinant colonies (inactivation of the *lacZ* gene), and replate for a second round of blue/white selection to avoid false positives. Collect the white colonies and grow them overnight in 2 ml of liquid selection media (50 μg/ml kanamycin, 7 μg/ml gentamycin, and 10 μg/ml tetracycline). Extract the viral DNA manually without the use of column plasmid kits, for example from Qiagen, because the large viral DNA will be sheared. Follow instead the instructions in the Gibco Bac-to-Bac user manual. Confirm by polymerase chain reaction (PCR) the presence of the recombinant gene using the BacA and BacB amplification primers

oligo(A) 5′ GTTTTCCCAGTCACGACGTTGTAAAACGAC3′
oligo(B) 5′ AGCGGATAACAATTTCACACAGGAAACAGC 3′

and the following PCR conditions (94° for 5 min, followed by 35 cycles of 94° for 1 min/55° for 1 min/72° for 6 min, and finally 1 cycle at 72° for 7 min).

Transfect separately by the CaCl$_2$ technique[21] 5 μl of the viral DNAs (Rabaptin-5 and Rabex-5) into one 25-cm^2 dish each of subconfluent *Spodoptera frugiperda* (Sf9) insect cells. The Sf9 cells are cultured in TNM-FH medium (Sigma, Deisenhofen), 10% fetal calf serum (FCS), and penicillin/streptomycin. No selection is necessary because only the released virus is of interest. Collect the P1 viruses 4 days later. Amplify the viruses twice by infecting Sf9 cells with a small amount of virus (for example, 100 μl of P1, then P2 into a 175-cm^2 flask of cells) to obtain large stocks of virus (P2, then P3). The precise time to harvest depends on the titer of the viruses and the culture conditions.

Protein Expression. The use of Sf9 cells is particularly suited for the expansion of baculoviruses. However, protein expression tends to be significantly better in general in High Five insect cells.[20] These latter cells are also easier to manipulate. Although it is recommended to titrate the viruses by plaque assay (plaque forming units/ml), it is also possible by protein

---

[21] F. L. Graham and A. J. v. d. Eb, *Virology* **52,** 456 (1973).

expression, an approach that we generally employ. It is important to optimize the infection conditions for each stock of baculovirus. Therefore, for an optimal protein expression, add different amounts of recombinant virus to High Five cells in 175-cm^2 flasks and titrate by SDS–PAGE and Coomassie blue staining, the amounts of virus and optimal time required to obtain maximal protein expression. This will also be a good starting point to evaluate the amounts of viruses to be used for coinfections. Follow the guidelines provided in the Gibco Bac-to-Bac user manual. Verify the degradation of the complex by Western analysis with a Rabaptin-5 antibody and if available one against Rabex-5. Once identified, perform a coinfection with the appropriate amounts of both Rabaptin-5 and Rabex-5 recombinant viruses and harvest the cells by scraping them in PBS at the predetermined optimal time (usually 40–50 hr postinfection). Wash the pellet twice with PBS to remove all traces of serum albumin and freeze the pellet at $-80°$ until ready to purify the Rabaptin-5 complex. Typically, 10–20 245 × 245 mm^2 plates of cells yield roughly 5–10 ml of pellet.

Purification of Complex by Affinity Chromatography. The key feature for the purification scheme is the presence of a histidine tag on Rabaptin-5 and the lack of tag on Rabex-5. This implies that the complex can be purified in a single purification step on a nickel agarose column. For this reason, a stringent wash regime is used to minimize contamination by insect proteins. To purify the complex, thaw the above pellet on ice and resuspend in 5 ml of lysis buffer per gram of pellet [lysis buffer: 50 mM Tris, pH 8.5, 4 mM 2-mercaptoethanol (2-ME), 5 μg/ml DNase I, and a cocktail of protease inhibitors (see above)]. Break the cells once in a French press (pressure of 800–1100 psi) and spin the lysate at 125,000g for 45 min at 4°. Meanwhile, preequilibrate 0.25–1 ml of nickel-NTA agarose beads (Qiagen, Karlsruhe)/20 plates of cells with lysis buffer. Filter the supernatant through a 0.22 μm filter and incubate in batch with the nickel beads for 1 hr at 4° on a rotator. Sediment the beads at low g force for 2–5 min and wash with 10 column volumes of buffer A [20 mM Tris, pH 8.0, 500 mM NaCl, 20 mM imidazole, 4 mM 2-ME, and 10% (v/v) glycerol] followed by 10 column volumes of buffer B [20 mM Tris, pH 8.5, 1 M NaCl, 4 mM 2-ME, and 10% (v/v) glycerol]. Resuspend the beads in buffer C (20 mM HEPES, pH 7.5, 100 mM NaCl, 20 mM imidazole, 4 mM 2-ME, and 10% glycerol) and load them in an empty 10-ml chromatographic column (Bio-Rad, Munich). Wash the beads with 10 column volumes of buffer C and elute the Rabaptin-5 complex with 10 column volumes of buffer D (20 mM HEPES, pH 7.4, 100 mM NaCl, 4 mM 2-ME, 200 mM imidazole, and 10% glycerol), collecting 10 fractions of 1 column volume each. Take aliquots to analyze by SDS–PAGE and Coomassie blue staining. Dilute the fractions to 20 mM HEPES, pH 7.4/100 mM NaCl/4 mM 2-ME/100 mM imidazole/10% glycerol to avoid precipitation and freeze aliquots of the fractions. It is best not to pool the fractions, because one of the fractions will be of much higher concentration than the other ones.

Analysis of the Purified Complex. Analyze the fractions by SDS–PAGE on a 10% polyacrylamide gel and by Coomassie blue staining. Figure 4 shows a typical purification preparation. The activity of the recombinant complex can be accessed as described above. It is as active as the native complex (data not shown).

Comments. The amount of nickel-NTA agarose beads influences both the concentration of the sample and the total yield. For instance, 0.25 ml of beads/20 plates of cells gives the highest concentration of complex, but also the lowest yield as some complex ends up in the flowthrough and washes. On average, the yield of complex is low overall ($\leq$70 μg/20 plates of cells), but nevertheless up to 20-fold better than for the native complex. In addition, the peak fraction of the recombinant complex is 5–15 times more concentrated than the native sample depending on the volume of nickel-NTA agarose beads used ($\leq$2700 NM). Furthermore, the purified complex is relatively pure ($\geq$70–80%) and devoid of the contaminating

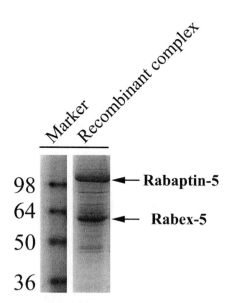

Fig. 4. Coomassie staining of the purified recombinant complex. On purification, 10 μl of the sample was loaded onto a 10% SDS–PAGE gel and stained with Coomassie blue. The molecular mass markers are indicated in kilodaltons. The identity of the Rabaptin-5 and Rabex-5 bands was confirmed by Western blot analysis.

mammalian proteins that are present in the native preparation and perhaps able to modulate the fusion of the early endosomes. Finally, this recombinant material is sufficient to perform many functional assays.

Acknowledgments

We thank Y. Takai (Department of Molecular Biology and Biochemistry, Osaka University Graduate School of Medicine, Japan) for the generous gift of several purified GTPases (Ki-Ras, RhoA, Rac1, and Rab3A). We also thank Heidi McBride and Marta Miaczynska for critical reading of the manuscript. This work was supported by the Max Planck Gesellschaft and by grants from the Human Frontier Science Program (RG-432/96), EU TMR (ERB-CT96-0020), and Biomed (BMH4-97-2410) to M.Z. R.L. was initially funded by an EMBO fellowship and subsequently by a fellowship from the Medical Research Council of Canada. H.H. was funded by an EMBO fellowship.

[16] Measurement of Rab5 Protein Kinase B/akt and Regulation of Ras-Activated Endocytosis

By Manuel A. Barbieri, A. Gumusboga, Richard L. Roberts, and Philip D. Stahl

Introduction

Endocytosis is a carefully orchestrated process required by all cells for nutrition and defense. Whereas an increasing number of Rab GTPases localize to the endocytic pathway, including Rab5, Rab4, Rab11, and Rab7,[1] the endocytic rate appears to be regulated by Rab5.[2] Rab5 is also rate limiting for endosome fusion reconstituted *in vitro*.[2–4] Earlier work has shown that Rab5, in turn, is regulated by phosphatidylinositol 3-kinase (PI3K) and Ras.[5,6] Ras is the prototype of a large family of 20- to 35-kDa monomeric GTPases that serve as molecular switches in regulating diverse

---

[1] P. Novick and M. Zerial, *Curr. Biol.* **9**, 496 (1997).

[2] C. Bucci, R. G. Parton, I. H. Mather, H. Stunnenberg, K. Simons, B. Hoflack, and M. Zerial, *Cell* **70**, 715 (1992).

[3] G. Li, M. A. Barbieri, M. I. Colombo, and P. D. Stahl, *J. Biol. Chem.* **269**, 14631 (1994).

[4] M. A. Barbieri, G. Li, M. I. Colombo, and P. D. Stahl, *J. Biol. Chem.* **269**, 18720 (1994).

[5] G. Li, C. D'Souza-Schorey, M. A. Barbieri, R. L. Roberts, and P. D. Stahl, *Proc. Natl. Acad. Sci. U.S.A.* **92**, 10207 (1995).

[6] G. Li, C. D'Souza-Schorey, M. A. Barbieri, J. A. Cooper, and P. D. Stahl, *J. Biol. Chem.* **272**, 10337 (1997).

cell functions[7,8] including cell proliferation and differentiation.[9–11] The best characterized Ras-activated pathway involves a cascade of protein kinases including Raf, MEK (mitogen-activated protein kinase), and ERK (mitogen-activated protein kinase).[12–14] The activation of MEK results in the expression of a specific set of target genes.[15] This signaling pathway is triggered by a direct interaction between Ras–GTP and Raf.[16–20] In addition to Raf, Ras–GTP also directly interacts with other effector molecules and activates multiple signal transduction pathways. In mammalian cells, the putative Ras effectors include Ras–GAP (GTPase-activating protein),[21] Ral–GDS (GDP dissociation stimulator),[22,23] and PI3K.[24] PI3K has multiple lipid products including phosphatidylinositol 3-phosphate (PI3P), phosphatidylinositol 3,4-bisphosphate (PI3,4P), and phosphatidylinositol 3,4,5-trisphosphate (PI3,4,5P). The first suggestion of PI3K involvement in membrane trafficking came from the study of VPS34p, a yeast protein essential for protein targeting to the yeast vacuole and vacuole morphogenesis.[25] VPS34p shares sequence homology with the catalytic subunits (p110) of mammalian PI3K and indeed exhibits PI3K activity.[26] Several isoforms of

[7] G. M. Bokoch and C. J. Der, *FASEB J.* **7,** 750 (1993).

[8] H. R. Bourne, D. A. Sanders, and F. McCormick, *Nature* **348,** 125 (1990).

[9] M. Barbacid, *Annu. Rev. Biochem.* **56,** 779 (1987).

[10] G. Bollag and F. McCormick, *Nature* **351,** 576 (1991).

[11] D. R. Lowy and B. M. Willumsen, *Annu. Rev. Biochem.* **62,** 851 (1993).

[12] P. Dent, W. Haser, T. A. J. Haystead, L. A. Vincent, T. M. Roberts, and T. W. Sturgill, *Science* **257,** 1404 (1992).

[13] L. R. Howe, S. J. Leevers, N. Gomez, S. Nakielny, P. Cohen, and C. J. Marshall, *Cell* **71,** 335 (1992).

[14] S. G. Macdonald, C. M. Crews, L. Wu, J. Driller, R. Clark, R. L. Erickson, and F. McCormick, *Mol. Cell. Biol.* **13,** 6615 (1993).

[15] C. S. Hill, R. Marais, S. John, J. Wynne, S. Dalton, and R. Treisman, *Cell* **73,** 395 (1993).

[16] S. A. Moodie, B. M. Willumsen, M. J. Weber, and A. Wolfman, *Science* **260,** 1658 (1993).

[17] L. V. Aelst, M. Barr, S. Marcus, A. Polverino, and M. Wigler, *Proc. Natl. Acad. Sci. U.S.A.* **90,** 6213 (1993).

[18] A. B. Vojtek, S. M. Hollenberg, and J. A. Cooper, *Cell* **74,** 205 (1993).

[19] P. H. Warne, P. R. Viciana, and J. Downward, *Nature* **364,** 352 (1993).

[20] X.-F. Zhang, J. Settleman, J. M. Kyriakis, E. Takeuchi-Suzuki, S. J. Elledge, M. S. Marshall, J. T. Bruder, U. R. Rapp, and J. Avruch, *Nature* **364,** 308 (1993).

[21] P. Polakis and F. McCormick, *J. Biol. Chem.* **268,** 9157 (1993).

[22] F. Hofer, S. Fields, C. Schneider and G. S. Martin, *Proc. Natl. Acad. Sci. U.S.A.* **91,** 11089 (1994).

[23] M. Spaargaren and J. R. Bischoff, *Proc. Natl. Acad. Sci. U.S.A.* **91,** 12609 (1994).

[24] P. Rodriguez-Viciana, P. H. Warne, R. Dhand, B. Vanhaesebroeck, I. Gout, M. J. Fry, M. D. Waterfield, and J. Downward, *Nature* **370,** 527 (1994).

[25] M. P. Ymann and L. Pirola, *Biochem. Biophys. Acta* **1436,** 127 (1998).

[26] P. K. Herman and S. D. Emr, *Mol. Cell Biol.* **10,** 6742 (1990).

the PI3K have already been subcloned and characterized.[27] Although a detailed mechanism has not been established, these lines of evidence strongly suggest the involvement of PI3K in several aspects of intracellular membrane trafficking.[28] Several findings also indicate that PI3K is involved in regulation of protein kinase B (PKB/akt) as well as other proteins.[24] Because PI3K is stimulated by activation of the small GTPase Ras via the p110 catalytic subunits, PKB/akt is also controlled by Ras.[29–33] In addition to the PI3K-mediated regulation of PKB/akt, other pathways exist for the regulation of PKB/akt that are apparently not sensitive to PI3K.[34,35] PKB/akt is a serine threonine kinase that has emerged as a key intermediate between signal transducing growth factor receptors, including insulin[36] and platelet-derived growth factor[37,38] and a variety of downstream effectors.[39,40] Recent work has linked PKB/akt to such diverse processes as cell survival (e.g., by suppressing apoptosis via phosphorylation of BAD[41]) and the metabolic response to insulin (e.g., via the regulation of intracellular trafficking of vesicles containing Glut4[42]). Activation of PKB/akt requires phosphorylation of the kinase by at least two phosphoinositide-dependent kinases that phosphorylate PKB/akt at Thr-308 and Ser-473.[43,44] Known downstream targets of PKB/akt include p70 S6 kinase[38,45] and glycogen

[27] P. V. Schu, *Science* **260,** 88 (1993).

[28] J. Downward, *Curr. Opin. Cell Biol.* **10,** 262 (1998).

[29] A. Klippel, *Mol. Cell Biol.* **16,** 4117 (1996).

[30] B. M. Marte, *Curr. Biol.* **7,** 63 (1997).

[31] K. Datta, *J. Biol. Chem.* **271,** 30835 (1996).

[32] A. Khwaja, *EMBO J.* **16,** 2783 (1997).

[33] W. G. King, *Mol. Cell Biol.* **17,** 4406 (1997).

[34] H. Konishi, *Proc. Natl. Acad. Sci. U.S.A.* **93,** 7639 (1996).

[35] H. Konishi, *Biochem. Biophys. Res. Commun.* **205,** 817 (1994).

[36] A. Barthel, S. T. Okino, J. Liao, K. Nakatani, J. Li, J. P. Whitlock, Jr., and R. A. Roth, *J. Biol. Chem.* **274,** 20281 (1999).

[37] A. D. Kohn, K. S. Kovacina, and R. A. Roth, *EMBO J.* **14,** 4288 (1995).

[38] T. F. Franke, S. I. Yang, T. O. Chan, K. Datta, A. Kazlauskas, D. K. Morrison, D. R. Kaplan, and P. N. Tsichlis, *Cell* **81,** 727 (1995).

[39] B. M. Marte and J. Downward, *Trends Biochem. Sci.* **22,** 355 (1997).

[40] D. R. Alessi and P. Cohen, *Curr. Opin. Genet. Dev.* **8,** 55 (1998).

[41] S. R. Datta, H. Dudek, X. Tao, S. Masters, H. Fu, Y. Gotoh and M. E. Greenberg, *Cell* **91,** 231 (1997).

[42] A. D. Kohn, S. A. Summers, M. J. Birnbaum, and R. A. Roth, *J. Biol. Chem.* **271,** 31372 (1996).

[43] D. R. Alessi, S. R. James, C. P. Downes, A. B. Holmes, P. R. J. Gaffney, C. B. Reese, and P. Cohen, *Curr. Biol.* **7,** 261 (1997).

[44] D. Stokoe, L. R. Stephens, T. Copeland, P. R. Gaffney, C. B. Reese, G. F. Painter, A. B. Holmes, F. McCormick, and P. T. Hawkins, *Science* **277,** 567 (1997).

[45] B. M. T. Burgering and P. J. Coffe, *Nature* **376,** 599 (1994).

synthase kinase,[46] although many other targets most likely exist. This conclusion is suggested by the fact that PKB/akt is present in at least three isoforms.[47-49] Moreover, multiple Ras genes are known to exist that impact mammalian cell proliferation.[50] Parallel to the activation of nuclear gene expression, Ras–GTP also triggers profound changes in the cytoplasm. The Ras signaling cascade involving the small GTPase Rac1 plays an important role in regulating the actin cytoskeleton and cell surface membrane ruffles.[51] Ras activation also stimulates pinocytosis and vesicular transport to endosomes and lysosomes.[6] This process involves the small GTPase, Rab5.[6] This chapter describes the methodology to express the small GTPases, Rab5 and H-Ras, and the regulatory kinase, PKB/akt, using the Sindbis virus transient expression system for biochemical and morphological analysis of the endocytic pathway. Functional data are also provided in order to assess the physiologic relevance of the GTP-binding proteins.

Experimental Procedure

Construction of Recombinant Sindbis Virus

Since preparation of recombinant virus stocks and the expression Sindbis virus encoding cDNAs share identical procedures, the following method is described only for the Rab5 recombinant Sindbis virus. Identical procedures can be applied for the expression of H-*ras* and PKB/akt recombinant viruses. All DNA manipulations were conducted according to standard methods.[52] cDNAs of H-Ras, Rab5, and PKB/akt were subcloned into the unique *Xba*I restriction site of the Sindbis vector Toto 1000 3' 2J (2JC1). The plasmid was then linearized by *Xho*I digestion and used as a template for *in vitro* transcription with SP6 RNA polymerase. The resulting RNA transcripts were used to prepare recombinant Sindbis viruses.

[46] D. A. Cross, D. R. Alessi, P. Cohen, M. Anjelkovich, and B. A. Hemmings, *Nature* **378,** 785 (1995).
[47] P. F. Jones, T. Jakubowicz, F. J. Pitossi, F. Maurer, and B. A. Hemmings, *Proc. Natl. Acad. Sci. U.S.A.* **88,** 4171 (1991).
[48] J. O. Cheng, A. K. Godwin, A. Bellacosa, T. Taguchi, T. F. Franke, T. C. Hamilton, P. N. Tsichlis, and J. R. Testa, *Proc. Natl. Acad. Sci. U.S.A.* **89,** 9267 (1992).
[49] H. Konishi, S. Kuroda, M. Tanaka, H. Matsuzaki, Y. Ono, K. Kameyama, T. Haga, and U. Kikkawa, *Biochem. Biophys. Res. Commun.* **216,** 526 (1995).
[50] M. A. White, C. Nicolette, A. Minden, A. Polverino, L. V. Aelst, M. Karin, and M. H. Wigler, *Cell* **80,** 533 (1995).
[51] A. J. Ridley, H. F. Paterson, C. L. Johnston, D. Diekmann, and H. Hall, *Cell* **70,** 401 (1992).
[52] J. Sambrook, E. F. Fristsch, and T. Maniatis, "Molecular cloning: A Laboratory Manual." Cold Spring Harbor Laboratory, Cold Spring Harbor, NY, 1989.

Method. cDNAs of Rab5 are subcloned into the unique *Xba*I site of the Sindbis vector, Toto 1000 3′ 2J (2JC1). The plasmid is then linearized by cutting with *Xho*I. This cut DNA is used as template for the *in vitro* transcription:

1. For one reaction, mix (all reagents must be RNase-free):
    5 μl 4 mM NTPs (nucleotide 5′-triphosphate, Pharmacia, Piscataway, NJ)
    1 μl 20 mM m^{7}G cap (New England BioLabs, Inc.)
    4 μl 5× transcription buffer (200 mM Tris-HCl, pH 7.9, 30 mM MgCl$_2$, 50 mM NaCl, 10 mM spermidine)
    0.5 μl 400 mM DTT (DL-dithiothreitol, Sigma, St. Louis, MO)
    0.5 μl RNase (Promega, Madison, WI)
    12.5 U SP6 polymerase (Epicentre)
    0.2 μl BSA (bovine serum albumin, New England BioLabs)
    1–1.5 μg DNA
    RNase-free H$_2$O to 20 μl
2. Incubate 1 hr at 37°.
3. Check 0.5 μl on agarose/ethidium bromide gel. The resulting transcripts (usually, one reaction can transfect 4 × 75-cm^2 flasks) are used to prepare recombinant Sindbis virus.

Preparation of Recombinant Sindbis Virus Stocks

The resulting RNA transcripts are used for transfection of confluent BHK-21 cell monolayers using a lipofectin-mediated procedure (Life Technology, Inc.). Cells are maintained at 37°, and the media containing released viruses are harvested 24–40 hr after transfection. Virus titers are generally between 10^8 and 10^9 plaque-forming units per ml. Virus stocks are aliquoted and kept frozen at −80° before use.

Method. The resulting transcripts are used to transfect BHK-21 cells in a 75-cm^2 flask.

1. Mix the RNA, Lipofectin (20 μl), and 1.5 ml ice-cold RNase-free PBS (150 mM NaCl, 50 mM Na$_2$PO$_4$, pH 7.2).
2. Incubate 5 min on ice.
3. Wash the 70% confluent BHK-21 cell monolayer 3 times with ice-cold PBS.
4. Add the Lipofectin mix to the cells.
5. Let stand 5 min (or until cells detach from the flask).
6. Remove Lipofectin mix.
7. Add 15 ml of α-MEM (α-minimum essential medium) with 3% fetal calf serum (FCS).

8. Incubate cells at 37°.
9. The media containing released virus are harvested 24–40 hr after transfection.

Expression of Small GTPase and PKB/akt with Recombinant Sindbis Virus

For expression of Rab5 proteins, the cells are grown to near confluence (~80%) in six-well plates, washed twice with 2 ml PBS, and then incubated with the appropriate dilution (10^6 pfu/ml) of Rab5-recombinant Sindbis virus in 500 μl serum-free α-MEM without bicarbonate containing 20 mM HEPES (Sigma), pH 6.8, for 30 min at room temperature, with intermittent agitation, to obtain a homogeneous distribution of the viruses. Subsequently, this medium is replaced with 3% fetal bovine serum α-MEM and the cells are incubated in a CO_2 incubator at 37° for several hours. The expression of Rab5 via recombinant Sindbis virus, which is detectable by Western blot analysis after 2 hr, increases up to sixfold over the endogenous level after 6 hr, and remains stable up to 12 hr after infection. Thus, this expression system offers the advantages of rapid and stable expression in essentially the whole cell population. Furthermore, this system allows for rapid expression of Rab GTPases at a very high level with little apparent cytophatic effects, which has not been possible using the vaccinia virus expression system. Vaccinia virus kills the cells within a short period of time.[53] In Fig. 1, we show the expression of Rab5: *wild type,* H-Ras: *wild type,* and PKB/akt: *wild type* as a function of time. We have also noted by Western blot analysis that a doublet is recognized by each of the specific antibodies used. In the case of Rab5 and H-Ras, the reason is that the Rab and Ras prenyltransferases, under conditions of high overexpression, may become limiting and modify only part of the expressed proteins.[54] In the case of PKB/akt, the PI3P-dependent kinases 1 and 2 (PDK1 and PDK2), kinases that activate PKB/akt, may be become limiting and only a part of the PKB/akt expressed is phosphorylated.[40]

Analysis of Sindbis-Infected Cells Overexpressing Small GTPases and PKB/akt

Cells expressing Rab GTPases (H-Ras or PKB/akt) can be analyzed in a number of ways, including immunofluorescence confocal microscopy, electron microscopy, Western blot analysis, metabolic labeling, and measurements of virus processing during intracellular transport. In addition,

---

[53] H. Stenmark, C. Bucci, and M. Zerial, *Methods Enzymol.* **257,** 155 (1995).
[54] G. Li and P. D. Stahl, *J. Biol. Chem.* **268,** 24475 (1993).

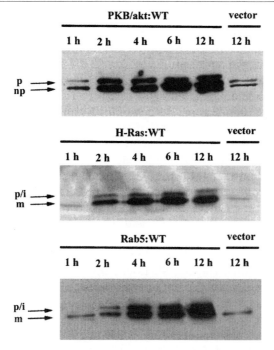

Fig. 1. Expression of Rab5, H-Ras, and PKB/akt in NIH 3T3 cells. Confluent NIH 3T3 cell monolayers were infected with recombinant viruses expressing PKB/akt, H-Ras, and Rab5 and the expressed proteins were determinated by immunoblot analysis using specific antibodies for each protein. The data are the mean of triplicate samples.

cytosol containing the overexpressed proteins can be isolated and used in cell-free assays. When analyzing the effects of the overexpressed Rab GTPase on morphology and transport, possible toxic effects of the Sindbis virus have to be considered, and infection times should be kept at a minimum. For experiments involving long infection times (12 hr or more), it is essential to include as a control the expression of the recombinant Sindbis virus alone (not encoding any foreign cDNA).

Measurement of Endocytosis in Sindbis-Infected Cells

In our laboratory, we have overexpressed several small GTPases and kinases (including Rab5, H-Ras, PKB, and PI4P5-kinase) to investigate the role of these proteins in the endocytic pathway. This chapter describes the methods we have used to measure the effect of the overexpression of these proteins on fluid-phase and receptor-mediated endocytosis. The same methodology can be applied to other small GTPases assumed to function in endocytosis and phagocytosis.

To study the effect of the overexpression of Rab proteins on fluid phase endocytosis, six-well plates of 80% confluent baby hamster kidney (BHK)-21, NIH 3T3, or CHO (chinese hamster ovary) cells are infected for 5 hr with Rab recombinant Sindbis virus as described earlier. Cells are then washed twice at room temperature with PBS containing 0.1% BSA, and 1500 μl α MEM containing 10 mg/ml horseradish peroxidase (HRP, type II, Sigma) is added to each well. The plates are incubated in a water bath at 37° for 1–60 min to allow HRP internalization. The cells are then washed extensively on ice with PBS/0.1% BSA, scraped from the dish, and washed three times by centrifugation in an Eppendorf centrifuge at 1500 rpm for 5 min with PBS. The cell pellet is then lysed in 200 μl of 0.1% (v/v) Triton X-100 and 10 mM HEPES, pH 7.2, and the nuclei are spun down at 3000 rpm for 10 min. Two milliliters of substrate solution containing 50 mM sodium phosphate, pH 5.5, 0.002% (v/v) H_2O_2, 0.1% (v/v) Triton X-100, and 0.01% dianisidine (Sigma, St. Louis, MO) is added to 20 μl of the supernatant. The reactions are kept in the dark from 10 to 30 min and are stopped with 100 μl of H_2SO_4. The absorbance of the samples at 455 nm is then read in a spectrophotometer. Ten microliters of the cell extract is used to determine the amount of protein, using the Bio-Rad (Hercules, CA) protein assay, and the measured HRP values are normalized to the protein concentrations. The time course of HRP endocytosis in control cells and in cells expressing a constitutively active form of PKB/akt is shown in Fig. 2A. Constitutively active forms of PKB/akt and H-Ras:V12[4,53] (Fig. 2A) enhanced both the rate of internalization and accumulation of HRP. In contrast, the expression of the PKB/akt "kinase dead" construct blocked HRP uptake (Fig. 2A). When the GTP-binding domain defective mutant of Rab5 (Rab5:N34) was coexpressed with the constitutively active form of PKB/akt and H-Ras, inhibition of endocytosis of HRP was observed.[6,55] Furthermore, the addition of wortmannin, a PI3K inhibitor, selectively blocked H-Ras:V12-stimulated HRP uptake[6,55]; wortmannin was unable to block PKB/akt-stimulated HRP uptake[55] nor Rab5:L79 stimulated uptake.[5] Taken together, these data indicate that H-Ras regulates endocytosis via PI3K, PKB/akt, and Rab5.

Ras is coupled to a variety of upstream tyrosine kinase-linked receptors, including EGF receptors, which are known to stimulate endocytosis. For measuring the effect of Rab5 overexpression on the epidermal growth factor (EGF) stimulated endocytosis, serum starved NIH 3T3 cells expressing human EGF receptor (in four-well Nunc plates, ca. 85% confluent) are

[55] M. A. Barbieri, A. D. Kohn, R. A. Roth, and P. D. Stahl, *J. Biol. Chem.* **273**, 19367 (1998).

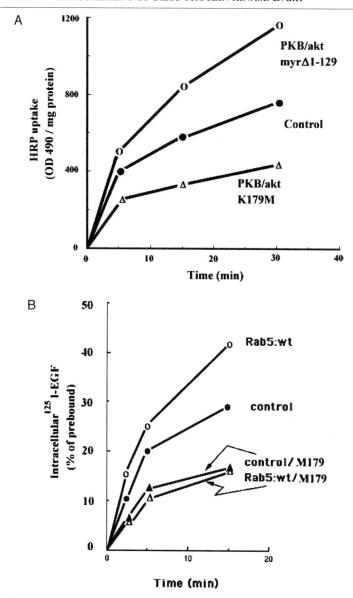

FIG. 2. Horseradish peroxidase and [^{125}I]EGF uptake in NIH 3T3 cells expressing Rab5 : WT and PKB/akt. (A) Horseradish peroxidase (HRP) uptake kinetics in cells
infected with either the vector virus or the recombinant virus encoding PKB/akt mutants.
(B) [^{125}I]EGF uptake kinetics in cells infected with either the vector as a control or the
recombinant virus encoding PKB/akt and Rab5 : WT (indicated). The data are the mean of
triplicate samples.

infected with Sindbis virus alone or with Sindbis virus encoding Rab5 or the cDNA of interest as described earlier. Four hours posttransfection, the cells are washed with serum-free medium and incubated further for 90 min in serum-free medium containing 0.2% BSA. Under these experimental conditions, most of the endogenous EGF is removed. The plates are placed on ice, then washed three times with ice-cold PBS, and 200 μl cold α MEM without bicarbonate containing 25 mM HEPES, pH 7.0, and 0.2% BSA; 100 pM[^{125}I]EGF (5 $\times$ 10^5 cpm/ng) is then added. The plates are kept on ice for 2 hr, then washed three times with ice-cold PBS, and 300 μl of prewarmed α MEM, without bicarbonate and containing 25 mM HEPES, pH 7.0, and 0.2% BSA, is added. Simultaneously, the plates are placed on a glass plate immersed in a water bath kept at 37°, and the cells are incubated for 1–90 min. Following this, the medium is collected rapidly, the cells are placed on ice, washed twice with 0.5 M NaCl/0.5 acetic acid, pH 3.0, and then twice with PBS. The washes are combined and the radioactivity is determined, indicating the amount of [^{125}I]EGF that remained at the cell surface. Finally, the cells are solubilized with 0.1% NaOH for 1 hr with intermittent agitation, and the radioactivity in the lysates is determined. The radioactivity in the lysate, obtained from the acid-neutral stripped, and that in the medium are then measured in a gamma counter. The measured values represent the internal, surface bound and released EGF, respectively. Control experiments are carried out to estimate the binding of [^{125}I]EGF in the presence of a 200-fold excess of unlabeled EGF and to Sindbis virus alone transfected and untransfected cells. PKB/akt activity and Rab5 function (M.A.B. and P.D.S., unpublished data) appear to be regulated by several growth factors, including EGF.[26] As shown in Fig. 2B, receptor-mediated endocytosis of [^{125}I]EGF was inhibited by ~40% in control cells as well as in cells expressing Rab5 : *wild type* (~79% inhibition) when the PKB/akt "kinase dead" construct was expressed. It is also important to note that the expression of Rab5 : *wild type* increased the rate as well as the accumulation of [^{125}I]EGF.

Confocal Microsocopy and Video Analysis

For morphologic analysis by immunofluorescence confocal microscopy, cells are grown on 12-mm round coverslips in 12-well microtest plates and transfected as described earlier. When the Sindbis system is used, significant amounts of unprenylated Rab proteins are often found in the cytosol. A possible reason is that the Rab-geranylgeranyltransferase, under conditions of high Rab overexpression, may become limiting and therefore the Rab proteins may not be fully prenylated. In such cases, the cytosolic Rab may complicate the detection of membrane-bound Rab, and it is therefore

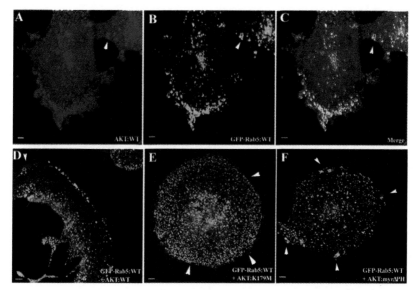

Fig. 3. Confocal microscopy of early endosomal structures in cells expressing PKB/akt and Rab5:*wild type*. Coexpression of PKB/akt:*wild type* (A–D) and the constitutively active form of PKB/akt(myrΔ1–129) (F) in cell line stably expressing GFP–Rab5:*wild type* induces the formation of enlarged endosomal structures. Coexpression of inactive PKB/akt in the GFP–Rab5:*wild type* cell line induced the fragmentation of endosomal structures (E). Arrowheads indicate the size of the endosomes. Bar: 1 μm.

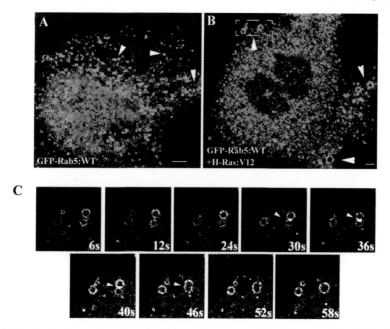

Fig. 4. Confocal microscopy of early endosomal structures in cells expressing H-Ras:V12 and Rab5:*wild type*. (A and B) Coexpression of the constitutively active form of H-Ras (H-Ras; V12) in a GFP–Rab5:*wild type* stable cell line induced the formation of enlarged endosomal structures. (C) The sequence shows an endosome docking reaction and fusion event. The time interval between frames is indicated in seconds. Note that the vesicles make contact in the 24-sec frame without an increase in pixel intensity, whereas the 30-sec frame (6 sec later), there is a nearly five-fold fluorescence intensity in the "contact" region compared to the rest of the membrane. Bar: 1.2 mm.

recommended to permeabilize the cells with 0.02% (w/v) saponin (Sigma) in 80 mM HEPES, pH 6.8, 1 mM MgCl$_2$, 0.25 M sucrose, and 5 mM EGTA (Sigma) for 5 min at 17° prior to fixation with 2.5% (w/v) paraformaldehyde. After 20-min fixation at room temperature, the cells are washed twice with PBS and then incubated for 20 min with 50 mM NH$_4$Cl in PBS. The coverslips are then placed with the cells facing down on Parafilm containing 30-μl droplets of 20% goat serum in PBS and incubated in a humidified chamber at room temperature for 30 min. Subsequently, the coverslips are placed on droplets containing antibody diluted in PBS containing 20% goat serum and are incubated at room temperature for 1 hr. The overexpressed Rab5 and tagged HA-PKB/akt proteins are visualized with an affinity purified antibody against a C-terminal peptide of Rab5, whereas the PKB/akt is visualized with the mouse anti-HA monoclonal antibody. After the incubation with the first antibody, the coverslips are washed three times for 15 min with 500 μl PBS and then incubated for 30 min at room temperature with rhodamine-labeled goat anti-mouse immunoglobulin G (IgG) antibodies and fluorescein isothiocyanate (FITC)-labeled goat anti-rabbit IgG antibodies (both from Sigma). After incubation with the second antibody, the coverslips are washed three times for 15 min with 500 μl PBS and mounted on 4 μl of mounting medium (10 mg/ml phenylenediamine, 45% glycerol in PBS) on a microscope slide. Control experiments are carried out to ensure that the antibodies do not cross-react. The cells are viewed in a confocal scanning beam fluorescent microscope at excitation wavelengths of 476 and 529 nm, respectively. In Fig. 3C (see color plate) we show extensive colocalization between the Rab5 (red) and the HA-PKB/akt (green) in endosomes. Also in cells coexpressing Rab5: *wild type* and PKB/akt: *wild type,* we observed the formation of enlarged endosomes (arrowhead in Figs. 3A–C). Interestingly, the expression of the constitutively active form of PKB/akt in stable cell lines expressing the GFP–Rab5: *wild type* also induced the formation of enlarged endosomes (arrowheads in Figs. 3D and 3F). In contrast, the expression of the PKB/akt "kinase dead" mutant induced the fragmentation of endosomes distributed throughout the cytoplasm (arrowheads in Fig. 3E). These data are consistent with previous observations, where the increase in the rate and accumulation of [^{125}I]EGF by the expression of Rab5: *wild type* was blocked by the coexpression of "kinase dead" PKB/akt. Similar inhibitory effects were also observed during Rab5-stimulated endocytosis of HRP.[55]

For video analysis by confocal microscopy, cells are grown on 22-mm^2 coverslips in six-well plates and transfected as described earlier. Cells stably transfected with pCDNA3 GFP–Rab5: *wild type* were coinfected with Sindbis virus alone or Sindbis virus encoding the GTPase-deficient H-

Ras:G12V mutant, which is a potent stimulator of fluid phase endocytosis.[3] Cotransfected cells grown on glass coverslips were inverted on glass slides made into a narrow flow cell by two strips of vacuum grease[56] and examined by either phase contrast or confocal microscopy. Mounted coverslips were warmed to 32–37° with heating lamps. Time-lapsed confocal microscopy was carried out on a Bio-Rad MRC1024 confocal microscope using a ∀63, 1.4-NA bright field objective and fluorescein and rhodamine filter sets. Confocal sequences were collected as Bio-Rad Pic files and converted to bitmaps for use in Photoshop 3.0 and pixel intensity was quantitated.[57] In cells expressing GFP–Rab5: *wild type,* labeled endosomal structures distributed throughout the cytoplasm were observed (Figs. 4A and 4B; see color plate). However, the endosomes in CHO cells coexpressing GFP–Rab5:*wild type* and H-Ras:V12 were singificantly enlarged (Fig. 4B). As expected the enlarged endosomes arose from the fusion between endosomes. Fusion between endosomes is shown in Fig. 4C. Note that the fusion between endosomes correlates with the localization of Rab5 in the docking region (arrowhead). This result clearly indicates that Rab5 is part of the fusion machinery and that the H-Ras signaling pathway plays a key, if not essential, role in the regulation of Rab5 function in endocytosis and endosome fusion.

Acknowledgments

We thank E. Peters, M. Levy, and L. La Rose for excellent technical assistance. This work was supported by NIH grants GM42259. AI20015, and AI35884 (P.D.S.) M.A.B. is a recipient of research fellowship 5T32CA09547-14 and R.L.R. is a recipient of research grant MH0162302.

---

[56] J. Heuser, Q. Zhu, and M. Clarke, *J. Cell Biol.* **121**, 1311 (1993).
[57] R. L. Roberts, M. A. Barbieri, K. M. Pryse, M. Chua, J. H. Morisaki, and P. D. Stahl, *J. Cell Sci.* **112**, 3667 (1999).

[17] Expression, Purification, and Biochemical Properties of Rabkinesin-6 Domains and Their Interactions with Rab6A

By Arnaud Echard, Ahmed El Marjou, and Bruno Goud

Introduction

The Rab6A GTPase is an ubiquitous Rab protein associated with the Golgi apparatus and the trans-Golgi network (TGN), as shown by light and immunoelectron microscopy.[1,2] Overexpression of GTP-bound ("active") Rab6A mutants in mammalian cells leads to the progressive microtubule-dependent redistribution of Golgi resident proteins to the endoplasmic reticulum (ER), and to an inhibition of the secretion.[3,4] Videomicroscopy experiments on live cells reveal that fluorescent Rab6A is localized to the Golgi region and labels highly dynamic tubulovesicular structures, which move along microtubules (MTs).[5] These transport intermediates define a novel, COP I-independent pathway from the Golgi apparatus to the ER, which was found to be specifically regulated by Rab6A and used by the B-fragment of Shiga toxin to enter the ER.[5,6] We have also found that Rab6A directly interacts with a Golgi-associated kinesin (Rabkinesin-6), a microtubule-dependent molecular motor expected to move cargo from the center to the cell periphery.[7]

A link between Rab machinery and the cytoskeleton has been docu-

[1] B. Goud, A. Zahraoui, A. Tavitian, and J. Saraste, *Nature* **345,** 553 (1990).

[2] C. Antony, C. Cibert, G. Geraud, A. Santa Maria, B. Maro, V. Mayau, and B. Goud, *J. Cell Sci.* **103,** 785 (1992).

[3] O. Martinez, A. Schmidt, J. Salamero, B. Hoflack, M. Roa, and B. Goud, *J. Cell Biol.* **127,** 1575 (1994).

[4] O. Martinez, C. Antony, G. Pehau-Arnaudet, E. G. Berger, J. Salamero, and B. Goud, *Proc. Natl. Acad. Sci. U.S.A.* **94,** 1828 (1997).

[5] J. White, L. Johannes, F. Mallard, A. Girod, S. Grill, S. Reinsch, P. Keller, B. Tzschaschel, A. Echard, B. Goud, and E. H. K. Stelzer, *J. Cell Biol.* **147,** 743 (1999).

[6] A. Girod, B. Storrie, J. C., Simpson, L. Johannes, B. Goud, L. M. Roberts, J. M. Lord, T. Nilsson, and R. Pepperkok, *Nat. Cell Biol.* **1,** 423 (1999).

[7] A. Echard, F. Jollivet, O. Martinez, J. J. Lacapere, A. Rousselet, I. Janoueix-Lerosey, and B. Goud, *Science* **279,** 580 (1998).

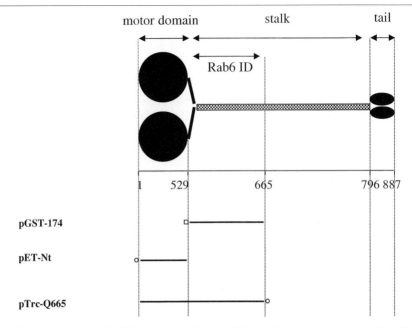

FIG. 1. A model for Rabkinesin-6 organization. The positions of the constructs described in this method are indicated: histidine-tagged (○) Nt or Q665 and GST-tagged (□) 174 domain (see text for details). Rab6 ID, Rab6A interaction domain.

mented for several Rabs, such as Rab8,[8,9] Rab3,[10] and Rab5.[11] Moreover, growing evidence suggests that direct interactions with molecular motors may be a general feature of Rab proteins: it has been reported that Rab33B binds to an Xklp2-related kinesin,[12] whereas Rab11A interacts with the actin-based myosin Vb motor.[13]

Based on homology and primary sequence analysis, Rabkinesin-6 can be divided into three structural domains (Fig. 1): an amino-terminal globular motor domain (amino acids 1–529) involved in ATP hydrolysis and MT

[8] D. Deretic, L. A. Huber, N. Ransom, M. Mancini, K. Simons, and D. S. Papermaster, *J. Cell Sci.* **108,** 215 (1995).

[9] J. Peränen, P. Auvinen, H. Virta, R. Wepf, and K. Simons, *J. Cell Biol.* **135,** 153 (1996).

[10] M. Kato, T. Sakaki, T. Ohya, H. Nakanishi, H. Nishioka, M. Imamura, and Y. Takai, *J. Biol. Chem.* **271,** 31775 (1996).

[11] E. Nielsen, F. Severin, J. M. Backer, A. A. Hyman, and M. Zerial, *Nat. Cell Biol.* **1,** 376 (1999).

[12] T. Koda, J. Y. Zheng, and M. Ishibe, *Mol. Biol. Cell* **10,** 214a (1999).

[13] R. Kumar, J. Navarre, and J. R. Goldenring, *Mol. Biol. Cell* **10,** 163a (1999).

binding, a predicted coiled-coil stalk region (amino acids 530–796) allowing homodimerization, and a carboxy-terminal globular tail domain (amino acids 797–887), possibly involved in membrane binding. The Rab6A interacting domain on Rabkinesin-6 (amino acids 529–665) is localized downstream to the motor domain, in the stalk region. This chapter describes the purification of several Rabkinesin-6 domains from bacteria and eukaryotic systems, and their biochemical characterization (ATPase, MT-binding, hydrodynamic properties). We also describe an assay to visualize the interaction between Rab6A and Rabkinesin-6.

Methods

Expression of Rabkinesin-6 Domains and Purification from Bacteria

Plasmids, Strains. Rabkinesin-6 motor domain cDNA (encoding amino acids 1–530, "Nt") is subcloned from pGEM-Nt plasmid[7] into pET* plasmid between *SmaI*/*KpnI* sites. pET* is derived from pET15b vector (Novagen, Madison, WI) by introducing the CATATGAACCCGGGGTCGAC-GGTACCGGATCC linker between *NdeI*/*BamHI* sites. This plasmid allows controlled expression in *Escherichia coli* bacteria of the protein fused at its NH_2 terminus to a hexahistidine (His_6) tag and a rapid purification by nickel chromatography.

Q665 construct (amino acids 1–665, "Q665") is purified as a carboxy-terminal fusion to a His_6 tag using the pTrcHis2B vector (Invitrogen, Grominger, The Netherlands): *SmaI*/*KpnI* cDNA Q665 insert from the pET-Q665 vector is subcloned into the blunted *BglII*/*KpnI* pTrcHis2B sites. Polymerase chain reaction (PCR) on Rabkinesin-6 cDNA with the primers AGCACAGTCCTCAGGTTGGCC and ATAGGTACCGTTGTGCCG-CTGGCTGCTG is digested by *Bsu*36I and *KpnI*, and introduced into the pET-Nt between these two sites, leading to the pET-Q665 plasmid.

Nt and Q665 domains are expressed in BL21 (DE3) *E. coli* strain (Novagen) carrying the bacteriophage T7 RNA polymerase gene under the control of the isopropyl-β-D-thiogalactopyranoside (IPTG) *lacUV5* promoter.

Protein Induction and Purification. *Escherichia coli* strain freshly transformed with pET-Nt or pTrc-Q665 plasmids are cultured overnight at 37° in 50 ml of Luria–Bertani (LB) broth (10 g of Bacto-tryptone, 5 g of Bacto-yeast extract, and 10 g NaCl) supplemented with 0.1 mg/ml ampicillin. These starter cultures are diluted in the same medium (1 liter) and grown at 37° until A_{600} reaches 0.6–0.8. Protein expression is then induced for

4 hr at 37° with 0.3 mM IPTG. Bacterial pellets are flash frozen into liquid nitrogen and stored at −80°.

Pellets are suspended in 10 ml H0 buffer [20 mM KOH–PIPES, pH 6.8, 1 mM MgCl$_2$, 1 mM EGTA, 1 mM dithiothreitol (DTT), protease inhibitors] supplemented with 0.5 mM Mg-ATP and 2 mg/ml lysozyme for 20 min on ice. Bacteria are lysed by sonication, and protein suspension is further incubated 10 min on ice with 1 μg/ml DNase I and 0.1% Tween 20. This lysate is clarified by centrifugation (20,000g, 20 min, 4°), and the supernatant is incubated for 1 hr with 1 ml Ni^{2+}-NTA agarose slurry (Qiagen, Valencia, CA) at 4° under stirring. Nickel beads are transferred on a column and washed at 4° with 20 ml H1 buffer (50 mM HEPES, pH 7.5, 1 mM MgCl$_2$, 10 mM imidazole, 300 mM NaCl, 0.1 mM Mg-ATP) and 20 ml H2 buffer (50 mM HEPES, pH 7.0, 1 mM MgCl$_2$, 10 mM imidazole, 0.1 mM Mg-ATP). Hexahistidine-tagged proteins are finally eluted by H3 buffer (50 mM HEPES, pH 7.5, 1 mM MgCl$_2$, 200 mM imidazole, 150 mM NaCl, 0.1 mM Mg-ATP). Peak fractions are pooled and directly stored in 10% glycerol at −80° after flash freezing into liquid nitrogen. For MT-binding experiments, fractions are first dialyzed against a low-salt buffer (15 mM imidazole, pH 7.8, 1 mM MgCl$_2$, 1 mM EGTA, 1 mM DTT) before storage.

Protein fractions are analyzed by SDS–PAGE and Western blot. This protocol leads to the purification of about 0.5–1 mg fusion proteins (purity >90%). Gel filtration chromatography and sucrose gradient sedimentation demonstrate that histidine-tagged Nt and Q665 display distinct hydrodynamic properties, as expected, since Q665 is predicted to be dimeric, whereas Nt is monomeric. For a sucrose gradient sedimentation, about 50 μg histidine-tagged Nt and Q665 are pooled and loaded on an 11-ml 7% (w/w)–20% (w/w) linear sucrose gradient. Proteins are sedimented in a SW41 rotor (Beckman) for 22 hr at 41,000 rpm, 4°. Twenty fractions (500 μl) are collected and analyzed on Western blot with an anti-His$_6$ antibody (Santa Cruz).

Microtubule Binding Assay. α/β-Tubulin is purified from bovine brain with two cycles of polymerization in the presence of GTP, and MT-associated proteins (MAPs) are removed by phosphocellulose chromatography essentially as described in Ref. 14. Tubulin concentration is estimated by the Bradford assay and MT concentration is expressed per tubulin heterodimer.

One mg/ml α/β-tubulin is polymerized at 37° for 30 min in BRB80 (80 mM KOH–PIPES, pH 6.80, 1 mM MgCl$_2$, 1 mM EGTA) supplemented with 1 mM GTP, 10% glycerol, and 20 μM taxol (paclitaxel, Sigma). MTs

[14] R. C. Williams, Jr., and J. C. Lee, *Methods Enzymol.* **85**, 376 (1982).

are centrifuged at 20,000g, 15 min, 25° and resuspended into BRB80, 1 mM GTP, 20 μM taxol to a final concentration of 1 mg/ml.

MT-binding assay is performed in 300-μl polyethylene centrifuge tubes (Beckman) coated with 1% gelatin and water-washed in order to avoid nonspecific motor domain interactions with plastic. Dialyzed motor domain is first centrifuged at 100,000g, 20 min, 4° to remove any aggregates, and then incubated for 20 min at 33° with taxol-stabilized MTs (motor domain: MT molar ratio 1.5 : 1). The complexes are centrifuged at 100,000g, 20 min, 33° on a 15% sucrose cushion, and equal amounts of pellet and supernatant are analyzed by SDS–PAGE. Typically, more than 80% of the motor domain is found associated with the MT pellets, whereas it remains in the supernatant in the absence of MTs. Moreover, excess of Mg-ATP (10 mM) for 20 min at 33° leads to the complete dissociation of the motor/MT complex. It is important to note that the Rabkinesin-6 motor domain–MT interaction is very sensitive to salts—150 mM KCl by itself dissociates this complex.

ATPase Assay. MT-induced ATPase activity of kinesins is often measured by using a NADH coupled enzymatic assay (for instance, see Ref. 15). Briefly, each hydrolyzed ATP leads to the oxidation of one NADH into one NAD^+, and the NADH concentration is determined at 340 nm by spectrophotometry. ATPase rates are measured under stirring at 25° in the reaction buffer (20 mM KOH–PIPES, pH 6.80, 5 mM $MgCl_2$, 1 mM phosphoenolpyruvate, 267 μM β-NADH, 0.1 mg/ml pyruvate kinase, 0.1 mg/ml lactate dehydrogenase, 1 mM Mg-ATP) in the presence of increasing concentrations of taxol-stabilized MT (0–100 μg), and either 55 nM purified motor domain or 675 nM purified Q665 construct. As shown in Fig. 2, the dimeric Q665 Rabkinesin-6 construct ATPase activity is enhanced by MT [0.1 μmol/min/mg, $k_{cat} = 16$ min^{-1} per dimer at maximum speed, $K_{0.5}(MT) = 5$ μg] and is reduced compared to the Rabkinesin-6 monomeric motor domain alone [1.2 μmol/min/mg, $k_{cat} = 70$ min^{-1}, $K_{0.5}(MT) = 10$ μg].

Direct Rab6A–Rabkinesin-6 Interaction Visualized by Overlay Experiment. Several methods have been used to show an interaction between Rab6A bound to GTP and Rabkinesin-6: two hybrid assay, coimmunoprecipitation after overexpression in HeLa cells or functional effect on Rab6A-induced intracellular defects.[7] Here we describe a rapid method that confirms the direct interaction between Rab6A and the "174 domain" (amino acids 529–665) localized just downstream of the motor domain (see Fig. 1). Equal amounts (0.3 nmol) of purified histidine-tagged motor domain,

[15] T. G. Huang, and D. D. Hackney, *J. Biol. Chem.* **269,** 16493 (1994).

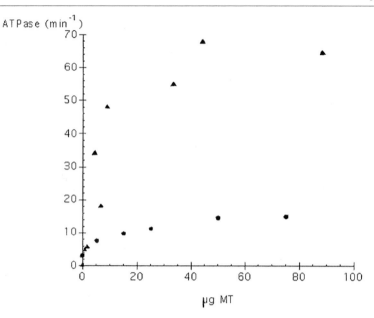

FIG. 2. MT-activated ATPase activity of a monomeric (Nt) or a dimeric (Q665) Rabkinesin-6. Purified histidine-tagged Nt (▲) or Q665 (●) Rabkinesin-6 was incubated with increasing concentration of taxol-stabilized MTs in the presence of 1 mM Mg-ATP. ATPase activities are measured by spectrophotometry using a NADH coupled enzymatic assay.

Q665 or glutathione S-transferase (GST)-tagged 174 domain[7] are separated on SDS–PAGE and transferred onto a nitrocellulose paper in a carbonate buffer: 3 mM Na_2CO_3, 10 mM $NaHCO_3$ (pH 9.8). Proteins are then renaturated for 4 hr at room temperature in the following buffer: 50 mM NaOH–HEPES, pH 7.2, 5 mM magnesium acetate, 100 mM potassium acetate, 3 mM DTT, 10 mg/ml bovine serum albumin (BSA), 0.1% Triton X-100, and 0.3% Tween 20. [α-^{32}P]GTP (50 μCi) is exchanged for 1 hr at 30° on bacterially purified histidine-tagged Rab6A (1.5 nmol) by 4 mM EDTA; 10 mM $MgCl_2$ is then added, and unincorporated nucleotide is removed by exclusion chromatography on Sephadex G-50. After the renaturation step, radiolabeled Rab6A is added to the nitrocellulose membrane in 10 ml of binding buffer (12.5 mM NaOH–HEPES, pH 7.2, 1.5 mM magnesium acetate, 75 mM potassium acetate, 1 mM DTT, 2 mg/ml BSA, 0.05 % Triton X-100, 0.05% CHAPS) for 1 hr at room temperature. The membrane is finally washed 3 times (15 min each) in the washing buffer (20 mM HCl-

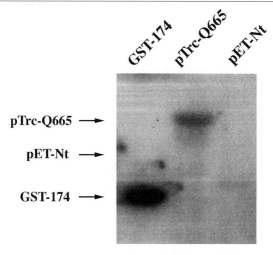

FIG. 3. Direct Rab6A–Rabkinesin-6 interaction visualized by an overlay experiment. Equal amounts of histidine-tagged Nt, histidine-tagged Q665, and GST-174 (Rab6A interacting domain) were analyzed on SDS–PAGE, then transferred on a nitrocellulose membrane. Renaturated membrane was incubated with Rab6A previously exchanged with [α-^{32}P]GTP and washed. Bound Rab6A was visualized after autoradiography.

Tris, pH 7.4, 100 mM NaCl, 20 mM MgCl$_2$, 0.005% Triton X-100) and autoradiographed. Rab6A binds to 174 and Q665 proteins, but not to the Rabkinesin-6 motor domain alone (Fig. 3), as already described by other means.[7]

Rabkinesin-6 Expression and Purification from High-Five Cells

Plasmids, Strains. Because histidine-tagged full-length Rabkinesin-6 is degraded when expressed in bacteria, we purified it with the baculovirus/ High-Five insect cell system. pET-Rabkinesin-6 was first constructed by subcloning blunted *Sma*I/*Hin*dIII fragment from pGEM-Rabkinesin-6[7] into pET* vector. Rabkinesin-6 cDNA was then transferred into the pFastBac plasmid, and the recombination step was performed as described (Life Technologies).

Protein Induction and Purification. Six T175-cm^2 flasks with 3×10^7 High-Five cells are infected with the Rabkinesin-6 recombinant baculovirus at a multiplicity of infection (MOI) of 10. Cells are collected by centrifugation 72 hr postinfection, and washed with PBS, pH 7.4. The cell pellet is

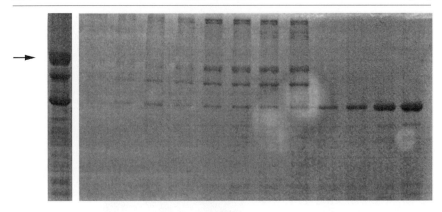

FIG. 4. Purification of the full-length Rabkinesin-6 from insect cells. The full-length histidine-tagged Rabkinesin-6 is purified from High-Five cells by chromatography on nickel beads. Eluted proteins (first lane) are separated by gel filtration (right panel). Full-length Rabkinesin-6 (arrow) is very sensitive to degradation, even during the gel filtration step.

lysed overnight at 4° in a lysis buffer (30 mM NaOH–MOPS, pH 6.7, 1 mM EGTA, 2 mM MgSO$_4$, 1 mM DTT, 1 mM Mg-ATP, 600 mM NaCl, 3% Triton X-100, protease inhibitor cocktail). The cell lysate is centrifuged at 50,000g for 45 min at 4°, the supernatant (8.7 ml) is diluted to 30 ml with 30 mM NaOH–MOPS, pH 6.7, 2 mM MgSO$_4$, 1 mM DTT, 0.1 mM Mg-ATP, 500 mM NaCl, 20 mM imidazole, pH 8.0, and protease inhibitor cocktail. A 1 ml NiNTA Fast Flow slurry column (Qiagen) is equilibrated with 30 mM NaOH–MOPS, pH 6.7, 2 mM MgSO$_4$, 1 mM DTT, 0.1 mM Mg-ATP, 500 mM NaCl, 20 mM imidazole, pH 8.0, 0.5% Triton X-100, 1 mM phenylmethylsulfonyl fluoride (PMSF), and protease inhibitor cocktail. The sample is loaded onto the column at a 0.5 ml/min flow rate with the gradiFrac system. Column is washed first with buffer Lav-1 (30 mM NaOH–MOPS, pH 6.7, 2 mM MgSO$_4$, 1 mM DTT, 0.1 mM Mg-ATP, 500 mM NaCl, 20 mM imidazole, pH 8.0, 0.5% Triton X-100, and protease inhibitor cocktail), then with buffer Lav-2 (30 mM NaOH–MOPS, pH 6.7, 2 mM MgSO$_4$, 1 mM DTT, 0.1 mM Mg-ATP, 500 mM NaCl, 20 mM imidazole, pH 8.0, and protease inhibitor cocktail). Histidine-tagged Rabkinesin-6 is eluted with 30 mM NaOH–MOPS, pH 6.7, 2 mM MgSO$_4$, 1 mM DTT, 0.1 mM Mg-ATP, 250 mM NaCl, 250 mM imidazole, pH 8.0, at a flow rate of 0.5 ml/min in 500-μl fractions. Rabkinesin-6 fractions are analyzed by SDS–PAGE and frozen in liquid nitrogen in the presence of 10% glycerol, and stored at −80°. Further purification is achieved by gel filtration chromatography on Superdex 200 10/30 (Pharmacia Biotech)

(Fig. 4). Rabkinesin-6 is localized on microtubules after infection and solubilized only in a combination of 3% Triton X-100, high salts, and ATP. Similar insolubility problems had already been observed in the case of CHO1, a kinesin closely related to Rabkinesin-6.[16]

[16] R. Kuriyama, S. Dragas-Granoic, T. Maekawa, A. Vassilev, A. Khodjakov, and H. Kobayashi, *J. Cell Sci.* **107,** 3485 (1994).

[18] Expression and Functional Analyses of Rab8 and Rab11a in Exocytic Transport from trans-Golgi Network

By WEI CHEN and ANGELA WANDINGER-NESS

Introduction

Both Rab8 and Rab11a belong to the Rab subfamily of Ras-related small GTPases. Rab8 has been shown to regulate the delivery of basolateral proteins from the trans-Golgi network (TGN) to the plasma membrane in both polarized cells and fibroblasts.[1–3] Rab11a has been localized to both the Golgi and recycling endosomes,[4–6] and we have established that Rab11a regulates Golgi-to-cell surface transport of a cognate basolateral protein, but not an apical marker in baby hamster kidney (BHK) cells.[6] Rab11a also controls the transit of internalized basolateral molecules through recycling endosomes in both epithelia and fibroblasts.[4,7] Data suggest that Rab11a controls passage from the Golgi through endosomes to the cell surface, whereas Rab8 functions in the direct transit from the Golgi to the surface (W. Chen and A. Wandinger-Ness, unpublished data, 1999). This chapter describes the morphologic and biochemical assays we have used to study the role of Rab8 and Rab11a in the exit from the TGN of cargo proteins destined for the plasma membrane.

[1] L. A. Huber, M. J. de Hoop, P. Dupree, M. Zerial, K. Simons, and C. Dotti, *J. Cell Biol.* **123,** 47 (1993).
[2] L. A. Huber, S. Pimplikar, R. G. Parton, H. Virta, M. Zerial, and K. Simons, *J. Cell Biol.* **123,** 35 (1993).
[3] J. Peränen, P. Auvinen, H. Virta, R. Wepf, and K. Simons, *J. Cell Biol.* **135,** 153 (1996).
[4] O. Ullrich, S. Reinsch, S. Urbé, M. Zerial, and R. G. Parton, *J. Cell Biol.* **135,** 913 (1996).
[5] M. Ren, G. Xu, J. Zeng, C. De Lemos-Chiarandini, M. Adesnik, and D. D. Sabatini, *Proc. Natl. Acad. Sci. U.S.A.* **95,** 6187 (1998).
[6] W. Chen, Y. Feng, D. Chen, and A. Wandinger-Ness, *Mol. Biol. Cell* **9,** 3241 (1998).
[7] J. E. Casanova, X. Wang, R. Kumar, S. G. Bhartur, J. Navarre, J. E. Woodrum, Y. Altschuler, G. S. Ray, and J. R. Goldenring, *Mol. Biol. Cell* **10,** 47 (1999).

Methods

Construction of Mutants and Expression Vectors

The functional analysis of Rab proteins is facilitated by the possibility of producing mutant forms that behave as dominant negative inhibitors of membrane transport. Mutants are generated by site-directed mutagenesis of conserved domains critical for nucleotide binding or hydrolysis. Substitution of T/S to N in the $GX_4GK(T/S)$ domain and N to I in the GNKXD domain results in proteins that bind GDP constitutively or are nucleotide free, respectively. Substitution of Q to L in the WDTAGQE domain generally results in a protein that hydrolyzes GTP at a greatly reduced rate due to poor interaction with its GTPase activating protein. In the case of Rab11a it was found that the Rab11aQ70L mutant behaved like the wild-type (wt) protein, but a slightly different mutant substituting S20 for V exhibited the expected reduction in GTP hydrolysis.[7] Therefore, it is advisable to screen all newly constructed Rab protein mutants in solution nucleotide binding and hydrolysis assays[7,8] or by ligand overlay blotting[9] to ascertain that they behave as expected.

Plasmid pRAB8-wt encoding wild-type Rab8 protein[3] was kindly provided by Johan Peränen (University of Helsinki, Helsinki, Finland). Plasmids pGEM1-rab8Q67L and -rab8T22N were generated by PCR mediated site-directed mutagenesis using the following primers: CAGCTGGTCTA-GAACGGTTTC, GTGGGGAAGAACTGTGTCCTG (changed bases are underlined), for rab8Q67L and rab8T22N, respectively. The nucleotide sequences were confirmed by sequencing. Plasmids pGEM1-rab11a, -rab11aQ70L and -rab11aS25N[4] were kindly provided by Marino Zerial (EMBL, Heidelberg, Germany). Rab11aS20V was generated by site-directed mutagenesis using the following primer: ATTGGAGATGT-TGGTGTTGG. Plasmids pEGFP-rab8, -rab8Q67L, -rab8T22N, -rab11a, -rab11aS25N, -rab11aQ70L, and -rab11aS20V encoding the GFP–Rab chimeric proteins under the control of a human cytomegalovirus (CMV) promoter were constructed for morphologic evaluations by subcloning the cDNA fragments in frame to pEGFP-C3 vector (Clontech, Palo Alto, CA) from the corresponding pGEM1-rab constructs.

Morphological Analyses

Expression and Localization of GFP–Rab Protein Chimeras. Green fluorescent protein (GFP) chimeras have been invaluable in cell biology

[8] S. Jones, R. J. Litt, C. J. Richardson, and N. Segev, *J. Cell Biol.* **130,** 1051 (1995).
[9] Y. Feng, B. Press, and A. Wandinger-Ness, *J. Cell Biol.* **131,** 1435 (1995).

for the localization of molecules in both live and fixed cells without requiring specific antibody reagents. The GFP–Rab constructs described above were expressed in cells together with various organelle markers for morphologic evaluation of their effects by immunofluorescence microscopy. For example, coexpression of the GFP-Rab8 chimeras and the TGN marker sialyltransferase (ST) in BHK cells revealed that Rab8 overexpression causes fragmentation of the TGN (Fig. 1). Similar experiments performed with Rab11a revealed that the Rab11aS25N mutant protein was preferentially colocalized with sialyltransferase in the TGN, while the wild-type protein was primarily associated with transferrin receptor positive early and recycling endosomes (Fig. 2, see color plate; see also Refs. 4 and 6).

For morphologic evaluation of Rab protein expression and localization, BHK cells are grown on 15-mm^2 coverslips in 35-mm dishes for 18 hr prior to transfection. Optimal transfection efficiencies are obtained if care is taken to ensure that cells are growing rapidly and in log phase, e.g., by maintaining subconfluent cultures prior to seeding for transfection. After washing once with serum-free medium, cells are transfected with equal amounts of pcDNA3-*myc*-ST[10] in combination with pEGFP-rab8, pEGFP-rab8Q67L, or pEGFP-rab8T22N plasmids. Lipofectamine is used for transfections according to the manufacturer's instructions (Gibco-BRL, Gaithersburg, MD); 1 μg DNA/9 μl Lipofectamine for each 35-mm dish BHK cells was optimal (DNA:Lipofectamine ratios should be optimized for different cell types). The DNA/Lipofectamine transfection mixture is removed 5 hr posttransfection and complete G-MEM is added. The transfected cells are cultured for an additional 12 hr at 37° in a 5% (v/v) CO$_2$ incubator and processed for immunofluorescence as described below.

In preparation for immunofluorescence staining, cells are washed once with phosphate-buffered saline (PBS) and permeabilized with 0.5% saponin (Sigma, St. Louis, MO) in 80 mM PIPES–KOH (pH 6.8), 5 mM EGTA, and 1 mM MgCl$_2$ for 5 min. Cells are then fixed with 3% paraformaldehyde in PBS$^+$ (PBS containing 1 mM CaCl$_2$ and 1 mM MgCl$_2$) for 15 min at room temperature. After fixation, cells are washed with 0.5% saponin in PBS for 5 min, and free aldehyde groups are quenched with 50 mM NH$_4$Cl in PBS for 10 min. Cells are washed with 0.5% saponin in PBS for 5 min and then incubated with mouse anti-*myc* antibody (for detection of *myc* epitope-tagged ST) in PBS containing 0.5% saponin for 20 min. After rinsing cells three times with 0.5% saponin in PBS (5 min/each) they are incubated with Texas Red-conjugated horse anti-mouse secondary antibody (Vector Laboratories, Burlingame, CA) in PBS containing 0.5% saponin for 20 min.

[10] J. Ma, R. Qian, F. M. 3rd. Rausa, and K. J. Colley, *J. Biol. Chem.* **272,** 672 (1997).

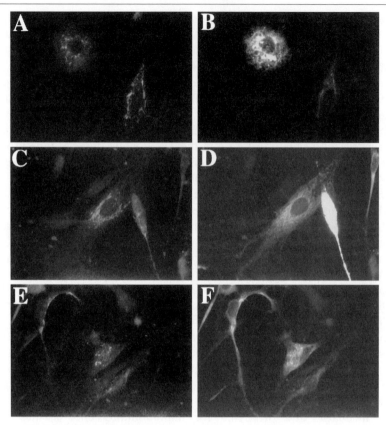

FIG. 1. Rab8 overexpression alters Golgi morphology. BHK cells were cotransfected with plasmids encoding *myc*-tagged sialyltransferase and various Rab8–GFP chimeras as detailed in the text. After fixation the trans-Golgi sialyltransferase was revealed by immunostaining with an anti-*myc* primary and a Texas Red-labeled secondary antibody (A, C, and E), while the Rab8–GFP chimeras were detected directly in the FITC channel (B, D, and F). Punctate structures positive for wild-type Rab8 (B) and rab8Q67L (F) were seen partially overlapping with disperse sialyltransferase positive trans-Golgi fragments (A and E). The rab8T22N mutant (D) appeared less membrane associated and the Golgi morphology remained more compact and perinuclear (C). Samples were imaged using a Zeiss Universal epifluorescence microscope outfitted with a 35-mm camera.

Cells are then washed once with PBS containing 0.5% saponin and three times with PBS (5 min/each). The coverslips are then mounted on glass slides in Mowiol 4-88 (Calbiochem, La Jolla, CA) and viewed with a fluorescence microscope. GFP can be detected using a fluorescein isothiocyanate (FITC) filter set.

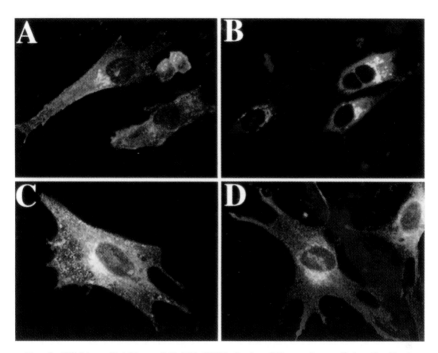

FIG. 2. Wild-type Rab11a and Rab11aS25N display different intracellular distributions. BHK cells were cotransfected with plasmids encoding either *myc*-tagged sialyltransferase or the human transferrin receptor and GFP–wild-type Rab11a or –Rab11aS25N chimeras as detailed in the text. (A and B) *trans*-Golgi sialyltransferase was revealed by immunostaining with an anti-*myc* primary and a Texas Red-labeled secondary antibody. (C and D) Human transferrin receptor was visualized with a goat anti-transferrin receptor primary antibody, biotinylated horse anti-goat antibody, and Texas Red-conjugated avidin for detection. GFP–Rab11a chimeras in (A–D) were detected directly in the FITC channel. Each image represents superimposed 0.4-μm sections of the GFP–Rab11a staining (green) and the sialyltransferase or transferrin receptor staining (red) taken in the same focal plane on a Zeiss LSM 410. Overlap in the two staining patterns is seen in yellow.

Morphologic Evaluation of GFP–Rab11a Function in Exocytosis. The wild-type vesicular stomatitis virus (VSV) G protein and the temperature-sensitive VSV G tsO45 variant proved to be useful transport markers to study the effects of the overexpression of GFP–Rab11a proteins on TGN-to-cell surface transport. Through selective temperature manipulations, VSV G or VSV G tsO45 proteins were accumulated in the TGN following synthesis. Subsequently their synchronous transport to the cell surface was initiated by return to permissive temperatures. Because cell surface and normal Golgi staining are readily distinguished by immunofluorescence staining, the progress of transport could be monitored morphologically. Expression of the mutant Rab11aS25N protein was found to prevent exit of both VSV G protein and its temperature-sensitive variant from the Golgi under these conditions, while wild-type Rab11a protein expression allowed normal delivery of VSV G protein to the cell surface.[6]

The impact of Rab11a expression on TGN-to-cell surface transport of VSV G tsO45 protein is morphologically evaluated as follows. BHK cells are grown on 15-mm^2 coverslips in 35-mm dishes for 18 hr prior to transfection. After washing once with serum-free medium, cells are transfected with equal amounts of pcDNA3-GtsO45[6] in combination with pEGFP-rab11a, -rab11aQ70L, or -rab11aS25N plasmids. Lipofectamine is used for transfections according to the manufacturer's instructions (Gibco-BRL); 1 μg DNA/9 μl Lipofectamine. The cells are then maintained at 39° (restrictive temperature) for 5 hr, at which time the DNA/Lipofectamine transfection mixture is removed and complete G-MEM is added. The transfected cells are cultured for an additional 12 hr at 39° in a 5% CO$_2$ incubator and processed for immunofluorescence as described above. Under these conditions the VSV G tsO45 protein accumulates in the endoplasmic reticulum. Cells are then incubated at 20° for 2 hr (to accumulate the VSV G tsO45 protein in the TGN) and either processed for immunofluorescence immediately or transferred to 31° (permissive temperature) for an additional hour (to allow the TGN export of the VSV G tsO45 protein) prior to permeabilization and fixation. Minor variations in the incubation temperatures are made if the wild-type VSV G protein is used as the transport marker (transfection, initial incubation to allow VSV G synthesis and cell surface transport are all performed at 37°; the 20° incubation is as above, see also Ref. 6). Care should be taken in evaluating the impact of Rab protein overexpression on the morphology of the Golgi prior to implementing this assay. For example, on account of the alterations in Golgi morphology induced by Rab8 overexpression (see Fig. 1), a biochemical rather than a morphologic assay was more effective for monitoring Rab8 function in TGN-PM transport.

Biochemical Analyses

Rab8 and Rab11a Functions in Exocytosis Monitored by Cell Surface Biotinylation. A recombinant vaccinia virus expression system was used to establish a biochemical assay useful for studying the function of Rab proteins in exocytic transport. This system was utilized on account of several important advantages. First, the high transfection efficiencies obtained with this system makes it particularly well suited for the development of biochemical assays that rely on having a uniform population of cells. Second, this system enables the coexpression of multiple proteins with relative ease. Thus, multiple Rab proteins could be overexpressed in combination together with relevant transport markers. Finally, the contributions of secondary effects that may be caused by mutant Rab protein expression are minimized in this system due to the narrow time frame (<6 hr) during which the Rab proteins are allowed to accumulate and the experiment is performed.

The system also has some disadvantages. For example, the cytotoxicity of vaccinia virus may interfere with normal cellular processes such as endocytosis, and assays requiring long time points for evaluation are not feasible due to cell death within 10 hr postinfection/transfection. Cytotoxicity may be minimized by inclusion of hydroxyurea (10 mM) to block viral replication, but assays requiring long-term evaluation necessitate the selection of inducible stable cell lines. The protocol below details the methods used for transient transfection to overexpress combinations of proteins, followed by radiolabeling to monitor expression and transport of newly synthesized proteins and biotinylation to quantify protein arrival at the plasma membrane. This assay was used successfully to demonstrate that expression of all forms of Rab8 is inhibitory, decreasing cell surface delivery of VSV G protein up to twofold. Wild-type Rab8 and the activating Rab8Q67L mutant, which acutely disrupt Golgi morphology and affect the cytoskeletal architecture,[3] are also most inhibitory in the exocytosis assay (Fig. 3A). In contrast, although the dominant negative rab11aS25N impairs exocytosis of VSV G protein by 50%, neither wild-type Rab11a nor the Rab11aS20V activating mutant exerts any inhibitory effect (Fig. 3B; see also Ref. 6).

Transient overexpression studies using the T7 RNA polymerase recombinant vaccinia virus expression system are based on published protocols.[9,11,12] Thirty minutes postinfection with vaccinia virus [performed with 5–10 plaque-forming units (pfu)/cell in serum-free medium], equal amounts

[11] T. R. Fuerst, E. G. Niles, F. W. Studier, and B. Moss, *Proc. Natl. Acad. Sci. U.S.A.* **83,** 8122 (1986).

[12] H. Stenmark and M. Zerial, *in* "Cell Biology: A Laboratory Handbook" (J. E. Celis, ed.), p. 201. Academic Press, San Diego, 1998.

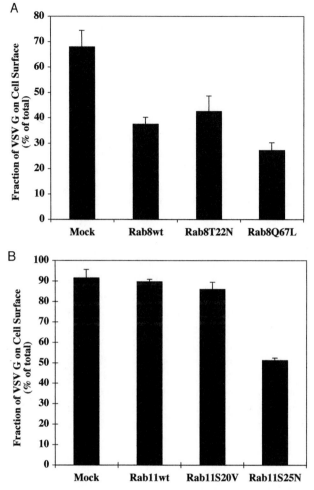

FIG. 3. A biochemical assay for Rab8 and Rab11a functions in exocytosis. (A) BHK cells were infected with recombinant vaccinia virus and transfected with plasmids encoding VSV G protein together with those encoding wild-type (wt) Rab8, Rab8T22N, or Rab8Q67L as detailed in the text. At 5 hr posttransfection cells were metabolically labeled for 10 min. After a chase period of 70 min, cell surface delivery of the VSV G protein was monitored by surface biotinylation as described in the text. Each column represents the mean ± SEM of triplicate samples from each of two independent experiments. (B) BHK cells were infected with recombinant vaccinia virus and transfected with plasmids encoding VSV G protein together with those encoding wild-type (wt) Rab11a, Rab11aS20V, or Rab11aS25N. At 5 hr posttransfection cells were metabolically labeled for 10 min. After a chase period of 120 min, cell surface delivery of the VSV G protein was monitored by surface biotinylation. Each column represents the mean ± SEM of triplicate samples from one of the three independent experiments.

of the plasmids encoding VSV G protein (pAR-G)[13] and Rab11a or Rab8 under the control of the bacteriophage T7 promoter are used for Lipofectamine-mediated cotransfection. A maximum of 1 μg total DNA and 9 μl Lipofectamine are used in the transfection of each 35-mm dish. Mock transfections are performed with pAR-G and a balancing amount of a pGEM1 vector lacking an insert to keep the amounts of transfected DNA constant and express VSV G alone. Under the experimental conditions used, transfection efficiencies are >90% and all cells are cotransfected. After 4.5 hr, transfected BHK cells are incubated for 30 min in medium without methionine and cysteine and metabolically labeled with Tran[35]S-label (ICN Biomedicals Inc., Irvine, CA) for 10 min (100 μCi per 35-mm dish) at 37°. Subsequently, the cells are transferred to complete G-MEM and incubated for various lengths of time (0–2 hr) at 37°. The cells are then transferred to ice and subjected to surface biotinylation using sulfo-NHS-SS-biotin (Pierce Biochemical, Rockford, IL) as described.[9] Briefly, cells are washed with cold PBS$^+$ and incubated three times for 10 min each with fresh 1 mg/ml sulfo-NHS-SS-biotin (in PBS$^+$) on ice with agitation. Incubating twice for 10 min each with 50 mM glycine in PBS$^+$ quenches the reaction. For analysis, cells are lysed in 500 μl RIPA buffer [1% Nonidet P-40, 0.5% deoxycholate, 0.1% sodium dodecyl sulfate (SDS), 50 mM Tris-HCl, pH 7.4, 150 mM NaCl] with a protease inhibitor cocktail PB/CLAP [1 mM phenylmethylsulfonyl fluoride (PMSF), 1 mM benzamidine, and 1 μg/ml each of chymostatin, leupeptin, antipain, and pepstatin A] and 10 mM iodoacetamide (Sigma, St. Louis, MO) added fresh.

RIPA lysates are precleared by centrifugation at 15,000 rpm in an Eppendorf microfuge for 15 min at 4°. A 250-μl aliquot of the supernatant is subjected to quantitative immunoprecipitation of the VSV G protein as follows. The lysate is incubated for 1 hr on ice with mouse monoclonal antibody (MAb) P5D4.[14] A polyclonal rabbit anti-mouse linker antibody is subsequently added (20 μg/ml, final concentration), and incubation continues for 1 hr followed by addition of 40 μl 50% slurry of protein A-Sepharose and rotation at 4° for an additional hour. The immune complexes are collected by centrifugation, washed twice with RIPA buffer, twice with high-salt RIPA buffer (500 mM NaCl), and twice more with 50 mM Tris-HCl, pH 7.4. The immunoprecipitates are boiled in 80 μl 10% SDS to release bound VSV G protein and clarified by a brief centrifugation at 15,000 rpm in an Eppendorf microfuge. Half of the immunoprecipitated sample is reserved for evaluation of the total VSV G protein and the

[13] M. A. Whitt, L. Chong, and J. K. Rose, *J. Virol.* **63**, 3569 (1989).
[14] T. E. Kreis and H. F. Lodish, *Cell* **46**, 929 (1986).

remainder is subjected to streptavidin precipitation to evaluate the amount delivered to the cell surface.

Biotinylated VSV G protein is precipitated from the released immuno-precipitates by diluting an aliquot (40 μl) of each sample 10-fold with RIPA buffer containing fresh PB/CLAP, 10 mM iodoacetamide, and 0.2% BSA and incubating for 1 hr at 4° with 40 μl 50% slurry of streptavidin-Sepharose (Pierce Biochemical, Rockford, IL). The biotinylated VSV G protein is released from the streptavidin-Sepharose by boiling in SDS–PAGE sample buffer and resolved by SDS–PAGE on 10% gels. These samples serve as a measure of the fraction of VSV G protein delivered to the cell surface, while the remaining 40 μl of each immunoprecipitate resolved directly by SDS–PAGE serves as a measure of the total VSV G protein. Rab protein overexpression is confirmed in each sample by immunoblot analysis using antibodies prepared as described.[3,4]

The exocytosis of FPV HA is monitored as outlined for VSV G protein. For these experiments the pTM3-FPV-HA plasmid[15] is used for cotransfections with Rab11a, and rabbit anti-FPV-HA antiserum is used for the immunoprecipitations.

Radiolabeled proteins are quantified by exposing gels to phosphoimager plates and analyzing the photostimulated luminescence on a Molecular Dynamics or Fuji Bioimager equipped with suitable software for quantification. The amount of cell surface VSV G protein is calculated as [biotinylated VSV G protein]/[total VSV G protein] × 100. The amount of cell surface HA is calculated as [biotinylated HA_1 + HA_2]/[total HA_0 + HA_1 + HA_2] × 100. The amount of cleaved HA is calculated as [total HA_1 + HA_2]/[total HA_0 + HA_1 + HA_2] × 100.

Rab11a Function in TGN–Endosome Meeting Assay. There is accumulating evidence that a fraction of newly synthesized molecules exits the Golgi and traverses through endosomes en route to the cell surface. This has been demonstrated to be the case for transferrin receptor, asialoglycoprotein receptor H1, and Semliki Forest virus p62 precursor using cofractionation or immunoisolation protocols,[16–18] thought to contribute to their endosomal-based sorting. We used horseradish peroxidase (HRP)-DAB-mediated cross-linking of endosomes to show that a fraction of newly synthesized VSV G, which also has a classical tyrosine coated pit internalization motif, is transported to the cell surface via endosomes and Rab11a

[15] K. Takeuchi and R. A. Lamb, *J. Virol.* **68,** 911 (1994).
[16] C. E. Futter, C. N. Connolly, D. F. Cutler, and C. R. Hopkins, *J. Biol. Chem.* **270,** 10999 (1995).
[17] B. Leitinger, A. Hille-Rehfeld, and M. Spiess, *Proc. Natl. Acad. Sci. U.S.A.* **92,** 10109 (1995).
[18] M. Sariola, J. Saraste, and E. Kuismanen, *J. Cell Sci.* **108,** 2465 (1995).

regulates transport along this pathway (W. Chen and A. Wandinger-Ness, unpublished data, 1999).

Transient overexpression studies are performed using the T7 RNA polymerase recombinant vaccinia virus expression system as described.[9,11] Equal amounts of the plasmids encoding VSV G protein (pAR-G) and Rab11a or Rab8 are used for cotransfection experiments. Mock transfections are performed with pAR-G and a balancing amount of a pGEM1 vector lacking an insert to keep the amounts of transfected DNA constant and express VSV G alone. After 4.5 hr, transfected BHK cells are incubated for 30 min in medium without methionine and cysteine and metabolically labeled with Tran[35]S-label (ICN Biomedicals Inc.) for 10 min (100 μCi per 35-mm dish) at 37°. Subsequently, the cells are transferred to complete G-MEM and incubated at 20° for 90 min to accumulate newly synthesized VSV G proteins in the TGN. The cells are then incubated at 37° for 30 min during which time HRP (5 mg/ml in complete G-MEM) is internalized. The cells are then transferred to ice and extensively washed with PBS[+]. Under the conditions used, the internalized HRP is confined to endocytic compartments and does not reach the Golgi.[19] The cells are scraped in PBS[+] using a windshield wiper blade to minimize cell damage. Cells from each dish are then divided into two equal aliquots, one aliquot is treated with DAB (0.4 mg/ml, final) and hydrogen peroxide (0.0225% final) for 1 hr at 4° in the dark to induce cross-linking,[20] and the second aliquot is left untreated as a control. Cells are pelleted by centrifugation and then lysed with the addition of Triton X-100. The cross-linked products are removed by centrifugation for 30 min at 15,000 rpm in an Eppendorf microfuge and the amount of VSV G protein in the supernatant fractions is analyzed by quantitative immunoprecipitation as described in the previous section. The immunoprecipitates are resolved by SDS–PAGE and the amount of radiolabeled VSV G protein in each sample is quantified by exposing gels to phosphoimager plates and analyzing the photostimulated luminescence. The amount of newly synthesized VSV G protein in endosomes is determined by comparing the DAB treated to the untreated samples. The endosomal fraction of VSV G protein is calculated as 100 − [VSV G protein in treated (HRP/DAB) aliquot]/[VSV G protein in untreated (HRP only) aliquot] × 100.

Conclusion

To assess the localizations and functions of Rab GTPases in exocytic membrane transport it is advisable to employ a combination of morphologic

[19] B. Press, Y. Feng, B. Hoflack, and A. Wandinger-Ness, *J. Cell Biol.* **140,** 1075 (1998).
[20] P. J. Courtoy, J. Quintart, and P. Baudhuin, *J. Cell Biol.* **98,** 870 (1984).

and biochemical analyses. Together such analyses discern alterations in Rab protein localization and disruptions in organelle or cytoskeletal architecture, identify the transport step for which a given Rab protein is required,and enable the quantitative evaluation of stimulatory or inhibitory effects exerted by mutant or wild-type Rab protein overexpression.

Acknowledgments

A.W.N. gratefully acknowledges support from an NSF CAREER grant (MCB-9507206, MCB-9996127), an award from NIDDK (R01-DK50141), start-up funds from a HHMI Research Resources for Medical Schools Award (76296-550501), and a NATO travel award (CRG 940772).

[19] Expression and Properties of Rab7 in Endosome Function

By Yan Feng, Barry Press, Wei Chen, Jay Zimmerman, and Angela Wandinger-Ness

Introduction

Rab7 is a small GTPase associated with late endosomes.[1] Our laboratory has devised a series of morphologic and biochemical assays to study the functions of Rab7 in late endocytic membrane transport. Initial morphologic evaluations demonstrated that Rab7 was required for transport from early to late endosomes in mammalian fibroblasts.[2] However, there remained some question regarding the function of Rab7 in later transport events.[3–5] A new biochemical assay based on the cleavage of vesicular stomatitis virus (VSV) G protein is detailed below and offers evidence that Rab7 functions exclusively in regulating transport from early to late endosomes with no apparent role in later steps. Rab7 regulates transport by interfacing with a phosphatidylinositol 3-kinase (PI3K) and the microtubule cytoskeleton as revealed by combinations of morphologic and biochemical assays. Because Rab7 mutants selectively inhibit late endocytic transport, systems for the

[1] P. Chavrier, R. G. Parton, H. P. Hauri, K. Simons, and M. Zerial, *Cell* **62**, 317 (1990).

[2] Y. Feng, B. Press, and A. Wandinger-Ness, *J. Cell Biol.* **131**, 1435 (1995).

[3] R. Vitelli, M. Santillo, D. Lattero, M. Chiariello, M. Bifulco, C. B. Bruni, and C. Bucci, *J. Biol. Chem.* **272**, 4391 (1997).

[4] S. Méresse, J. P. Gorvel, and P. Chavrier, *J. Cell Sci.* **108**, 3349 (1995).

[5] A. Mukhopadhyay, K. Funato, and P. D. Stahl, *J. Biol. Chem.* **272**, 13055 (1997).

conditional expression of mutant Rab7 proteins proved invaluable for determining the consequences of blocking Rab7-mediated transport.[6] Thus, the importance of Rab7 function in lysosome biogenesis and the transit of newly synthesized lysosomal hydrolases to lysosomes were revealed. This chapter describes the methodologies we have devised for dissecting late endocytic membrane transport and identifying the requisite regulatory molecules involved.

Methods and Assays

Construction of Mutants and Expression Vectors

Plasmid vectors encoding wild-type or mutant forms of Rab7 were constructed as described: pGEMRAB7-wt and -rab7N125I[1] and pGEM-rab7T22N.[2] pGEMrab7Q67L was generated by PCR-mediated site-directed mutagenesis using the following primer: AGCAGGCCTGGAACGGTT-CCA (changed base is underlined). Green fluorescent protein (GFP)–Rab7 chimeric proteins under the control of a human cytomegalovirus (CMV) promoter were made by subcloning the cDNA fragments from the corresponding pGEM-rab7 constructs in frame into the pEGFP-C3 vector (Clontech, Palo Alto, CA). For the generation of stable cell lines expressing Rab7, the entire coding regions of wild-type Rab7 and the various mutants were cloned into the XbaI site of the tetracycline-inducible expression plasmid pUHD10-3.[7]

Morphologic Assays for Monitoring Late Endocytic Membrane Transport and Interconnections with Cytoskeleton

Background. GFP chimeras have proven extremely useful for monitoring molecular dynamics in live cells.[8] GFP–Rab7 chimeras were generated as probes for purposes of analyzing late endosome–cytoskeleton connections. The GFP–Rab7 chimeras are concurrently being used to study late endocytic transport in real time by tracing the flux of fluorescently tagged endocytic tracers through Rab7-positive endosomes. It is well established that intact actin and microtubule networks are critical for endocytosis.[9] Although the Rab proteins have classically described roles in membrane docking and fusion (reviewed in Ref. 10), recent studies using GFP–Rab5

[6] B. Press, Y. Feng, B. Hoflack, and A. Wandinger-Ness, *J. Cell Biol.* **140,** 1075 (1998).
[7] M. Gossen and H. Bujard, *Proc. Natl. Acad. Sci. U.S.A.* **89,** 5547 (1992).
[8] K. F. Sullivan and S. A. Kay, eds., "Green Fluorescent Proteins," *Meth. Cell Biol.,* Vol. 58. Academic Press, London, 1999.
[9] M. P. Sheetz, *Eur. J. Biochem.* **262,** 19 (1999).
[10] J. S. Rodman and A. Wandinger-Ness, *J. Cell Sci.* **113,** 183 (2000).

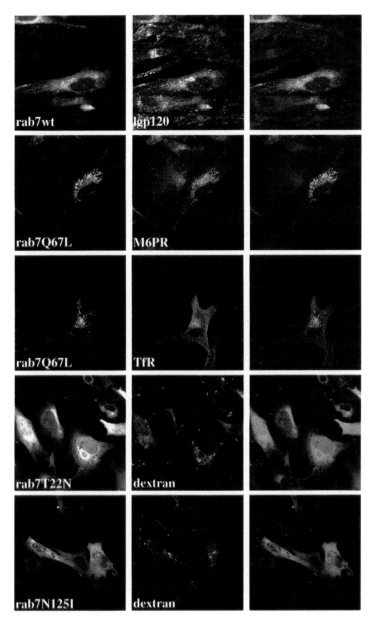

Fig. 1. Expression and localization of GFP–Rab7 chimeras. BHK cells were transfected with plasmids encoding GFP–Rab7 chimeric proteins and in some cases a plasmid encoding human transferrin receptor.[34] Six hours posttransfection some cells were allowed to internalize Texas Red-dextran for 12 hr. They were then washed and incubated for an additional 6 hr in medium without Texas Red-dextran to allow the dextran to accumulate in late endosomes and lysosomes. Eighteen to 24 hr after transfection, cells were fixed and processed for immuno-fluorescence microscopy. Images are 0.4-μm confocal sections taken on a Zeiss LSM 410. Left-hand panels show GFP–Rab7 wild-type (Rab7wt) and mutant protein (Rab7Q67L, Rab7T22N, Rab7N12I) localization visualized in the FITC (green) channel. Middle panels

(over)

show Texas Red-dextran (dextran, Molecular Probes, Eugene, OR) labeling or staining with antibodies against cation independent mannose 6-phosphate receptor (M6PR[35]), lysosomal membrane glycoprotein 120 (lgp120, monoclonal antibody a gift from J. Gruenberg, Université de Genève, Switzerland), or human transferrin receptor (TfR, polyclonal antibody a gift from S. Pfeffer, Stanford University, Stanford, CA) and Texas Red-labeled secondary antibodies (Vector Laboratories, Burlingame, CA). Right-hand panels show the extent of colocalization, seen as yellow in the merged images.

chimeras have uncovered a direct role for rab proteins in the regulation of membrane transport along the cytoskeleton.[11] The known dependence of early to late endocytic transport on functional Rab7[2] and an intact microtubule network[12] warranted an evaluation of the translocation of Rab7-positive endosomes along the cytoskeleton as detailed below.

Analyses of GFP Chimeras. GFP–Rab7 chimeras were constructed and immunofluorescence studies confirmed that both the wild-type and mutant chimeras exhibited a localization similar to that of their nonchimeric counterparts (not shown). For all analyses baby hamster kidney (BHK) cells were transiently transfected[12a] with the GFP-rab7 chimeras using Lipofectamine according to the manufacturer's instructions (Gibco-BRL, Gaithersburg, MD).[12b] Wild-type and Rab7Q67L chimeras displayed partial overlap with late endocytic and lysosomal markers (mannose 6-phosphate receptor and lysosomal membrane glycoprotein 120) and no colocalization with transferrin receptor on recycling endosomes as expected (Fig. 1, top three rows; see color plate). Due to poor prenylation the GFP–Rab7T22N protein exhibits diffuse staining (Fig. 1, left panel, row four), while the GFP–Rab7N125I protein is often present in large punctate structures (Fig. 1, left panel, row five). Although the chimeras between GFP and the dominant negative forms of Rab7 (T22N and N125I) were not obviously colocalized with internalized dextran in late endocytic compartments (Fig. 1, bottom two rows), subcellular fractionation studies suggest that the Rab7N125I protein is preferentially associated with low-density early endosomes or endocytic carrier vesicles.[6]

Late endosome dynamics in living cells was evaluated using the GFP-rab7 wild-type chimera. Live cell viewing requires a constant temperature microscope stage. Therefore, a novel stage warming device was built that rapidly changes sample chamber temperature using Peltier technology.[13] Eighteen hours following transfection GFP–Rab7 is sufficiently expressed for video microscopic analyses. At this timepoint, transfected cell samples on coverslips are mounted on the stage warming device and imaged at 2-sec intervals with a digital camera. The acquired images are analyzed using

[11] E. Nielsen, F. Severin, J. M. Backer, A. A. Hyman, and M. Zerial, *Nature Cell Biol.* **1,** 376 (1999).

[12] J. Gruenberg, G. Griffiths, and K. E. Howell, *J. Cell Biol.* **108,** 1301 (1989).

[12a] For detailed transfection and immunofluorescence staining protocols, see W. Chen and A. Wandinger-Ness, *Methods Enzymol.* **329,** [18], (2001) (this volume).

[12b] FuGene (Roche Diagnostics Corp., Indianapolis, IN) offers a new alternative to Lipofectamine-mediated transfection that does not require serum-free conditions and may be preferable on account of ease of use.

[13] Y. Feng and A. Wandinger-Ness, *Elsevier Trends Journals Technical Tips Online,* T01426 (1998).

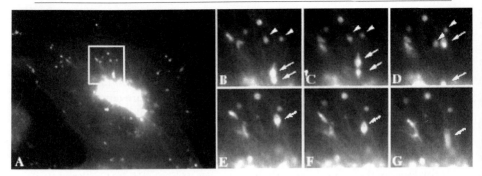

FIG. 2. Late endosome motility on microtubules. BHK cells were cotransfected with expression plasmids encoding GFP–Rab7 wild-type and GFP–tubulin (faint signal). Live cells were imaged at 2-sec intervals with a digital camera. (B)–(G) focus on inset box in (A). In the 4-sec sequence shown in (B)–(D) two endosomes pull apart and one moves rapidly toward the cell periphery (arrows). Arrowheads denote a second pair of endosomes moving in opposite directions. In the 2-sec sequence shown in (E)–(G) the large endosome, which first moved rapidly in the anterograde direction (D), stops and moves in the retrograde direction.

MetaMorph software (Universal Imaging Corp, West Chester, PA). Initial analyses indicate that Rab7-positive endosomes are transported bidirectionally on microtubule tracks (Fig. 2). Microtubules are visualized by cotransfecting cells with GFP–tubulin (Clontech, Palo Alto, CA). Endosome motility is sensitive to microtubule depolymerizing agents and Rab7 mutant protein expression (not shown). Work is in progress to identify the molecular motor(s) controlling transport and to discern the underlying regulation.

Biochemical Assay to Monitor Late Endocytic Membrane Transport and Identify Key Regulatory Factors

Background. VSV G protein is a well-characterized viral glycoprotein that has been widely used for studying exocytic membrane transport processes,[14] but it is also useful as an endocytic tracer. As the primary component of the viral envelope, VSV G protein can be implanted in the cellular plasma membrane by low pH mediated fusion.[12,15] The low pH induces a conformational change in the VSV G protein and causes fusion of the VSV viral envelope with the plasma membrane simultaneously releasing the viral nucleocapsid proteins to the cytoplasm. If the cell surface VSV G protein is subsequently cross-linked with antibody, the protein is synchro-

---

[14] J. E. Bergmann, *Meth. Cell Biol.* **32**, 85 (1989).
[15] J. White, K. Matlin, and A. Helenius, *J. Cell Biol.* **89**, 674 (1981).

nously internalized into the cell and serves as an endocytic tracer.[2,12] Under these circumstances internalized VSV G protein is targeted to lysosomes where it is completely degraded. Prior to complete degradation, specific cleavage fragments have been observed.[12] Such limited proteolytic cleavages have often been associated with the arrival of molecules in late endosomes. For example, proteolytic activation of lysosomal enzyme precursors such as cathepsin D and clipping of major histocompatibility complex (MHC) class II invariant chain are initiated in late endosomes.[16–18] Our laboratory has now fully characterized the sequential degradation of VSV G protein during transit along the endocytic pathway and demonstrated its utility as a quantitative diagnostic assay for discrete late endocytic transport events, as detailed below. Preliminary trials suggest that the cleavage assay may also be conducted on streptolysin O permeabilized cells, making it suitable for elucidating the molecular requirements for late endocytic transport.

VSV G Cleavage Assay. The assay is performed with radiolabeled VSV, obviating the need for immunoprecipitation since VSV G protein, its degradation fragments, and the two viral nucleocapsid proteins, M and N, are the predominant labeled proteins in the cell lysates and these are readily resolved by SDS–PAGE (Fig. 3). To begin, radiolabeled VSV G is allowed to bind to cells and implanted in the cell plasma membrane by brief exposure to low pH. Implanted VSV G protein is then cross-linked by antibody binding and its internalization and degradation monitored after increasing incubation times (0–90 min) at 37°. On internalization, two partial degradation products of VSV G protein are detected. The larger one, termed Ga, appears as a 62,000–60,000 molecular weight doublet, while the smaller Gb fragment is detected at 29,000 molecular weight. Ga is formed within 15 min of internalization, peaks at 30 min, and is consumed by 90 min. Gb is detected within 30 min of internalization and continues to accumulate during a 90-min internalization period. A typical time course of cleavage is shown in Fig. 3. The requirements for each of the two cleavage events have been carefully dissected and shown to occur sequentially in distinct endocytic compartments (Y. Feng and A. Wandinger-Ness, unpublished, 1998). The Ga fragments are formed following exit from early endosomes, likely beginning in endocytic carrier vesicles. Their formation is blocked by conditions that cause the accumulation of endocytic tracers in early endosomes, including the addition of nocodazole or incubation at 15°. Gb is formed by the action of a cysteine protease in lysosomes. This assay is

[16] R. Delbrück, C. Dessel, K. von Figura, and A. Hille-Rehfeld, *Eur. J. Cell Biol.* **64,** 7 (1994).
[17] M. A. Maric, M. D. Taylor, and J. S. Blum. *Proc. Natl. Acad. Sci. U.S.A.* **91,** 2171 (1994).
[18] G. R. Richo and G. E. Conner, *J. Biol. Chem.* **269,** 14806 (1994).

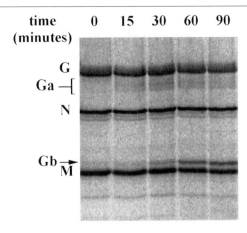

time   0   15   30   60   90
(minutes)

FIG. 3. VSV G cleavage to assay discrete endocytic transport steps. Radiolabeled VSV G protein was implanted in the plasma membrane of BHK cells as described in the text. Following antibody cross-linking, VSV G protein was internalized at 37° for 0–90 min. The degradation of VSV G protein and formation of the Ga (doublet) and Gb fragments were analyzed by resolving samples by SDS–PAGE and autoradiography. The relative migrations of VSV G protein, its cleavage fragments, and two viral nucleocapsid proteins M and N are as indicated.

novel in comparison to other established assays that measure lysosomal arrival based on proteolysis[3] because it is uniquely able to distinguish transport events occurring between intermediate endocytic compartments.

Based on the kinetics of cleavage shown in Fig. 4, Rab7 appears exclusively required for transport from early to late endosomes (measured as Ga formation). A PI3K activity, on the other hand, is necessary for transport both from early to late endosomes and also for transit from late endosomes to lysosomes (measured as Gb formation). There is mounting evidence for a lysosomal cycle, where delivery of cargo to lysosomes is thought to occur through the transient fusion between late endosomes and lysosomes.[19–21] Although, Luzio and colleagues have demonstrated the requirement for a small GTPase in the fusion event, Rab7 was unable to support late endosome–lysosome fusion in vitro.[22] The presence of wild-type Rab7 and the activated Rab7Q67L on lysosomes[4] (Fig. 1) does not imply an active role in this fusion event, because it may simply reflect a population of Rab7 molecules that were present on late endosomes prior to lysosomal fusion.

[19] N. A. Bright, B. J. Reaves, B. M. Mullock, and J. P. Luzio, *J. Cell Sci.* **110**, 2027 (1997).
[20] C. E. Futter, A. Pearse, L. J. Hewlett, and C. R. Hopkins, *J. Cell Biol.* **132**, 1011 (1996).
[21] I. Mellman, *Annu. Rev. Cell Dev. Biol.* **12**, 575 (1996).
[22] B. M. Mullock, J. H. Perez, T. Kuwana, S. R. Gray, and J. P. Luzio, *J. Cell Biol.* **126**, 1173 (1994).

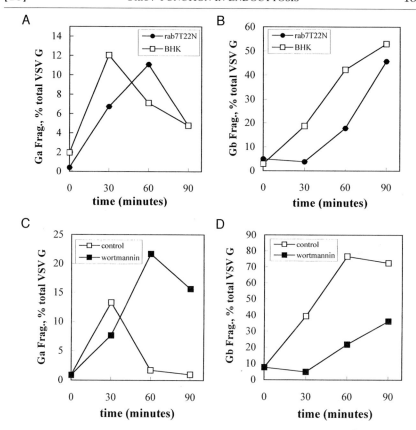

Fig. 4. Rab7 and PI3K requirements in late endocytic membrane transport. Radiolabeled VSV G protein was implanted on the plasma membrane and endocytic transport was monitored by the appearance of specific cleavage fragments. Ga is diagnostic of transport to late endosomes, and Gb measures transport from late endosomes to lysosomes. Expression of the dominant negative Rab7T22N delayed both (A) Ga and (B) Gb cleavages to a similar extent, suggesting that only the first transport step was impaired. Wortmannin (C) delayed Ga formation by 30 min and caused it to accumulate, while (D) Gb cleavage was delayed by nearly 100 min, implying an effect on both transport steps.

Although the identity of the GTPase regulating late endosome to lysosome transport still remains enigmatic, it is anticipated that the above assay will facilitate the testing and identification of further regulatory factors controlling individual late endocytic transport events.

VSV Cleavage Assay Methodologies. The assay is performed with radiolabeled VSV, which is prepared as described.[12] Briefly, four 75-cm^2 flasks of cells (we routinely use BHK) are infected with 10 plaque-forming units

(pfu) VSV for 1 hr. Subsequently, the cells are starved for 1.5 hr in 5 ml/
flask D-MEM containing low methionine concentrations (1.5 mg/liter) and
metabolically labeled with 1 mCi/flask of [^{35}S]methionine and [^{35}S]cysteine
(Tran^{35}S-label, ICN Biomedicals, Inc., Costa Mesa, CA). The medium containing radiolabeled virus is collected after 7.5 hr, clarified by a low-speed
centrifugation to remove cell debris, and loaded onto a glycerol step gradient (consisting of 3 ml 50% glycerol overlaid with 10 ml 10% glycerol,
prepared in 50 mM Tris-HCl, pH 7.4, and 2 mM EDTA). VSV bands at
the 10%–50% glycerol interface and is collected after centrifugation for 90
min at 25,000 rpm and 4° in a Beckman SW28 rotor.

The VSV G cleavage assay is performed as follows. BHK cells (although
not specifically tested, the assay is unlikely cell type specific) are grown to
90% confluence on 3.5-cm dishes. Radiolabeled VSV is added to cells
(2×10^6 cpm/dish) in 500 μl binding medium (G-MEM containing 0.35
g/liter NaHCO$_3$, 60 mg/liter KH$_2$PO$_4$, and 10 mM HEPES–NaOH, pH
7.2) and incubated for 30 min on ice. After two washes with binding medium, the cells are incubated with fusion medium (G-MEM, 10 mM MES,
pH 5.2) for 90 sec at 37°. Cells are then washed once with binding medium and incubated with an excess of monoclonal II (8G5F11)[23] directed
against the ectoplasmic domain of VSV G protein for 30 min on ice.
Unbound antibodies are removed by washing cells three times on ice with
cold binding medium. Subsequently, the cells are warmed to 37° for various lengths of time to allow internalization and degradation of the VSV
G-antibody complexes. For cleavage product analysis, cells are suspended
directly in SDS–PAGE sample buffer and the degradation fragments of
VSV G protein resolved by SDS–PAGE using 12.5% polyacrylamide gels.

Quantification of ^{35}S-labeled samples is accomplished by exposure of
the gels to phosphoimager plates and analysis of the photostimulated luminescence on a Molecular Dynamics phosphoimager or Fuji Bioimager
equipped with appropriate software. The formulas for determining the
percentages of the Ga and Gb fragments are as follows: total = $G + Ga\times$
(67/62) + Gb (67/29); %Ga = Ga$\times$ (67/62)/total $\times$ 100; %Gb = Gb$\times$ (67/
29)/total $\times$ 100.

Cleavage of implanted VSV G protein can also be monitored following
streptolysin O permeabilization. Radiolabeled VSV G is implanted on the
cell surface and cross-linked by antibody binding as above. Cells are then
incubated at 20° for 30 min, allowing accumulation of VSV G protein in
early endosomes. Cells are then incubated with 1.5 U/ml of streptolysin O
for 15 min on ice in IT$^+$ buffer [115 mM KC$_2$H$_3$O$_2$, 20 mM HEPES–KOH,
1 mM Mg(C$_2$H$_3$O$_2$)$_2$, 1$\times$ EGTA-Ca^{2+} (from 10$\times$ stock: 20 mM HEPES–

---

[23] L. Lefrancois and D. S. Lyles, *Virology* **121,** 157 (1982).

KOH, 50 mM EGTA-KOH, 18 mM CaCl$_2$, pH 7.2), and 1 mM DTT], pH 7.2. The cells are washed three times and incubated in prewarmed IT$^+$ buffer at 37° for 7 min. Cytosolic proteins are released by removing the permeabilization buffer and washing cells two times with cold IT$^+$ buffer on ice. The extent of cytosolic protein release is measured as a function of loss of cell associated lactate dehydrogenase (using LDH 500 kit from Sigma, St. Louis, MO). The release of lactate dehydrogenase serves a useful predictor for optimal *in vitro* reconstitution. Maximal dependence on exogenously added cytosol is achieved when no more than 60–70% lactate dehydrogenase is released, conditions under which all cells are permeabilized as judged by vital staining.

For *in vitro* transport studies, streptolysin O permeabilized cells are released from the dishes by scraping into IT$^+$ buffer (1 ml/10 cm dish). Aliquots (250 μl) of the cell suspension are used for each trial. To reconstitute endocytic transport *in vitro,* samples are supplemented with bovine brain cytosol (5 mg/ml) and 1$\times$ ATP regeneration system (1 mM ATP, 10 mM creatine phosphate, 10 U/ml creatine phosphate kinase). Samples are incubated at 37° for 60 min, after which time cells are pelleted by centrifugation and the cell pellets are resuspended in SDS–PAGE sample buffer. The degradation fragments of VSV G protein are resolved by SDS–PAGE on 12.5% gels and quantified as above.

Stable Rab7 Cell Lines to Evaluate Interconnections between Endo- and Exocytic Pathways

Background. Mutant Rab GTPases that inhibit transport at a particular point in the flow of exo- or endocytic traffic are useful for examining their downstream consequences and identifying interconnected pathways. For example, to identify whether newly synthesized lysosomal proteins were transported via a Rab7-regulated pathway a series of stable baby hamster kidney (BHK) cell lines exhibiting inducible expression of wild-type or dominant negative mutant forms of Rab7 were generated.[6] Studies using these cell lines revealed that a majority of cation-independent mannose 6-phosphate receptor and associated ligands were initially transported from the trans-Golgi to early endosomes and functional Rab7 was critical for their transport to later compartments.[6] In contrast, a lysosomal membrane protein was found to transit along a different route. A similar strategy could also be useful for dissecting the requisite pathways involved in MHC class II-mediated antigen presentation and for discerning the interconnections between specialized pathways in polarized neurons and epithelia.

Generation of Stable Cell Lines. Endocytic membrane transport has been well characterized in the BHK cell line and therefore it was chosen for the generation of stable cell lines that conditionally express Rab7.[6] The tetracycline inducible system originally described by Gossen and Bujard[7] was used on account of the tight regulatory control it offers, with nondetectable expression under noninducing conditions (3 μg/ml tetracycline).

Initially a parental cell line was generated that expressed the chimeric tetracycline-regulated transcription transactivator tTA by standard calcium phosphate-mediated transfection of subconfluent BHK21 cells with pUHD 15-1 (6 μg per 6-cm dish). After transfection, the cells were allowed to recover for 24 hr before passage at 1:2.6 and transfer to G-MEM containing 800 μg/ml Geneticin. Viable clones were isolated using cloning rings, expanded in medium containing Geneticin and conditioned medium from subconfluent BHK cells, and tested for their ability to induce the expression of luciferase under the control of a tetracycline-sensitive operator (using pUHC 13-3[7]). Limiting dilution cloning was used to subclone a BHK-tTA cell line that exhibited a 10-fold increase in luciferase after transfection and growth in the absence of tetracycline and <2% of the maximal luciferase activity when cultured in the presence of tetracycline. Obtaining a tightly regulated parental cell line is a critical first step to the successful isolation of subsequent cell lines where expression of the gene of interest is tightly controlled. A number of other cell types expressing tTA and useful as parental lines are now commercially available from Clontech.

To isolate stable lines expressing recombinant Rab7 proteins, subconfluent BHK-tTA cells were subjected to calcium phosphate-mediated transfection with 10 μg pUHD 10-3-Rab7 per 6-cm dish and 1 μg of the pMiwhph plasmid bearing the hygromycin resistance gene.[24] Cells were allowed to recover for 24 hr posttransfection in G-MEM containing 400 μg/ml Geneticin and 3 μg/ml tetracycline prior to a 1:2.6 split and the addition of 400 μg/ml hygromycin B (Calbiochem-Novabiochem, La Jolla, CA). Individual BHK-tTA/R7 clones were isolated and analyzed for overexpression of Rab7 in a tetracycline-regulated manner by immunoblot analysis and immunofluorescence staining. Those clones with levels of recombinant Rab7 protein expression below detection in the presence of tetracycline were subcloned twice by limiting dilution. These stable BHK-tTA/R7 cell lines were routinely maintained in normal G-MEM (5% FCS, 2.6 mg/ml tryptose phosphate broth, glutamine, and penicillin/streptomycin) supplemented with 200 μg/ml hygromycin B, 3 μg/ml tetracycline, and 400 μg/ml Geneticin. Under these conditions recombinant Rab7 protein expression could not be detected. G-MEM containing the selective agents is prepared as needed from single-use aliquots of frozen stock solutions due to the unstable

[24] A. Matsuda, H. Sugiura, K. Matsuyama, H. Matsumoto, S. Ichikawa, and K. Komatsu, *Biochem. Biophys. Res. Commun.* **186,** 40 (1992).

nature of tetracycline in aqueous solution. For some applications it may be advantageous to substitute doxycycline for tetracycline due to its higher stability and effectiveness at lower dosages (20 ng/ml).

Analyses of Cell Lines. Recombinant Rab7 expression was induced by transferring the BHK-tTA/R7 clones to selective GMEM lacking tetracycline or doxycycline. The inclusion of sodium butyrate (5 m*M,* final) on removal of tetracycline or doxycycline can further boost expression of the recombinant protein. Because some lots of fetal calf serum may contain high levels of tetracycline, it is advisable to use serum that has been screened and certified to contain no tetracycline (Clontech). Rab7 protein expression is detectable within 6 hr after transfer to tetracycline-free medium, reaching maximal levels after 48 hr (Fig. 5 and Ref. 6). For all analyses, cells were used 18 hr after transfer, a time point when overexpression levels were still modest (threefold above endogenous).

Rab7 function in lysosome biogenesis was scored using both morphologic and biochemical evaluations of the cell lines.[6] Immunofluorescence staining of cells overexpressing wild-type Rab7 exhibited normal distributions of cation-independent mannose 6-phosphate receptor and the lysosomal membrane glycoprotein (lgp) 120. In contrast, when the dominant

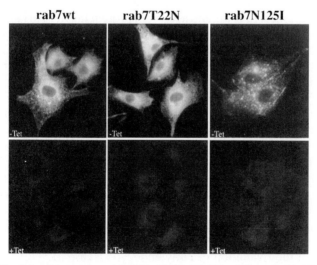

FIG. 5. Conditional expression of Rab7 in BHK-tTA/R7 cell lines. Stable BHK-tTA/R7 cell lines encoding wild-type or mutant forms of Rab7 were cultured in the absence of tetracycline (−Tet) for 24 hr. Duplicate control samples were maintained continuously in media containing 3 μg/ml tetracycline (+Tet). Shown are immunofluorescence micrographs taken on a Zeiss axiophot of representative fields of cells stained with an affinity-purified anti-Rab7 antibody.[36] The antibody was diluted to detect only the overexpressed Rab7 protein.

negative Rab7N125I protein was overexpressed, a significant fraction of the cation-independent mannose 6-phosphate receptor was localized in large peripheral vesicles, while the distribution of lgp 120 was not noticeably altered.

Cell fractionation and pulse-labeling studies were conducted to further evaluate the impact of the mutant rab7N125I expression on the trafficking of molecules to lysosomes. A Percoll gradient fractionation protocol successfully resolved early endosomes from late endosomes and lysosomes[6,25] and in combination with endosome cross-linking experiments revealed that the mutant Rab7N125I protein caused newly synthesized mannose 6-phosphate receptor and its ligand cathepsin D to accumulate in early endosomes.[6] Detailed protocols for both fractionation and DAB cross-linking of HRP-containing endosomes along with numerous control experiments are given in Ref. 6 and, therefore, they have not been included here. Using similar criteria, delivery of newly synthesized lgp 120 was shown to be insensitive to the mutant Rab7N12I. These data were indicative of a post-Golgi divergence in the routes followed by different lysosome-directed molecules. Given the interest in discerning the interconnections between the Golgi and endocytic compartments in specialized cell types, similar strategies could also prove useful in other systems.

The Rab7 cell lines have facilitated a number of other important observations with respect to Rab7 and accessory protein function in late endocytic transport. Coupled with the VSV G cleavage assay (see previous section) the Rab7 cell lines enabled us to pinpoint the precise site of Rab7 function in endocytic transport (see Fig. 4). The cells have also been useful for the identification of other regulatory factors governing late endocytic transport. For example, cell lines expressing wild-type or mutant forms of Rab7 were differentially sensitive to the PI3K inhibitor wortmannin. Expression of the dominant negative Rab7N125I exacerbated the wortmannin-induced dilation of endosomes, while the activated Rab7Q67L alleviated endosome swelling (Y. Feng and A. Wandinger-Ness, unpublished, 1998). This observation provided an initial clue that Rab7 might function synergistically with a PI3K to regulate late endocytic transport, which was further supported by coimmunoprecipitation studies. The details of a molecular interaction between Rab7 and the vps34 PI3K are currently under investigation. A similar interaction between vps34 and rab5 has been demonstrated to be important in the regulation of early endocytic transport.[26]

[25] R.-P. Czekay, R. A. Orlando, L. Woodward, M. Lundstrom, and M. G. Farquhar, *Mol. Biol. Cell* **8**, 517 (1997).
[26] S. Christoforidis, M. Miaczynska, K. Ashma, M. Wilm, L. Zhao, S.-C. Yip, M. D. Waterfield, J. M. Backer, and M. Zerial, *Nature Cell Biol.* **1**, 249 (1999).

Conclusion

Late endocytic membrane transport is a complex process involving sorting of molecules internalized from the cell surface or delivered from the exocytic pathway, recycling of molecules to the trans-Golgi, and flux of molecules to and from lysosomes.[10] It appears that there are multiple intersection points between the exocytic and endocytic pathways,[27] molecules may transit to and from lysosomes by both vesicular and nonvesicular routes,[19,28] and protein and lipid sorting depends on the dynamic formation of multivesicular bodies.[19,29,30] The elucidation of these events at the molecular level is being realized through the combined application of genetic, biochemical, and morphologic approaches in both yeast and mammalian cells.[20,31–33] Assays described in this chapter are useful for monitoring individual late endocytic events and elucidating their regulation by Rab7 and associated factors. Thus, they contribute to the growing arsenal of tools available for pinpointing the site of action of regulatory factors or testing the involvement of particular membrane compartments in a given biologic process.

Acknowledgments

A.W.N. gratefully acknowledges support from an NSF CAREER grant (MCB-9507206, MCB-9996127), an award from NIDDK (R01-DK50141), start-up funds from a HHMI Research Resources for Medical Schools Award (76296-550501), and a NATO travel award (CRG 940772). Mr. Jay Zimmerman (currently a student at University of Chicago Medical School) prepared the GFP–Rab7 chimeras, while conducting an undergraduate research project supported by a NSF REU supplement linked to MCB-9507206.

[27] R. Le Borgne and B. Hoflack, *Biochim. Biophys. Acta* **1404,** 195 (1998).
[28] L. M. Traub, S. I. Bannykh, J. E. Rodel, M. Aridor, W. E. Balch, and S. Kornfeld, *J. Cell Biol.* **135,** 1801 (1996).
[29] F. Gu and J. Gruenberg, *FEBS Lett.* **452,** 61 (1999).
[30] T. Kobayashi, F. Gu, and J. Gruenberg, *Semin. Cell Dev. Biol.* **9,** 517 (1998).
[31] C. G. Burd, M. Babst, and S. D. Emr, *Semin. Cell Dev. Biol.* **9,** 527 (1998).
[32] J. P. Gorvel, P. Chavrier, M. Zerial, and J. Gruenberg, *Cell* **64,** 915 (1991).
[33] F. Aniento, N. Emans, G. Griffiths, and J. Gruenberg, *J. Cell Biol.* **123,** 1373 (1993).
[34] W. Chen, Y. Feng, D. Chen, and A. Wandinger-Ness, *Mol. Biol. Cell* **9,** 3241 (1998).
[35] T. Ludwig, H. Munier-Lehmann, U. Bauer, M. Hollinshead, C. Ovitt, P. Lobel, and B. Hoflack, *EMBO J.* **13,** 3430 (1994).
[36] Y. Qiu, X. Xu, A. Wandinger-Ness, D. P. Dalke, and S. K. Pierce, *J. Cell Biol.* **125,** 595 (1994).

[20] Expression, Purification, and Properties of Rab8 Function in Actin Cortical Skeleton Organization and Polarized Transport

By JOHAN PERÄNEN and JOHANNA FURUHJELM

Introduction

Membrane traffic in eukaryotic cells is regulated by Rab proteins that form the largest family of Ras-like small GTPases.[1] Each Rab protein has a specific localization in the cell according to the transport route it controls.[2,3] Rab8 differs from other Rab proteins in its ability to drastically alter the cell shape when expressed in a variety of cells.[4,5] Rab8 was originally implicated in the regulation of constitutive membrane traffic between trans-Golgi network (TGN) and the basolateral surface domain in epithelial cells.[6] However, recent studies suggest that Rab8 has only a minor effect on the transport of newly synthesized membrane proteins to the cell surface in fibroblasts.[5,7] Although Rab8 did not affect the transport kinetics, it promoted a polarized transport of these proteins to the cell surface through reorganization of actin and microtubules.[5] This suggests that Rab8 participates in a membrane traffic route that mediates cell polarization. Aside from Rab8, animal cells also contain a closely related protein called Rab8b.[4] Interestingly, these two proteins show a distinct expression pattern among different tissues and organs.[4]

To better understand the functions of Rab8 in cell morphogenesis we have constructed different expression vectors containing mutants of Rab8 and Rab8b. We show here that expression of wild-type enhanced green fluorescent protein (EGFP)-Rab8b promotes reorganization of actin as previously reported for Rab8.[5] This is associated with a change in the cell shape as compared to pEGFP-C1 expressing cells. Moreover, we also show that coexpression of EGFP-Rab8b with Rab8-67L results in colocalization

[1] V. M. Olkkonen and H. Stenmark, *Int. Rev. Cytol.* **176,** 1 (1997).

[2] P. Chavrier, R. G. Parton, H. P. Hauri, K. Simons, and M. Zerial, *Cell* **62,** 317 (1990).

[3] P. Chavrier, J. P. Gorvel, E. Stelzer, K. Simons, J. Gruenberg, and M. Zerial, *Nature* **353,** 769 (1991).

[4] J. Armstrong, N. Thompson, J. H. Squire, J. Smith, B. Hayes, and R. Solari, *J. Cell Sci.* **109,** 1265 (1996).

[5] J. Peränen, P. Auvinen, H. Virta, R. Wepf, and K. Simons, *J. Cell Biol.* **135,** 153 (1996).

[6] L. A. Huber, S. Pimplikar, R. G. Parton, H. Virta, M. Zerial, and K. Simons, *J. Cell Biol.* **123,** 35 (1993).

[7] W. Chen, Y. Feng, D. Chen, and A. Wandinger-Ness, *Mol. Biol. Cell* **9,** 3241 (1998).

of these proteins on membrane structures. In addition, we describe new bacterial expression vectors that were used to express Rab8/Rab8b and selected deletions of these proteins. Several technical modifications and improvements are also reported that might be of general use in the study of small GTPases.

Construction of Bacterial Expression Vectors

We have described a set of T7 expression vectors useful for expression of toxic proteins in *Escherichia coli*.[8] However, these vectors are not generally suitable for expression of short peptide sequences. To circumvent this problem we constructed His-GST (glutathione *S*-transferase) fusion vectors (pGAT-1 and pGAT-2) that facilitate dual purification and high-level expression of a peptide sequence linked to the C terminus of the His-GST fusion. They also contain a thrombin sequence adjacent to the molecular cloning site to facilitate cleavage of the peptide or the protein from the fusion protein. These vectors have already been used in multiple Rab studies.[5,9,10] However, their construction is for the first time described here.

The pGAT-1 vector is constructed by first cloning an adapter (composed of oligonucleotides)

a1: 5' CT AGT GGA TCT GGT GGT GGT GGC CGC CTG GTT CCG AG CCT ATC GGA TCC ATG GC 3'

a2: 5' TCG AGC CAT GGA TCC GAT AGG CCT CGG AAC CAG GCG GCC ACC ACC ACC AGA TCC A 3'

into the *SpeI/XhoI* cleaved pHAT-1,[8] creating the vector pHAT-3. The adapter provides pHAT-3 with a serine glycine linker and a thrombin cleavage site. The GST gene is amplified by polymerase chain reaction (PCR) with two specific oligonucleotides

5' AGC ACT AGT CAT GTC CCC TAT ACT AGG TTA TTG 3'
5' TCC ACT AGT TGG AGG ATG GTC CGG ACC 3'

from pGEX-3X (Pharmacia, Piscataway, NJ) using Dynazyme polymerase according to the manufacturer (Finnzymes, Espoo, Finland). This fragment is cloned into the *SpeI* site of pHAT3 to create pGAT-1. We also construct another variant of this vector called pGAT-2 by cloning an adapter

5' GGA TCC ATG GAT ATC GAA TTC GTC GAC CTC GAG GGA TCC GGG CCC TCT AGA TGC GGC CGC ATG CAT A 3'
5' AG CTT ATG CAT GCG GCC GCA TCT AGA GGG CCC GGA TCC CTC GAG GTC GAC GAA TTC GAT ATC CAT GGA TCC 3'

[8] J. Peränen, M. Rikkonen, M. Hyvönen, and L. Kääriäinen, *Anal. Biochem.* **236,** 371 (1996).
[9] H. Stenmark, G. Vitale, O. Ullrich, and M. Zerial, *Cell* **83,** 423 (1995).
[10] F. A. Barr, *Curr. Biol.* **9,** 381 (1999).

into the *Stu*I–*Hin*dIII sites of pGAT-1. The pGAT-2 vector has a larger MCS and a thrombin cleavage site identical with that of pGEX-2T (Pharmacia).

We also construct a HA-tag vector based on pGEM-3 (Promega, Madison, WI). An adapter of two oligos

H1: 5′ AA TTC ACC ATG GGC TAC CCA TAC GAC GTC CCA GAC
     TAC GCC AGC GGC CAT ATG G 3′
H2: 5′ GA TCC CAT ATG GCC GCT GGC GTA GTC TGG GAC GTC
     GTA TGG GTA GCG CAT GGT G 3′

encoding the HA-tag is made and this is cloned into the *Eco*RI/*Bam*HI site of pGEM-3 to create the vector pHA-7. This vector is suitable for preliminary characterization of cloned genes, because it can be used to check the open reading frame by *in vitro* translation as well as the cellular localization by the T7-based vaccinia expression system.[11]

Cloning and Mutagenesis of Mouse Rab8b

Total RNA from mouse NIH 3T3 cells is isolated using the Qiagen Total RNA kit. This RNA was used to make cDNA (20 μl) with an oligo(dT) primer using the Superscript II reverse transcriptase according to the manufacturer (Life Technologies, Rockville, MD). Two oligonucleotides are synthesized: start1

5′ ACA ACA CAT ATG GCG AAG ACG TAC GAT TAT C 3′

and stop1

5′ AAC CAA GCT TAA AGC AGA GAA CAC CGG AAG AA 3′

based on the published sequence of rat Rab8b.[4] These oligonucleotides are used in PCR to amplify the mouse Rab8b coding region from the NIH 3T3 cDNA. The PCR mix contains 2 μl cDNA, 800 μM dNTP, 1× Dynazyme buffer (10 mM Tris-HCl, pH 8.8, 1.5 mM MgCl$_2$, 50 mM KCl, 0.1% Triton X-100) and 2 U of Dynazyme II polymerase (Finnzymes Ltd.). After 1 min of denaturation at 95°, 35 cycles are done under the following conditions: 1 min, 94°; 1 min, 55°; 3 min, 72°. The last elongation is done for 10 min at 72°. The obtained fragment is cloned into the *Nde*I/*Hin*dIII site of pHA-7 to create pHA-Rab8b. The sequence is verified from three independent clones by an automated DNA sequencer (A.L.F.; Pharmacia).

We found altogether seven differences in the coding sequence of mouse

---

[11] T. R. Fuerst, E. G. Niles, F. W. Studier, and B. Moss, *Proc. Natl. Acad. Sci. U.S.A.* **83,** 8122 (1986).

Rab8b compared to rat Rab8b (rat Rab8b/mouse Rab8b; 78 C/A; 222 A/G; 402 A/G; 459 A/G; 525 A/G; 570 A/G; 579 A/G). Nonetheless, the amino acid sequence of mouse Rab8b is identical with that of rat. Thereafter, we create a dominant negative mutant (T22N) and an activated mutant (Q67L) of the Rab8b coding region by site-directed mutagenesis. For Rab8b-T22n we make oligonucleotide 22N

5′ GGA ACA GGA GGC AGT TCT TGC CAA CGC 3′

and for Rab8b-Q67L oligonucleotide 67L

5′ GTT CGG AAT CTT TCT AGA CCC GCC GTG TCC 3′

These oligonucleotides are used in combination with start1 and stop1 oligonucleotides to create corresponding mutant genes according to a megaprimer-based PCR method.[12] The mutated genes are thereafter cloned into *Nde*I/*Hin*dIII site of pHA-7 to create pHA-Rab8b-22N and pHA-Rab8b-67L. Both genes are verified by sequencing.

Cloning Rab8b into pEGFP Vectors

To express GFP-fused Rab8b, we first clone Rab8b into the *Bgl*II/*Eco*RI site of pEGFP-C1 (Clontech). This construct gives a nonspecific localization of the protein and, hence, is not functional. We reasoned that Rab8b was joined too close to the EGFP protein. To avoid this we made a new construct that contained an alanine linker region between these two proteins. Oligo ala5

5′ AAC AAG ATC TGC AGC GGC AGC TGC ACA TAT GGC GAA
GAC GTA CGA TTA TC 3′

and oligo stop1 (see above) are used to amplify the ala5-Rab8 gene from pHA-Rab8b by PCR. The fragment is then cloned into the *Bgl*II/*Hin*dIII site of pRAT-5[8] to create pRAT-ala5-Rab8b. This vector is cleaved with *Nde*I/*Hin*dIII to delete the Rab8b, leaving the alanine linker. The original sequenced Rab8b and corresponding mutants are then cloned back into this *Nde*I/*Hin*dIII cleaved pRAT-ala5-Rab8b vector to create (pala5-Rab8b). The idea of replacing the PCR product with the original Rab8b genes avoids the sequencing of the whole gene. Only the alanine encoding region must be sequenced. This strategy is very useful when large genes are to be provided with new tags or restriction sites by PCR, because DNA sequencing can be minimized. Next ala5-Rab8b is excised by *Bgl*II and *Hin*dIII from pala5-Rab8b and cloned into the *Bgl*II/*Hin*dIII site of

---

[12] O. Landt, H. P. Grunert, and U. Hahn, *Gene* **96,** 125 (1990).

pEGFP-C1 to obtain pEGFP-a5-Rab8b. When this vector is used for expression in animal cells, EGFP-Rab8b localized as Rab8 proteins expressed without tags. Moreover, it also had a clear effect on the cell shape and actin dynamics (see Figs. 1 and 2).

The good experience of introducing an alanine linker prompted us to modify the original pEGFP1-C1. Two oligonucleotides:

G1:

5′ GA TCA GCA GCT GCA GCT GCA GGA GAT ATC AGA TCT
   GAA TTC ACT AGT CTC GAG AAG CTT C 3′

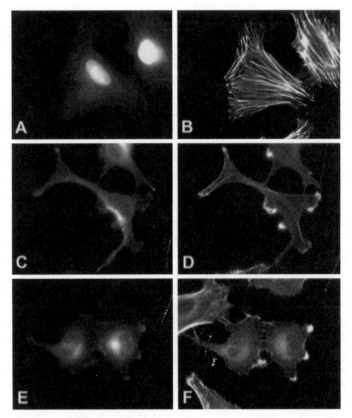

Fig. 1. Expression of Rab8 and Rab8b results in reorganization of actin. (A, B) HeLa cells transfected with the pEGFP-C1 control vector contained actin stress fibers and showed nuclear accumulation of EGFP (C, D), whereas contrast expression of Rab8-67 (C) or EGFP-Rab8b (E) leads to reorganization of actin in HeLa cells (D, F).

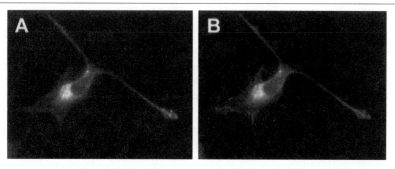

Fig. 2. Coexpression of Rab8-67L and EGFP-Rab8b. Cells were cotransfected with pCMV-Rab8-67L and pEGFP-Rab8b, and labeled with affinity purified anti-Rab8 antibodies. The transfected cells had long protrusions that contained vesicles where Rab8 (A) and EGFP-Rab8b (B) colocalized.

G2:

5′ AA TTG AAG CTT CTC GAG ACT AGT GAA TTC AGA TCT
GAT ATC TCC TGC AGC TGC AGC TGC T 3′

are synthesized to create an adapter. This adapter, which contains an alanine linker and new cloning sites, is introduced into the *Bgl*II-*Eco*RI sites of pEGFP-C1 to create pEGFP-C1A.

Production of Rab8b Hypervariable Peptide for Immunization

To produce antibodies against Rab8b we chose to express the hypervariable region of it in *E. coli*. The hypervariable region is amplified by PCR using oligo C

5′ CAA GGA TCC GAT ATA ATG ACA AAA CTC AAC 3′

and oligo stop1 (see above) from plasmid pHA-Rab8b. The amplified fragment is cleaved with *Bam*HI and *Hin*dIII and cloned into the *Bam*HI/*Hin*dIII sites of pGAT-2 to obtain pGAT-Rab8b-ct. For expression of recombinant proteins we use either the JM109(DE3) (Promega) or the BL21(DE3) (Novagen) strain of *E. coli*. However, the recombinant protein His-GST-Rab8b-ct is unstable in JM109(DE3). Thus we use only the BL21(DE3) strain for expression studies described here.

An overnight culture of BL21(DE3) containing pGAT-Rab8b-ct is diluted 1/50 in 200 ml Luria–Bertani (LB) medium containing ampicillin (100 μg/ml). The cells are grown at 37° to $A_{600\,\text{mn}}$ 0.6–0.8 and induced by adding 0.5 mM isopropylthio-β-D-galactoside (IPTG). After 3 hr they are harvested by centrifugation at 8000 rpm for 20 min at 4°. The pellet is then dissolved in 7 ml lysis buffer [20 mM Tris-HCl, pH 8.0, 0.1% Triton X-100, 0.5 mM phenylmethylsulfonyl fluoride (PMSF)] and lysed by a French press. The lysate is centrifuged at 12,000 rpm for 30 min at 4°. The supernatant is adjusted by adding NaCl to a final concentration of 400 mM. The His-GST-Rab8b-ct protein is affinity purified by using Talon beads according to the manufacturer (Clontech). The purity of His-GST-Rab8b-ct is checked by SDS–PAGE and the yield is 5–10 mg. The protein is dialyzed overnight against phosphate-buffered saline (PBS). The protein is aliquoted and stored at −80°.

Two rabbits are injected subcutaneously on four different sites. The first injection contains 400 μg His-GST-Rab8b-ct in 1 ml Freund's complete adjuvant (Sigma, St. Louis, MO). After 4 weeks each rabbit obtains 300 μg of antigen in Freund's incomplete adjuvant (Sigma). This injection is repeated 2 weeks later. One week after this, the rabbits are tested for antibody production by screening the antisera by immunofluorescence microscopy against pEGFP-Rab8b transfected cells (see below). The rabbits are boosted twice at 2-week intervals and bled 1 week after the final boost. The blood is incubated for 1 hr at 37° and then at 4° overnight in a 50-ml Falcon tube containing a long wooden stick. After removing the coagulated clot around the stick, the serum is spun at 4000g for 15 min at 4°. The serum is then stored in aliquots at −20°. Characterization of the anti-Rab8b antibodies show that they react with Rab8b, but not with Rab8.

Affinity Purification of Antisera

The buffer of purified His-GST, His-GST-Rab8ct, and His-GST-Rab8b-ct is changed to binding buffer (0.1 M NaHCO$_3$, pH 8.3, 0.5 M NaCl) by passing it twice through a PD-10 column (Pharmacia). The proteins are then bound to 330 mg CNBr-activated Sepharose according to the manufacturer (Pharmacia). We dilute 1 ml of antisera with 4 ml PBS. Because the antisera contains antibodies against GST, we preincubate it with GST beads for 1 hr at 4°. Unabsorbed antibodies are then incubated with His-GST-Rab8ct or His-GST-Rab8b-ct beads by rotation overnight at 4°. The beads are put into a column and washed twice by passing 10 ml PBS through the column. The antibodies are eluted with 0.1 M glycine, pH 2.4. One-milliliter fractions are collected and neutralized by adding 30 μl 3 M Tris-HCl, pH 8.8. We also add 20 μl 5 M NaCl, 100 μl of 10 mg/ml bovine serum albumin (BSA), and glycerol to a final concentration of 50%. The affinity purified antibodies are divided into aliquots and stored at −20°. When soluble proteins are not available for making an affinity column, antibodies are affinity purified by proteins blotted onto nitrocellulose strips.[13]

Cell Transfection

High-efficiency transfections (40–90%) of HeLa and HT1080 cells are obtained by using the transfection reagent FuGENE 6 Reagent (Roche, Mannheim, Germany). Cells are plated onto glass coverslips in 35-mm petri dishes 1 day prior to transfection. For HT1080 cells we use coverslips coated with 5 μg/ml fibronectin. The next day 3 μl of FuGENE 6 is mixed with 93 μl of Optimem media (GIBCO, Grand Island, NY) and incubated for 5 min at room temperature. The FuGENE 6 solution is then mixed with 1 μg of appropriate Rab8 vector (pEGFP-Rab8b) or, in the case of double transfection, 0.6 μg of each vector. After a 15-min incubation, the transfection mix is transferred directly onto the 2 ml of growth media overlaying the cells. About 3–4 hr later, the transfection mix is replaced by fresh growth medium and the transfection continued for 16–20 hr. Longer transfection times lower the number of Rab8 expressing cells, probably due to toxicity.

We have shown that Rab8 promotes polarized transport of newly synthesized membrane proteins.[5] The T7 vaccinia system is used to simultaneously introduce both the Rab8 gene and the VSV-G gene. However, a rapid high expression of Rab8 is difficult to control due to drastic changes in cell shape. Thus, we have changed the original protocol, so that we now use cells that have been transfected with a Rab8 construct 1 day prior to the experiment (see above). The cells, which express modest levels of Rab8, are transfected by the T7 vaccinia system or the Semliki Forest virus system to introduce the genes encoding the membrane proteins to be studied as previously described.[5,11,14]

Analysis of Rab8 by Confocal and Immunofluorescence Microscopy

After transfection the cells are fixed by freshly made 4% paraformaldehyde in PBS buffer for 15 min at room temperature. Free aldehyde groups are blocked by incubating the cells for 10 min in PBS containing 20 mM glycine. The cells are permeabilized with 0.1% Triton X-100 for 10 min at room temperature. Then the cells are washed twice with DPBS (Dulbecco's PBS containing 0.5% BSA). Affinity purified anti-Rab8 (1:300), together with phalloidin Texas Red-X (1:300; Molecular Probes, Eugene, OR) or

---

[13] J. Peränen, *Biotechniques* **13**, 546 (1992).
[14] P. Liljeström and H. Garoff, *Biotechnology* **9**, 1356 (1991).

phalloidin Oregon Green 488 (1:300; Molecular Probes), is diluted in DPBS. This mix is applied to the cells for 30 min at room temperature. The cells are then washed four times during 30 min with DB buffer. Secondary antibodies, either anti-rabbit Lissamine rhodamine (Jackson, West Grove, PA) or anti-rabbit FITC (Jackson) is diluted according to the manufacturer in DPBS buffer. The secondary antibodies are applied to the cells for 30 min at room temperature. Cells are washed free of secondary antibodies as above. After longer periods of storage, antibody dissociation may occur. To avoid this, we postfix the cells with 4% paraformaldehyde in PBS for 5 min at room temperature, after which the cells are washed twice with DPBS. We have noticed that postfixation not only preserves the samples better, but also gives a stronger signal and less background.

After a rinse in distilled water the coverslips are mounted with 50% (v/v) glycerol in PBS onto object glasses. In some experiments we use 0.1% saponin instead of Triton X-100 to permeabilize the cells, but in such cases saponin is also included in all steps where the DPBS was used except for the final wash before postfixation. Saponin has also been used prior to primary fixation to get rid of background staining of overexpressed Rab proteins.[6,15] However, we have noticed that saponin pretreatment drastically alters cell and organelle structure, especially the plasma membrane and the reticular endoplasmic reticulum. Thus, we think that this procedure must be used with caution, and only in cases where other methods do not work. Paraformaldehyde fixation is superior in preserving cell architecture compared to methanol and acetone-based fixatives. However, paraformaldehyde may mask antigenic sites. This problem can be partially circumvented by a short incubation in guanidine hydrochloride as previously described.[16]

[15] M. Zerial, R. Parton, P. Chavrier, and R. Frank, *Methods Enzymol.* **219,** 398 (1992).
[16] J. Peränen, M. Rikkonen, and L. Kääriäinen, *J. Histochem. Cytochem.* **41,** 447 (1993).

[21] Properties of Rab13 Interaction with Rod cGMP Phosphodiesterase δ Subunit

By Anne-Marie Marzesco, Thierry Galli, Daniel Louvard, and Ahmed Zahraoui

Introduction

Rab proteins constitute a distinct branch (more than 40 Rab proteins are known) of the Ras superfamily. They are closely related to the yeast Ypt1 and Sec4 gene products. They bind GDP/GTP and possess a weak intrinsic GTPase activity. Rab proteins are involved in the regulation of endocytic and exocytic vesicular transport.[1] Although the Rab proteins share highly conserved domains, in particular regions involved in GDP/GTP switch, they may have diverse functions. Several putative effectors that preferentially bind to the GTP-bound form of Rab proteins have been identified. These effectors are structurally distinct from each other and are involved in diverse cellular processes.[2–7] To fulfill their function, Rab proteins cycle between cytosolic and membrane-bound forms. Association to membranes/dissociation is strictly controlled by Rab accessory proteins (GDI, GEF, GAP, etc.).

This chapter deals with Rab13 and its interaction with the phosphodiesterase (PDE) delta subunit (δPDE). The small GTPase Rab13 protein is highly homologous to the mammalian Rab8, and Rab10 and to the yeast Sec4 protein involved in polarized secretion during the budding process.[8,9] Rab13 associates with vesicles in fibroblasts and accumulates in tight junctions of polarized epithelial cells. This subcellular localization depends on

[1] P. Novick and M. Zerial, *Curr. Opin. Cell Biol.* **9**(4), 496 (1997).

[2] A. Echard, F. Jollivet, O. Martinez, J. J. Lacapere, A. Rousselet, I. Janoueix-Lerosey, and B. Goud, *Science* **279,** 580 (1998).

[3] M. Kato, T. Sasaki, T. Ohya, H. Nakanishi, H. Nishioka, M. Imamura, and Y. Takai, *J. Biol. Chem.* **271,** 31775 (1996).

[4] H. Stenmark, G. Vitale, O. Ullrich, and M. Zerial, *Cell* **83**(3), 423 (1995).

[5] M. Ren, J. Zeng, L. De, C. Chiarandini, M. Rosenfeld, M. Adesnik, and D. D. Sabatini, *Proc. Natl. Acad. Sci. U.S.A.* **93,** 5151 (1996).

[6] Y. Wang, M. Okamoto, F. Schmitz, K. Hofmann, and T. C. Sudhof, *Nature* **388,** 593 (1997).

[7] S. Christoforidis, M. Miaczynska, K. Ashman, M. Wilm, L. Zhao, S.-C. Yip, M. D. Waterfield, J. M. Backer, and M. Zerial, *Nature Cell Biol.* **1,** 249 (1999).

[8] A. Zahraoui, G. Joberty, M. Arpin, J. J. Fontaine, R. Hellio, A. Tavitian, and D. Louvard, *J. Cell Biol.* **124,** 101 (1994).

[9] W. Guo, D. Roth, C. Walch-Solimena, and P. Novick, *EMBO J.* **18**(4), 1071 (1999).

the integrity of tight junctions,[8] which constitute a specialized subdomain of the lateral plasma membrane separating the apical from the basolateral cell surface in epithelial cells. In budding yeast, Sec4 may control the assembly of the exocyst, which targets secretory vesicles to specific sites on the plasma membrane at the bud tip.[9] In mammalian cells, the Sec6/8 subunits of the exocyst complex associate with cell–cell contact in epithelial cells.[10] Our previous results suggested that Rab13 may be involved in docking/fusion of a subset of transport vesicles at specialized microdomains of the lateral plasma membrane.[8] This is reminiscent of early studies showing that aminopeptidase N is recruited to the cell–cell contact before its final localization to the apical membrane.[11] Taken together, these results indicate that the intercellular junctions may constitute a specific targeting site on the plasma membrane for delivery of specialized cargo vesicles. To investigate the function of Rab13, we have searched for its protein partners. We present here different techniques we have used to identify δPDE, a Rab13 interacting protein, and to show that it is required for the recycling of Rab13 from membranes to the cytosol.

Two-Hybrid Screen

There are several well-defined procedures for the identification of proteins interacting with a protein of interest. The yeast two-hybrid system was developed to detect protein–protein interactions and was applied to study a variety of biological functions, such as signal transduction, intracellular transport, and oncogenesis. In recent years, the yeast two-hybrid assay has successfully been used to isolate many Rab interacting proteins.

Strains and Media

The genotype of the *Saccharomyces cerevisiae* L40 is *Mat a his 3Δ200 trp 1-901 leu 2-3112 ade2 LYS2*::(*4lexAop-HIS 3*) *URA 3*::(*8lexAop-lac Z*) *GAL4.* Yeast strains are grown at 30° in different media:
   YPD: For 1 liter (20 g peptone Y, 10 g yeast extract Y, 20 g dextrose + 17 g agar for solid medium)
   DO: For 1 liter (1.7 g yeast nitrogen base, 20 g dextrose, 5 g ammonium sulfate, 17 g agar for solid medium), and amino acid mixture
   DO W⁻L⁻: DO without tryptophan and leucine
   DO W⁻L⁻H⁻: DO minus tryptophan, leucine, and histidine

[10] K. K. Grindstaff, C. Yeaman, N. Anandasabapathy, S. C. Hsu, E. Rodriguez-Boulan, R. H. Scheller, and W. J. Nelson, *Cell* **93**(5), 731 (1998).
[11] D. Louvard, *Proc. Natl. Acad. Sci. U.S.A.* **77**, 4132 (1980).

Plasmids

Two yeast expression vectors are used for the two-hybrid screen: pLex10 plasmid contains the LexA DNA binding domain (LexDBD) under the control of ADH promoter, and pGAD.GH contains the Gal4 activation domain (pGAD) and nuclear localization signal that localizes the hybrid protein to the nucleus for activation of the reporter gene. The reporter genes commonly used encode the yeast His3p and *Escherichia coli* β-galactosidase enzyme. These plasmids contain the origin of replication *trp* (pLex10) or *leu* (pGAD) gene for selection in yeast. The DBD or GAD domains are followed by a multiple cloning site for the in-frame generation of fusion proteins and a nuclear localization signal.

pLexA-Rab13wt, pLexA-Rab13T22N, pLexA-Rab13Q67L, and pLexA-Rab13ΔCaaX plasmids are constructed by inserting the *Eco*RI/*Bam*HI fragments of pCR3.1-Rab13wt, pCR3.1-Rab13T22N, pCR3.1-Rab13Q67L, and pCR3.1-Rab13ΔCaaX, respectively, into the yeast pLex10 vector. The resulting plasmids express Rab13 as fusion proteins to the DNA binding domain of LexA (LexA-Rab13).

Method

The yeast two-hybrid system used in this study has already been described.[12] Briefly, the pLexA-Rab13wt construct is used to screen 3×10^6 clones from a HeLa cDNA library (a gift from Dr. J. Camonis, Institut Curie, Paris, France) fused to the Gal4 activation domain (GAD). The *S. cerevisiae* L40 strain, which contains the reporter genes His3 and LacZ downstream of the binding sequence of LexA, is transformed with pLexA-Rab13wt plasmid. LexA-Rab13 fusion protein expression is checked by Western blot with anti-Rab13 antibodies. Then it is transformed with a HeLa cDNA library fused to Gal4 activation domain by the lithium acetate method:

1. Inoculate 20 ml of an overnight preculture of L40 strain transformed with pLexA-Rab13wt in 80 ml of YPD. Incubate for 3 hr at 30° to get 4×10^7 cell/ml.
2. Centrifuge and wash the pellet with 50 ml of sterile water and 50 ml of 0.1 *M* lithium acetate, pH 7.5. Then resuspend cells in 10 ml of 0.1 *M* lithium acetate and shake carefully for 60 min at 30°. Lithium acetate should be prepared in TE (10 m*M* Tris-HCl, pH 7.5, 1 m*M* EDTA) buffer.

---

[12] M. D. Rose, F. Winston, and P. Hieter, "Laboratory Course Manual for Methods in Yeast Genetics." Cold Spring Harbor Laboratory, Cold Spring Harbor, NY, 1990.

3. Centrifuge and resuspend the yeast pellet in 2 ml of 0.1 M lithium acetate. Mix with 60 μg library cDNA and 40 μg of fish sperm DNA. The total volume of DNA mixture should not exceed 11 μl.
4. After 10 min of incubation at room temperature, add 20 ml of 50% (w/v) polyethylene glycol (PEG)3350/0.1 M lithium acetate. Incubate 1 hr at 30° and 30 min at 42°.
5. Spin down and wash twice with 10 ml of DO W⁻ medium.
6. Resuspend the cells in 2 ml of DO W⁻ and spread 100 μl on large selective plates.

Double transformants are plated on selective medium lacking trypto-phan, leucine, and histidine for selection of His⁺ colonies. The plates are incubated at 30° for 3 days. His⁺ colonies are patched on solid selective medium lacking Trp/Leu/His and assayed for β-galactosidase activity by filter assay using 5-bromo-4-chloro-3-indolyl-β-D-galactoside (X-Gal) as substrate. Library positive clones from transformed yeast are recovered using the bacteria HB101 as a recipient strain, selected on M9 medium minus leucine. The library positive clones are tested for specificity by cotransformation into L40 strain with pLexA-Rab6T27N, pLexA-Rab6Q72L, pLexA-Rab8wt, and pLexA-lamin. (pLexA-Rab6T27N and pLexA-Rab6Q72L are a gift from Dr. B. Goud, Institut Curie, Paris, France.) The specific cDNA clones are sequenced using the Sanger dideoxy-termination method. Using this technique we could identify eight identical clones encoding the δPDE (an 18-kDa protein), which interacts specifically with Rab13. The δPDE/Rab13 interaction requires the geranyl-geranylation of Rab13[13,14] (Table I).

In Vitro Interaction of Human δPDE with Rab13

To confirm the interaction of human δPDE with Rab13, we also studied it biochemically.

Expression of Human δPDE in Escherichia coli

Large amounts of purified δPDE protein are required for biochemical studies and for analyzing its interaction with Rab13. Different methods have been employed to express proteins of interest either in bacteria or in insect cells using baculovirus expression vectors. We use the T7 expression system to produce native δPDE. This system, originally developed by

[13] G. Joberty, A. Tavitian, and A. Zahraoui, *FEBS Lett.* **330,** 323 (1993).
[14] A. M. Marzesco, T. Galli, D. Louvard, and A. Zahraoui, *J. Biol. Chem.* **273**(35), 22340 (1998).

TABLE I
ANALYSIS BY YEAST TWO-HYBRID SYSTEM OF
INTERACTION OF HUMAN δPDE PROTEIN WITH Rab13[a]

Fused to LexA DBD	Fused to Gal4 AD	His +
Ras	Raf	+
Rab13Q67L	None	−
Rab13Q67L	Lamin	−
Rab13WT	δPDE	+
Rab13Q67L	δPDE	+
Rab13T22N	δPDE	+
Rab13ΔCaaX	δPDE	−
Rab8WT	δPDE	−
Rab6Q72L	δPDE	−
Rab6T27N	δPDE	−

[a] An L40 yeast strain is cotransformed with two plasmids, one expressing Rab13 protein as a fusion with LexA DNA binding domain (LexADBD) and the second encoding δPDE wild-type protein as a fusion with Gal4 activating domain (Gal4AD). The Ras/Raf two-hybrid interaction is used as a positive control, the Rab13Q67L/lamin two-hybrid interaction is used as a negative control. Only the activation of the His reporter gene (+) is shown.

Studier *et al.*,[15] allows cloning of target cDNAs at sites where they are selectively transcribed by T7 RNA polymerase *in vitro* and in *E. coli.* Transcription is controlled by the strong ϕ10 promoter of T7 RNA polymerase. The coding region of the protein of interest is cloned downstream of a T7 RNA polymerase promoter and then introduced into an *E. coli* strain containing a chromosomal copy of the T7 RNA polymerase gene under control of the inducible *lacUV5* promoter. Expression of T7 RNA polymerase in this bacteria is induced by addition of isopropylthio-β-D-galactoside (IPTG). Two oligonucleotides have been used to create by polymerase chain reaction (PCR) an *Nde*I restriction site that includes the ATG initiator codon of δPDE coding sequence and a *Bam*HI site downstream of δPDE termination codon. The *Nde*I–*Bam*HI insert containing the complete δPDE coding sequence is cloned into pET-15b (Novagen Inc., Madison, WI). This vector carries a stretch of six consecutive histidine residues adjacent to the cloning site and allows the expression of an N-terminal histidine–δ PDE fusion protein. It also contains a thrombin cleavage site that allows removal

[15] A. H. Rosenberg, B. N. Lade, D. S. Chui, J. J. Dunn, and F. W. Studier, *Gene* **56,** 125 (1987).

of the His tag. However the presence of His tag does not appear to affect the biological activity of most target proteins.

The His tag binds to divalent cations immobilized on metal chelation resin, such as nickel resin Ni-NTA (Qiagen GmbH, Germany) or cobalt resin TALON (Clontech, GmbH, Germany). Under our purification conditions (see below) cobalt beads give better results than nickel beads.

1. The recombinant plasmid pET-15b/ δPDE is introduced into *E. coli* strain DH5α, which is convenient to maintain the plasmid and allows for a high-level plasmid yield.
2. Identify the correct recombinant colonies by PCR using the appropriate primers. Plasmid DNA is prepared from positive colonies and sequenced.
3. Transform with the resulting plasmid *E. coli* strain BL21(DE3)plysS and select for ampicillin (50 μg/ml) and chloramphenicol (12.5 μg/ml)-resistant transformants. This strain is a lysogen bearing the IPTG-inducible T7 RNA polymerase gene. It also bears the plasmid plysS, which carries bacteriophage T7 lysozyme gene and confers resistance to chloramphenicol. T7 lysozyme is a specific inhibitor of T7 RNA polymerase; its presence in a host that carries the inducible gene for T7 RNA polymerase increases the tolerance for toxic target proteins and allows us to efficiently lyse the cells (e.g., freezing and thawing). The analysis of transformants for expression of His-δPDE is carried out by a rapid screening.
4. Pick single colonies of transformants into 5 ml of prewarmed Luria–Bertani (LB) media containing 100 μg/ml of ampicillin and 12.5 μg/ml of chloramphenicol. Grow the cultures overnight at 37°.
5. Inoculate 5 ml of prewarmed LB media containing 100 μg/ml of ampicillin with 500 μl of the overnight culture, and grow 60 min at 37° with vigorous shaking. Inoculate an extra culture to serve as the uninduced control.
6. Induce expression 3 hr by adding 0.4 m*M* IPTG. The cells are harvested by centrifugation 3 min at 12,000*g* at 4°, and resuspended in 100 μl of sodium dodecyl sulfate (SDS) sample buffer. Extracts from uninduced and induced culture are boiled 5 min, and 10 μl of each sample is analyzed for expression of His–δPDE protein on a 15% SDS–polyacrylamide gel. Clones expressing the human His–δPDE protein are identified by Coomassie blue staining.

Large-Scale Production of His–δPDE

1. Inoculate 50 ml of LB media containing 100 μg/ml of ampicillin and 12.5 μg/ml chloramphenicol. Grow the culture overnight at 37°.

2. Inoculate 2 liters culture of 2× YT (16 g tryptone-B, 10 g yeast extract-B, 5 g NaCl per liter) medium containing 100 μg/ml ampicillin with 40 ml of an overnight culture. Grow at 37° for 3 hr with vigorous shaking.

3. Induce the production of the His–δPDE protein with 0.4 mM IPTG for 1 hr at 37°. The cells are harvested by centrifugation at 7000g for 8 min at 4°, washed once with phosphate-buffered saline (PBS), and stored at −80°.

4. Thaw the cell pellet on ice for 15 min and resuspend it in 20 ml of ice-cold lysis buffer [20 mM Tris-HCl, pH 8.0, 100 mM NaCl, 10 mM imidazole, 1% Triton X-100, and protease inhibitor cocktail (Sigma, St. Louis, MO)]. The endogenous lysozyme allows us to efficiently lyse the cells.

5. Sonicate on ice 3× for 1-min intervals with cooling on ice for 30 sec between each sonication. Centrifuge lysate at 30,000 rpm for 30 min at 4° in a 70Ti rotor (Beckman, Palo Alto, CA) to remove unbroken bacteria and debris.

6. A 1-ml bed volume of nickel or cobalt TALON resin is equilibrated with 10 volumes of lysis buffer. The clarified extract is mixed with the resin and gently rotated at room temperature for 30 min, allowing the protein to bind to the resin.

7. Centrifuge at 700g for 5 min and remove as much supernatant as possible. Wash the resin four times with 10 bed volumes of wash buffer (20 mM Tris-HCl, pH 8, 100 mM NaCl, 15 mM imidazole, 1% Triton X-100, and protease inhibitor cocktail). The resin is poured into an empty column, and allowed to pack by gravity flow.

8. Elute the bound His–δPDE fusion protein with a step-gradient of elution buffer (wash buffer with increasing concentration of imidazole 50, 100, 150, 200, and 250 mM). The elute fractions are analyzed by SDS–PAGE and Coomassie blue staining. The fractions containing pure His–δPDE protein are pooled and dialyzed against 200 volumes of buffer [20 mM Tris-HCl, pH 7.5, 100 mM NaCl, protease inhibitor cocktail, 10% (v/v) glycerol] overnight at 4° and stored at −80°.

The His–δPDE protein with a purity exceeding 98% is obtained (Fig. 2A). The protein concentration is determined by the method of Bradford using bovine serum albumin (BSA) for calibration.

Rab 13-Coupled Transcription/Translation in Vitro

To analyze the interaction of Rab13 with δPDE protein, it is necessary to produce the isoprenylated form of the small GTPase Rab13. Because bacteria expressed Rab13 protein lacks the C-terminal geranylgeranyl lipid, we use the TNT T7 quick coupled transcription/translation system (Promega, Madison, WI) to produce *in vitro* a geranylgeranylated Rab13 protein. Indeed, rabbit reticulocyte lysate has been reported to contain a variety of posttranslational processing activities including isoprenylation.[16,17]

Rab13 and Rab3A cDNAs are cloned in pCR3.1 plasmid (Invitrogen, Carlsbad, CA) under the control of a T7 RNA polymerase promoter. Transcription and translation in the presence of [^{35}S]methionine (Amersham Pharmacia Biotech, Europe GmbH, Germany) are performed *in vitro* using the TNT T7 quick coupled transcription/translation system. This system incorporates transcription in the translation mix, which reduces the time necessary for *in vitro* protein synthesis. To avoid RNase contamination, all materials used should be RNase free. Mix the following compounds in an Eppendorf tube:

TNT T7 Quick Master Mix	40 μl
[^{35}S]Methionine (1000 Ci/mmol) at 10 mCi/ml	2 μl
pCR3.1/Rab13 DNA template	1 μg
Nuclease-free water to final volume	50 μl

The reaction is incubated at 30° for 90 min and 1–5 μl of translation reaction is subjected to SDS–PAGE. The gel is stained in Coomassie blue solution, destained in 10% (v/v) acetic acid and 10% (v/v) ethanol, then washed twice with water and incubated 1 hr with 1 *M* sodium salicylate (which increases the detection sensitivity of fluorography). The gel is dried and the detection is carried out by X-ray film. As a control for the specificity of Rab13/δPDE binding, we have also synthesized *in vitro* Rab3A. Under these conditions, radiolabeled Rab13 protein translated *in vitro* fractionated into the detergent phase with Triton X-114, suggesting it is isoprenylated.

In vitro Interaction Assay

The capacity of δPDE to bind Rab13 protein is assayed *in vitro*. [^{35}S]Methionine *in vitro* translated Rab13 or Rab3A are incubated with the His–δPDE protein immobilized onto Ni^{2+} beads.

1. A 60-μl bed volume of beads immobilizing His–δPDE fusion protein is equilibrated with buffer A [20 m*M* Tris-HCl, pH 7.5, 150 m*M* NaCl, 5 m*M* MgCl$_2$, 1 m*M* Pefabloc, 0.5% Nonidet P-40 (NP-40)].
2. Then 30 μl of His–δPDE beads is incubated overnight at 4° with 7.5 μl of radiolabeled Rab13 or Rab3A protein in the same buffer, then

---

[16] B. T. Kinsella and W. A. Maltese, *J. Biol. Chem.* **267**, 3940 (1992).
[17] O. Ullrich, H. Stenmark, K. Alexandrov, L. A. Huber, K. Kaibuchi, T. Sasaki, Y. Takai, and M. Zerial, *J. Biol. Chem.* **268**, 18143 (1993).

washed with 3 volumes of buffer A. The proteins retained on the His–δPDE beads are resuspended in SDS sample buffer, boiled 5 min, and analyzed by SDS–PAGE and autoradiography as indicated above. We observed that His–δPDE binds Rab13 and not Rab3A.[14]

Solubilization Assay of Membrane-Bound Rab13

According to the functional Rab model, Rab proteins associate with the membrane of transport vesicles at the donor compartment, which move to, dock, and fuse with the acceptor membrane. On GTP hydrolysis, Rab proteins are recycled back to the donor compartment via a cytosolic inter-mediate.[1] The bovine δPDE protein has been reported to solubilize the isoprenylated α and β catalytic subunits of the retinal rod phosphodiester-

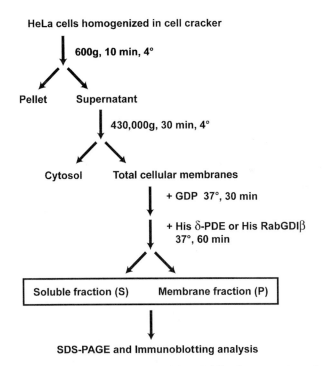

FIG. 1. Flow diagram of cell fractionation and the solubilization assay of membrane-bound Rab13. HeLa cells are homogenized, and cytosolic and pellet fractions are prepared. The pellet is incubated with δPDE or Rab GDIβ isoform and centrifuged. The resulting soluble (S) and membranous fractions (P) are analyzed by SDS–PAGE and immunoblotting as indicated in the text.

A

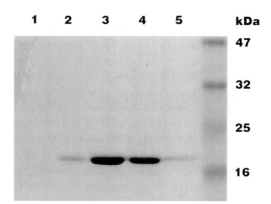

B

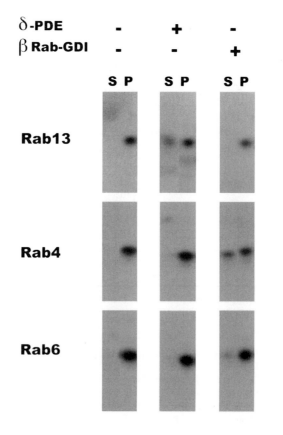

ase.[18] We tested whether the recombinant His–δPDE was able to retrieve Rab13 protein from membranes. As a positive control, we used Rab GDI, which has been shown to dissociate several membrane-bound Rab proteins.[16,19–22]

Preparation of Cellular Membrane Fractions

HeLa cells are grown in Dulbecco's modified Eagle's medium (DMEM) supplemented with 10% fetal calf serum, 2 mM glutamine, 100 U/ml penicillin, and 10 mg/ml streptomycin. The culture is incubated at 37° under 10% CO_2 atmosphere. HeLa cells grown to confluency on 10-cm plastic dishes are washed twice with cold PBS and scraped with a rubber policeman. Cells are homogenized on ice in buffer A (10 mM Tris-HCl, pH 7.4, 0.25 M sucrose, 1 mM MgCl$_2$, 5 mM CaCl$_2$ and protease inhibitor cocktail containing 10 μg/ml of leupeptin and of aprotinin, 1 μg/ml of pepstatin, and 0.5 mM Pefabloc) using a cell cracker. The homogenate is centrifuged at 600g for 10 min to remove nuclei and unbroken cells. The supernatant is centrifuged at 430,000g in a TL 100 rotor (Beckman) for 30 min at 4° to separate a membrane pellet from a cytosol fraction. The protein concentration is determined by the method of Bradford using BSA for calibration. The membrane pellet is used for the solubilization assay of membrane bound Rab13 (Fig. 1).

1. 100 μg of membrane are incubated for 60 min at 30° with either 1 mM GDP or GTPγS in the solubilization buffer (25 mM HEPES,

[18] S. K. Florio, R. K. Prusti, and J. A. Beavo, *J. Biol. Chem* **271**(39), 24036 (1996).
[19] S. Araki, K. Kaibuchi, T. Sasaki, Y. Hata, and Y. Takai, *Mol. Cell. Biol.* **11**(3), 1438 (1991).
[20] T. Soldati, M. A. Riederer, and S. R. Pfeffer, *Mol. Biol. Cell* **4**, 425 (1993).
[21] R. Regazzi, A. Kikuchi, Y. Takai, and C. B. Wollheim, *J. Biol. Chem.* **267**, 17512 (1992).
[22] L. I. Janoueix, F. Jollivet, J. Camonis, P. N. Marche, and B. Goud, *J. Biol. Chem.* **270**, 14801 (1995).

FIG. 2. (A) Purification of recombinant His–δPDE protein on cobalt-affinity chromatography resin (TALON resin). The bound His–δPDE fusion protein is eluted with the elution buffer containing 50 (lane 1), 100 (lane 2), 150 (lane 3), 200 (lane 4), and 250 (lane 5) mM imidazole. The His–δPDE (18-kDa) protein is eluted at 150–200 mM imidazole. The eluted fractions are analyzed by SDS–PAGE and Coomassie blue staining. (B) 100 μg of crude membrane fraction prepared from HeLa cells is incubated with 10 μg of purified recombinant δPDE or Rab GDIβ isoform proteins. Soluble fraction (S) and pellets (P) are then separated and analyzed by SDS–PAGE and immunoblotting using polyclonal anti-Rab13 (*upper*), anti-Rab4 (*middle*), and Rab6 (*lower*) antibodies.

20 m*M* Tris-HCl, pH 7.5, 1 m*M* dithiothreitol, 1 m*M* MgCl$_2$, 5 m*M* EDTA, 150 m*M* NaCl, 0.3 *M* sucrose, 0.5 m*M* ATP, protease inhibitor cocktail).

2. Add different amounts (0, 2, 5, 10, and 15 μg) of the purified recombinant δPDE or (0, 2, 5, 10, and 15 μg) of Rab GDIβ isoform. After a further incubation for 60 min at 37°, the reactions are stopped on ice and samples are separated into membranes and soluble fractions by centrifugation at 4° for 30 min at 430,000*g* and analyzed by SDS–PAGE and immunoblotting (Fig. 2B).

3. For immunoblotting, proteins from membrane and cytosolic fractions are separated on 15% SDS–PAGE and transferred to Immobilon-P membrane (Millipore, Bedford, MA). The filter is blocked for 1 hr in TBS-T buffer (20 m*M* Tris-HCl, pH 7.5, 150 m*M* NaCl, 0.1% Tween 20) + 5% milk, and then incubated in the same buffer with rabbit anti-Rab13, anti-Rab6, and anti-Rab4 polyclonal antibodies at 1:1000 for 1 hr at room temperature. After washing 4 times for 10 min each with TBS-T, the filter is incubated in TBS-T buffer with horseradish peroxidase-conjugated goat anti-rabbit antibodies. Finally, after 4 washes of 10 min each, bound antibodies are revealed by the enhanced chemiluminescence detection reagent (Roche Diagnostics, Meylan, France).

The results indicate that δPDE, but not Rab GDI, is capable of dissociating Rab13 from the membranes. GDIβ successfully released Rab4 and Rab6, but not Rab13, from membranes (Fig. 2). Similar results are obtained with Rab GDIα isoform (not shown).

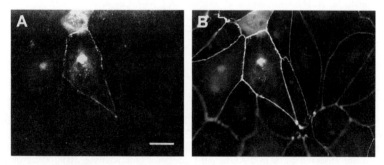

Fig. 3. Immunofluorescence microscopy showing the distribution of δPDE protein. The VSV-G epitope tag is inserted at the N terminus of δPDE. VSV–δPDE is transiently overproduced in the epithelial cell line LLC-PK1 by transfection. Cells are (A) double stained for VSV–δPDE and (B) for the tight junction protein, ZO-1. Bar: 10 μm.

Immunofluorescence Localization Studies

Although we have not succeeded in obtaining specific antibodies against δPDE, we tagged δPDE with the vesicular stomatitis virus glycoprotein G (VSV) epitope at the N terminus. The resulting VSV–δPDE construct, verified by sequencing, is cloned into the eukaryotic expression vector pCB6 under control of the CMV promoter. For these experiments, epithelial LLC-PK1 cells are split at 3×10^5 cells per 3-cm coverslip 18 hr before transfection. The day of transfection, cells are washed twice with Opti-MEM (Life Technologies, Rockville, MD) medium and transfected with VSV–δPDE/pCB6 construct using lipofectin reagent according to the instructions of the manufacturer (Life Technologies). Cells are incubated for 6 hr at 37° with the transfection mixture in a 10% CO_2 incubator. They are then washed with PBS, incubated in DMEM containing 10% fetal calf serum, and 2 mM glutamine in a 10% CO_2 incubator for 18 hr, and processed for immunofluorescence microscopy. Cells transiently transfected with VSV–δPDE/pCB6 express a high level of the protein. This prompted us to permeabilize the cells with 0.05% saponin as detergent prior to fixation. This procedure allows for the removal of most of the cytosolic pool and the detection of the membrane pool. Cells are washed with PBS containing 1 mM $CaCl_2$, 0.5 mM $MgCl_2$, and then permeabilized with 0.05% saponin in PBS for 2 min. After 15 min of fixation with 4% paraformaldehyde (PFA) in PBS, free aldehyde groups are quenched with 50 mM NH_4Cl in PBS for 15 min. Cells are washed and incubated in 0.05% saponin, 0.2% BSA in PBS for 15 min. All subsequent incubations with antibodies and washes are performed in this buffer. Cells are double stained with the mouse monoclonal anti-VSV P5D4 and rabbit polyclonal anti-ZO-1 (a tight junction marker) antibodies for 1 hr at room temperature. After rinsing three times for 10 min, the primary antibodies are visualized with a mixture of rabbit FITC anti-mouse and goat TRITC anti-rabbit immunglobulins (Cappel, Organon Teknika Corp, West Chester, PA). Immunofluorescence analysis is performed with a Leica TLS confocal microscope (Leica Microsystems Inc., Buffalo, NY).

The VSV–δPDE exhibited a patchy staining pattern outlining the periphery of the cells (Fig. 3). The immunofluorescence observations suggest that δPDE may be localized in vesicular structures in close contact with the lateral plasma membrane of epithelial cells.

[22] Subcellular Localization of Rab17 by Cryo-Immunogold Electron Microscopy in Epithelial Cells Grown on Polycarbonate Filters

By Peter J. Peters and Walter Hunziker

Introduction

Rab17 is one of several Rab GTPases associated with vesicular compartments of the endocytic pathway. Expression of Rab17 message and protein seems to be restricted to organs rich in epithelial tissues (i.e., intestine, liver, kidney) and the GTPase is neither detected in organs lacking epithelial cells nor in fibroblasts.[1] In kidney proximal tubule, Rab17 expression has been shown to be restricted to epithelial cells by immunogold electron microscopy (EM).[1] Despite the epithelial specific expression of Rab17, its expression may be induced in fibroblast under particular physiologic conditions. Rab17 expression, for example, was induced in human peridontal ligament fibroblasts subjected to mechanical stretching.[2]

At the subcellular level, Rab17 has been localized by immunogold EM to the basolateral plasma membrane as well as tubules and vesicles in the apical cytoplasm of mouse proximal tubule kidney cells.[1] In Madin-Darby canine kidney (MDCK) epithelial cells, transfected mouse Rab17 was also present on an apical vesicular–tubular compartment but could not be detected at the basolateral plasma membrane.[3] Endogenous Rab17 and mouse Rab17 expressed from an inducible vector were also absent from the basolateral plasma membrane of EpH4 mouse mammary epithelial cells.[4] In contrast, the GTPase was detected along the basolateral membrane of enterocytes in mouse small intestinal explants.[5] Although the GTPase was present on apical vesicles in enterocytes, it was more prominent on vesicles and tubules in the basolateral cytoplasm. These observations suggest that

[1] A. Lutcke, R. G. Parton, C. Murphy, V. M. Olkkonen, P. Dupree, A. Valencia, K. Simons, and M. Zerial, *J. Cell Sci.* **107,** 3437 (1994).

[2] E. K. Basdra, A. G. Papavassiliou, and L. A. Huber, *Biochim. Biophys. Acta* **1268,** 209 (1995).

[3] W. Hunziker and P. J. Peters, *J. Biol. Chem.* **273,** 15734 (1998).

[4] P. Zacchi, H. Stenmark, R. G. Parton, D. Orioli, F. Lim, A. Giner, I. Mellman, M. Zerial, and C. Murphy, *J. Cell Biol.* **140,** 1039 (1998).

[5] G. H. Hansen, L. L. Niels-Christiansen, L. Immerdal, W. Hunziker, A. J. Kenny, and E. M. Danielsen, *Gastroenterol.* **116,** 610 (1999).

the subcellular localization of Rab17 can differ from one epithelial cell type to another.

Based on the accessibility of the Rab17 positive tubular and vesicular cytosolic structures to internalized ligands, Rab17 associates with early compartments of the endocytic pathway. In EpH4 cells transfected with a chimeric receptor encoding a mutant cytosolic domain of the LDLR (FcLR 5-27), antibodies internalized via FcLR 5-27 were found in Rab17 positive endosomes.[4] Also transferrin taken up by these cells colocalized with Rab17. In MDCK cells coexpressing murine Rab17 and the rabbit polymeric immunoglobulin receptor (pIgR), a significant fraction of Rab17 positive structures carried dimeric immunoglobulin A (dIgA).[3] Similarly, in small intestinal explants, dIgA secreted by plasma cells of the lamina propria was taken up by enterocytes via the endogenous pIgR into Rab17 positive vesicles and tubules.[5] Furthermore, Rab17 was detected in transcytotic vesicles immunoisolated from rat liver using antibodies to the cytosolic tail of the pIgR.[6]

Interestingly, at least a fraction of the Rab17 positive endosomes appear to conform morphologically and functionally to a subclass of early endosomes termed recycling endosomes[7] or apical recycling compartment in epithelial cells.[8,9] In epithelial cells, recycling endosomes represent a vesicular and tubular early endocytic compartment clustered around the centrosome, which is accessible to receptors internalized either from the apical or the basolateral surface.[8–12] Importantly, protein sorting must occur in this compartment as both recycling and transcytosing receptors colocalize in these structures before being selectively transported to opposite plasma membrane domains. Indeed, in MDCK cells at least a subset of the Rab17 positive compartment may receive ligands internalized from the apical and basolateral domain.[3]

Consistent with the localization of Rab17 to recycling endosomes and their possible role in polarized sorting is the observation that overexpressing wild-type or mutant forms of the GTPase affects endocytic traffic of transcytotic proteins. In MDCK cells overexpressing murine Rab17, pIgR-mediated basolateral to apical transcytosis of dIgA was inhibited and basolateral recycling of dIgA concomitantly increased.[3] In EpH4 cells, the transfection of Rab17 mutants defective in GTP binding or hydrolysis stimulated baso-

[6] M. Jin, L. Saucan, M. G. Farquhar, and G. E. Palade, *J. Biol. Chem.* **271,** 30105 (1996).
[7] J. Gruenberg and F. R. Maxfield, *Curr. Opin. Cell Biol.* **7,** 552 (1995).
[8] G. Apodaca, L. A. Katz, and K. E. Mostov, *J. Cell Biol.* **125,** 67 (1994).
[9] M. Barroso and E. S. Sztul, *J. Cell Biol.* **124,** 83 (1994).
[10] E. J. Hughson and C. R. Hopkins, *J. Cell Biol.* **110,** 337 (1990).
[11] A. Knight, E. Hughson, C. R. Hopkins, and D. F. Cutler, *Mol. Biol. Cell* **6,** 597 (1995).
[12] D. R. Sheff, E. A. Daro, M. Hull, and I. Mellman, *J. Cell Biol.* **145,** 123 (1999).

lateral to apical transcytosis of FcLR 5-27 and transferrin and also increased apical recycling.[4]

Although the localization to recycling endosomes and the effect of overexpressing wild-type or mutant Rab17 on transcytotic receptors implies a role for the GTPase in transcytosis, the precise step(s) regulated by Rab17 remain elusive. In MDCK cells not all Rab17 positive membranes appear to contain internalized dIgA, suggesting that dIgA may not reach a subset of the Rab17 compartment or that Rab17 is associated with a particular domain or region of interconnected endosomal tubules. Furthermore, other Rab proteins are present on early endosomes of which Rab11, Rab15, and Rab25 have been localized to recycling endosomes.[13-15] It will therefore be important to elucidate how the different Rab proteins coordinate traffic to and from recycling endosomes. This will require a combination of cell biological, biochemical, and morphological assays. Here we provide a protocol for the high-resolution localization of Rab17 in MDCK cell monolayers grown on permeable polycarbonate filters by cryo-immunogold EM. This protocol should also be applicable to the localization of other Rab proteins in polarized epithelial cells and allow comparative topological biochemistry of different Rab proteins at the ultrastructural level.

High-resolution cryo-immunogold EM is the most sensitive procedure for immunodetection of antigens on ultrathin sections prepared from chemically fixed cells since aldehyde fixation is the only denaturation step. The omission of harsh organic solvents (such as those used for plastic embedding) ensures better preservation of protein antigenicity. Fixed material is embedded in gelatin, cryosectioned, mounted on Formvar-coated grids, and labeled.

Below we describe a typical labeling protocol based on the work by Slot and Geuze[16] during which the sectioned material is exposed first to antibody, then to protein A-gold particles. If the primary antibody does not bind to protein A, incubation with a secondary antibody that does bind protein A can be added. The precise reaction conditions used for immunogold labeling may vary depending on the antibody and are determined by trial and error.[17]

[13] O. Ullrich, S. Reinsch, S. Urbe, M. Zerial, and R. G. Parton, *J. Cell Biol.* **135**, 913 (1996).
[14] P. A. Zuk and L. A. Elferink, *J. Biol. Chem.* **274**, 22303 (1999).
[15] J. E. Casanova, X. Wang, R. Kumar, S. G. Bhartur, J. Navarre, J. E. Woodrum, Y. Altschuler, G. S. Ray, and J. R. Goldenring, *Mol. Biol. Cell* **10**, 47 (1999).
[16] J. W. Slot and H. J. Geuze, *J. Cell Biol.* **90**, 533 (1981).
[17] G. Raposo, M. J. Kleijmeer, G. Posthuma, J. W. Slot, and H. J. Geuze, "Handbook of Experimental Immunology," 5th ed., p. 208. Blackwell Science, Cambridge, Massachusetts, 1996.

Fixation of Cell Monolayers on Transwell Polycarbonate Filters for
Immunogold Labeling

The fixation step is critical for successful cryo-immunogold EM. The
purpose of fixation is to immobilize all subcellular structures and antigens
in their most native form. It is important, however, to maintain maximum
antigenicity. Three different fixatives are commonly used: paraformaldehyde (PFA), glutaraldehyde (GA), or a combination of both (PFA/GA).
It is advantageous to first determine which of these fixatives is best suited
for maintaining optimal antigenicity within the proper subcellular compartment.

Materials

     Cultures of epithelial cells, for example, MDCK cells
     Culture medium
     2× fixative: 2× PFA (see "Reagents and Solutions"), 2× GA (see
          "Reagents and Solutions"), *or* 2× PFA/GA (see "Reagents and
          Solutions"), room temperature
     Storage solution (see "Reagents and Solutions")
     Transwell polycarbonate filter units, 0.4-mm pore size, 6 or 12 mm in
          diameter (Costar, Cambridge, MA)
     Screw-top microcentrifuge tubes
     Razor blade

Procedure

     1. Seed and grow MDCK cells onto Transwell polycarbonate filter units
        as described elsewhere.[18] Add fresh medium to the cells on Transwell
        units the day before the experiment.
     2. Add an equal volume of 2× fixative to the medium present in the
        apical and basolateral compartment of Transwell units immediately
        after removing from the incubator. Incubate the cells 24 hr (for PFA)
        or 1 hr (for GA or PFA/GA) at room temperature. Do not disturb
        cells during the fixation period. It is advisable not to wash cells in
        phosphate-buffered saline (PBS) or any other medium before fixa-
        tion, in order to avoid artifacts.
     3. Using a razor blade, cut out the polycarbonate filter from the plastic
        holder of the Transwell unit. The filter with the cells will roll up.
        Transfer the filter with the cells into a screw-top microcentrifuge

[18] W. Hunziker and I. Mellman, *J. Cell Biol.* **109,** 3291 (1989).

tube containing 1 ml storage solution for storage. Store at 4° in an airtight container.

Embedding Cells for Immunogold Labeling

Cells are embedded in gelatin to facilitate handling during the rest of the procedure.[19] It is advisable to have a dissecting microscope with "cold" light optics permanently present in a cold room. The gelatin will melt once the ultrathin cryosections are prepared for immunolabeling (see below). Sucrose infiltration prevents damage to the material by ice–crystal formation during freezing in liquid nitrogen.[20,21] Care should be taken to prevent the samples from drying out.

Materials

Fixed cells in storage solution
PBS containing 0.15 M glycine
10% (w/v) gelatin (see "Reagents and Solutions"), 37°
2.3 M sucrose in 0.1 M sodium phosphate buffer, pH 7.4
Petri dishes
Razor blades, ethanol-rinsed and air-dried
Dissecting microscope with cold light optics (Leica)
1-ml vials
End-over-end rotator
High-precision tweezers
Aluminum specimen holders (Leica), roughened with sandpaper, soap-cleaned, and dust-free

Procedure

1. Pipette off the storage solution and wash the filters twice, each time for 5 min, in 1 ml PBS/0.15 M glycine.
2. Cut the filter (0.2 × 0.2 cm) and incubate 30 min each in 2.5% and 12% (w/v) gelatin (prewarmed to 37°).
3. With a pipette transfer the filter/gelatin mixture on top of a lid from a petri dish, taking care that the filter lies parallel to the surface of the petri dish. You may position two to four filters on top of each other.

[19] P. J. Peters, J. Borst, V. Oorschot, M. Fukuda, O. Krahenbuhl, J. Tschopp, J. W. Slot, and H. J. Geuze, *J. Exp. Med.* **173,** 1099 (1991).
[20] K. T. Tokuyasu, *J. Cell Biol.* **57,** 551 (1973).
[21] W. Liou, H. J. Geuze, and J. W. Slot, *Histochem. Cell Biol.* **106,** 41 (1996).

4. Place the dish on ice to solidify the drop.
5. Just before the gelatin solidifies, place a small petri dish on top of the droplet to create a flattened piece of solidified gelatin enclosing the filter.
6. Remove the top dish and add cold PBS/0.15 M glycine to the solidified material and keep at 4° to prevent the gelatin from melting.
7. Cut the filter roll/gelatin into slices 0.5 mm thick and then cut the slices into 0.5-mm cubic blocks, using clean (ethanol-rinsed, air-dried) razor blades, under a dissecting microscope. This step is best carried out in a cold room to ensure that the gelatin does not melt and that the sample does not dry out.
8. Transfer the cubic block to a 1-ml vial containing 2.3 M sucrose.
9. Rotate sample in sucrose for at least 2 hr in an end-over-end rotator in the cold room.
10. Working under the dissecting microscope and using high-precision-grade tweezers with a minimum of force, transfer the samples, along with a minimum of sucrose, from the vial to very clean aluminum specimen holders that have been roughened with sandpaper.
11. Remove excess sucrose by capillary action on the forceps or using clean filter paper, and transfer holders to small containers filled with liquid nitrogen.

Samples can be stored in liquid nitrogen and are ready for ultrathin cryosectioning (see below).

Cryosectioning

The procedure for ultrathin cryosectioning is time consuming and can still be considered an art. The preparation of glass knives for ultrathin cryosectioning has been described in detail elsewhere[22]; one may also consult the instructions provided by the manufacturer of the ultramicrotome. High-quality diamond knives are now also available. The sample blocks that are stored in liquid nitrogen can be used repeatedly.

Materials

    Embedded cells on filter (see above)
    2.3 M sucrose in 0.1 M sodium phosphate buffer, pH 7.4, *or* 3 : 2 mixture of 2.3 M sucrose and 2% (v/v) methyl cellulose
    2% gelatin (see "Reagents and Solutions") in a small petri dish

[22] G. Griffiths, "Fine Structure Immunocytochemistry" pp. 144–151. Springer Verlag, New York, 1993.

Ultramicrotome (Leica) with cryochamber and antistatic devices
(Diatome, Biel, Switzerland)
Glass or diamond knife (Drukker, Luijk, The Netherlands, and Dia-
tome, Biel, Switzerland)
Trimming knife
Eyelash mounted on a wooden stick
1.5-mm-diameter stainless steel loop mounted on 15-cm wooden stick
Carbon- and Formvar-coated copper grids (see below)

Procedure

1. Set the ultramicrotome to $-100°$. Insert and secure specimen and
   trimming knife in the ultramicrotome as described by the manufacturer. Make sure that the filter is inserted in a right angle to the
   knife. Trimming of the sample takes most of the work and is a
   prerequisite for making ultrathin sections. The end result is a per-
   fectly trimmed rectangular block from which ultrathin cryosections
   are made.
2. Trim the sample with a glass knife or trimming device by making
   sections from the short side of the sample block. Maintain the temperature setting of the microtome at $-100°$.
3. Trim sections of 0.1-μm thickness at a speed of 50 mm/sec. Rotate
   the sample at $90°$ angles and trim along each side to end up with a
   shiny rectangular block of $\sim 0.2 \times 0.3$ mm. At this step, it is advisable
   to check whether the trimmed block is suitable for further sectioning,
   by cutting a few 0.2-μm thin cryosections. These can then be picked
   up (step 6) and evaluated by light microscopy. Toluidine blue staining
   or immunocytochemistry may be performed on these sections. If
   necessary, one can trim the sample even further after the light microscope examination.
4. For ultrathin sectioning, set and maintain the temperature setting of
   the cryochamber, knife holder, and specimen holder of the microtome
   at $-120°$. Fluctuations in temperature should be $<1°$.
5. Cut ultrathin sections of 50-nm thickness (for PFA fixed material)
   or 65-nm thickness (for PFA/GA-fixed material) at a speed of 1.6
   mm/sec using clean glass or diamond knives. Sections should be lined
   up adjacent to each other so as to form "ribbons" of five to six
   sections. During sectioning, sections, can be manipulated with an
   eyelash mounted on a wooden stick. This can be used to carefully
   position sections onto the mounting material of the diamond knife.

The mounting material seems to have the best conductivity or any other unknown characteristic and is the best location for picking up sections.

6. Retrieve sections from the cryochamber using a droplet of 2.3 M sucrose on a 1.5-mm stainless steel loop attached to a 15-cm wooden stick. Draw the droplet toward the face of the ribbon sections until the sections attach. Just before freezing, the droplet should be very gently pushed onto the section so that the sections will gently attach to the droplet. It is preferable, but more difficult, to use a 1 : 1 mixture of 2.3 M sucrose/2% methyl cellulose to retrieve the sections.[21]

7. Quickly remove the loop carrying the ribbon of sections from the cryochamber and let the sections thaw.

8. As soon as the sections are thawed, place them onto carbon- and Formvar-coated copper grids.

9. Place grids coated side down onto 2% gelatin in a small petri dish, which is itself in a wet petri dish on ice ("gelatin storage plate"). Alternatively, store sections in a closed box at 4°.

Immunogold Labeling of Ultrathin Cryosections with Epithelial Cells on Transwell Polycarbonate Filters

Materials

Gelatin storage plate with cryosections mounted on Formvar-coated grids

PBS containing 0.15 M glycine

PBS containing 1% (w/v) BSA

Primary antibody in PBS/1% BSA

PBS/0.1% (w/v) BSA (see "Reagents and Solutions" for 10% BSA stock)

Secondary antibody in PBS/1% BSA (optional)

10-nm protein A-gold particles[16] at OD 0.1 in PBS/1% BSA (see "Reagents and Solutions" for 10% BSA stock)

PBS containing 1% (v/v) glutaraldehyde

Uranyl oxalate solution (see "Reagents and Solutions"; optional)

Methyl cellulose/uranyl acetate solution (see "Reagents and Solutions")

37° hot plate or incubator

Forceps

Stainless steel loop slightly larger than grids attached to 1000-μl pipette tips

Procedure

All procedures are carried out at room temperature, except 1.

1. Place the gelatin storage plates containing the cryosections (see above, step 9) for 20 min on a 37° hot plate or in an incubator maintained at 37°. The gelatin will melt and the grids will start floating. Also, residual 10% gelatin that was introduced into the sample during the embedding step is melted away at this point.

2. Lift the floating grids with a pair of forceps and transfer them onto 100-ml droplets of PBS/0.15 M glycine laid out on a sheet of Parafilm. Wash four times, each time by floating the grids for 3 min on fresh 100-μl droplets of PBS/0.15 M glycine. All subsequent treatments (washings and incubations) involve the transfer of the grids onto fresh droplets of medium that are laid out on Parafilm; 100-μl droplets should be used for all solutions except the antibody and protein A-gold, for which 10-μl droplets are used.

3. Wash grids twice, each time by floating for 5 min on droplets of PBS/1% BSA.

4. Incubate grids for the appropriate period of time at room temperature on 10-μl droplets of primary antibody diluted to an appropriate concentration in PBS/1% BSA. The appropriate dilution and incubation times vary with the antibody used. Incubation time ranges from 30 to 100 min. Antibody titer is usually in the range of 0.2 μg/ml but must be titered for each antibody, at the EM level, in pilot experiments.

5. Wash the grids four times, each time by floating for 2 min on droplets of PBS/0.15 M glycine, then perform a final wash in PBS/0.1% BSA.

6. *Optional:* When a primary antibody that does not react with protein A is used, incubate grids with a secondary (bridging) antibody that binds protein A, diluted in PBS/0.1% BSA, for 20 min.

7. Wash grids five times, each time by floating for 2 min in droplets of PBS/0.1% BSA.

8. Incubate grids for 20 min on 10-μl droplets of 10-nm protein A-gold particles (at an OD of 0.1) in PBS/1% BSA.

9. Wash grids seven times, each time by floating for 2 min on droplets of PBS/0.1% BSA.

10. Incubate grids for 5 min on droplets of PBS/1% glutaraldehyde. For double immunogold labeling: after this step wash the grids with PBS/0.15 M glycine (step 2) and repeat the entire procedure from

that point on with the second primary antibody.[23] For double-labeling experiments, use protein A with different sizes of gold particles (10 and 15 nm).

11. Wash grids seven times, each time by floating for 2 min on droplets of distilled water.

12. *Optional:* Incubate grids for 5 min on droplets of uranyl oxalate solution.

13. Quickly wash grids by floating for 1 min on droplets of methyl cellulose/uranyl acetate solution, then floating 5 min in fresh droplets of methyl cellulose/uranyl acetate solution.

14. Retrieve the grids with a stainless steel loop of diameter slightly larger than the grid, attached to the tip of a 1000-μl pipette.

15. Drain the excess methyl cellulose/uranyl acetate solution by touching the loop at an angle of 45° to a filter paper (coffee filter works perfect).

16. Air dry grids at room temperature or under warm air by placing the pipette tips upside-down in a pipette tip rack. A film of methyl cellulose/uranyl acetate is left on the grid after drying. This is important to give optimal contrast and to preserve the integrity of membrane structures.[24]

17. Examine grid by electron microscopy at 80 kV with a small aperture.

Preparation of Carbon- and Formvar-Coated Copper Grids

Carbon- and Formvar-coated copper grids are used for mounting sections of samples for immunogold labeling.

Materials

    Formvar (Merck, Darmstadt, Germany)
    Chloroform, analytical grade
    Glass microscope slides
    Glass-stoppered Erlenmeyer flask accommodating 100 ml
    Coplin jar
    100-mesh copper grids
    Address labels
    Carbon-coating device (BOC Edwards, Crawley, UK)

---

[23] J. W. Slot, H. J. Geuze, S. Gigengack, G. E. Lienhard, and D. E. James, *J. Cell Biol.* **113,** 123 (1991).

[24] G. Griffiths, R. Brands, B. Burke, D. Louvard, and G. Warren, *J. Cell Biol.* **95,** 781 (1982).

Procedure

1. Rub the fingertip along the surface of a glass microscope slide to transfer "body grease," and then gently rub the surface with lens paper. Make sure that the slide is dust-free.
2. Dissolve 1.1 g Formvar in 100 ml chloroform in a glass-stoppered Erlenmeyer flask while stirring. Transfer the Formvar solution gently into a Coplin jar.
3. Place the slide for 70% of its length into Formvar solution for 2 min.
4. Remove the slide and allow it to air dry for 1 min. The rate at which the slide is removed will determine the thickness of the Formvar coat (optimum thickness ~50 nm).
5. Insert the slide slowly into a glass container of distilled water and allow the film to float on the surface. Formvar films will float on the surface and should have a grayish (interference color) appearance (due to the 50-nm thickness).
6. Overlay clean 100-mesh copper grids onto the Formvar film, with the rough surface of the grid facing the Formvar film.
7. Remove Formvar-coated grid from the water surface by laying the grids over the surface of a glass slide, one side of which is covered with a white "address label," which allows the grids to adhere onto the slide.
8. Let the grids air dry.
9. Evaluate the quality of the Formvar-coated grids by viewing them under an electron microscope. The film should be dust free, without crevices, holes, or tears, and should not collapse under a high-energy beam (80 kV without objective aperture).
10. Coat the grids the next day with a layer of evaporated carbon under high vacuum according to the equipment manufacturer's instructions. The desired thickness is critical and can be judged by evaluating the color of the address label relative to a noncoated address label. There should be only a minor change in color.

Reagents and Solutions

Deionized or distilled water is used in all recipes and protocol steps.

BSA Stock Solution, 10% (w/v)

Prepare 10% (w/v) bovine serum albumin (BSA, fraction V; Sigma) in distilled water containing 0.02% sodium azide. Adjust to pH 7.4 and centrifuge 1 hr at 100,000g, 4°. Retain supernatant and store up to 1 month in 1-ml aliquots at 4° in closed tubes.

Gelatin, 2% and 10% (w/v)

Prepare 2% and 10% (w/v) gelatin (Merck) solutions in 0.1 *M* sodium phosphate buffer, pH 7.4, containing 0.02% azide. Store 2% gelatin in small petri dishes in a closed box; store 10% gelatin in small vials up to 1 month at 4°.

Glutaraldehyde (GA) Fixative, 2×

Mix 2.5 ml 0.4 *M* PHEM buffer (see "Reagents and Solutions") and 2.5 ml 8% glutaraldehyde (EM grade; commercially available). Prepare immediately before use. The final concentrations are 0.2 *M* PHEM and 4% glutaraldehyde.

Methyl Cellulose/Uranyl Acetate

Methyl Cellulose Stock. Add 2 g methyl cellulose (Sigma, 25 centipoise), with stirring, to 98 ml distilled water that has been prewarmed to 90°. Cool on ice while stirring, until temperature has dropped to 10°. Stir overnight at low speed in the cold room. Let the solution mature for 3 days in the cold room. Centrifuge 95 min at 97,000g (29,000 rpm in a Beckman 45 Ti rotor), 4°, then divide the supernatant into 10-ml aliquots and store up to 3 months at 4°.

To Prepare Methyl Cellulose/Uranyl Acetate. Carefully mix 1 ml of 4% uranyl acetate (see "Reagents and Solutions") and 9 ml of methyl cellulose stock. Store up to 1 month at 4° in a dark container.

Paraformaldehyde (PFA) Fixative, 2×

Thaw an aliquot of 16% paraformaldehyde stock solution (see "Reagents and Solutions") in 60° water. When solution is clear, mix the following:

  10 ml 0.4 *M* PHEM buffer (see "Reagents and Solutions")
  5 ml 16% paraformaldehyde stock (see "Reagents and Solutions")
  5 ml H$_2$O

Prepare immediately before use. The final concentrations are 0.2 *M* PHEM buffer and 4% PFA.

Paraformaldehyde/Glutaraldehyde (PFA/GA) Fixative, 2×

Thaw an aliquot of 16% paraformaldehyde stock solution (see "Reagents and Solutions") in 60° water. When solution is clear, mix the following:

10 ml 0.4 M PHEM buffer (see "Reagents and Solutions")
5 ml 16% paraformaldehyde stock solution (see "Reagents and Solutions")
5 ml 8% glutaraldehyde (EM grade)

Prepare immediately before use. The final concentrations are 0.2 M PHEM, 4% PFA, and 0.4% GA.

Paraformaldehyde (PFA) Stock Solution, 16%

Dissolve 32 g paraformaldehyde in 200 ml water. To help dissolution, add 0.34 g Na_2CO_3 to the PFA solution and stir until the solution is clear. Heat the solution on a hot plate until the paraformaldehyde is completely dissolved, but do not heat above 60°. Divide into aliquots in 10-ml tubes and store up to several years at −20° in a sealed box. **CAUTION:** *PFA is toxic, so it is necessary to prepare the solution in a fume hood.*

PHEM Buffer, 0.4 M

100 ml 240 mM PIPES
100 mM HEPES
8 mM $MgCl_2$
40 mM EGTA

Adjust pH to 6.9 with NaOH. Store up to several years at −20°.

Storage Solution

Thaw an aliquot of 16% paraformaldehyde stock solution (see "Reagents and Solutions") in 60° water. When solution is clear, mix the following:

5 ml 0.4 M PHEM buffer (see "Reagents and Solutions")
0.6 ml 16% PFA stock solution (see "Reagents and Solutions"), thawed and clear
14 ml H_2O

The final concentrations are 0.1 M PHEM and 0.5% PFA.

Uranyl Acetate, 4%

Prepare a 4% (w/v) stock of uranyl acetate in distilled water and adjust pH to 4 with HCl. Store up to 1 month at room temperature; filter through Millipore filter before use. **CAUTION:** *Uranyl acetate is radioactive.*

Uranyl Oxalate Solution

Mix 1 vol of 4% uranyl acetate (see "Reagents and Solutions") and 1 vol of 0.15 M oxalic acid. Adjust to pH 7 with 25% ammonium hydroxide.

Troubleshooting

The most commonly encountered problems in immunogold staining of ultrathin cryosections are low-quality sections, absence of specific labeling, high background labeling, colabeling artifacts, dirty grids, and weak contrast. Unfortunately, the problems will only become apparent at the end of the procedure when the specimen is analyzed in the electron microscope.

Bad sectioning is often due to unsuitable preparation of the specimen or instability in the cryomicrotome. Dilutions of fixatives must be prepared just before they are used for specimen fixation. Specimens that have dried out during preparation, or the presence of residual sucrose around the gelatin block, may cause sectioning problems. Also, loose attachment of the specimen holder or knife holder to the cryomicrotome will cause problems. Contamination of specimen and knife, which can result from condensation of ice crystals, must be avoided.

Nonspecific labeling is encountered mainly because the antibody does not recognize the antigen after aldehyde fixation. This can be tested by evaluating the antibody on aldehyde-fixed material by FACS (fluorescence cell sorter analysis) or immunofluorescence microscopy. Overexpression of the antigen (protein) in a control cell that does not express the antigen endogenously is the best quality control for specificity.

Background labeling often results because a too large concentration of protein A-gold was used or because the antibody titer was too high. For the protein A-gold concentration it is best to follow the instructions of the manufacturer. For most applications, the antibody titer is used in the range of 0.2–2 μg/ml. It is best to try different dilutions and evaluate the specific signal-to-background ratio. Polyclonal, monoclonal, and recombinant Fab antibodies can be successfully employed. It is important to use a bridging antibody that binds specifically and efficiently to protein A-gold as an intermediate step if the primary antibody does not bind protein A efficiently.

Artifacts are often encountered when double-labeling procedures are carried out. In this case, the two sizes of gold particles are always adjacent to each other, at <20 nm apart, and this is often mistaken for colocalization. The cause of colabeling is not always understood, but often the antibody that is used for the detection of the second antigen binds to the antibody that was used to detect the first antigen. Also, protein A-gold that is used in the second step may still bind to the antibody used in the first step. Glutaraldehyde incubation after the first immunolabeling should overcome the later problem.[23] It is important to always check using a single label first.

Dirt on the grid is usually caused by phosphate contamination of solutions. Also, partial drying of droplets during the antibody incubation can result in dirty grids. Always keep the grids under a petri dish with a wet tissue during incubation and never touch the Parafilm where droplets are being applied.

A screening of antibodies by FACS or light microscopy before and after aldehyde fixation and an evaluation of specificity by immunoblots and immunoprecipitation may speed successful labeling on cryosections. Con-

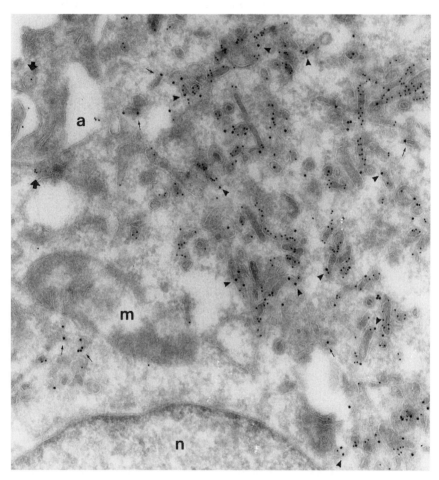

FIG. 1. Electron micrograph of a MDCK cell grown on Transwell filters that has been labeled with anti-Rab17 antibodies followed by 15-nm protein A-gold. Internalized dIgA was detected using an anti-IgA antibody followed by protein A conjugated to 10 nm gold. (a) Apical plasma membrane; m, mitochondria; n, nucleus. Arrows indicate membranes carrying Rab17 only (small arrows), dIgA only (large arrows), or vesicles and tubules with dIgA and Rab17 (arrowheads).

trol cells that are transfected with or without the specific protein are very instrumental as a positive control.

Protein A-gold particles of 10 and 15 nm are routinely used. Transmission electron microscopy is carried out at 80 kV with the smallest objective aperture. Exposure time of film is between 0.5 and 1 sec. Damage to sections can be reduced by never evaluating sections without objective aperture. For more information on the background, see Peters.[25]

Results

Successful immunogold labeling should allow the detection of specific antigens at the subcellular level, as shown in the electron micrograph of a MDCK cell in Fig. 1.

Acknowledgments

We thank Dr. Crislyn D'Souza-Schorey for help with the writing of this manuscript.

[25] P. J. Peters, Cryo-immunogold electron microscopy, in "Current Protocols in Cell Biology" (J. S. Bonifacino, M. Dasso, J. B. Harford, J. Lippincott-Schwartz, and K. M. Yamada, eds.), pp. 4.7.1–4.7.12. John Wiley & Sons, New York, 1999.

[23] Expression and Properties of Rab25 in Polarized Madin–Darby Canine Kidney Cells

By James R. Goldenring, Lorraine M. Aron, Lynne A. Lapierre, Jennifer Navarre, and James E. Casanova

Introduction

Rab proteins are a family of small GTPases with molecular masses of 23–28 kDa. The family includes almost 50 eukaryotic members.[1] Within the family of Rab proteins are a number of subfamilies of more closely related proteins. The Rab11 family consists of three members: Rab11a, Rab11b, and Rab25. Rab11a sequences have been cloned from a number

[1] P. Novick and M. Zerial, Curr. Opin. Cell Biol. 9, 496 (1997).

of species including mouse, dog, rabbit, and human.[2–5] Rab11b has been cloned from mouse.[6] Rab11b differs from Rab11a only in its terminal 30 amino acids within the hypervariable domain. Rab25 was originally cloned from rabbit parietal cells,[7] but sequences from human and mouse are also now available (GenBank accession numbers R28405 and AF119675). Rab25 differs from Rab11a in both its amino and carboxyl termini. In addition, unique among Rab proteins, Rab25 contains the sequence WDTAGLE in its P3 nucleotide binding domain, compared to WDTAGQE in all other Rab proteins reported to date.[7] The WDTAGLE motif is present in Rab25 sequences from rabbit, mouse, and human. This Q to L change in other small GTP-binding proteins has been implicated as a dominant active mutation, substantially reducing the GTPase activity.[8] However, in the case of Rab25, the wild-type protein with a leucine in position 71 has an active intrinsic GTPase activity that is also stimulated by a cytosolic GAP.[9] In addition, the analogous Q70L mutation in Rab11a had no effect on either intrinsic or GAP-stimulated GTPase activity.[9] Most importantly, while Rab11a and Rab11b appear to be ubiquitous, Rab25 expression is limited to epithelial cells.[7]

Two important considerations limit much of the research on Rab proteins. First, while some Rab proteins, such as Rab11a, are abundant in particular cells types (e.g., the high level of Rab11a in parietal cells), relatively low amounts of particular Rab proteins exist in most cells.[10] Thus, it is difficult to study endogenous Rab proteins without very high sensitivity antibodies, and even if these are available, detection may be limited to Western blot analysis. Second, the high level of similarity among Rab proteins has limited the production of specific antibodies, especially murine

[2] P. Chavrier, M. Vingron, C. Sander, K. Simons, and M. Zerial, *Mol. Cell. Biol.* **10**, 6578 (1990).

[3] P. Chavrier, K. Simons, and M. Zerial, *Gene* **112**, 261 (1992).

[4] G. T. Drivas, A. Shih, E. E. Coutevas, P. D'Eustachio, and M. G. Rush, *Oncogene* **6**, 3 (1991).

[5] J. R. Goldenring, C. J. Soroka, K. R. Shen, L. H. Tang, W. Rodriguez, H. D. Vaughan, S. A. Stoch, and I. M. Modlin, *Am. J. Physiol.* **267**, G187 (1994).

[6] F. Lai, L. Stubbs, and K. Artzt, *Genomics* **22**, 610 (1994).

[7] J. R. Goldenring, K. R. Shen, H. D. Vaughan, and I. M. Modlin, *J. Biol. Chem.* **268**, 18419 (1993).

[8] H. Adari, D. R. Lowy, B. M. Willumsen, C. J. Der, and F. McCormick, *Science* **240**, 518 (1988).

[9] J. E. Casanova, X. Wang, R. Kumar, S. G. Bhartur, J. Navarre, J. E. Woodrum, G. S. Ray, and J. R. Goldenring, *Mol. Biol. Cell.* **10**, 47 (1999).

[10] J. R. Goldenring, J. Smith, H. D. Vaughan, P. Cameron, W. Hawkins, and J. Navarre, *Am. J. Physiol.* **270**, G515 (1996).

monoclonal antibodies. In Madin–Darby canine kidney (MDCK) cells, Rab25 levels are very low, whereas endogenous Rab11a levels are relatively high. Therefore, to study Rab25 in MDCK cells, we were forced to isolate cell lines stably transfected with the wild-type sequence of rabbit Rab25. To detect Rab25, we also developed monoclonal antibodies specific for Rab25, and nonreactive against Rab11a.

Procedures

MDCK Lines Stably Transfected with Rab25

The rabbit Rab25 sequence is cloned into pCB6 for expression under the control of the CMV promoter. Rab25-pCB6 is then transfected into MDCK II cells by a standard calcium phosphate precipitation/glycerol shock technique using 20 μg of supercoiled plasmid for transfection of 2 million cells. The transfected cells are allowed to grow to confluence for 48 hr in 110-mm petri dishes in nonselecting media. The cells are then trypsinized in trypsin/EDTA and replated in 110-mm dishes in serial dilutions from 1 : 40 to 1 : 1280 in media containing G418 for neomycin resistance selection. Cell colonies are allowed to grow for 7–14 days until they reach a size of 2–4 mm. Plates with lower dilutions are used to ensure easy separation of isolated colonies.

Two techniques are used to isolate colonies without the use of cloning rings and are found to be equally effective. In both cases, cells are visualized under an inverted phase contrast microscope with a 20× objective. Using the first method, under microscopic visualization colonies are isolated by gentle scraping and aspiration into a 200-μl wide-mouth micropipette tip. The aspirated colony is then transferred into 0.25 ml of MDCK media with G418 in a 24-well plate. A second method can also be used and is technically somewhat easier. A sterile cotton applicator is used to pick up the colony under direct visualization under the microscope. Cells are then transferred to 0.5 ml of media with G418 in a 24-well plate and dispersed into the well with rapid rotation of the applicator tip. Both of these techniques can be performed outside of a cell culture hood as long as the air flow in the room is clean and the investigator works expeditiously. In our experience we have had negligible contamination using these techniques.

Colonies are grown in 24-well plates to confluence and cell lines are picked for analysis based on the maintenance of a normal tight cobblestone appearance of the monolayer. These colonies are digested and replated into two T25 flasks, which are again grown to confluence. One flask is used for passage of the line, and the second flask is used for preparation of RNA using a standard lysis with RNAZOL.[11] Northern blots are then performed with 20 μg of total RNA using a full-length rabbit Rab25 as a probe.

Production of Monoclonal Antibodies against Rab25

Rab11a and Rab25 differ primarily only in their amino- and carboxyl-terminal sequences (overall 68% identity). In the past, a number of investigators and commercial firms have isolated polyclonal antibodies against specific Rab proteins utilizing synthetic peptide fragments derived from the carboxyl-terminal variable regions. In our experience, however, it has been difficult to raise Rab-specific anti-peptide monoclonal antibodies that are useful for both Western blot and immunofluorescence applications. We therefore have utilized a double-screening enzyme-linked immunosorbent assay (ELISA) protocol to isolate monoclonal antibodies against Rab25 that do not react with Rab11a.

As an immunogen, we utilize histidine-tagged Rab25[7] since the polyhistidine sequence is a poor antigen. Five-week-old female BALB/c mice are injected at multiple sites subcutaneously with 165 μg of recombinant Rab25 in Freund's complete adjuvant as an initial immunization. Four weeks later, mice are injected intraperitoneally with 100 μg of Rab25 in Freund's incomplete adjuvant. Titers of antisera are determined from retro-orbital bleeds using a Rab25 ELISA as detailed below. When its half-maximal titer reaches 1 : 30,000, one mouse is injected intraperitoneally with 50 μg of Rab25 without adjuvant. Three days later, the mouse is sacrificed and the spleen harvested and dispersed mechanically. The isolated splenocytes are fused with Sp2/0 myeloma cells and the cells plated out across twenty 96-well microtiter plates. Wells are monitored for colony growth, and the supernates from wells containing active hybridoma colonies are screened using ELISA for both Rab11a and Rab25.

Rab ELISA

Solutions

ELISA borate saline (EBS): 200 mM boric acid, 500 mM $Na_2B_4O_7$, 75 mM NaCl, pH 8.5

ELISA phosphate-buffered saline (EPBS): 8.24 mM Na_2HPO_4, 1.77 mM NaH_2PO_4, 140 mM NaCl, pH 7.4

ELISA wash buffer: 10 mM Tris, pH 8.0, 0.05% Tween 20, 0.02% NaN_3

Alkaline phosphatase substrate buffer: 0.5 mM $MgCl_2$, 0.9 M diethanolamine, pH 9.6

---

[11] P. Chomczybski and N. Sacchi, *Anal. Biochem.* **162,** 156 (1987).

ELISA Protocol. Wells of 96-well Immulon-2 plates are coated with 50 μl of 5 μg/ml of either Rab11a or Rab25 in EBS or no antigen (EBS alone) overnight at 4°. The plates are washed three times in ELISA wash buffer and then blocked for 30 min at room temperature with 200 μl of 10 mg/ml BSA in EPBS. Following blocking the plates are washed three times with ELISA wash buffer and then 50 μl of media from the hybridomas to be tested is added to the wells and incubated at room temperature for 1 hr. Following the primary incubation, the plates are washed three times with ELISA wash buffer, and then 50 μl of alkaline phosphatase-conjugated rabbit anti-mouse IgG (diluted 1:500 in EPBS) is added for 1 hr at room temperature. The plates are then washed three times with ELISA wash buffer and *p*-nitrophenyl phosphate color substrate is added in alkaline phosphatase buffer (5 mg/8.33 ml alkaline phosphatase substrate buffer). The color reaction is allowed to continue for 30 min. The yellow color development is assessed spectrophotometrically by absorbance at 405 nm (corrected for background by subtraction of absorbance at 492 nm). Wells showing a greater than fivefold difference between immunoreactivity for Rab25 and Rab11a are subcloned by dilution. Supernates from clones that maintain their selectivity for Rab25 are then screened for utility in Western blot and immunofluorescence applications.

Confocal Fluorescence Microscopy. To visualize the three-dimensional distribution of Rab proteins in polarized cells, the stably transfected Rab25-expressing 2A3 cell line is grown to confluence on Transwell clear filters and maintained for 3 days following confluence, a point where resistances are stable. The cells are then fixed in 4% paraformaldehyde for 30 min at 4°. Cells are permeabilized and blocked simultaneously with 5% goat serum, 0.3% Triton X-100 in phosphate-buffered saline (PBS) for 0.5 hr at room temperature. Cells are then incubated for 2 hr at room temperature with a combination of murine monoclonal anti-Rab25 (12C3, 1:50), rat anti-ZO-1 (Chemicon 1:300), and one of the following rabbit polyclonal antibodies: anti-Rab11a (Zymed, 1:200), anti-Rab5 (QCB, 1:100), or anti-Rab4 (QCB, 1:100). After washing in PBS, cells are incubated for 0.5 hr at room temperature with fluorochrome-conjugated multiply absorbed, species-specific secondary antibodies: Cy2 anti-mouse IgG, Cy3 anti-rabbit IgG, and Cy5 anti-rat IgG (Jackson Immunoresearch). Following a final washing in PBS, the cells are mounted in Prolong Antifade solution (Molecular Probes, Eugene, OR). In some experiments, synthetic blocking peptides at a concentration of 5 μM are included with the primary incubation. These peptide sequences correspond to the variable carboxyl termini of Rab11a and Rab25. To block Rab11a binding, the peptide QKQMSDRRENDMSPSNNVVPIHVPPTTENKPKVQ is used,

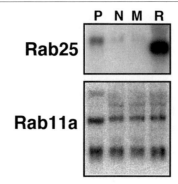

FIG. 1. Rab11a and Rab25 expression in MDCK cell lines. 20 μg of total RNA from nontransfected MDCK cells (N), mock-transfected cells stably transfected with pCB6 (M), the 2A3 Rab25-transfected cell line (R), and rabbit parietal cells (P) were probed with either the rabbit Rab25 or Rab11a cDNAs.

whereas KVSKQIQNSPRSNAIALGSAQAGQEPGPGQKR is used for Rab25.[12]

Cells are examined using scanning confocal fluorescence microscopy (Molecular Dynamics, Sunnyvale, CA). To image triple labeled samples, sections series containing forty 0.3-μm optical sections are performed twice for each sample, once with dual imaging with 488/647-nm excitation laser lines to visualize Cy2 and Cy5 and a second series with the 568-nm laser excitation to visualize Cy3. The section series are then rendered using the look-through projections of Imagespace software (Molecular Dynamics) using a Silicon Graphics workstation.

Results

Characterization of Rab25-Transfected Cell Lines

Out of 20 lines screened, 6 lines were found to contain a transfected Rab25 message. The 2A3 line utilized for characterization of the association of Rab11a and Rab25[9] was one of the two lowest expressing cell lines. To compare the effects of stable overexpression of Rab25 on endogenous message levels, Rab25 and Rab11a mRNA levels were compared between mock pCB6-transfected cells and the Rab25-transfected 2A3 line (Fig. 1). The results demonstrate that while both nontransfected (N) and mock-transfected (M) cells do demonstrate a small amount of Rab25 mRNA, the levels are far below those observed in gastric parietal cells (P). Endogenous

[12] B. C. Calhoun and J. R. Goldenring, *Biochem. J.* **325,** 559 (1997).

Rab25 message in MDCK cells was slightly larger than the Rab25 mRNA in rabbit parietal cells (Fig. 1). The 2A3 cell line (R) has a greater than 10-fold increase in Rab25 mRNA expression. Interestingly, overexpression of Rab25 had no effect on endogenous Rab11a expression (Fig. 1).

Specificity of Rab11a and Rab25 Antibodies

Out of the isolated clones, one hybridoma line, 12C3, showed strong immunoreactivity with recombinant Rab25 in ELISA, but no immunoreactivity for recombinant Rab11a. Similarly, 12C3 saw only Rab25 in Western blots.[12] To establish further the specificity of monoclonal anti-Rab25 (12C3), the 2A3 Rab25-transfected cell line was stained with a polyclonal anti-Rab11a antibody raised against the carboxyl-terminal region of Rab1a (Zymed) or the 12C3 monoclonal anti-Rab25 in the absence or presence of synthetic peptides corresponding to the carboxyl termini of either Rab11a or Rab25. Figure 2 demonstrates that Rab11a and Rab25 antibodies co-

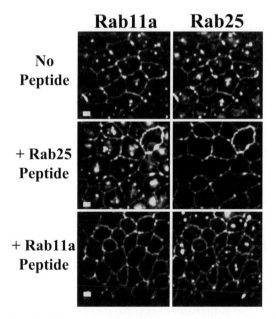

Fig. 2. Specificity of Rab11a and Rab25 antibodies. The 2A3 Rab25-transfected cells were grown to confluence on Transwell permeable supports and fixed in 4% formaldehyde for 30 min at 4°. Cells were triple stained with polyclonal anti-Rab11a, monoclonal anti-Rab25 (12C3), and rat anti-ZO-1 in the absence or presence of either 5 μM Rab11a carboxyl-terminal oligopeptide or 5 μM Rab25 carboxyl-terminal oligopeptide. Images are shown as double labels of Rab11a or Rab25 staining with ZO-1 staining as a reference for cell boundaries. Bar: 4 μm.

stained centrally located vesicles in the 2A3 cell line. The Rab11a carboxyl-terminal peptide abolished staining with the antibody against Rab11a, but had no effect on Rab25 staining. Conversely, the Rab25 carboxyl-terminal peptide abolished Rab25 staining without affecting Rab11a staining. The results establish the specificity of the 12C3 monoclonal antibody for Rab25 and demonstrate that the antibody is directed against the carboxyl variable domain of the protein.

Distribution of Rab25 in Comparison with Other Rab Proteins

To establish the distribution of Rab25 in transfected cells, we compared the Rab25 immunostaining with that for endogenous Rab11a, Rab4, and Rab5 in cells polarized on permeable supports. Endogenous Rab11a immunostaining colocalized with Rab25 in the 12C3 cell line in vesicles localized beneath the apical plasma membrane (Figs. 3a and 3b). In contrast, endogenous Rab4 and Rab5 were distributed in vesicles distributed throughout the cytoplasm of the cells and there was no evidence of significant overlap with Rab25 (Fig. 3). These results in polarized MDCK cells contrast with the results of Trischler et al.[13] in nonpolarized CHO cells where they observed Rab4 in immunoisolated Rab11a-containing vesicles. This discrepancy may reflect differences in recycling system pathways in nonpolarized versus polarized cells.

Functional Impact of Rab25 Transfection

To assess the impact of Rab25 transfection on trafficking in MDCK cells, the trafficking of immunoglobulin A (IgA) was assessed in a stable cell line transfected with both the polymeric IgA receptor and Rab25.[9] Rab25 inhibited both basolateral to apical transcytosis and apical recycling by approximately 40%. Similar results were also obtained using adenoviral infection of MDCK cells with a tetracycline-inhibitable Rab25 expression construct.[9] These results suggested that Rab25 may function as an inhibitory Rab protein along the apical recycling pathways.

Concluding Remarks

The apical recycling system of polarized cells appears to contain at least three Rab proteins: Rab11a, Rab25, and Rab17.[9,14] These proteins appear to have complex and perhaps cell-specific effects on the processes of transcytosis and apical recycling. Both Rab17 and Rab25 are specific to the

[13] M. Trischler, W. Stoorvogel, and O. Ullrich, J. Cell Sci. 112, 4773 (1999).
[14] W. Hunziker and P. J. Peters, J. Biol. Chem. 273, 15734 (1998).

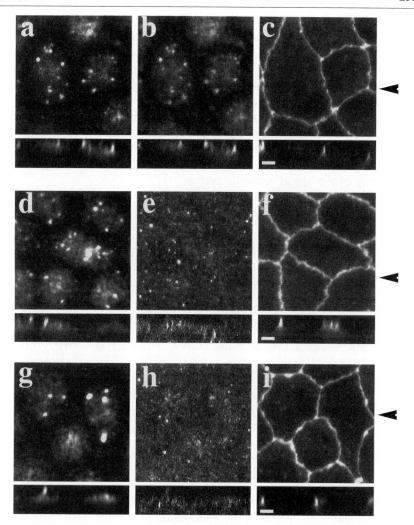

FIG. 3. Distribution of Rab25 in 2A3 cell line compared with endogenous Rab11a, Rab5, and Rab4. The 2A3 Rab25-transfected cells were grown to confluence on Transwell permeable supports and fixed in 4% formaldehyde for 30 min at 4°. Cells were triple stained with monoclonal anti-Rab25 (a, d, g) and rat anti-ZO-1 (c, f, i) along with polyclonal anti-Rab11a (b), anti-Rab5 (e), or anti-Rab4 (h). Images represent look-through projections assembled from forty 0.29-μm optical sections. Upper panels are X-Y projections, while the lower panels are X-Z assembled from points indicated by arrowheads on each of the X-Y projections. Bar: 4 μm.

epithelia,[7,15] while Rab11a is ubquitous. Rab11a and Rab25 show considerable sequence similarity, and thus they may share some upstream and downstream effectors. It remains to be determined whether the functions of Rab11a in regulating transferrin receptor plasma membrane recycling in nonpolarized cells[16] will also be maintained in polarized cells. In polarized cells, the inhibitory action of Rab25 may reflect either a true inhibitory function of Rab25 in these pathways or competition of these proteins for effectors required for both proteins to function in separable but parallel trafficking pathways. Identification of common and unique effectors for Rab11a and Rab25 as well as for Rab17 should allow dissection of the complexity of the apical recycling system.

Acknowledgments

This work was supported by NIH grants DK48370 and DK43405 and a Veterans Administration Merit Award to J.R.G. and grants from the NIH (AI32991, DK33506) and the Good Samaritan Foundation to J.E.C.

[15] A. Lutke, S. Jansson, R. G. Parton, P. Chavrier, A. Valencia, L. A. Huber, E. Lehtonen, and M. Zerial, *J. Cell Biol.* **121,** 553 (1993).
[16] O. Ullrich, S. Reinsch, S. Urbe, M. Zerial, and R. G. Parton, *J. Cell Biol.* **135,** 913 (1996).

[24] Purification of TRAPP from *Saccharomyces cerevisiae* and Identification of Its Mammalian Counterpart

By Michael Sacher and Susan Ferro-Novick

Introduction

Membrane traffic between the endoplasmic reticulum (ER) and Golgi complex has been well studied in the budding yeast *Saccharomyces cerevisiae.* Many genes whose products play a role at this stage of the secretory pathway have been identified and characterized. To date, however, a molecular understanding of how a secretory vesicle is properly targeted to the Golgi is not understood. Several components have recently been reported to be involved in vesicle docking at the Golgi including TRAPP,[1,2] the

[1] M. Sacher, Y. Jiang, J. Barrowman, A. Scarpa, J. Burston, L. Zhang, D. Schieltz, J. R. Yates III, H. Abeliovich, and S. Ferro-Novick, *EMBO J.* **17,** 2494 (1998).
[2] J. Barrowman, M. Sacher, and S. Ferro-Novick, *EMBO J.* **19,** 862 (2000).

Sec34p/Sec35p complex,[3,4] and Uso1p.[5] One common feature shared between these components is their strong genetic interactions with *YPT1*, the gene that encodes the small GTPase involved in the late stages of ER to Golgi traffic.[6] TRAPP appears to play a pivotal role in docking vesicles to the Golgi because 7 of its 10 subunits are essential for the growth of yeast[7] and, unlike Uso1p and the Sec34p/Sec35p complex, overexpression of *YPT1* cannot bypass the loss of this complex.[2] Therefore, identification of the TRAPP subunits (*TRS* gene products) as well as a functional characterization of this complex is necessary to better understand the details of vesicle docking.

Construction of Bet3p-Protein A Fusion Used to Purify TRAPP

TRAPP is initially purified using a strain in which the sole copy of Bet3p is tagged at its carboxy terminus with three c-*myc* epitopes.[1] The Bet3p-associated proteins are then purified from this strain by precipitating the complex from a lysate. This is done by incubating a clarified detergent extract with affinity purified anti-c-*myc* antibody that is bound to Affi-Gel. The bound protein is eluted and analyzed by SDS–PAGE. This affinity purification led to the identity of several (Trs20p, Trs23p, and Trs33p; see Table I) TRAPP subunits. Two other subunits, Bet3p and Bet5p, were identified in genetic screens.[8,9]

Because we could not quantitatively precipitate TRAPP from a lysate by this method, and the preparations were not sufficiently pure to identify the other subunits, a new method of purification was required. The new method employs a different carboxy-terminal tag on Bet3p, protein A (PrA). Precipitation of Bet3p-PrA resulted in a quantitative depletion of the TRAPP complex from a lysate and ultimately led to the identification of the remaining subunits.

To tag the carboxy terminus of Bet3p with PrA, the PrA coding sequence is inserted before the *BET3* stop codon using several polymerase chain

[3] D.-W. Kim, M. Sacher, A. Scarpa, A. M. Quinn, and S. Ferro-Novick, *Mol. Biol. Cell* **10**, 3317 (1999).

[4] S. M. VanRheenen, X. Cao, S. K. Sapperstein, E. C. Chiang, V. V. Lupashin, C. Barlowe, and M. G. Waters, *J. Cell Biol.* **147**, 729 (1999).

[5] C. Barlowe, *J. Cell Biol.* **139**, 1097 (1997).

[6] N. Segev, J. Mulholland, and D. Botstein, *Cell* **52**, 915 (1988).

[7] M. Sacher, J. Barrowman, and S. Ferro-Novick, *Eur. J. Cell Biol.* **79**, 71 (2000).

[8] G. Rossi, K. Kolstad, S. Stone, F. Palluault, and S. Ferro-Novick, *Mol. Biol. Cell* **6**, 1769 (1995).

[9] Y. Jiang, A. Scarpa, L. Zhang, S. Stone, E. Feliciano, and S. Ferro-Novick, *Genetics* **149**, 833 (1998).

TABLE I
IDENTIFICATION OF TRAPP SUBUNITS

TRAPP subunit	Yeast ORF/gene name
Bet5p	YML077w/*BET5*
Trs20p	YBR254c
Bet3p	YKR068c/*BET3*
Trs23p	YDR246w
Trs31p	YDR472w
Trs33p	YOR115c
Trs65p	YGR166w/*KRE11*
Trs85p	YDR108w/*GSG1*
Trs120p	YDR407c
Trs130p	YMR218c

reaction (PCR)/primer extension reactions. First (reaction 1), the *BET3* promoter and open reading frame (until the stop codon) are amplified with the following primers:

primer A   5′-ATACTAGTCGACTCTAGAAAGAGAAAAACGTA
           TAAAATGATTCAAG-3′
primer B   5′-TTGGATAAAACCATTGCGTTGATCCCAATCTTCG
           CCGATCGGTATTTCGTC-3′

Primer B contains the first 20 bases of the PrA coding sequence. Next (reaction 2), the PrA coding sequence is amplified from the vector pRIT2T (Pharmacia) using the following primers:

primer C   5′-GACGAAATACCGATCGGCGAAGATTGGGATCAA
           CGCAATGGTTTTATCCAA-3′
primer D   5′-AAAAAGCTTACATGTCCACTAGTCGACGGATCC
           CCGGGAATTC-3′

Primer C contains the last 20 bases of *BET3* (before the stop codon), whereas primer D contains the stop codon and the subsequent 20 bases from the 3′ UTR of *BET3*. Finally (reaction 3), the *BET3* 3′ UTR is amplified with the following primers:

primer E   5′-GAATTCCCGGGGATCCGTCGACTAGTGGACATG
           TAAGCTTTTT-3′
primer F   5′-ATACTAAGAAGCCGCTCTTCTTCTATACATTC-3′

Primer E contains the last 22 bases of PrA. The products of reactions 2 and 3 are mixed and amplified (reaction 4) using primers C and F above to give PrA fused to the 3′ UTR of *BET3*. Subsequently, the product of reaction 4 is mixed with reaction 1 and amplified with primers A and F above to give the *BET3*-PrA fusion construct. The 2.3-kb *XbaI*/*NheI* fragment from this construct is inserted into the *XbaI* site of pRS305 (*LEU2* integrating vector), linearized with *AflII* and integrated at the *LEU2* locus in SFNY472 (*MATa*/α *ura3-52/ura3-52 BET3/bet3Δ::URA3 leu2-3,112/leu2-3,112*). Transformants are sporulated, dissected, and the Ura⁺ Leu⁺ colonies selected. One transformant is saved and called SFNY737. Western blot analysis with anti-Bet3p antiserum confirmed the loss of wild-type Bet3p and the presence of a new immunoreactive band at ~50 kDa, the expected size of the Bet3p-PrA fusion protein. This clone, which is used for the purification of TRAPP, is sequenced along its entire length to verify that no PCR-generated mutations were present.

Large-Scale Purification of TRAPP and Identification of TRAPP Subunits

TRAPP is purified from 15,000 OD_{600} units of cells that were converted to spheroplasts during a 60-min incubation at 37° in 5 liters of spheroplasting buffer [1.4 M sorbitol, 50 mM KP_i, pH 7.5, 50 mM 2-mercaptoethanol, 33 μg/ml zymolyase 100T (Seikagaku, Tokyo, Japan)] and lysed in 250 ml of buffer A [150 mM KCl, 20 mM HEPES, pH 7.2, 2 mM EDTA, 1% Triton X-100, 0.5 mM dithiothreitol (DTT), 1× protease inhibitor cocktail (PIC[10])] using a Wheaton Dounce homogenizer (tight pestle). All subsequent steps are performed at 4° unless otherwise noted. The lysate is centrifuged at 33,000 rpm in a Type 70 Ti rotor (Beckman, Fullerton, CA) for 1 hr and protein is measured by the method of Bradford.[11] A total of 3.5 g of lysate is diluted to 10 mg/ml with buffer A and the lysate incubated with 6 ml of a 50% slurry of Sepharose CL-4B for 4 hr with rotation. The beads are removed using an EconoColumn (1.5 cm × 50 cm; Bio-Rad, Hercules, CA) and the flow-through is incubated for 16 hr with 6 ml of a 50% slurry of IgG-Sepharose (Pharmacia, Piscataway, NJ). The beads are gently pelleted (~2000g) and loaded into an EconoColumn (1 cm × 10 cm) where they are washed with 15 ml of buffer B (same as buffer A with 500 mM KCl) and 15 ml of buffer A. The bound protein is eluted from the beads by passing 9 ml of 0.2 M glycine, pH 2.8, over the column. Protein is precipitated from the eluate by adding 100% (w/v) trichloroacetic acid (TCA) to the sample to give a final TCA concentration of 10%. After a 1-hr incubation on ice, the sample is centrifuged at 15,000g in an SS34 rotor for 30 min. The precipitate is washed 3 times with ice cold acetone and air dried. The

[10] M. G. Waters and G. Blobel, *J. Cell Biol.* **102,** 1543 (1986).
[11] M. M. Bradford, *Anal. Biochem* **72,** 248 (1976).

protein is solubilized in 1.2 ml SDS–PAGE sample buffer by boiling for 5 min with intermittent vortexing and fractionated on an SDS–8% polyacrylamide gel (for Trs65p, Trs85p, Trs120p, and Trs130p; note that the TRAPP subunits have been designated *TRS* followed by their apparent molecular size on SDS–PAGE) or an SDS–13% polyacrylamide gel (for Trs20p, Trs23p, Trs31p, and Trs33p). The well size for loading the sample is 1.5 mm × 2.5 cm × 2.5 cm, and the sample is loaded twice in aliquots of 600 μl using a 1-ml syringe with an 18-gauge needle. Polypeptides are visualized with Coomassie Brilliant Blue and those which are specific to the Bet3p-PrA tagged strain (Fig. 1) are excised and sequenced as described previously.[1] The identity of each subunit is shown in Table I.

Many laboratory yeast strains contain two distinct species of double-stranded viral RNA. One species, referred to as L-A, encodes a protein called gag of ~85 kDa. Because the gag protein is a common contaminant of the TRAPP preparation and masked Trs85p, SFNY737 can be cured of the virus as follows. The cells are grown on YPD agar plates at 40° and single colonies are purified by three more passages on YPD plates at the same temperature. Clones are tested for the absence of the gag protein by Western blot analysis using a monoclonal anti-gag antibody (kind gift of Dr. D. Toft, Mayo Clinic, MN). Those that no longer contain the gag protein are saved as SFNY904.

Small-Scale Precipitation of TRAPP

TRAPP has also been purified on a smaller scale from labeled yeast lysates to examine its stability in salt and for pilot experiments prior to scaling up. To do this, yeast strains are radiolabeled at 2 OD_{600} units/ml for 2 hr at 25° in synthetic medium (Difco, Detroit, MI) with the appropriate supplements and 100 μCi/ml ProMix (Amersham, Arlington Heights, IL). Aliquots of 15 OD_{600} units of cells are spheroplasted in 5 ml of spheroplast buffer (see above) and lysed in 600 μl of buffer A. Lysates are clarified by centrifugation in an SW 50.1 rotor at 30,000 rpm for 1 hr at 4° (all subsequent steps are performed at 4° unless otherwise noted). IgG-Sepharose precipitations are performed on 50 × 10^6 cpm of lysate (25–90 μl) by first diluting the sample to 1 ml with buffer A and then incubating the samples with 60 μl of a 50% slurry of Sepharose CL-4B (preequilibrated in buffer A) for 1 hr. The beads are pelleted at 3000g for 30 sec and the supernatant is transferred to a fresh tube containing 60 μl of a 50% slurry of IgG-Sepharose prepared in buffer A. Following a 2-hr incubation, the beads are pelleted as above and washed 3 times with ice-cold buffer B and 2 times with ice-cold buffer A (1 ml/wash). After the final wash, protein is eluted from the beads by incubating them with 100 μl of 0.2 M glycine, pH 2.8, at room

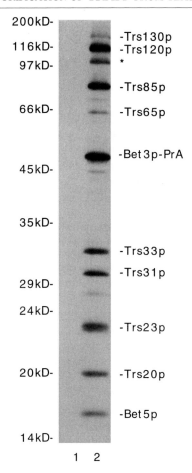

FIG. 1. A Bet3p-PrA fusion protein can be used to purify TRAPP. Untagged (lane 1) and Bet3p-PrA-tagged (lane 2) strains were radiolabeled, converted to spheroplasts, and lysed as described in the text. The lysates (200 × 10⁶ cpm) were fractionated on a Superdex-200 gel filtration column (conditions are described in the "Identification of Mammalian TRAPP from HeLa Lysates" section of the text), Triton X-100 was added to a final concentration of 1%, and the fractions (1 ml) were treated with IgG-Sepharose as described in the "Small-Scale Precipitation of TRAPP" section. The fraction (fraction 9) containing Bet3p-PrA with all the TRAPP subunits is shown. TRAPP subunits are listed on the right and the positions of the molecular weight markers are shown on the left. The asterisk (∗) band is a breakdown product of Trs120p and the unmarked band immediately below Trs31p is a breakdown product of Trs130p.

temperature for 5 min. The beads are pelleted and the supernatant is removed and neutralized with KOH (pH was tested with pH paper). For Bet3p-*myc* precipitations, samples (50 × 10⁶ cpm) are diluted with buffer A to 1 ml and rotated with 30 μl of a 50% slurry of PrA-Sepharose CL-4B for 1 hr. The beads are pelleted as above and the supernatant transferred to a fresh tube containing 2 μl of anti-c-*myc* ascites fluid and rotated for 1 hr. At that time, 30 μl of a 50% slurry of PrA-Sepharose is added, and the incubation is continued for an additional hour. Beads are washed as described above and protein is eluted by boiling in 70 μl of SDS–PAGE sample buffer. The precipitates are analyzed on an SDS–13% polyacrylamide gel followed by fluorography.

Identification of Mammalian TRAPP from HeLa Lysates

To examine whether an analogous complex exists in mammalian cells, antibodies are raised to recombinant human orthologs of Bet3p and Trs20p and used to probe samples from a gel filtration column. HeLa cells used for gel filtration are purchased from the National Cell Culture Center. Aliquots (60 × 10⁸) of cells are lysed by homogenization in a Wheaton Dounce homogenizer (10 strokes with pestle A) in 2 ml of 20 mM HEPES, pH 7.2. The lysate is transferred to a fresh tube and NaCl, EDTA, DTT, and PIC were added to final concentrations of 150 mM, 2 mM, 1 mM, and 1×, respectively. The sample is clarified in an SW 50.1 rotor at 100,000g for 1 hr at 4°. A Superdex-200 gel filtration column (25-ml bed volume; HR 10/30; Pharmacia) is equilibrated with buffer (150 mM NaCl, 20 mM HEPES, pH 7.2, 2 mM EDTA, 1 mM DTT, and 1× PIC) and loaded with 8 mg HeLa lysate. A flow rate of 0.4 ml/min is used while 1 column volume

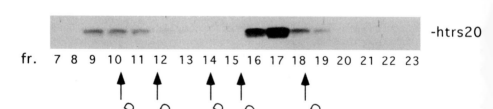

FIG. 2. HeLa cells contain htrs20 in a high molecular weight fraction. A lysate from HeLa cells was prepared and fractionated by gel filtration chromatography as described in the text. Samples from the column were boiled in SDS–PAGE sample buffer, fractionated on an SDS–15% polyacrylamide gel, transferred to nitrocellulose, and probed with anti-htrs20 antiserum. The positions of size standards used for gel filtration are shown on the bottom of the panel.

of buffer is passed through and collected as 1-ml fractions. Samples are boiled in SDS–PAGE sample buffer and proteins are detected by Western blot analysis using anti-human bet3 and anti-human trs20 antiserum (both antisera were used at 1 : 1000 dilution).

As seen in Fig. 2, the majority of human trs20 was found in a low molecular weight pool likely corresponding to monomers. However, a second higher molecular weight pool was seen which, at ~670 kDa, is somewhat smaller in size compared to the yeast complex of ~1000 kDa.[6] Note that the human orthologs are smaller in size compared to their yeast counterparts,[1,7,9] possibly explaining the size difference noted above between yeast and mammalian TRAPP.

Section II

ADP-Riboylation Factor (ARF) GTPases

[25] Structural and Functional Organization of ADP-Ribosylation Factor (ARF) Proteins

By Sophie Béraud-Dufour and William E. Balch

ADP-ribosylation factors (ARFs), which were originally identified as cofactors for cholera toxin,[1] are small G proteins (21 kDa) implicated in the maintenance of organelle structure, formation of coated vesicles in the secretory and endocytic pathways, and activation of enzymes that modify phospholipids such as phospholipase D or phosphatidylinositol-4-OH kinase. Six ARF proteins have been identified in mammals, three in yeast, and several ARF-related proteins have been found in mammals (ARL, ARD1, ARP). ARF1, which is very closely related to ARF3 (seven differences, exclusively in the N-terminal and C-terminal extensions), is the best studied and controls the formation of at least two different coats (coatomer and clathrin through the binding of AP-1 and AP-3 adapter complexes) that act at many steps in intracellular trafficking, including endoplasmic reticulum (ER)–Golgi retrograde transport, ER–Golgi anterograde transport, intra-Golgi transport, and post-Golgi transport. ARF4 and ARF5 are closely related to each other in terms of sequence and, although ARF5 appears to be localized in the Golgi and endoplasmic reticular golgi intermediate compartment (ERGIC), their function is mostly unknown. ARF6, the least conserved ARF protein, is found at the plasma membrane and regulates endocytic and recycling pathway and cytoskeleton organization.

Like all G proteins, ARFs bind guanine nucleotides and adopt two different conformations: an inactive GDP-bound state and an active GTP-bound state. The crystal structure of ARF1 and ARF6 have been determined in several forms[2–4] (Fig. 1). ARF1 shares structural homologies with the Ras core domain but presents specific characteristics. ARF1 has a supplementary myristoylated amphipatic α helix situated in the N terminal, which controls the interaction of ARF1 with membrane.[5] ARF1–GDP is quasi-soluble; the helix (but not the myristate) is buried in the core domain of the protein.

[1] R. A. Kanh and G. A. Gilman, *J. Biol. Chem.* **259,** 6235 (1984).

[2a] T. Amor, D. H. Harrison, R. A. Kahn, and D. Ringe, *Nature* **372,** 704 (1994).

[2b] S. E. Greaslet, H. Jhoti, C. Teahan, R. Solari, A. Fensome, G. M. H. Thomas, S. Cockcroft, and B. Bax, *Nature Structural Biology* **2,** 797 (1995).

[3] J. Menetrey, E. Macia, S. Pasqualeto, M. Franco, and J. Cherfils, *Nat. Struct. Biol.* **7,** 466 (2000).

[4] J. Goldberg, *Cell* **95,** 237 (1998).

[5] B. Antonny, S. Béraud-Dufour, P. Chardin, and M. Chabre, *Biochemistry* **36,** 4675 (1997).

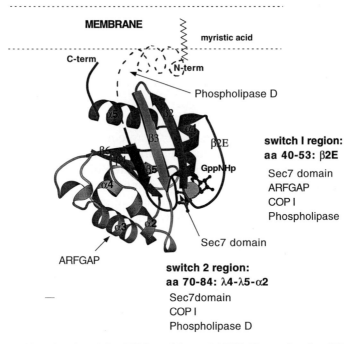

FIG. 1. Ribbon drawing of the GTP-bound form of ARF1. The myristoylated N-terminal helix of ARF1 has been modeled in the potential membrane-bound conformation. The switch 1 and switch 2 regions (regions modified during GDP-GTP exchange) are indicated. The binding sites for membranes,[5] the Sec7 domain of GEF,[10,4] ARF-GAP,[13b] phospholipase D,[15] and subunits of the COPI proteins[16] are indicated. Most contacts have been probed by site-directed mutagenesis, but only the interfaces between ARF1 and GEFs or GAPs have been described in 3D structures.

ARF1–GTP interacts strongly with membrane as the helix has swung and is now oriented to interact with the membrane. Indeed, following GDP–GTP exchange, the switch regions of ARF1 are restructured, inducing the movement of one loop (between $\beta2$ and $\beta3$ strands), which now occupies the position of the N-terminal helix. This helix is shorter in the case of ARF6.

Specific classes of guanine nucleotide exchange factors (GEFs) and GTPase activating proteins (GAPs) control the GDP–GTP cycle of ARFs proteins. All GEFs for ARFs share a common catalytic domain: the Sec7 domain.[6] Several GEFs have been identified in mammals: the ARNO family (which are *in vitro* GEFs for ARF1[6,7]), p200, GBF1, and EFA6 (specific

[6] P. Chardin, S. Paris, B. Antonny, S. Robineau, S. Béraud-Dufour, C. L. Jackson, and M. Chabre, *Nature* **384,** 481 (1996).

[7] M. Franco, J. Boretto, S. Robineau, S. Monier, B. Goud, P. Chardin, and P. Chavrier, *Proc. Nat. Acad. Sci. U.S.A.* **95,** 9926 (1998).

for ARF6[8]) and in yeast (Sec7p, Gea1, and GEa2[9]). The "large" GEFs are sensitive to brefeldin A, a fungal metabolite that induces the fusion of the Golgi apparatus with the ER, and that acts by stabilizing an abortive complex between ARF1–GDP and the Sec7 domain of GEFs.[10] The structure of a complex between a truncated form of ARF1, [Δ17]ARF1, and the Sec7 domain of Gea2 has also been solved recently.[4] The Sec7 domain catalyzes the GDP–GTP exchange by destabilizing the nucleotide with a glutamate residue (the "glutamic finger"[11]) and by restructuring the switch regions and the phosphate-binding loop of ARF1. Membranes are necessary for the GDP–GTP exchange catalyzed by ARNO proteins.[12a,12b] Several ARF–GAP proteins for mammals and and yeast ARFs have been identified. They all show a characteristic zinc finger motif.[13b] The structure of a complex between ARF1–GDP and the catalytic domain of ARF1–GAP has been crystallized.[13a] An arginine from the GAP (the "arginine finger") is essential to promote the hydrolysis of the third phosphate of GTP. In this case, it might also be possible that a component of the coat helps to position properly this arginine.[13a,14]

[8] M. Franco, P. J. Peters, J. Boretto, E. van Donselaar, A. Neri, C. D'Souza-Schorey, and P. Chavrier, *EMBO J.* **18**, 1480 (1999).
[9] A. Peyroche, S. Paris, and C. L. Jackson, *Nature* **384**, 479 (1996).
[10] P. Chardin and F. McCormick, *Cell* **97**, 153 (1999).
[11] S. Béraud-Dufour, S. Robineau, P. Chardin, S. Paris, M. Chabre, J. Cherfils, and B. Antonny, *EMBO J.* **17**, 3651 (1998).
[12a] S. Paris, S. Béraud-Dufour, S. Robineau, J. Bigay, B. Antonny, M. Chabre, and P. Chardin, *J. Biol. Chem.* **272**, 22221 (1997).
[12b] S. Béraud-Dufour, S. Paris, M. Chabre, and B. Antonny, *J. Biol. Chem.* **274**, 37629 (1999).
[13a] J. Goldberg, *Cell* **96**, 893 (1999).
[13b] V. Mandiyan, J. Andreev, J. Schlessinger, and S. R. Hubbard, *EMBO J.* **18**, 6890 (1999).
[14] E. Szafer, E. Pick, S. Rotman, I. Huber, and D. Cassel, *J. Biol. Chem.* **275**, 23615 (2000).
[15] D. Jones, C. Morgan, and S. Cockcroft, *Biochim. Biophys. Acta* **1439**, 299 (1999).
[16a] L. Zhao, J. B. Helms, B. Brugger, C. Harter, B. Martoglio, R. Graf, J. Brunner, and F. T. Wieland, *Proc. Natl. Acad. Sci. U.S.A.* **94**, 4418 (1997).
[16b] L. Zhao, J. B. Helms, J. Brunner, and F. T. Wieland, *J. Biol. Chem.* **274**, 14198 (1999).

[26] Expression and Properties of ADP-Ribosylation Factor (ARF6) in Endocytic Pathways

By JULIE G. DONALDSON and HARISH RADHAKRISHNA

Introduction

ADP-ribosylation factor 6 (ARF6) is a member of the family of ARF GTPases and is the most distantly related to the well-studied ARF1 protein.

ARF6 is distinct from ARF1 in being peripherally localized to the plasma and endosomal membranes, and not associated with the Golgi complex.[1] Furthermore, ARF6 is activated in cells by guanine nucleotide exchange factors (GEFs) insensitive to brefeldin A(BFA),[2,3] unlike the BFA-sensitive activation of ARF1 at the Golgi complex. ARF6 protein is expressed ubiquitously[4,5] and in many cells remains associated with membrane whether bound to GTP or GDP.[4,6] Others, however, have evidence that ARF6, like ARF1, may cycle on and off membranes when GTP and GDP bound, respectively.[7] ARF6 functions in cells to regulate a plasma membrane (PM)–endosomal recycling pathway that in many cells is separate from the clathrin-facilitated endocytosis pathway.[8] Additionally, accumulation of ARF6–GTP at the PM is associated with various cortical actin-rich structures including protrusions[9] and membrane ruffling[10] and ARF6 activity has been shown to be required for cell spreading.[4] Here we describe approaches to studying ARF6 roles in membrane trafficking, which is integral to the additional effects of ARF6 observed on the actin cytoskeleton.

Expression of Wild-Type ARF6 and Mutants in Mammalian Cells

A convenient method for studying ARF6, and novel GTPases in general, is to use transient expression in cells of wild-type and mutant GTPases. In this way, the consequences of expression of GTP-binding defective (i.e., inactive, GDP-bound form) and GTPase-defective (i.e., constitutively active, GTP-bound form) mutants can be observed in cells and compared with that observed by expression of the wild-type ARF6. We typically transfect cells with expression vectors containing wild-type and mutant proteins using the calcium phosphate coprecipitation method.[11] Alterna-

[1] P. J. Peter, V. W. Hsu, C. E. Ooi, D. Finazzi, S. B. Teal, V. Oorschot, J. G. Donaldson, and R. D. Klausner, *J. Cell Biol.* **128,** 1003 (1995).

[2] S. Frank, S. Upender, S. H. Hansen, and J. E. Casanova, *J. Biol. Chem.* **273,** 23 (1998).

[3] M. Franco, P. J. Peters, J. Boretto, E. van Donselaar, A. Neri, C. D'Souza-Schorey, and P. Chavrier, *EMBO J.* **18,** 1480 (1999).

[4] J. Song, Z. Khachikian, H. Radhakrishna, and J. G. Donaldson, *J. Cell Sci.* **111,** 2257 (1998).

[5] C. Z. Yang, H. Heimberg, C. D'Souza-Schorey, M. M. Mueckler, and P. D. Stahl, *J. Biol. Chem.* **273,** 4006 (1998).

[6] M. M. Cavenagh, J. A. Whitney, K. Carroll, C. Zhang, A. L. Boman, A. G. Rosenwald, I. Mellman, and R. A. Kahn, *J. Biol. Chem.* **271,** 21767 (1996).

[7] J. Gaschet and V. W. Hsu, *J. Biol. Chem.* **274,** 20040 (1999).

[8] H. Radhakrishna and J. G. Donaldson, *J. Cell Biol.* **139,** 49 (1997).

[9] H. Radhakrishna, R. D. Klausner, and J. G. Donaldson, *J. Cell Biol.* **134,** 935 (1996).

[10] C. D'Souza-Schorey, R. L. Boshans, M. McDonough, P. D. Stahl, and L. vanAelst, *EMBO J.* **16,** 5445 (1997).

[11] J. S. Bonifacino, P. Cosson, and R. D. Klausner, *Cell* **63,** 403 (1990).

tively, cells can be transfected using commercially available lipid-based reagents (Lipofectamine, Gibco-BRL Gaithersburg, MD; Fugene-6, Roche Molecular Biochemicals, Indianapolis, IN). The use of inducible expression systems or retroviral expression is also quite popular.

The expression of epitope-tagged ARF proteins has some advantages and disadvantages. Because ARF proteins must be myristoylated at the amino terminus to retain biological activity, epitope tagging must be confined to the carboxyl end of the molecule. Small tags appended to the carboxyl terminus (i.e., influenza hemagglutinin, HA) still allow the expressed ARF proteins to localize correctly (ARF1-HA to Golgi complex; ARF6-HA to PM and endosomes) and function reasonably well in cells.[1,8] The epitope tagging also allows for identification of transfected ARFs by immunolocalization and immunoblotting, and facilitates the immunoprecipitation of the ARFs. Also, the wide availability of commercial antibodies to these peptide tags facilitates studies investigating the roles of different ARF proteins in cells.

A significant drawback, however, of the use of epitope-tagged ARFs is that the presence of the tag does impair to some extent the "potency" of the protein. We have noticed this especially to be a problem for seeing effects of the "dominant, negative" ARF6 mutant, T27N. These mutants in small GTPases are believed to exert their effects by binding to and sequestering the guanine nucleotide exchange factors.[12] Although we have observed that the Rac1 mutant T17N is remarkably effective at inhibiting wild-type Rac1 activation even when the wild-type protein is in excess, ARF6-T27N must be in excess of wild-type ARF6 to effectively inhibit activation of the wild-type ARF6.[13,14] This potency problem is exacerbated when ARF6-T27N-HA is expressed. In HeLa cells, ARF6-T27N is much more effective at inhibiting cell spreading than ARF6-T27N-HA.[14] To sum, in cases where inhibition with ARF6-T27N is not observed, the situation should be assessed with regard to the presence of epitope tags and the levels of endogenous ARF6 in the cells. Surprisingly, ARF1 and ARF6 tagged with the green fluorescent protein (GFP) show reasonable localization in cells,[14,15] however, whether the dominant negative and constitutively active mutants of these chimeras are fully functional is not clear.

The expression of untagged ARF6 proteins can be detected by ARF6-specific antibodies. Although commercial sources of these antibodies are

[12] M. S. Boguski and F. McCormick, *Nature* **366,** 543 (1993).

[13] H. Radhakrishna, O. Al-Awar, Z. Khachikian, and J. G. Donaldson, *J. Cell Sci.* **112,** 855 (1999).

[14] H. Radhakrishna and J. G. Donaldson, unpublished observations (1998).

[15] C. Vasudevan, W. P. Han, Y. D. Tan, Y. M. Nie, D. Q. Li, K. Shome, S. C. Watkins, E. S. Levitan, and G. Romero, *J. Cell Sci.* **111,** 1277 (1998).

not available, several investigators have generated such antibodies.[4–6,16] We have successfully made antipeptide antibodies in rabbits using peptides from the ARF6 sequence. Peptides 94–111 (DRIDEAR..) and 164–175 (EGLTWLT..) conjugated to keyhole limpet hemocyanin (KLH) have made reasonable immunogens for generating antibodies specific for ARF6.[4] These peptides likely encompass the α_3 and α_5 helicies based on the crystal structure of the ARF1 protein.[17] Pan-ARF antibodies like the hybridoma 1D9, which will detect the overexpressed protein, are also available from commercial sources (Affinity Bio Reagents, Golden, CO).

ARF6 in Endocytosis in HeLa Cells

We have been studying ARF6 function in HeLa cells and our experiences convince us that these cells can serve as a model for studying alternative endocytic pathways. The ARF6 endosomal compartment in these cells is separate and distinct from the well-studied transferrin receptor endosomal system.[8] This is also the case in primary human fibroblasts.[14] However, in Chinese hamster ovary (CHO) cells, there is partial overlap between the ARF6 endosome and transferrin receptor-containing endosomal compartment.[16]

Our studies in HeLa cells have been facilitated by the identification of reversible, pharmacologic reagents that mimic the effects of the GTP-binding defective ARF6-T27N and GTPase-defective ARF6-Q67L mutants. This has allowed us to study the trafficking from the PM through the ARF6 endosome in cells expressing the wild-type protein. In particular, we use low concentrations of cytochalasin D (CD) to accumulate and trap membrane from the PM into the endosome. This accumulation and trapping is also observed on expression of ARF6-T27N in cells indicating that ARF6 activation and actin filaments are required for recycling of this membrane back to the PM. The advantage of CD treatment over ARF6-T27N expression is that after loading the endosomal compartment, we can remove CD and observe the recycling of membrane back to the PM. In cells expressing ARF6-T27N, membrane is internalized even in the absence of CD treatment, but it does not recycle.[8] Alternatively, we can treat cells with aluminum fluoride (AlF) (mixtures of sodium fluoride and aluminum chloride) to mimic the effects of expression of the activated mutant of ARF6, Q67L, in cells.[9] In this case, actin-containing protrusions are formed and a block in internalization into the ARF6 endosomal compartment is observed.[8]

To study this novel endosomal membrane recycling pathway, we also

[16] C. D'Souza-Schorey, G. Li, M. I. Colombo, and P. D. Stahl, *Science* **267**, 1175 (1995).
[17] J. C. Amor, D. H. Harrison, R. A. Kahn, and D. Ringe, *Nature* **372**, 704 (1994).

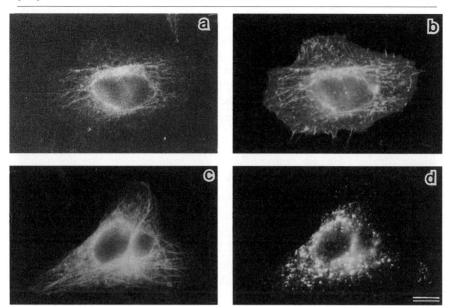

FIG. 1. Differential endocytic pathways marked by Tac and Tac-DKQTLL. HeLa cells cotransfected with ARF6 and either Tac (a, b) or Tac-DKQTLL (c, d) were incubated in media containing 7G7 anti-Tac antibody in the presence of CD (0.1 μM) for 30 min. Cells were then fixed and labeled with antibodies to ARF6 (a, c) and internalized Tac antibody detected by fluorescent secondary antibody labeling (b, d). Note that Tac is internalized from the PM into the ARF6-labeled endosome, whereas Tac-DKQTLL is internalized into other endosomal/lysosomal structures. Bar: 10 μm.

have made use of the Tac antigen, the IL2 receptor α subunit, to mark the ARF6 endosomal pathway in cells. Tac, a type I membrane protein, does not contain any tyrosine or dileucine motifs in its cytoplasmic tail and hence is not internalized via AP2-clathrin facilitated endocytosis. However, chimeras of Tac can be made with cytoplasmic tails encoding clathrin-internalization motifs such as the dileucine motif DKQTLL.[18] Hence the Tac and Tac-DKQTLL proteins can be used to mark the different pathways in the cells. Conveniently, the hybridoma cell line 7G7B6 (HB-8764) available from the American Type Culture Collection (Manassas, VA) produces an excellent monoclonal antibody that is useful for immunofluorescence and immunoprecipitation.[19] As an example, in Fig. 1 we show that in the presence of CD, which enhances the loading and visualization of the endoso-

[18] F. Letourneur and R. D. Klausner, *Cell* **69,** 1143 (1992).
[19] L. A. Rubin, C. C. Kurman, W. E. Biddison, N. D. Goldman, and D. L. Nelson, *Hybridoma* **4,** 91 (1985).

mal compartment, Tac antibody internalized into Tac-expressing cells colocalizes with ARF6 on the endosome. However, Tac antibody uptake into Tac-DKQTLL-expressing cells does not colocalize with the ARF6 endosome, but rather labels early and late endosomes and lysosomes, representative of pathways taken by many proteins internalized via clathrin-mediated mechanisms. Bonifacino and colleagues have made many Tac chimeras that have been used in subsequent studies to reveal the further complexity of the endosomal pathways in cells.[20,21]

Internalization and Recycling of Anti-Tac Antibodies to Monitor Tac Trafficking along ARF6 Endocytic Pathway

Overview

The following two assays provide a convenient way to follow both the internalization and recycling of Tac through the ARF6 endosomal compartment in cells using indirect immunofluorescence methods.[8] Below is a brief overview of these assays, followed by a more detailed description of the methods in latter sections.

To monitor Tac internalization, cells expressing Tac are chilled to 4° and incubated with anti-Tac antibodies on ice to label the surface pool of Tac. Warming to 37° permits the antibodies bound to Tac to be cointernalized into these cells. Subsequent immunofluorescence localization allows the visualization of both the internalized antibodies as well as those remaining at the cell surface. Acid washing prior to fixation enables one to selectively view only the internalized anti-Tac antibodies.

To monitor Tac recycling from ARF6 endosomes to the PM, cells are first permitted to internalize anti-Tac antibodies as described above. Acid stripping is used to remove uninternalized antibodies from the cell surface followed by subsequent warming at 37° to initiate the recycling of both Tac and the bound antibodies back to the PM. Immunofluorescence localization of the anti-Tac antibodies in *fixed and permeabilized* cells allows the visualization of both the recycled antibodies and those still remaining in the ARF6 endosomes. In contrast, localization of anti-Tac antibodies in *fixed but nonpermeabilized* cells allows the visualization of only anti-Tac antibodies that have recycled back to the plasma membrane.

Internalization of Anti-Tac Antibodies

HeLa cells grown on 12-mm circular glass coverslips are transfected with plasmids encoding Tac either alone or with other plasmids (e.g., ARF6

[20] M. S. Marks, L. Woodruff, H. Ohno, and J. S. Bonifacino, *J. Cell Biol.* **135**, 341 (1996).
[21] W. G. Mallet and F. R. Maxfield, *J. Cell Biol.* **146**, 345 (1999).

WT) using the calcium phosphate coprecipitation method.[11] We routinely use 2 μg of each plasmid when transfecting the cells in 35-mm dishes or six-well dishes. The cells are ready for experimentation 30–36 hr later. Typically, six coverslips are needed per transfection sample per time point for a complete analysis of antibody uptake.

Individual coverslips are transferred to 12 well dishes containing ice-cold medium [Dulbecco's modified Eagle's medium (DMEM) + 10% (v/v) fetal bovine serum (FBS)] and placed in an ice water slurry in the cold room. The cells are rinsed twice with cold medium and covered with approximately 0.5 ml of culture supernatant containing anti-Tac antibodies and supplemented with 10 mM HEPES, pH 7.4. Depending on the antibody titer, the culture supernatant can be diluted. These dishes are covered and incubated for 30 min on ice. The cells are rinsed with ice-cold medium three times to remove excess antibodies. The washes must be gently aspirated and added since the cells are generally fragile at this step. Two coverslips are removed and fixed in ice-cold 2% formaldehyde in phosphate-buffered saline (PBS), pH 7.4. These represent the initial bound samples and all of the Tac antibody staining is confined to the plasma membrane. This can be confirmed by labeling the surface of cells with fluorescently conjugated concanavalin A (ConA) in the absence of saponin. We routinely use fluorescent ConA purchased from Molecular Probes (Eugene, OR).

The remaining four coverslips are then warmed to 37° for 30 min in the presence or absence of pharmacologic reagents [e.g., 1 μM CD, aluminum fluoride (30 mM NaF + 50 μM AlCl$_3$), etc.] to allow the bound antibody to be internalized. As an alternative to the 4° binding and internalization protocol described above, cells can be incubated at 37° in the hybridoma cell supernatant for simultaneous antibody binding and internalization. At the end of the 37° internalization, two coverslips are transferred to a 10-cm dish containing 10 ml of acid wash (0.5% acetic acid, 0.5 M NaCl, pH 3.0) and swirled at room temperature for approximately 15–30 sec to remove surface bound antibody. The coverslips are then transferred to a second 10-cm dish containing 10 ml of medium, swirled as above, and then fixed along with the remaining two non-acid-washed coverslips in 2% formaldehyde in PBS. At this point there will be three sets of duplicate coverslips: (1) two initial bound samples, (2) two uptake samples (no acid wash), and (3) two uptake samples (acid washed). All of the coverslips are incubated with PBS containing 10% FBS (PBS/serum) for 5 min to quench any residual fixative.

To label the total pool of cell-associated Tac and ARF6, the fixed cells are first incubated with rabbit antibody to ARF6 in PBS/serum containing 0.2% saponin, followed by incubation in PBS/serum/0.2% saponin containing fluorescently conjugated anti-rabbit secondary antibodies to detect

ARF6 and anti-mouse antibodies to visualize the cell-associated Tac antibody. To visualize only the surface-associated Tac antibody, the remaining coverslips are incubated with anti-mouse secondary antibodies in the absence of saponin. This can be followed by a brief period of fixation (3 min in 2% formaldehyde), followed by immunolabeling for ARF6 in the presence of saponin.

Observations

The ARF6 endosomal compartment in HeLa cells exhibits multiple tubular extensions that arise from a juxtanuclear center (see Fig. 1).[8] Tac normally cycles between the PM and this compartment. In cells treated with low concentrations (0.1–1 μM) of CD, internalized Tac accumulates in this compartment due to a block in recycling (see below). The tubular extensions align along microtubules and are disrupted on treatment with nocodazole.[8]

In cells expressing Tac and wild-type ARF6, Tac antibody is mostly internalized via large vesicular structures that do not colocalize with proteins internalized via clathrin-mediated endocytosis, such as transferrin or Tac-DKQTLL. Although some Tac antibody can be observed in small vesicles that colabel with transferrin, it is quickly sorted away into the ARF6 endosome. Significant labeling of the ARF6 tubular endosomal compartment can be observed within 10–15 min. The accumulation is enhanced in the presence of CD because the membrane becomes trapped there, but can also be observed in cells loaded with Tac antibody in the absence of CD.

Recycling of Anti-Tac Antibodies from ARF6 Endosomes to Plasma Membrane

HeLa cells expressing Tac are "loaded" with Tac antibodies at 37° for at least 30 min as described above either in the presence or absence of CD. "Loading" in the presence of CD facilitates the accumulation of Tac antibody in the ARF6 compartment. The antibodies remaining at the surface are removed by acid washing the cells as described above. Two coverslips should be fixed after the acid-washing step. This represents the starting point for the recycling experiment. The remaining samples are warmed to 37° under various conditions (e.g., drug treatment, time course) and then fixed with formaldehyde as above.

Immunofluorescence localization of Tac antibodies in the absence of detergent permeabilization permits the visualization of only the Tac antibodies that have recycled to the cell surface. At time zero of recycling, there should be no surface staining in nonpermeabilized cells. Localization in the presence of detergent permits the visualization of both recycled antibodies and those remaining in the endosomal compartment.

Observations

When the cells are warmed in the absence of CD, Tac antibodies recycle to discrete sites at the PM within 15 min.[8,13] The small GTPase, Rac1, also associates with ARF6 endosomes and is similarly transported to the PM.[13] These sites of antibody reappearance at the PM are enriched in F-actin suggesting that Tac recycles to sites of actin polymerization.[14] In cells coexpressing ARF6, these sites exhibit enhanced F-actin staining within 5–10 min after removal of CD and this effect is slightly diminished thereafter. This suggests that the initial delivery to the PM of membrane and/or contents (e.g., Rac1) from the ARF6 endosome results in a transient "burst" of actin polymerization that diminishes afterward.

Tac recycling is critically dependent on intact actin filaments since warming the cells in the presence of even low concentrations (e.g., 0.1 μM) of CD blocks the recycling of Tac antibodies.[8] At these concentrations of CD, actin stress fibers are not extensively depolymerized, suggesting that perhaps it is the cortical actin meshwork that is critical for recycling from ARF6 endosomes. In addition, a chemically and mechanistically unrelated inhibitor of actin polymerization, latrunculin B, also blocks recycling and causes the accumulation of ARF6 and Tac in ARF6 endosomes.[14] In contrast, treatment of cells with the actin filament-stabilizing compound, jasplakinolide, does not affect recycling to the PM. These results suggest that it is the presence of intact filaments rather than the dynamic polymerization/depolymerization of actin that is important for recycling from ARF6 endosomes. Membrane recycling from the ARF6 endosomal compartment is somewhat reminiscent of melanosome transport to the PM in melanocytes. Wu et al. have shown that melanosomes are transported from the perinuclear region to the edges of cells along microtubules and the subsequent delivery to the PM is dependent on both actin filaments and the actin motor protein, myosin V.[22] The role of myosin proteins in recycling along the ARF6-regulated endocytic pathway is not known.

Concluding Remarks

We have described here a protocol for visualizing the internalization of PM proteins into the ARF6 endosome and their recycling back to the PM. These techniques can be used to look at trafficking of other Tac

---

[22] X. Wu, B. Bowers, K., Rao, Q. Wei, and J. A. Hammer III, *J. Cell Biol.* **143,** 1899 (1998).

chimeras as well.[20,21] For a quantitative variation on these assays, surface biotinylation can be utilized to follow quantitatively the fate of internalized Tac in the cells.[23]

[23] M. P. Lisanti, A. LeBivic, M. Sargiacomo, and E. Rodriguez-Boulan, *J. Cell Biol.* **109**, 2117 (1989).

[27] Expression and Analysis of ARNO and ARNO Mutants and Their Effects on ADP-Ribosylation Factor (ARF)-Mediated Actin Cytoskeletal Rearrangements

By Lorraine C. Santy, Scott R. Frank, and James E. Casanova

Introduction

ARNO is a member of a family of guanine nucleotide exchange factors (GEFs) with specificity for the ADP-ribosylation factor (ARF) GTPases.[1] Although divergent in overall sequence, these proteins share a common catalytic domain with homology to the yeast protein Sec7, and are thus designated as the Sec7 family. The Sec7 family can be readily divided into two subsets. The first are all large (>150 kDa), localize to the Golgi apparatus, and with one exception (GBF1[2]) are inhibited by the fungal metabolite brefeldin A (BFA). ARNO is a member of the second group, which has members that are smaller in size and resistant to BFA. This group also includes cytohesin-1,[3] GRP-1/ARNO3,[4,5] and EFA6.[6] ARNO, cytohesin-1, and GRP-1/ARNO3 are closely related and share a similar domain structure (Fig. 1). The amino terminus of these proteins contains an ~60 residue coiled-coil domain, thought to function in homodimeriza-

[1] C. L. Jackson and J. E. Casanova, *Trends Cell Biol.*, in press (2000).
[2] A. Claude, B. P. Zhao, C. E. Kuziemsky, S. Dahan, S. J. Berger, J. P. Yan, A. D. Armold, E. M. Sullivan, and P. Melanon, *J. Cell Biol.* **146**, 71 (1999).
[3] W. Kolanus, W. Nagel, B. Schiller, L. Zeitlmann, S. Godar, H. Stockinger, and B. Seed, *Cell* **86**, 233 (1996).
[4] J. K. Klarlund, L. E. Rameh, L. C. Cantley, J. M. Buxton, J. J. Holik, C. Sakelis, V. Patki, S. Corvera, and M. P. Czech, *J. Biol. Chem.* **273**, 1859 (1998).
[5] M. Franco, J. Boretto, S. Robineau, S. Monier, B. Goud, P. Chardin, and P. Chavrier, *Proc. Natl. Acad. Sci. U.S.A.* **95**, 9926 (1998).
[6] M. Franco, P. J. Peters, J. Boretto, E. van Donselaar, A. Neri, C. D'Souza-Schorey, and P. Chavrier, *EMBO J.* **18**, 1480 (1999).

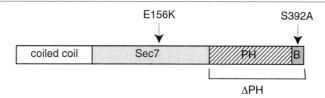

Fig. 1. Domain organization of ARNO. The positions of mutations discussed in this chapter are indicated. B, Polybasic domain.

tion. The central ~200 amino acid catalytic domain is immediately followed by a PH domain; this tandem arrangement is also found in the *dbl* family of GEFs for Rho GTPases.[7] The pleckstrin homology (PH) domain of each of these proteins has been shown to bind polyphosphoinositides and to, at least in part, mediate their transient recruitment to the plasma membrane. ARNO, cytohesin-1, and GRP-1 have all been shown to undergo translocation from the cytosol to the plasma membrane in response to phosphatidylinositol-3-kinase (PI-3K) activation, suggesting that they function downstream of PI-3K in signaling pathways originating at the cell surface (for review, see Ref. 1). The C terminus is enriched in basic amino acids, which appear to enhance the affinity of the PH domain for negatively charged lipids. In ARNO and cytohesin, but not GRP1, this polybasic region is interrupted by a protein kinase C (PKC) site whose phosphorylation may regulate the association of these proteins with the plasma membrane.[8]

Using *in vitro* exchange assays, we have determined that ARNO is capable of stimulating nucleotide exchange on both ARF1 and ARF6.[9] It is therefore possible that the substrate specificity of the ARNO family GEFs is determined by their subcellular localization. Among the mammalian ARFs, ARF6 is unique in at least two respects: first, it is localized predominantly to the plasma membrane and endosomes rather than to the Golgi; second, it can regulate aspects of actin cytoskeleton dynamics.[10] To determine whether ARNO can activate ARF6 *in vivo*, we have transiently expressed either wild-type ARNO or a panel of mutants in HeLa cells and have examined the morphology of the actin cytoskeleton. HeLa cells are chosen for a number of reasons: (1) the effects of ARF6 on cytoskeletal assembly have been previously characterized and (2) high transfection efficiencies can be achieved.

[7] R. A. Cerione and Y. Zheng, *Curr. Opin. Cell Biol.* **8,** 216 (1996).
[8] L. C. Santy, S. R. Frank, J. C. Hatfield, and J. E. Casanova, *Curr. Biol.* **9,** 1173 (1999).
[9] S. Frank, S. Upender, S. H. Hansen, and J. E. Casanova, *J. Biol. Chem.* **273,** 23 (1998).
[10] H. Radhakrishna, R. D. Klausner, and J. G. Donaldson, *J. Cell Biol.* **134,** 935 (1996).

Constructs

To facilitate detection of ARNO mutants in the presence of the endogenous protein, an N-terminal *myc* epitope tag (EQKLISEEDL) is introduced by polymerase chain reaction (PCR) mutagenesis as previously described.[9] Constructs tagged at the C terminus behave identically to those with N-terminal tags. (Plasmids encoding forms of ARF1 and ARF6 with HA epitope tags at their C termini are kindly provided by Dr. Victor Hsu, Department of Rheumatology, Brigham and Women's Hospital, Boston, MA.) These tagged forms of ARF have been extensively characterized and behave identically to the native proteins in all aspects tested to date.

Generation of ARNO Mutants

Mutations in each of the functional domains of ARNO are generated by PCR mutagenesis as described previously[11] and are shown in Fig. 1.

1. E156K is a point mutation in a conserved residue of the catalytic domain that completely abrogates exchange activity.
2. ΔPH introduces a stop codon C-terminal to the Sec7 domain, in place of Leu-269, deleting the PH and polybasic domains.
3. S392A converts phosphorylated serine in PKC site to alanine.

Each of the constructs was subcloned into the mammalian expression vector pCB7 in which transcription of the inserted gene is driven by a cytomegalovirus (CMV) immediate early promoter.

Culture of HeLa cells

HeLa cells are maintained in Dulbecco's modified Eagle's medium (DMEM) containing 0.11 g/liter sodium pyruvate, 2 mM L-glutamine, 4.5 g/liter glucose, 10% fetal bovine serum (FBS), 100 U/ml penicillin, and 100 U/ml streptomycin in a humidified incubator at 5% (v/v) CO_2. Cells are passaged twice weekly and maintained for approximately 24 passages after thawing. For transient transfection, cells are seeded onto glass coverslips for morphologic analysis or directly into six-well culture plates at a concentration of 5×10^4 cells/well, and allowed to spread for 24 hr prior to transfection.

Transient Transfection Using Lipofection

Although HeLa cells are readily transfectable by the $CaPO_4$ method (described in detail in Ref. 12), we have found that we achieve more

[11] S. R. Frank, J. C. Hatfield, and J. E. Casanova, *Mol. Biol. Cell* **9**, 3133 (1998).
[12] H. Damke, M. Gossen, S. Freundlieb, H. Bujard, and S. L. Schmid, *Methods Enzymol.* **257**, 209 (1995).

consistent and reproducible transfection efficiencies using the cationic lipid reagent Fugene6 (Roche/Boehringer, Indianapolis, IN). This reagent is nontoxic to cells and can be used in serum-containing medium. For each well of a six-well plate, 3 μl Fugene6 reagent is diluted into 100 μl serum-free medium and incubated at room temperature for 5 min. The lipid suspension is then added dropwise to 1 μg plasmid DNA and incubated an additional 15 min. *Note:* If cotransfecting with more than one plasmid, the total DNA should not exceed 1 μg. When cotransfecting ARF with ARNO, 0.5 μg of each plasmid is included in the transfection mixture, but if ARF is transfected alone, 0.5 μg of ARF plasmid is mixed with 0.5 μg of empty vector plasmid. The DNA/lipid mixture is then added to cells in a total volume of 2 ml complete (serum-containing) medium and incubated for 24–48 hr before further analysis. We routinely achieve >50% transfection efficiencies using this protocol.

Analysis of ARF Nucleotide Exchange *in Vivo*

Unlike Ras or Rho family GTPases, the intrinsic (non-GAP-mediated) rate of GTP hydrolysis among ARF family members is quite low. It is therefore possible to quantitate the GTP–GDP ratio of individual ARFs by extraction of bound nucleotide from immunoprecipitates from metabolically labeled cells.[13] A sample of such an analysis is shown in Fig. 2.

1. Transfect cells as described above, in six-well plates or 35-mm culture dishes.
2. Twenty-four hours after transfection, wash cells 2× in phosphate-free MEM (GIBCO, Grand Island, NY). Then to each well add 0.5 ml of the same medium supplemented with 0.3 mCi/ml ortho [^{32}P]phosphate, and incubate overnight.
3. The following day, aspirate labeling medium and dispose of appropriately.
4. Place cells on ice and wash 2× with ice-cold phosphate-buffered saline (PBS).
5. To each well add 1 ml ice-cold lysis buffer [20 mM Tris-HCl, pH 8.0, 100 mM NaCl, 1 mM MgCl$_2$, 1% Triton X-100, 0.05% cholate, 0.005% sodium dodecyl sulfate (SDS), 1 mM dithiothreitol (DTT), 1 mM NaF, 1 mM Na$_2$VO$_4$, 100 $\mu$$M$ ATP and a cocktail of protease inhibitors including 1 mM phenylmethylsulfonyl fluoride (PMSF), 5 μg/ml leupeptin, 10 μg/ml pepstatin, and 10 μg/ml antipain].

---

[13] S. E. Langille, V. Patki, J. K. Klarlund, J. M. Buxton, J. J. Holik, A. Chawla, S. Corvera, and M. P. Czech, *J. Biol. Chem.,* in press (1999).

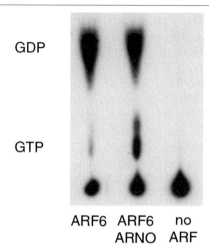

GDP

GTP

ARF6   ARF6   no
       ARNO   ARF

Fig. 2. Thin-layer chromatography (TLC) analysis of ^{32}P-labeled nucleotides extracted from ARF6. Cells expressing HA-tagged ARF6 in the absence (lane 1) or presence of coexpressed ARNO (lane 2) were metabolically labeled and ARF recovered by immunoprecipitation. An immunoprecipitate from mock-transfected cells is shown in lane 3. Bound nucleotides were extracted as described in the text and resolved by TLC.

6. Place lysates in 1.5-ml microfuge tubes and spin 5 min at 4° at maximum speed in a microfuge to pellet nuclei and other detergent-insoluble material.

7. After transfer of supernatants to clean tubes, immunoprecipitate HA-tagged ARF by adding 2 μl rabbit anti-HA antibody (BABCO, Richmond, CA) and 20 μl of a 50% slurry of protein A-sepharose. Incubate on a rotating platform for 2 hr at 4°. *Note:* A mock-transfected control should be included to determine the background level of free nucleotide brought down by the beads (see Fig. 2).

8. Following incubation with antibody, spin 30 sec in microfuge to pellet beads. The pellets should then be washed extensively to remove free nucleotide. Wash 2× in buffer A (50 mM HEPES-NaOH, pH 7.4, 0.5 M NaCl, 5 mM MgCl$_2$, 0.1% Triton X-100, 0.05% cholate, 0.005% SDS, 100 μM ATP), then 5× in buffer A lacking ATP.

Thin-Layer Chromatography (TLC) Analysis of Bound Nucleotides

1. Aspirate all liquid from beads using a Hamilton syringe.

2. Add 20 μl elution buffer (75 mM KH$_2$PO$_4$, pH 3.4, 5 mM EDTA, 0.5 mM GTP, and 0.5 mM GDP). Heat to 85° for 3 min to extract nucleotide.

3. To resolve eluted nucleotides, spot the entire eluate onto PEI cellulose fluorescent indicator TLC plates (Selecto Scientific, F-254, 100
   μm, 20 × 20 cm, obtained from Fisher Scientific, Pittsburgh, PA).
   These are plastic-backed plates that can be cut to fit the number of
   samples applied. Spot samples 1 cm apart. In one lane spot a standard
   containing 0.5 mM GTP and 0.5 mM GDP.
4. Place TLC plate(s) in a glass tank containing developing buffer
   (0.65M KH$_2$PO$_4$, pH 3.4) and develop at room temperature until the
   buffer reaches the top of the plate.
5. After air-drying, subject plates to autoradiography.
6. To quantitate nucleotide levels, place plates under UV illumination
   to identify GTP and GDP spots. Inclusion of nucleotides in elution
   buffer helps to precisely locate spots on the TLC plate. Mark spots
   with a pencil.
7. Cut spots from plate with either scissors or a scalpel and quantitate
   by scintillation counting. Pieces of plate can be placed directly into
   scintillant without scraping from the plastic backing.

Morphologic Analysis of Cells Expressing ARNO and ARNO Mutants

HeLa cells are cultured on glass coverslips as described above, and
transfected 24–48 hr prior to analysis, depending on the level of expression
desired. As mentioned above, ARNO contains a consensus site for phosphorylation by PKC at its C terminus. To determine the effects of PKC
activation on ARNO function, a subset of cultures is treated with phorbol
13-myristate acetate (PMA) for 30 min prior to fixation. PMA is maintained
as a 1 mM stock in DMSO at −20°. In these experiments, final concentration
of PMA was 1 μM; however, concentrations as low as 20 nM are effective.

Fixation and Staining of Cells

Prior to fixation, cells are washed twice with phosphate-buffered saline
(PBS) to remove serum proteins and nonadherent cells. Cells are fixed in
either 2% paraformaldehyde or 3.7% formaldehyde in PBS for 10 min at
room temperature. Paraformaldehyde is prepared in advance and frozen
in aliquots at −20°. Individual aliquots are thawed once and discarded.
Formaldehyde is diluted fresh from a 37% stock solution immediately before
use. Cells are fixed for 10 min at room temperature, then washed twice in
PBS to remove excess fixative. Blocking and permeabilization are achieved
simultaneously by a 10-min incubation with PBS containing 10% normal

goat serum (NGS) and 0.2% saponin (PBS–NGS). Coverslips are then inverted onto 100-μl drops of PBS–NGS into which primary antibody has been diluted. In this case, we used the monoclonal antibody 9E10, which recognizes the *myc* epitope tag. After a 30-min incubation at room temperature in a humidified chamber (we use a plastic box with a damp Kimwipe rolled up at each end), the coverslips are returned to a multiwell plate and washed 3× 5 min with 2 ml PBS to remove excess primary antibody. They are then inverted onto 100-μl drops of PBS–NGS containing a mixture of diluted secondary antibody (in this case Cy2-conjugated donkey anti-mouse IgG, Jackson Immunoresearch, West Grove, PA) and Texas Red-conjugated phalloidin (Molecular Probes, Engene, OR) to detect filamentous actin. Phalloidin is used at a final concentration of 0.1 μg/ml; optimal working dilutions of primary and secondary antibodies should be determined empirically, but 10 μg/ml of purified immunoglobulin G (IgG) is generally sufficient. After 30 min at room temperature, coverslips are returned to a six-well plate and washed 3× 5 min with PBS and once with water. They are then mounted in a 10-μl drop of Fluoromount G (Southern Biotechnology, obtained from Fisher Scientific, Pittsburgh, PA) containing 2.5 mg/ml n-propyl gallate to prevent photobleaching.

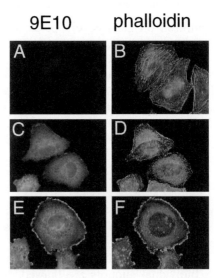

FIG. 3. Effects of ARNO expression on the actin cytoskeleton. HeLa cells were either mock transfected (A, B) or transfected with plasmid encoding wild-type ARNO (C–F). Cells in E and F were pretreated with 1 μM PMA 30 min prior to fixation. Cells were double labeled with antibody 9E10 to localize *myc*-ARNO (A, C, E), and phalloidin for filamentous actin (B, D, F).

Effects of Wild-Type and Mutant ARNO on Actin Cytoskeleton

The most prominent feature of the actin cytoskeleton in nontransfected HeLa cells is the presence of stress fibers (Fig. 3B). Expression of wild-type ARNO in these cells induces a loss of stress fibers and the appearance of more punctate actin-containing structures throughout the cell (Fig. 3D and Ref. 11). ARNO staining is distributed diffusely throughout the cytoplasm, but appears to be somewhat concentrated at the margins of the cells (Fig. 3C). Interestingly, treatment of ARNO-expressing cells with PMA for 30 min induces a dramatic redistribution of both ARNO (Fig. 3E) and actin (Fig. 3F) into lamellar structures at the cell periphery. Induction of lamellae was observed as early as 5 min after PMA treatment and was dependent on PKC catalytic activity, as it did not occur in cells pretreated with 10 μM bisindoylmaleimide, a specific inhibitor of the PKC catalytic domain (not shown, but see Ref. 11). A catalytically inactive ARNO mutant, E156K, did not support this cytoskeletal reorganization indicating that ARF nucleotide exchange is an integral component of this process (Figs. 4A and 4B). Similarly, an ARNO mutant lacking the PH domain, ARNOΔPH, did not support actin redistribution (Figs. 4C and 4D), in agreement with *in vitro* evidence that interaction of ARNO with membrane phosphoinositides

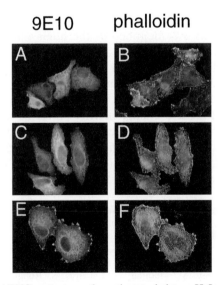

FIG. 4. Effects of ARNO mutants on the actin cytoskeleton. HeLa cells were transfected with ARNOE156K (A, B), ARNOΔPH (C, D), or ARNOS392A (E, F) and treated with 1 μM PMA 30 min prior to fixation. Cells were double labeled to localize ARNO (A, C, E) and filamentous actin (B, D, F).

is required for its exchange activity.[14] However, although ARNO is itself a substrate for PKC, mutation of the phosphorylated serine to alanine (S392A) did not prevent either its membrane recruitment or the formation of lamellar structures (Figs. 4E and 4F). We have subsequently determined that phosphorylation of S392 negatively regulates the association of ARNO with polyphosphoinositides *in vitro*.[8] This apparent paradox can be accommodated by a model in which ARNO recruitment and ARF activation precedes phosphorylation of S392 by PKC, which then displaces ARNO from the membrane, terminating the activation signal.

[14] P. Chardin, S. Paris, B. Antonny, S. Robineau, S. Beraud-Dufour, C. L. Jackson, and M. Chabre, *Nature* **384,** 481 (1996).

[28] Expression, Purification, and Measurements of Activity of ARNO1, a Guanine Nucleotide Exchange Factor for ADP-Ribosylation Factor 1 (ARF1)

By Sophie Béraud-Dufour and Sylviane Robineau

Introduction

ARNO1 (47 kDa) was identified as the human homolog of the yeast ADP-ribosylation factor (ARF) exchange factors Gea1 and Gea2[1,2] (in this volume[2a]). Gea1, Gea2, and ARNO share a common central domain of 200 amino acids, called the Sec7 domain. This domain was first identified in Sec7p,[3] the product of one of the genes involved in secretion in yeast, and is responsible for the exchange activity on ARF1 and other members of the ARF family. The crystal structure of the Sec7 domain of ARNO1 has been solved.[4,5] It consists of 10 helices, and shows a striking hydrophobic groove, responsible for the interaction with ARF1.[6] The structure of a complex between a truncated form of ARF1: [Δ17]ARF1 and the Sec7

[1] A. Peyroche, S. Paris, and C. L. Jackson, *Nature* **384,** 479 (1996).
[2] P. Chardin, S. Paris, B. Antonny, S. Robineau, S. Béraud-Dufour, C. L. Jackson, and M. Chabre, *Nature* **384,** 481 (1996).
[2a] A. Peyroche and C. L. Jackson, *Methods Enzymol.* **329** [31] (2000) (this volume).
[3] T. Achstetter, A. Frankzusoff, C. Field, and R. Schekman, *J. Biol. Chem* **263,** 11711 (1988).
[4] J. Cherfils, J. Menetrey, M. Mathieu, G. Le Bras, S. Robineau, S. Beraud-Dufour, B. Antonny, and P. Chardin, *Nature* **392,** 101 (1998).
[5] E. Mossessova, J. M. Gulbis, and J. Goldberg, *Cell* **92,** 415 (1998).
[6] S. Béraud-Dufour, S. Robineau, P. Chardin, S. Paris, M. Chabre, J. Cherfils, and B. Antonny, *EMBO J* **17,** 3651 (1998).

domain of Gea2 has also been solved.[7] The Sec7 domain catalyzes the GDP–GTP exchange on ARF1 by destabilizing the nucleotide with a glutamate residue, the "glutamic finger,"[6] and by restructuring the switch regions and the phosphate-binding loop of ARF1.[6,7] Two homologs of ARNO1 have been cloned in mammalian cells: cytohesin-1 (ARNO2)[8] and Grp1 (ARNO3).[9] These three proteins also contain an amino-terminal coiled-coil motif and a carboxyl-terminal pleckstrin-homology (PH) domain responsible for the interaction with phosphoinositides such as phosphatidylinositol triphosphate (PIP$_3$) or phosphatidylinositol biphosphate (PIP$_2$).[10] The Sec7 domain of ARNO proteins has a low sensitivity to brefeldin A.

The expression in *Escherichia coli* and purification of ARNO1 and its Sec7 domain, ARNO1-Sec7, are described in this chapter, as well as assays to monitor the exchange activity of ARNO1 on ARF1. Note that although both ARF1 and ARNO1 are soluble proteins, they interact with membrane lipids.[10,11] More importantly, these membrane interactions are necessary for the functional interaction between the two proteins.[12] Therefore membrane lipids (for instance, artificial lipid vesicles) must be included in functional assays. However, this requirement can be bypassed when ARNO1-Sec7 and [Δ17]ARF1, an N-terminal truncated form of ARF1, are used. The isolation of a soluble stoichiometric complex between ARNO1-Sec7 and [Δ17]ARF1 by gel filtration is described here. The purification schemes for ARF1 and [Δ17]ARF1 have been described elsewhere.[13,14]

Methods

Expression and Purification of ARNO1

The cDNA of full-length ARNO1 is introduced in a pET-11d vector (*Nco*I–*Bam*HI) (Novagen, Madison, WI) and *E. coli* BL21DE3 strains are transformed with the modified vector. Expression is checked on minicultures in Luria-Bertani (LB) with ampicillin (60 mg/liter) induced by addition of 0.1 mM isopropylthiogalactoside (IPTG). Large-scale cultures are

[7] J. Goldberg, *Cell* **95,** 237 (1998).

[8] L. Liu and B. Pohajdak, *Biochim. Biophys. Acta* **132,** 75 (1992).

[9] J. K. Klarlund, A. Guilherme, J. J. Holik, J. V. Virbasius, A. Chawla, and M. P. Czech, *Science* **275,** 1927 (1997).

[10] S. Paris, S. Béraud-Dufour, S. Robineau, J. Bigay, B. Antonny, M. Chabre, and P. Chardin, *J. Biol. Chem.* **272,** 22221 (1997).

[11] B. Antonny, S. Béraud-Dufour, P. Chardin, and M. Chabre, *Biochemistry,* **36,** 4675 (1997).

[12] S. Béraud-Dufour, S. Paris, M. Chabre, and B. Antonny, *J. Biol. Chem.* **274,** 37629 (1999).

[13] M. Franco, P. Chardin, M. Chabre, and S. Paris, *J. Biol. Chem.* **270,** 1337 (1995).

[14] B. Antonny, S. Béraud-Dufour, P. Chardin, and M. Chabre, *Biochemistry* **36,** 4675 (1997).

carried out in rich medium containing Bacto-tryptone 50 g, yeast extract 25 g, KH_2PO_4 11 g, K_2HPO_4 37.5 g, and glycerol 90 ml. *E. coli* cells are precultured overnight in 500 ml of the medium and ampicillin (60 mg/liter) at room temperature (OD_{600} up to 0.4). Five liters of medium containing ampicillin (60 mg/liter) are autoclaved in a fermenter, cooled, and inoculated with all the preculture in order to control the oxygenation, the agitation, and the temperature of the culture. The culture is grown at 37° to an OD_{600} of 0.8–1, with agitation. The temperature is then reduced to 35° and the induction is obtained by addition of 0.1 mM IPTG. The cells are then grown for 3 hr. The cells are collected and then centrifuged for 30 min at 4200 rpm at 4°. The pellets are washed with 400 ml of 100 mM NaCl, 50 mM Tris, pH 8.0. The cells are centrifuged again, and the pellets are conserved at −20° (or −80° for long-term storage). About 20 g of cells is lysed, in 35 ml of buffer A [Tris 50 mM, pH 8, $MgCl_2$ 5 mM, dithiothreitol (DTT) 1 mM] plus one tablet of Complete protease inhibitors (Boehringer Mannheim), using the French press at 1200 psi. The cells are then centrifuged for 30 min at 9000 rpm in a Beckman JA20 rotor and the supernatant is recentrifuged for 45 min at 25000 rpm in a SW28 rotor. The soluble extract is loaded on a 200-ml QAE-Sepharose FF column (XK-50/20 Pharmacia, Piscataway, NJ) equilibrated with buffer B (Tris 50 mM, pH 8, $MgCl_2$ 1 mM, DTT 1 mM) at a flow rate of 5 ml/min and eluted with a linear gradient of 0–1 M NaCl (buffer C: Tris 50 mM, pH 8, $MgCl_2$ 1 mM, DTT 1 mM, NaCl 1 M) in the cold room. The fractions are analyzed on 12% SDS–PAGE gel: ARNO1 elutes at about 250 mM NaCl. Then the fractions are precipitated by ammonium sulfate (50% saturated), resuspended in 5 ml of buffer B plus 100 mM NaCl and loaded onto a gel-filtration column: 2 liters of Sephacryl S-200 HR (XK-50/100 Pharmacia), and run in buffer B (flow 6 ml/min) in the cold room. The major peak containing ARNO1 is concentrated by ammonium sulfate precipitation (70% saturated) and dialyzed against buffer B. The fractions are analyzed on a 12% SDS–PAGE gel (Fig. 1). The purified protein represents about 70% of the total protein content (1.7 mg/ml; total of 5 mg). The proteins are stored at −80°. The activity of the protein decreases on thawing and refreezing.

Expression and Purification of ARNO1-Sec7

The cDNA of ARNO1-Sec7 (residues 50–252 of ARNO1) is introduced in a pET-11d vector (*Nco*I–*Bam*HI) (Novagen), and *E. coli* BL21DE3 are transformed with this construct. One colony of *E. coli* BL21DE3 is cultured overnight in 50 ml Luria–Bertani (LB) medium containing ampicillin (60 mg/liter) at 37° with agitation (OD_{600} up to 0.4). Then 1 liter of LBA is inoculated with 10 ml of the preculture and grown at 37° with agitation to

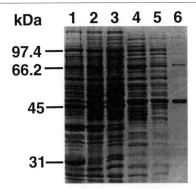

FIG. 1. Purification of ARNO1. Coomassie blue stained SDS–polyacrylamide gel (12%). Lane 1, uninduced *E. coli* extracts; lanes 2–3, induced *E. coli* extracts; lane 4, soluble extract loaded on the QAE column; lane 5, fractions pooled from QAE column; lane 6, purified ARNO1 after the gel filtration column and ammonium sulfate precipitation.

an OD_{600} of 0.8–1 induced with 0.1 mM IPTG. The cells are grown for 3 hr. The cells are then collected by centrifugation for 15 min at 7000 rpm in JA 14 rotor. The pellet is washed with 100 ml of Tris 50 mM, pH 8.0, EDTA 1 mM. About 3 g of cells are lysed for 30 min at 4° in 30 ml of Tris 50 mM, pH 8.0, EDTA 1 mM, DTT 1 mM supplemented with lysozyme (1 mg/ml) and one tablet of Complete protease inhibitor (Boehringer Mannheim). Then 5 mM MgCl$_2$ and 0.02% deoxycholate are added for 30 min, followed by the addition of 100 μg/ml DNase for an additional 30 min. The suspension is centrifuged for 60 min at 400,000g at 4°. The clear supernatant is applied to a QAE-Sepharose F.F. in the same conditions as for ARNO1. The fractions are analyzed on a 15% SDS–PAGE gel (Fig. 2). The fractions containing ARNO1-Sec7 are pooled and concentrated on ice with an Amicon (Danvers, MA) cell (filter YM10) to 5 ml and loaded onto the same gel-filtration column as used for ARNO1 and run in buffer B (flow 6 ml/min). The fractions are analyzed on a 15% SDS–PAGE gel and the major peak containing ARNO1-Sec7 is concentrated with Amicon cell (filter YM10). The proteins are stored at −80°.

Vesicle Preparation

Unilamellar vesicles are prepared by the extrusion method.[15] All lipids are purchased from Sigma (St. Louis, MO). Solutions of phosphatidylcholine (PC), phosphatidylglycerol (PG) in chloroform and PIP$_2$ in chloroform : methanol : H$_2$O : HCl (1 N) (20:9:1:0.1), stored at −20° in glass

[15] M. J. Hope, M. B. Bally, G. Webb, and P. R. Cullis, *Biochim. Biophys. Acta* **812,** 55 (1985).

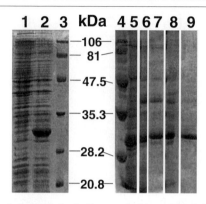

FIG. 2. Purification of ARNO1-Sec7. Coomassie blue stained SDS–polyacrylamide gel (15%). Lane 1, uninduced *E. coli* extracts; lane 2, induced *E. coli* extracts; lane 3, molecular weight markers; lane 4, molecular weight markers; lane 5, cell lysis; lane 6, fraction loaded on the QAE column; lane 7, fractions pooled from the QAE column; lane 8, fraction loaded on the gel filtration column; lane 9, purified ARNO1-Sec7 after the gel filtration column.

ampules, are mixed in the required proportion: 65:35:5 (w/w/w). The chloroform:methanol 2:1 ratio is maintained in the mixture to allow homogeneous mixing of the PIP_2 and the other lipid solutions. The solution is placed in a Rotavapor in order to evaporate all solvents: a film is formed and resuspended at 4 mg/ml in 50 mM HEPES–KOH pH 7.5, and 100 mM KCl. The suspension is then vortexed for 20 min and freezed–thawed (using liquid nitrogen and a bath at 42°) five times in order to form large vesicles. Vesicles can be stored at this step, at −20°, after the last freezing. Unilamellar vesicles are produced by extrusion through a 0.1-μM pore size polycarbonate filter (Millipore, Bedford, MA) using an extrusion apparatus (Avanti Polar Lipids, Birmingham, AL).[16] The phospholipid concentration is determined using a Fiske–Subbarow phosphorus assay.[17] The lipid can be stored for 1 week at 4° or for 1 day at room temperature.

Measure of Activity of ARNO1 and ARNO1-Sec7

Two different kinds of assay can be used to monitor the exchange activity of ARNO1 or ARNO1-Sec7 on ARF1: classical nucleotide binding measurements with radiolabeled nucleotides (described in this volume[17a]

[16] R. C. MacDonald, R. I. MacDonald, B. P. Menco, K. Takeshita, N. K. Subbarao, and L. R. Hu, *Biochim. Biophys. Acta* **1061,** 297 (1991).
[17] C. H. Fiske and Y. Subbarow, *J. Biol. Chem.* **66,** 375 (1925).
[17a] P. Chavrier and M. Franco, *Methods Enzymol.* **329** [29] (2000) (this volume).

for the exchange activity of EFA6 on ARF6) or a real-time assay based on the large difference between the tryptophan fluorescence of ARF1-GDP and the tryptophan fluorescence of ARF1-GTP. The obvious advantage of the fluorescence assay is its time resolution. However this assay requires purified proteins for a good signal-to-noise ratio, and cannot detect futile nucleotide exchanges (for instance, the exchange of GDP by GDP), since such exchanges are spectroscopically silent.

The dramatic increase in the fluorescence of ARF1 that accompanies the exchange of GDP by GTP is due to the relative movement of two tryptophan residues (66 and 78). The signal is so large ($+200\%$) that it can be easily detected with concentration of ARF1 in the range of 10^{-8} to 10^{-6} M. All fluorescence measurements are performed with a Shimadzu RF-5000 fluorimeter containing a cuvette holder with stirring and temperature regulation facilities. Although tryptophan residues show a maximum of absorption at 280 nm, the excitation is usually set at 297–300 nm to prevent light absorption by added nucleotides. A small excitation bandwidth is used (1.5–5 nm) to minimize photobleaching. The emission is measured at 340 nm with a large emission bandwidth (up to 30 nm) to enhance the signal. We use a cylindrical (6-mm-diameter) cuvette in which the sample is stirred with a small magnetic bar. This cell requires a sample volume of 0.6 ml as compared to 1.5 ml for a classical 10×10 cell. Injections are done with Hamilton syringes, and the cover of the sample compartment contains a guide for the syringes. This setup allows manual injection of proteins and nucleotides from stock solutions without interrupting the fluorescence recording; hence, completion of injection and mixing takes about 1 sec and kinetics in the range of few seconds can be resolved.

Kinetics measurements are performed at $37°$ in a buffer containing 50 mM HEPES, pH 7.5, 120 mM KCl, 1 mM MgCl$_2$, and 2 mM DTT. The buffer must be filtered and degased before use to avoid light scattering artifacts. In a typical experiment, the fluorescence cuvette contains initially 600 μl of buffer supplemented with vesicles of defined lipid composition at concentration of 0.1–1 mM. ARF-GDP (0.2–1 μM), GTP (10–100 μM) and ARNO (10–50 nM) are added in stepwise manner (for instance, each minute) from stock (50–100 X) solutions. On the addition of GTP onto ARF-GDP, a slow fluorescence increase is observed that reflects the spontaneous exchange of GDP for GTP on ARF. The subsequent addition of a catalytic amount of ARNO accelerates this fluorescence change whose kinetics reflects the catalysis of GDP/GTP exchange on ARF1 by ARNO (Fig. 3). Similar experiments can be done with ARNO1-Sec7 and [Δ17]ARF1, in which case, lipid vesicles are omitted.

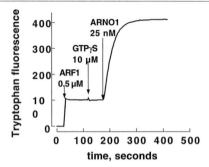

Fig. 3. Stimulation of GDP/GTP exchange on ARF1 by ARNO1. The GDP–GTP exchange is monitored by the correlated variation in tryptophan fluorescence. ARF1 (0.5 μM) is injected in a fluorescence cuvette in HEPES 50 mM, KCl 100 mM, MgCl$_2$ 1 mM, 2 mM DTT, and 0.3 g/liter vesicles PC-PG-PIP2 (65:30:5). At the indicated times, GTPγS (10 μM) and ARNO1 (25 nM) were added to promote the GDP release and its substitution by GTPγS.

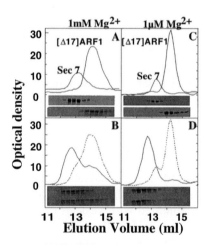

Fig. 4. Interaction between [Δ17]ARF1 and ARNO1-Sec7 on a gel filtration column. [Δ17]ARF1 (10 μM), ARNO-Sec7 (10 μM), or a mixture of the two proteins is applied to a gel filtration column. Protein absorbancy is monitored at 280 nm, and a sample of each fraction is analyzed by SDS–polyacrylamide gel electrophoresis and Coomassie blue staining. The concentration of free Mg^{2+} is 1 mM (A and B) or 1 μM (C and D). (A and C) Elution profiles obtained when [Δ17]ARF1 and ARNO1-Sec7 are loaded separately on the column. (B and D) Elution profiles obtained with the mixture of the two proteins. The dotted line represents the calculated sum of the two absorbancy profiles obtained for the separate proteins under the same Mg^{2+} condition, shown in the panel above. The upper band on the gels corresponds to the ARNO1-Sec7 domain, and the lower band corresponds to [Δ17]ARF1.

Measurement of Interaction between [Δ17]ARF1 and ARNO-Sec7 by Gel Filtration Experiment

A complex in solution between the truncated form of ARF1, [Δ17]ARF1, and ARNO-Sec7 could be formed.[18] The formation of the complex is analyzed by gel filtration experiment. All experiments are performed on a Superose 12 column (HR 10/30, Pharmacia) at room temperature. The running buffer contains 20 mM Tris, pH 7.5, 100 mM NaCl, 1 mM MgCl$_2$, 5 mM 2-mercaptoethanol, and 0.1 mM phenylmethylsulfonyl fluoride (PMSF). This buffer can be supplemented with 2 mM EDTA to reduce the concentration of free magnesium to 1 μM in order to favor the release of the nucleotide. [Δ17]ARF1 (10 μM) is incubated with or without ARNO-Sec7, also 10 μM, in 200 μl of the running buffer at room temperature. Then the sample is applied to the column at a flow rate of 0.5 ml/min and the optical density is recorded continuously at 280 nm. Fractions of 300 μl are collected and 60 μl of each fraction is concentrated by evaporation and analyzed on 15% SDS–PAGE gel (Fig. 4). It is also possible to analyze the presence of the nucleotide in ARF1 by using a [³H]GDP-labeled [Δ17]ARF1. [Δ17]ARF1 (40 μM) is incubated with a trace of [³H]GDP (final concentration: 10 μCi/ml) at 25° in 50 mM HEPES, pH 7.5, 100 mM NaCl, 1 mM MgCl$_2$, 2 mM EDTA (free MgCl$_2$: 1 μM) for 15 min. Then the concentration of free MgCl$_2$ is increased to 1 mM. The gel filtration is performed in the same conditions and the radioactivity each fraction is determined by counting 50 μl of the 300-μl fractions. We used this technique in order to determine the binding sites on ARF1 and on ARNO-Sec7, by making different mutants of the two proteins.[6]

Acknowledgments

We thank Bruno Antonny and Pierre Chardin for helpful comments on the manuscript.

[18] S. Paris, S. Béraud-Dufour, S. Robineau, J. Bigay, B. Antonny, M. Chabre, and P. Chardin, *J. Biol. Chem.* **272,** 22221 (1997).

[29] Expression, Purification, and Biochemical Properties of EFA6, a Sec7 Domain-Containing Guanine Exchange Factor for ADP-Ribosylation Factor 6 (ARF6)

By PHILIPPE CHAVRIER and MICHEL FRANCO

Introduction

In higher eukaryotes, ADP-ribosylation factors (ARFs) form a group of six small (20-kDa) ubiquitous GTP-binding proteins related to Ras that are required for maintaining the integrity of organelle structure and intracellular transport.[1] ARF6, the least conserved ARF protein, controls the early endocytic pathway and actin cytoskeleton organization.[2–4]

Like all Ras-like proteins, ARF6 cycles between two conformations, a GDP-bound inactive form and a GTP-bound active conformation. Activation (GDP–GTP exchange) is catalyzed by guanine nucleotide exchange factors (GEFs). A combination of genetic and biochemical approaches has led to the identification of a family comprised of several ARF-specific GEFs.[1] These proteins share a conserved 200 amino acid domain homologous to a region of the yeast *Saccharomyces cerevisiae* Sec7 protein. The Sec7 domain is sufficient both for guanine nucleotide exchange activity[5] and substrate specificity.[6] So far only two GEFs have been shown to activate ARF6. ARNO, a Sec7 domain-containing protein initially described as an ARF1 GEF,[5] is also capable of promoting to some extent nucleotide exchange on ARF6.[7] More recently, a new Sec7 domain factor called EFA6 has been identified as an ARF6-specific GEF.[8] EFA6 promotes efficient nucleotide exchange on ARF6 but not ARF1 *in vitro*. In addition, the localization of EFA6 to the plasma membrane, together with its effects on

[1] P. Chavrier and B. Goud, *Curr. Opin. Cell Biol.* **11**, 466 (1999).

[2] C. D'Souza-Schorey and P. D. Stahl, *Exp. Cell Res.* **221**, 153 (1995).

[3] C. D'Souza-Schorey, E. van Donselaar, V. W. Hsu, C. Yang, P. D. Stahl, and P. J. Peters, *J. Cell Biol.* **140**, 603 (1998).

[4] H. Radhakrishna, R. D. Klausner, and J. G. Donaldson, *J. Cell Biol.* **134**, 935 (1996).

[5] P. Chardin, S. Paris, B. Antonny, S. Robineau, S. Béraud-Dufour, C. L. Jackson, and M. Chabre, *Nature* **384**, 481 (1996).

[6] M. Franco, J. Boretto, S. Robineau, S. Monier, B. Goud, P. Chardin, and P. Chavrier, *Proc. Natl. Acad. Sci. U.S.A.* **95**, 9926 (1998).

[7] S. Frank, S. Upender, S. H. Hansen, and J. E. Casanova, *J. Biol. Chem.* **273**, 23 (1998).

[8] M. Franco, P. J. Peters, J. Boretto, E. van Donselaar, A. Neri, C. D'Souza-Schorey, and P. Chavrier, *EMBO J.* **18**, 1480 (1999).

endocytosis and actin cytoskeleton organization support a prominent role of EFA6 as an ARF6-GEF *in vivo.*

The methods described herein outline the different techniques we have used to produce and analyze the biochemical activity of EFA6 as an ARF6-specific GEF and to assess its intracellular localization in mammalian cells.

Production of Recombinant Proteins in *Escherichia coli*

Unmyristoylated and Myristoylated ARF6 (myrARF6)

Human ARF6 cDNA is inserted into pET3a (Novagen, Madison, WI) and the resulting construct is transformed into the *E. coli* strain BL21(DE3). Transformed bacteria are grown to an OD_{600} of ~0.8 and protein expression is induced by addition of 0.2 mM isopropyl-1-thio-β-D-galactopyranoside (IPTG) for 2 hr at 37°. Following centrifugation, the bacterial pellet (13 g) is resuspended in 150 ml of lysis buffer [50 mM Tris-HCl, pH 8.0, 1 mM EDTA, 1 mM dithiothreitol (DTT), 200 μM GTP, protease inhibitor cocktail, Boehringer Mannheim-Roche, Meylan, France]. All subsequent steps are performed at 4°. The bacterial suspension is incubated for 1 hr in the presence of lysozyme (0.5 mg/ml). Then, 0.05% sodium deoxycholate, 3 mM MgCl$_2$, and 0.1 mg/ml DNase I are added for 30 min. The lysate is cleared by centrifugation at 10,000g for 20 min, followed by ultracentrifugation at 120,000g for 90 min (Ti45 rotor, Beckman, Fullerton, CA). The supernatant is loaded onto a Q-Sepharose Fast Flow column (XK50, 150-ml bed volume, Pharmacia) previously equilibrated in buffer A (50 mM Tris-HCl, pH 8.0, 1 mM MgCl$_2$, 1 mM DTT). ARF6 is recovered in the flow-through and precipitated at 70% saturation ammonium sulfate. After centrifugation, the protein pellet is resuspended in 30 ml of buffer A and loaded onto a Sephacryl S200 HR gel filtration column (XK50/100, Pharmacia, Piscataway NJ). ARF6-containing fractions, as determined by protein separation by SDS–PAGE and Coomassie blue staining, are pooled and concentrated using an Amicon (Danvers, MA) cell. The protein is more than 90% pure and the yield is about 20 mg of ARF6 per liter of bacterial culture.

Recombinant ARF6 can be myristoylated by coexpression with yeast *N*-myristoyltransferase.[9] Protein expression is induced with IPTG as above except that bacteria are cultured at 27° for 3 hr in medium supplemented with 50 μM myristate mixed with 10 μM bovine serum albumin (BSA).

---

[9] R. J. Duronio, E. Jackson-Machelski, R. O. Heuckeroth, P. O. Olins, C. S. Devine, W. Yonemoto, L. W. Slice, S. S. Taylor, and J. I. Gordon, *Proc. Natl. Acad. Sci. U.S.A.* **87,** 1506 (1990).

After centrifugation, the bacterial pellet is resuspended in lysis buffer as described above and myrARF6 in the lysate is precipitated at 35% saturation ammonium sulfate. The protein pellet is resuspended in 15 ml buffer A and dialyzed overnight against buffer A supplemented with 2 μM GTP. The sample is loaded onto a DEAE Sepharose Fast Flow column (XK16, 25-ml bed volume, Pharmacia) in buffer A and myrARF6 is recovered in the flow-through. After gel filtration on a Sephacryl S200 HR column as above, peak fractions are pooled and myrARF6 is concentrated using an Amicon cell. The protein is judged to be 60% pure with an estimated yield of 1 mg of myrARF6 per liter of culture.

EFA6

Human EFA6 cDNA inserted in pET3a[8] is transformed into BL21 (DE3). Bacteria are grown to an OD_{600} of ~0.8 and protein expression is induced by addition of 0.2 mM IPTG for 3 hr at 27°. Bacteria are lysed as described above except that GTP is omitted from the lysis buffer. After ultracentrifugation, the supernatant that contained about 10–20% of recombinant EFA6 (as judged by SDS–PAGE and Coomassie blue staining) is used directly as a source of GEF.

Analysis of Bound Nucleotides on Recombinant ARF6

Recombinant ARF1 is purified as a GDP-bound form (M. Franco, 1999, unpublished). Tryptophan fluorescence is used to determine the nature of the nucleotide bound to purified ARF6. A large increase in the intrinsic fluorescence of ARF occurs when GDP is exchanged for GTP.[10] Fluorescence measurements are performed at 37° with a Shimadzu (Kyoto, Japan) RF-5000 fluorimeter with 297.5- and 340-nm excitation and emission wavelengths, respectively. At the beginning of the experiment 0.6 ml of 50 mM HEPES, pH 7.5, 120 mM KCl, 1 mM MgCl$_2$ is injected into a 6-mm-diameter cylindrical quartz cell and continuously agitated. After 30 sec, recombinant ARF6 (1 μM final concentration) is added, inducing a strong fluorescence increase (Fig. 1). Excess GDP or GTP (20 μM) is subsequently injected and spontaneous exchange is accelerated by the addition of 2 mM EDTA (free [Mg^{2+}] ~ 1 μM). Addition of GDP induces a large fluorescence decrease, whereas in the presence of excess GTP, only a small fluorescence increase is observed. These results indicate that recombinant ARF6 is predominantly loaded with GTP that could exchange with GDP, resulting in the observed fluorescence decrease. Of note, recombinant ARF6 is also recovered in the GTP-bound form even when GDP is added during purification.

[10] R. A. Kahn and A. G. Gilman, *J. Biol. Chem.* **261**, 7906 (1986).

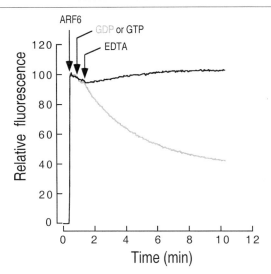

FIG. 1. Purified recombinant ARF6 is bound to GTP. Following purification from *E. coli* lysate, recombinant ARF6 was incubated in the presence of an excess of GDP or GTP, and EDTA was added as described in the text. Nucleotide exchange was monitored by online measurement of tryptophan fluorescence.

Analysis of Myristoylated ARF6

Activation of ARF1 by GEFs occurs only when the ARF/GEF complex is associated with membranes.[11,12] In the case of ARF1, the myristoyl moiety that increases membrane association of both the GDP- and GTP-bound forms is required for catalyzed nucleotide exchange.[13,14] Therefore, myristoylated ARF6 is produced with the aim of analyzing the nucleotide exchange activity of EFA6. To assess the extent of myristoylation, modified ARF6 is discriminated from the unmodified form by its increased affinity for membrane phospholipids. Large unilamellar vesicles of azolectin are prepared as described.[15] Briefly, 20 mg of azolectin (soybean lipids, Sigma, St. Louis, MO) is dissolved in 6 ml diethyl ether, and 1 ml of aqueous buffer (50 mM HEPES, pH 7.5) is added. The mixture is then sonicated for 2 min at 0° and the solvent removed under reduced pressure. The aqueous vesicle suspension is filtered through a 0.8-μm Millipore (Bedford,

[11] M. Franco, P. Chardin, M. Chabre, and S. Paris, *J. Biol. Chem.* **271,** 1573 (1996).
[12] S. Paris, S. Beraud-Dufour, S. Robineau, J. Bigay, B. Antonny, M. Chabre, and P. Chardin, *J. Biol. Chem.* **272,** 22221 (1997).
[13] M. Franco, S. Paris, and M. Chabre, *FEBS Lett.* **362,** 286 (1995).
[14] B. Antonny, I. Huber, S. Paris, M. Chabre, and D. Cassel, *J. Biol. Chem.* **272,** 30848 (1997).
[15] F. Szoka Jr. and D. Papahadjopoulos, *Proc. Natl. Acad. Sci. U.S.A.* **75,** 4194 (1978).

MA) filter to produce vesicles of homogeneous size. ARF6 or myrARF6 (1 μM) is incubated in the presence of 3 mg/ml azolectin vesicles in 50 mM HEPES, pH 7.5, 120 mM KCl, 1 mM DTT, 1 mM MgCl$_2$, 2 mM EDTA, 100 μM GTP for 1 hr at 37° in a total volume of 100 μl. The free Mg^{2+} concentration is then raised to 1 mM by adding 2 mM MgCl$_2$. After ultracentrifugation (400,000g for 10 min at 20° in a TL100.1 Beckman rotor), the pellet (P) and the supernatant (S) are analyzed by 15% SDS–PAGE and Coomassie blue staining. As shown in Fig. 2, under these conditions unmodified ARF6 is soluble while myrARF6 is membrane bound, indicating that the protein is fully myristoylated.

[^{35}S]GTPγS Binding Assay

The ability of recombinant EFA6 to catalyze the binding of guanosine 5'-[γ-thio]triphosphate ([^{35}S]GTPγS) is monitored on myrARF6. Myr-ARF6 (1 μM) is incubated at 37° in 50 mM HEPES-NaOH, pH 7.5, 1 mM MgCl$_2$, 1 mM DTT, 100 mM KCl, 10 μM [^{35}S]GTPγS (~1000 cpm/pmol, NEN, Boston, MA) supplemented with azolectin vesicles (1.5 g/liter). Soluble bacterial protein lysate containing EFA6 is added to a final concentration of 40 μg/ml of total proteins. The estimated GEF concentration in the assay is ~100–200 nM. At indicated time intervals, aliquots of 25 μl are removed, diluted in 2 ml ice-cold buffer (20 mM HEPES, pH 7.5, 100 mM NaCl, 10 mM MgCl$_2$), and filtered through 25-mm BA85 nitrocellulose filters (Schleicher & Schuell, Keene, NH). Filters are washed twice with 2 ml of the same buffer, dried, and counted for radioactivity. As shown in

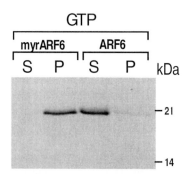

FIG. 2. Analysis of membrane association of recombinant ARF6 and myrARF6. Recombinant myrARF6 and ARF6 were loaded with GTP in the presence of azolectin vesicles. After centrifugation, the supernatant (S) and the pellet (P) were analyzed by SDS–PAGE for the presence of the proteins.

Fig. 3, in the presence of EFA6, nucleotide exchange on myrARF6 is rapid
and reaches a maximal rate within 5 min.

Intracellular Localization of EFA6 in Mammalian Cells

To determine the cellular localization of EFA6, human EFA6 cDNA
is inserted into pSRα expression vector in front of an amino-terminal
vesicular stomatitis virus glycoprotein (VSV-G) tag. Baby hamster kidney
(BHK-21) or Chinese hamster ovary (CHO)-derived cells (TRVb-1) plated
on 11-mm round glass coverslips are transfected with the VSV-G-EFA6
vector using Fugene-6 reagent according to the manufacturer's instructions
(Roche). Forty hours after transfection, cells are washed with phosphate-

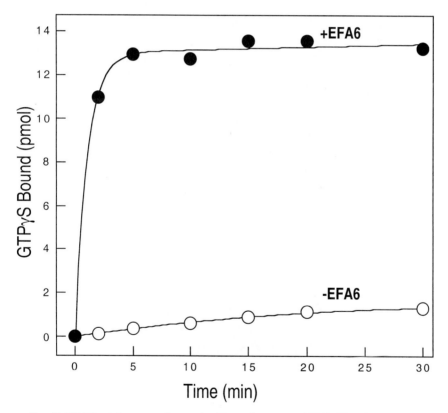

FIG. 3. EFA6 catalyzes guanine nucleotide exchange on ARF6. The slow kinetics of
spontaneous [35]GTPγS-binding to ARF6 (open circles) is dramatically increased in the presence of EFA6 (closed circles).

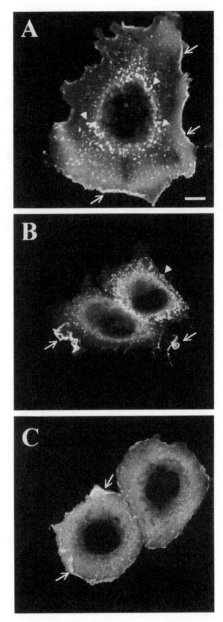

Fig. 4. Localization of overexpressed EFA6 by indirect immunofluorescence microscopy. BHK-21 (A) or TRVb-1 cells (B and C) were grown on coverslips and tansfected with expression plasmids encoding VSV-G-tagged EFA6 (A and B) or HA-tagged ARF6 (C). Cells were fixed and labeled with either anti-VSV-G (A and B) or anti-HA (C) epitope antibody. Arrows indicate labeling of plasma membrane ruffles at cell edges. Arrowheads point to labeling of EFA6 on microvilli at the dorsal cell surface. Bar: 10 μm.

buffered saline (PBS) and fixed in 3% paraformaldehyde. After two washes in PBS, free aldehyde groups are quenched with 50 mM NH$_4$Cl in PBS for 10 min. After one wash in PBS, cells are permeabilized with 0.5% (w/v) saponin in PBS for 5 min. To reduce nonspecific antibody binding, cells on coverslips are incubated in PBS containing 10% horse serum and 0.05% saponin and then incubated for 20 min at 25° with anti-VSV-G tag mouse monoclonal antibody (clone P5D4) diluted in the same buffer (5 μg/ml). After rinsing the cells three times, primary antibody binding is visualized with donkey anti-mouse fluorescein isothiocyanate (FITC)-conjugated antibodies diluted as above. After three washes in PBS–0.05% saponin, and one wash in PBS, the coverslip is mounted on a glass slide in Mowiol (Calbiochem, LA Jolla, CA). Samples are viewed by confocal microscopy with a Leica TCS 4D confocal microscope equipped with a mixed gas argon/krypton laser (Leica Laser Teknik, Heidelberg, Germany). Overexpressed VSV-G-tagged EFA6 localizes to the plasma membrane and is enriched in membrane ruffles (arrows) and microvilli-like structures (arrowheads) that are visible at the dorsal surface of the cells (Fig. 4).

Acknowledgments

This work was supported by INSERM and CNRS institutional fundings and a grant from the Association pour la Recherche sur le Cancer to P.C. We thank Dr. W. Hempel and E. Macia for critical reading of the manuscript.

[30] Isolation and Properties of GRP1, an ADP-Ribosylation Factor (ARF)–Guanine Nucleotide Exchange Protein Regulated by Phosphatidylinositol 3,4,5-Trisphosphate

By JES K. KLARLUND and MICHAEL P. CZECH

Introduction

GRP1 is a member of a family of proteins that possess three cognate domains: an N-terminal coiled-coil region of about 50 amino acids, a Sec7 homology domain of about 180 residues, and a pleckstrin homology (PH) domain containing about 120 amino acids.[1] These proteins also contain a

---

[1] J. K. Klarlund, A. Guilherme, J. J. Holik, J. V. Virbasius, A. Chawla, and M. P. Czech, *Science* **275,** 1927 (1997).

short COOH-terminal basic amino acid sequence that may be important in protein–protein or protein–lipid interactions. The Sec7 homology domain is able to catalyze guanine nucleotide exchange of ADP-ribosylation factor (ARF) proteins *in vitro* and *in vivo*,[2] indicating a role in membrane budding, trafficking, or interaction with the cytoskeleton.[3,4] The GRP1 PH domain binds phosphatidylinositol 3,4,5-trisphosphate [PtdIns(3,4,5)P$_3$] with unusually high affinity and selectivity over PtdIns(3,4)P$_2$ and PtdIns(4,5)P$_2$.[5,6] The high abundance of PtdIns(4,5)P$_2$ over the 3'-polyphosphoinositides in intact cells suggests this high binding selectivity is required in order for GRP1 to be responsive to the appearance of PtdIns(3,4,5)P$_3$. These considerations suggest that GRP1 is a protein that can be recruited to cell membranes in response to the activation of phosphatidylinositol-3-kinase (PI3K), which generates PtdIns(3,4,5)P$_3$, and this has been confirmed.[2] Thus, GRP1 apparently acts to activate ARF proteins at the cell surface and possibly other cell membranes in response to growth factors and other agents that signal through the PI3K pathway. Other family members that exhibit the same domain structure as GRP1 and that have been characterized in some detail are ARNO[7] and cytohesin-1.[8]

It is still not clear what specific biological functions are regulated by GRP1. Recent work has led some investigators to suggest that Golgi structure or function may be a target, with ARF1 as mediator for such effects.[7] However, endogenous GRP1 appears to colocalize with ARF6 proteins at the cell surface membrane rather than in intracellular membranes.[2] Golgi localization may only be observed at high levels of expression of epitope-tagged constructs of GRP1 or other related proteins. Brefeldin A disrupts Golgi structure and inhibits the GTP loading of ARF1 but not ARF6,[2] presumably based on the action of the drug to inhibit endogenous exchange factors for ARF1. However, the ability of expressed GRP1 to catalyze ARF1 guanine nucleotide exchange is not compromised by brefeldin A, consistent with the hypothesis that it does not act as the endogenous ARF1

[2] S. E. Langille, V. Patki, J. S. Klarlund, Buxton, J. M., J. J. Holik, A. Chawla, S. Corvera, and M. P. Czech, *J. Biol. Chem.* **274,** 27104 (1999).

[3] J. Moss and M. Vaughan, *J. Biol. Chem.* **273,** 21431 (1998).

[4] J. Song, Z. Khachikian, H. Radhakrishna, and J. G. Donaldson, *J. Cell. Sci.* **111,** 2257 (1998).

[5] J. K. Klarlund, L. E. Rameh, L. C. Cantley, J. M. Buxton, J. J. Holik, C. Sakelis, V. Patki, S. Corvera, and M. P. Czech, *J. Biol. Chem.* **273,** 1859 (1998).

[6] J. M. Kavran, D. E. Klein, A. Lee, M. Falasca, S. J. Isakoff, E. Y. Skilnik, and M. A. Lemmon, *J. Biol. Chem.* **273,** 30497 (1998).

[7] M. Franco, J. Boretto, S. Robineau, S. Monier, B. Gould, P. Chardin, and P. Chavrier, *Proc. Natl. Acad. Sci. U.S.A.* **95,** 9926 (1998).

[8] W. Kolanus, W. Nagel, B. Schiller, L. Zeitlmann, S. Godar, H. Stockinger, and B. Seed, *Cell* **86,** 233 (1996).

exchange factor in the Golgi. Many studies have further shown that GRP1-like proteins are specifically recruited to plasma membranes in reponse to growth factors.[9-11] Taken together, the data are consistent with the concept that plasma membrane ARF proteins are targets of GRP1. ARF6 may be the major, although not exclusive, ARF protein at the cell surface, and may act in processes that include macropinocytosis, vesicular trafficking, and cell spreading, as well as macrophage phagocytosis.[4,12-14] These processes in turn are good candidates for regulation by GRP1 and related proteins.

Isolation of GRP1

GRP1 was isolated based on the high affinity binding to PtdIns(3,4,5)P$_3$ of its PH domain,[1] using a screen of a cDNA expression library. We have also isolated the phosphoinositide-dependent protein kinase-1 by this procedure, and the method should be successful in identifying other PtdIns(3,4,5)P$_3$ binding proteins.[15] The method used follows.

The 3-phosphatidylinositol probes labeled at the 3' position are generated with GST-p110 PI3K purified from recombinant baculovirus-infected Sf9 (*Spodoptera frugiperda* ovary) cells[16] and unfractionated bovine brain lipid (Sigma, St. Louis, MO). Phospholipid (20 μg) in chloroform is dried under N$_2$, resuspended in 30 μl of buffer containing 20 mM Tris-HCl (pH 7.4) and 1 mM EDTA by sonication and incubated in 200 μl of phosphorylation medium containing 20 mM Tris-HCl (pH 7.4), 100 mM NaCl, 10 mM MgCl$_2$, 0.5 mM EGTA. 0.2 mM adenosine, glutathione S-transferase (GST)-p110 PI3K, and 5–10 mCi of [γ-^{32}P]ATP (2.7–3.9 Ci/μmol). The reaction is incubated for 2 hr at room temperature, quenched by the addition of 0.2 ml of 1M HCl, followed by 0.5 ml of a chloroform : methanol (1 : 1) mixture. The organic phase is washed four times with 0.4 ml of methanol : HCl (1M) (1 : 1) and stored at $-70°$. Just before use, the lipid is dried under a stream of nitrogen with phosphatidlyserine corre-

[9] K. Venkateswarlu, P. B. Oatley, J. M. Tavare, and P. J. Cullen, *Curr. Biol.* **8,** 463 (1998).

[10] U. Ashery, H. Koch, V. Scheuss, N. Brose, and J. Rettig, *Proc. Natl. Acad. Sci. U.S.A.* **96,** 1094 (1999).

[11] K. Ventateswarlu, F. Gunn-Moore, P. B. Oatley. J. M. Tavare, and P. J. Cullen, *Biochem. J.* **335,** 139 (1998).

[12] H. Radhakrishna and J. G. Donaldson, *J. Cell Biol.* **139,** 49 (1997).

[13] H. Radhakrishna, R. D. Klausner, and J. G. Donaldson, *J. Cell Biol.* **134,** 935 (1996).

[14] D. Cox, C. C. Tseng, G. Bjekic, and S. Greenberg, *J. Biol. Chem.* **274,** 1240 (1999).

[15] V. R. Rao, M. N. Corradetti, J. Chen, J. Peng, J. Yuan, G. D. Prestwich, and J. S. Brugge, *J. Biol. Chem.* **274,** 37893 (1999).

[16] J. V. Virbasius, A. Guilherme, and M. P. Czech, *J. Biol. Chem.* **271,** 13304 (1996).

Positive Plaque

  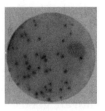

Primary   Secondary   Tertiary

FIG. 1. Cloning of GRP1. Nitrocellulose filters were incubated with ^{32}P-labeled phosphoino-sitides, and autoradiographs are shown after exposure for 2 days at $-70°$ with an intensifying screen. The primary spot is fairly faint, but the presence of a genuine positive is demonstrated by the presence of both plaques that bind and plaques that do not bind the probe in the secondary screen. At the tertiary purification, all plaques were positive. The primary screen was performed in 15-cm dishes, and the following in 10-cm dishes.

sponding to a final concentration of 20 μg/ml. The identities of the PtdIns(3)P, PtdIns(3,4)P$_2$, and PtdIns(3,4,5)P$_3$ are confirmed by thin-layer chromatography and high-performance liquid chromatography (HPLC) analysis.[17]

A mouse brain cDNA expression library in λ-ZAP vectors (Stratagene, La Jolla, CA) is plated and protein expression induced by standard techniques: 40,000 plaque-forming units (pfu) of the cDNA libraries are plated on each of eighteen 15-cm plates and incubated for 4 hr at 42°. Nitrocellulose filters that have been soaked in 10 mM isopropylthio-β-D-galactoside and subsequently dried at room temperature are placed on the plates and incubated for 14–16 hr at 37°. The plates are cooled to 4° and filters are removed and washed three to four times in 300 ml of assay buffer [25 mM Tris, pH 7.4, 100 mM NaCl, 0.25% Nonidet P-40 (NP-40), 0.1% sodium cholate, 1 mM MgCl$_2$, and 0.5 mM dithiothreitol (DTT)] under constant agitation. The filters are then incubated for 30 min in a crystallization bowl with the dissolved lipid in 30 ml of assay buffer at room temperature with the labeled mixed brain lipid (1–2 μCi/ml) and shaken vigorously. The filters are washed with five changes of the same buffer for a total of approximately 20 min, dried, and subjected to autoradiography. The phage is purified by removing an agar plug corresponding to a positive plaque, followed by elution and replating, and the procedure is repeated. Usually the phage is pure after two rounds of purification.

The signal from the primary screen is fairly weak (Fig. 1), and false

---

[17] L. A. Serunian, K. R. Auger, and L. C. Cantley, *Methods Enzymol.* **198,** 78 (1991).

positives are common. The presence of a bona fide positive is clearly seen at the level of the secondary screen because both positive and negative plaques will be present. Also the signals tend to be stronger because the plaques are larger. The success of the procedure appears to depend critically on the buffer conditions. The probe is expected to be present in micelles that carry negative charge from the cholate molecules. This may contribute to the binding since no binding to GRP1 is detectable if cholate is omitted. Possibly, other proteins may be isolated by varying the composition of the micelles.

ARF Exchange by GRP1 Sec7 Homology Domain

The Sec7 homology domain of GRP1 appears to be capable of catalyzing GTP–GDP exchange on all known ARF proteins. Direct demonstration[2,5] of such activity in cell-free systems has been reported for ARF1, ARF5, and ARF6 proteins, which represent each of the three classes of known ARF proteins.[3] This suggests that the related ARF2, ARF3, and ARF4 are also targets of GRP1. However, initially it was observed that ARF6 was apparently not a substrate for GRP1[5] or ARNO.[7] Subsequently, using a different assay method, two groups including our laboratory showed that ARF6 was indeed a good substrate for these proteins as well.[2,18] Both assays used are excellent, however, when used with ARF1, and therefore both are provided here.

ARF Exchange Assays

All solutions are prepared in assay buffer (50 mM HEPES, pH 7.5, 1 mM MgCl$_2$, 100 mM KCl, 1 mM DTT), and ARF and GRP1 proteins are transferred to this buffer either by washing beads containing the proteins or, if in solution, by the use of Centricon 30 microconcentrators (Amicon, Danvers, MA). Glutathione S-transferase (GST) fusion constructs of GRP1 or its individual domains are constructed by standard procedures to obtain fusions at precise junctions that we have previously described.[1] To obtain soluble protein for assay of exchange activities, they are cloned into pGEX-4T, and the recombinant protein is bound to glutathione agarose beads (Sigma), as described by the manufacturer of the cloning vector (Pharmacia, Piscataway, NJ). The GST fusion proteins are cleaved by incubation with 5 μg/ml thrombin in 200 μl of 20 mM Tris, pH 8.0, 2.5 mM CaCl$_2$, 150 mM NaCl overnight at 4°. Complete cleavage is verified by SDS–polyacrylamide gel electrophoresis. The proteins are stored at $-20°$ in 50% (v/v) glycerol in assay buffer.

[18] S. Frank, S. Upender, S. H. Hanson, and J. E. Cassanova, *J. Biol. Chem.* **273**, 23 (1998).

Assay Using Immobilized ARF1. Recombinant baculovirus encoding ARF proteins fused at the C terminus to a 9-amino-acid sequence corresponding to the major antigenic determinant of influenza virus hemagglutinin (YPYDVPDYA) are constructed. Sf9 cells are infected with the recombinant baculovirus, the cells are harvested 3 days later by centrifugation for 5 min at 3000 rpm, and the pellets are stored at −70°. A cell pellet corresponding to 50 ml of culture medium is dissolved in 1 ml of assay buffer supplemented with 1% Triton X-100, 1 mM benzamidine, 5 μg/ml leupeptin, 5 μg/ml aprotinin, 1 mM phenylmethylsulfonyl fluoride (PMSF), and 50 μM GDP. After clarification by centrifugation at 20,000 rpm for 5 min, 50 μl of a rabbit antiserum that has been produced by immunization with a peptide (YPYDVPDYA) conjugated to hemocyanin is added and incubated overnight on ice. The following day, 75 μl of protein A conjugated to Sepharose CL-4B (Sigma) is added and incubated on an end-over-end mixer for 1 hr. The beads are collected by centrifugation and washed five times with 1 ml of assay buffer. An additional 500 μl of Sepharose CL-4B is added as a carrier. An alternative source of myristoylated ARF protein is from *Escherichia coli* coexpressing N-myristoyltransferase.[19] It is important to estimate the level of modification by myristic acid because the modification is necessary for the activity of ARF proteins. This is most easily accomplished by labeling with [³H]myristic acid.[19]

The required amount of dimyristoylphosphatidylcholine (final concentration in assay 3 mM) is dried down in a Speed-Vac (Savant) and dissolved in assay buffer containing 0.4% cholate by vigorous agitation. The presence of negatively charged lipid or detergent is required for the exchange reaction, and their concentrations greatly influence the reaction.[5] The effects of other lipids on the exchange reaction may be examined by drying them down together with the dimyristoylphosphatidylcholine. The dissolved lipids are combined with an equal volume of 10 μM GTPγS, [³⁵S]GTPγS (typically 0.75 μCi per assay), 1% bovine serum albumin (BSA) in assay buffer, and the required amount of GRP1. Reactions are performed at room temperature, and started by adding 5 μl of the mix to 5-μl beads containing immunoabsorbed ARF. After 40 min, the beads are washed quickly four times with 1 ml of assay buffer, and bound ³⁵S is quantitated by liquid scintillation counting. A typical result from this assay is depicted in Fig. 2.

The results of such assays of GRP1 show a marked activation of ARF1 exchange activity in the presence of PtdIns(3,4,5)P₃ compared to the standard conditions described above. This is presumably due to the higher degree of recruitment of the PH domain of GRP1 to the phospholipid/

[19] P. A. Randazzo and R. A. Kahn, *Methods Enzymol.* **250**, 394 (1995).

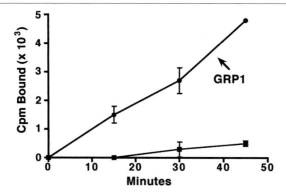

FIG. 2. Time course of a typical exchange reaction of ARF1 immobilized on beads. Incubations were in the absence or presence of 0.15 mM GRP1. Note that a small, but detectable, basal exchange rate is evident in the absence of GRP1.

detergent micelles when the polyphosphoinositide is present. We have studied the specificity of this activation and found it to be dependent on the detergent conditions.[5] Interestingly, when a relatively high concentration of negative charge is present in the micelles (0.1% cholate), PtdIns(3,4,5)P$_3$, PtdIns(4,5)P$_2$, and PtdIns(3,4)P$_2$, all stimulated ARF1 exchange activity in the presence of GRP1. In the absence of charge (0.1% CHAPS, no cholate), none of these phosphoinositides was able to enhance the ARF exchange activity of GRP1. In contrast, PtdIns(3,4,5)P$_3$ selectively stimulated [^{35}S]GTPγS binding to ARF1 in the presence of GRP1 when a low level of negative charge (0.05% cholate) was present with the phosphatidylcholine in our assay. Under these conditions PtdIns(4,5)P$_2$ and PtdIns(3,4)P$_2$ had significantly less effect. These experiments establish an *in vitro* assay system that reveals selective regulation of GRP1-catalyzed ARF1 guanine nucleotide exchange activity by PtdIns(3,4,5)P$_3$.

Assay of ARF Exchange Activities in Solution. In this assay free, recombinant myriostoylated ARF protein is used in purified form. Methods to prepare these proteins have been described in detail.[20] Azolectin (Sigma) is dried in a Speed-Vac (Savant), and a 6 mg/ml suspension of azolectin is prepared by sonication. This is combined with one volume of 20 μM GTPγS, [^{35}S]GTPγS (typically 0.75 μCi per assay), and one volume of GRP1 at the desired concentration. Reactions are started by adding 15 μl of this reaction mix to 5 μl of 4 μM ARF at room temperature. They are stopped by adding 1 ml ice-cold assay buffer to each reaction tube. Samples are then filtered through nitrocellulose membranes using a Bio-Rad (Hercules, CA) Bio-Dot apparatus and washed extensively with cold buffer to remove unbound

[20] P. A. Randazzo, O. Weiss, and R. A. Kahn, *Methods Enzymol.* **257,** 128 (1995).

[^{35}S]GTPγS from the ARF6 retained on the membranes. ARF6-bound [^{35}S]GTPγS is quantitated by counting the washed nitrocellulose membranes in a beta counter.

Assay of Guanosine Nucleotide Loading of ARF in Intact Cells. To assess the ability of expressed GRP1 or other proteins to cause activation of ARF proteins in intact cells, we developed an assay that measures labeled GTP loading of HA-epitope-tagged ARF proteins expressed in cultured cells.[2] COS-1 cells were transfected with HA-tagged wild-type and mutant ARF6 constructs or pCMV5 control vector, and then labeled with [^{32}P]orthophosphate. The level of GTP-bound ARF was determined by measuring the ratio of radiolabeled GTP–GDP bound to ARF protein immunoprecipitated from cell lysates using anti-HA antiserum. Using this method, cell lysates from pCMV5 transfected cells yielded no labeled guanine nucleotide after immunoprecipitation with anti-HA antiserum.[2] The wild-type, T27N (constitutively inactive), and Q67L (constitutively active) ARF6 constructs expressed well and on their immunoprecipitation yielded the expected changes in ratios of labeled GTP–GDP.[2] This assay also provides an estimate of the endogenous exchange activity and its properties for a given ARF protein. For example, we found that GTP loading of expressed HA-ARF1, but not HA-ARF6, was inhibited by brefeldin A in COS-1 cells, consistent with previous data obtained with cell free assays of endogenous exchange activity. The assay is performed as follows.

COS-1 cells are seeded in six-well tissue culture plates at a concentration of 5×10^4 cells/well. The following day, cells are transfected with varying amounts of HA-tagged ARF together with the exchange factor to be tested, or empty vector. Then, 24–48 hr posttransfection, cells are transferred to phosphate and serum-free Dulbecco's modified Eagle's medium (DMEM) supplemented with 25 mM HEPES (pH 7.2), 2 mM pyruvate, and 375 μCi/ml [^{32}P]orthophosphate for 16 hr. The presence of HEPES allows placement of cultures in incubators with no CO_2 supply. Spent medium is aspirated from each well and replaced with 500 μl of lysis buffer [20 mM Tris, pH 8.0, 100 mM NaCl, 1 mM MgCl$_2$, 1% Triton X-100, 0.05% cholate, 0.005% sodium dodecyl sulfate (SDS), 1 mM DTT, 1 mM phenylmethylsulfonyl fluoride (PMSF), 1 mM benzamidine, 5 μg/ml aprotinin, 5 μg/ml leupeptin, 1 mM NaF, 1 mM vanadate]. Cell lysates are scraped from each well and cleared by centrifugation at 20,000g for 5 min. Aliquots of supernatants should be analyzed by Western blotting for determination of protein expression. Anti-HA polyclonal antiserum (5 μl) and 10 μl of protein A-Sepharose beads are added to the supernatants, and immunoprecipitation is conducted at 4° for 2 hr on an end-over-end mixer. This is followed by seven to eight washes of the beads with ice-cold wash buffer (50 mM HEPES, pH 7.4, 0.5 M NaCl, 5 mM MgCl$_2$, 0.1% Triton X-100, 0.05% cholate, 0.005% SDS).

These extensive washes are required to avoid contaminants. After careful aspiration of the supernatant, 20 μl of elution buffer (75 mM KH$_2$PO$_4$ pH 3.4, 5 mM EDTA, 0.5 mM GTP, 0.5 mM GDP) is added to each tube, heated to 85° for 3 min, and spotted on polyethyleneimine-cellulose thin-layer chromatography plates containing a fluorescent indicator (Merck). Chromatography is conducted for approximately 1 hr using 0.65 M KH$_2$PO$_4$ buffer (pH 3.4). After autoradiography, separated guanine nucleotides are visualized with UV light, cut from the chromatography plates, and counted in a beta counter.

Polyphosphoinositide Binding by GRP1 PH Domain

PH domains consist of about 120 residues and have been identified in more than 100 proteins.[21,22] Most but not all PH domains bind polyphospho-inositides, which target these domains to cell membranes. The structures of several PH domains bound to the polar head groups Ins(1,4,5)P$_3$ or Ins(1,3,4,5)P$_4$ of the relatively high abundant PtdIns(4,5)P$_2$ or the PI3K product PtdIns(3,4,5)P$_3$, respectively, have been reported. These structures reveal a fold comprised in general of seven beta sheets with connecting loops that resemble structures of several other ligand binding domains that exhibit little sequence similarity with PH domains. Thus, it has been suggested that PH domains represent the prototype of a superfamily or superfold.[22] Of the many PH domains that have been characterized with respect to polyphosphoinositide binding, the PH domain of GRP1 appears to exhibit not only high affinity for PtdIns(3,4,5)P$_3$, but high selectivity as well.[5,6] Thus, affinities for these two lipids vary by more than two orders of magnitude.

One of the major problems in measuring binding affinities of PH domains for phospholipids is the insolubility of the latter. To overcome this problem, Rameh et al.[23] developed a binding assay that employed the dioctanoyl form of PtdIns(3,4,5)P$_3$, which is soluble in aqueous buffer. This method works well, but should be used with the understanding that binding affinities may differ from those that characterize the longer fatty acyl chain polyphosphoinositides.[23] Also, PH domains may bind to other membrane components in the intact cell. Nonetheless, this assay is very valuable in

[21] A. Musacchio, T. Gibson, P. Rice, J. Thompson, and M. Saraste, *Trends Biomed. Sci.* **18**, 343 (1993).

[22] N. Blomberg, E. Baraldi, M. Nilges, and M. Saraste, *Trends Biomed. Sci.* **24**, 441 (1999).

[23] L. E. Rameh, A. Arvidsson, K. L. Carraway III, A. D. Couvillon, G. Rathbun, A. Crompton, B. VanRenterghem, M. P. Czech, S. Ravichandran, S. J. Burakoff, D. Wang, C. S. Chen, and L. C. Cantley, *J. Biol. Chem.* **272**, 22059 (1997).

estimating the relative selectivity of binding to these lipids by GRP1-related proteins and for their predicted relative responsiveness to PI3K activation.

Polyphosphoinositide Binding and Competition Assays

Direct Binding Assay. GST–PH fusion proteins are immobilized on glutahione agarose beads and transferred to HNE (30 mM HEPES, pH 7.0, 100 mM NaCl, and 1 mM EDTA) supplemented with 0.02% Nonidet P-40. Gluthathione beads bound to GST alone are added as carrier. For the binding determinations, 160 μl of serial dilutions of unlabeled PtdIns(3,4,5)P$_3$ (Matreya) is added to 2 ml straight-walled screw-cap centrifuge tubes, followed by addition of 20 μl [^{3}H]dioctanoyl-C$_8$ immobilized PtdIns(3,4,5)P$_3$; approximately 2000 cpm are used per incubation. Then 20

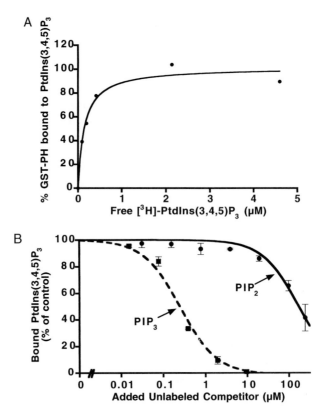

FIG. 3. Binding of [H^3]C$_8$PtdIns(3,4,5)P$_3$ to the PH domain of GRP1 and competition by C$_8$PtdIns(3,4,5)P$_3$ and C$_8$PtdIns(4,5)P$_2$ (A) Binding and (B) competition assays of the GRP1 PH domain.

μl of a 50% slurry of the GST–PH domains is added. For GRP1, approximately 200 pmol of GST–PH domain is used, but more may be needed with lower affinity binders. The tubes are placed on a rocking platform at room temperature. Care should be taken not to allow beads to adhere to the sides of the tubes, which will generate variability. After 1 hr, the tubes are centrifuged and the supernatants are collected and counted in a beta counter. The amount of [^{3}H]C$_8$PtdIns(3,4,5)P$_3$ bound to the PH domains is calculated by subtracting the amount of free ^{3}H present in the GST–PH domain supernatant from the amount of free ^{3}H present in control incubations with GST alone. The concentrations of free C$_8$PtdIns(3,4,5)P$_3$ remaining in the supernatants are calculated, and the data are fit to the equation [bound] = $B_{max}{}^\times$ [free]/(K_D + [free]) by the method of least squares (Kaleidagraph), where [bound] is the concentration of [^{3}H]C$_8$PtdIns(3,4,5)P$_3$ bound, [free] is the concentration of [^{3}H]C$_8$PtdIns-3,4,5-P$_3$ free in solution, B_{max} is the saturation binding, and K_D is the dissociation constant. The equation assumes a single class of binding sites. Figure 3A shows results from a typical binding assay.

Competition Assay. Incubations are set up as above. After 1 hr of incubation at room temperature, the beads are washed twice quickly with 1 ml of HNE–0.5% Nonidet P-40, and the washed beads counted in a beta counter. The data are plotted as a percentage of the control with no unlabeled lipid added, and it is fitted to the equation % bound = 100 − $n \times L/[K_I(app) + L]$, where n is the percent specific binding, L is the concentration of unlabeled lipid added, and $K_I(app)$ is the apparent competitive dissociation constant. This assumes simple competitive binding. As discussed by Rameh *et al.*,[23] the ratios of the measured dissociation constants accurately reflect the ratios of the true dissociation constants under these experimental conditions. Figure 3B shows the results from a typical competition assay.

Acknowledgment

Supported by NIH grants DK30898 and DK30648.

[31] Functional Analysis of ADP-Ribosylation Factor (ARF) Guanine Nucleotide Exchange Factors Gea1p and Gea2p in Yeast

By Anne Peyroche *and* Catherine L. Jackson

Introduction

Proteins of the ADP-ribosylation factor (ARF) family of small G proteins bind guanine nucleotides (either GDP or GTP), and their activity in cells depends on the nature of the nucleotide bound. Guanine nucleotide exchange factors (GEFs) catalyze the conversion of ARF–GDP to ARF–GTP in cells, and hence are key regulators of the biological functions of ARF proteins. A genetic approach in the yeast *Saccharomyces cerevisiae* led to the identification of the first family of GEFs for ARF, and the demonstration that the "Sec7 domain," a protein module of approximately 200 amino acids, is the catalytic domain for exchange.[1,2] The genetic selection used was a search for multicopy suppressors of the dominant negative growth defect conferred by ARF2[T31N], and the yeast Gea1p and Gea2p proteins were identified using this approach.[1] Chardin and colleagues found a human protein, ARNO, with sequence homology to the Gea1p and Gea2p proteins, but only in the Sec7 domain [so-called because it is also found in the Sec7p protein, necessary for transport in the endoplasmic reticulum (ER)–Golgi system in yeast[3]]. ARNO was found to have very potent ARF GEF activity, and the purified Sec7 domain alone was sufficient to catalyze exchange.[2] A biochemical approach aimed at purification of an ARF exchange activity identified several members of the Sec7 family, including mammalian homologs of Sec7p itself, and twelve Sec7 domain proteins have now been characterized.[4] The Sec7 domain carries out exchange by catalyzing nucleotide dissociation on ARF, the rate of which is very slow under physiologic conditions. Because the concentration of free GTP in cells is much higher than that of free GDP, release of nucleotide from a GTP-binding protein is sufficient to accomplish nucleotide exchange. Antonny and colleagues have demonstrated that the ARNO Sec7 domain accelerates the rate of nucleotide dissociation on ARF by an impressive

[1] A. Peyroche, S. Paris, and C. L. Jackson, *Nature* **384,** 479 (1996).

[2] P. Chardin, S. Paris, B. Antonny, S. Robineau, S. Beraud-Dufour, C. L. Jackson, and M. Chabre, *Nature* **384,** 481 (1996).

[3] A. Franzusoff, K. Redding, J. Crosby, R. S. Fuller, and R. Schekman, *J. Cell Biol.* **112,** 27 (1991).

[4] J. Moss and M. Vaughan, *J. Biol. Chem.* **273,** 21431 (1998).

4–5 orders of magnitude.[5] The yeast Gea1p protein has not been purified to homogeneity but useful information has been obtained as to its function from partial purification of a His$_6$-tagged version of the protein from yeast, as described below. We also describe purification of myristoylated yeast ARF2 from *Escherichia coli* and an assay for measuring exchange of GTP for GDP on myristoylated yeast ARF2. The yeast ARF1 and ARF2 proteins (which share approximately the same level of sequence homology to both class I and class II mammalian ARFs) are 96% identical and are functionally interchangeable.[6]

Purification of His$_6$-Gea1p from *Saccharomyces cerevisiae*

Gea1p is a protein of relatively low abundance in yeast, and it is not possible to overexpress the protein to a high level, making biochemical analysis difficult. We first constructed plasmid pCLJ202, in which the hexa-histidine (His$_6$)-modified *GEA1* gene is cloned into the vector pYES2 (Invitrogen, San Diego, CA) under the control of the strong *GAL1* promoter (which can lead to as much as a 1000-fold increase over the basal level for some proteins[7]). However, Gea1p cannot be overexpressed to more than 10 times its normal endogenous level in cells carrying this plasmid, even when expressed as the sole source of Gea1p or Gea2p in the cell (in strain CJY52-10-2 *MATα ura3-52 leu2 his3Δ200 lys2-801 ade2-101 trp1-Δ63 gea1::HIS3 gea2::HIS3*/pCLJ202). A similar level of overexpression can be obtained by simply placing the *GEA1* gene on a multicopy 2-μm vector (such as pCLJ92HT). For strain 52-10-2/pCLJ202, a 3-liter culture is grown in YPGal medium [2% (w/v) Bacto-peptone, 1% (w/v) yeast extract, and 2% (w/v) galactose] at 30° to an absorbance at 600 nm of 0.8. For strain 52-10-2/pCLJ92HT, a 1-liter culture is grown at 30° to an absorbance at 600 nm of 1.5 in YPD medium [2% (w/v) Bacto-peptone, 1% (w/v) yeast extract, and 2% (w/v) glucose]. Cells are collected by centrifugation, washed once with 300 ml of NaN$_3$ (10 mM), once with cold water, once with 100 ml buffer A-50 (20 mM HEPES, pH 7.5, 50 mM NaCl, 10% glycerol) and resuspended in 10 ml of buffer A-50 containing protease inhibitors [2 μg/ml leupeptin, aprotinin, chymostatin, and pepstatin, and 1 mM phenylmethylsulfonyl fluoride (PMSF)]. Cells are frozen at −80°, and lysed using a French press (pressure of 6 tons). The crude extract is then centrifuged [20,000 rpm (36,000g), 15 min, 4°] in a Beckman Ti50.2 ultracentrifuge rotor to remove cellular debris. The supernatant is collected and centrifuged

[5] S. Béraud-Dufour, S. Robineau, P. Chardin, S. Paris, M. Chabre, J. Cherfils, and B. Antonny, *EMBO J.* **17,** 3651 (1998).

[6] T. Stearns, R. A. Kahn, D. Botstein, and M. A. Hoyt, *Mol. Cell Biol.* **10,** 6690 (1990).

[7] J. C. Schneider and L. Guarente, *Methods Enzymol.* **194,** 373 (1991).

in a Ti50.2 rotor at 100,000g (40,000 rpm), 1 hr, 4°. Ammonium sulfate is added to the S100 to 40% saturation (4.5 g into 15 ml S100), the mixture incubated with stirring at 4° for 1 hr, then centrifuged in a Beckman Ti50.2 rotor at 100,000g, 30 min, 4°. The supernatant is carefully removed and the pellet is centrifuged again to allow more complete removal of the supernatant, then resuspended in 3 ml of buffer A-0 (20 mM HEPES, pH 7.5, 10% glycerol) containing protease inhibitors, and dialyzed against buffer A-50. The dialyzed protein solution is loaded onto a 1 ml Ni^{2+}-HiTrap chelating FPLC (fast protein liquid chromatography) column (Pharmacia, Piscataway, NJ) equilibrated with buffer A-50, and loading is carried out at a rate of 0.2 ml/min. The column is washed with 10 ml of buffer A-50 containing 20 mM imidazole. Proteins are eluted with a 20-ml linear gradient from 20 to 150 mM imidazole in buffer A-50 at a rate of 0.5 ml/min. Fractions of 300 μl are collected. Selected fractions are analyzed by Western blotting (Fig. 1). This procedure does not lead to a high yield of Gea1p protein, but results in a preparation that is pure enough to be used in *in vitro* ARF exchange assays.[1,8] It is useful for comparing the exchange activity of a mutant version of the Gea1p protein to that of the wild type. The procedure can also be used to identify proteins that copurify with Gea1p. An example is shown in Fig. 1. We identified Scp160[9,10] as a multicopy suppressor of a temperature-sensitive *gea1-4 gea2Δ* mutant. Scp160p expressed from a multicopy vector also suppresses the hypersensitivity of a *gea1-4 gea2Δ* strain to brefeldin A, a drug that acts as an uncompetitive inhibitor of the ARF exchange reaction catalyzed by sensitive Sec7 domain proteins (see below).[8] These results suggest that the two proteins both function in a common process. Scp160 coelutes with Gea1p, indicating that they exist in a complex in the cell, and hence providing evidence for a direct functional link (Fig. 1).

Expression and Purification of Myristoylated Yeast ARF2

For production of myristoylated recombinant yeast ARF2 (myr-yARF2), BL21(DE3) *E. coli* cells are cotransformed with plasmids pCLJ74 (pET14 carrying the yeast *ARF2* gene) and pBB131 containing the yeast *NMT1* gene encoding *N*-myristoyltransferase.[11] Cotransformants are selected for both ampicillin resistance (selecting for pCLJ74) and kanamycin

[8] A. Peyroche, B. Antonny, S. Robineau, J. Acker, J. Cherfils, and C. L. Jackson, *Mol. Cell* **3**, 275 (1999).

[9] U. Wintersberger, C. Kuhne, and A. Karwan, *Yeast* **11**, 929 (1995).

[10] V. Weber, A. Wernitznig, G. Hager, M. Harata, P. Frank, and U. Wintersberger, *Eur. J. Biochem.* **249**, 309 (1997).

[11] R. J. Duronio, E. Jackson-Machelski, R. O. Heuckeroth, P. O. Olins, C. S. Devine, W. Yonemoto, L. W. Slice, S. S. Taylor, and J. I. Gordon, *Proc. Natl. Acad. Sci. U.S.A.* **87**, 1506 (1990).

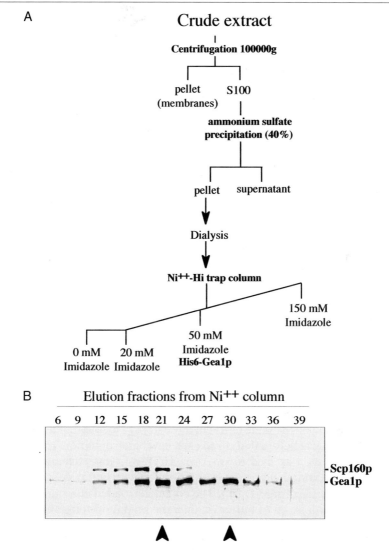

FIG. 1. Coelution of His$_6$-Gealp and Scp160p from the Ni^{2+} column. (A) Flow diagram of the purification scheme is shown. (B) Western blot analysis of fractions from the Ni^{2+} column. One-tenth the volume of each of the indicated fractions was loaded onto a 6% SDS–polyacrylamide gel, and analyzed by Western blot using polyclonal antisera against Scp160p (kindly provided by Dr. Ulrike Wintersberger, University of Vienna) and Gealp. Arrows indicate two peaks of elution of the Gealp protein.

resistance (to select for pBB131). Six liters of transformed cells are grown to an optical density at 600 nm of 0.8 in Luria–Bertani (LB) medium supplemented with ampicillin and kanamycin. Myristate ($50 \mu M$) bound to BSA ($6 \mu M$) is then added to the culture as a 100-fold stock solution.[12] After incubation at 37° for 20 min, the expression of both yARF2 and N-myristoyltransferase is induced by adding 0.6 mM isopropylthio-β-D-galactoside (IPTG). Cultures are incubated with shaking for 3.5 hr at 30°; the lower temperature increases the yield of myristoylated ARF2 protein. Cells are collected by centrifugation for 5 min at 5000 rpm ($4420g$) in a Beckman centrifuge using a JA10 rotor, frozen at −80°, then resuspended in 100 ml lysis buffer [Tris (pH 8) 20 mM, EDTA 1 mM, GDP 200 μM, dithiothreitol (DTT) 1 mM, PMSF 0.4 mM]. The cell suspension is transferred to a 40-ml hand-operated Dounce homogenizer, and the piston passed 15–20 times. Lysozyme is added to a final concentration of 0.38 mg/ml. The lysate is stirred for 30 min at 4° using a magnetic bar, then subjected to a second round of homogenization as before. Benzonase (Merck, Darnstadt, Germany) is added to a final concentration of 60 units/ml using a stock solution at 250 units/μl, and MgCl$_2$ to a final concentration of 2 mM. The solution is incubated at 4° with mild agitation for 30 min, then centrifuged at 14,000 rpm ($30,000g$) in a JA14 rotor for 20 min at 4°. The supernatant is transferred to two Ti45 ultracentrifuge tubes, and centrifuged at $100,000g$ for 1 hr, 30 min at 4°. The supernatant is recovered, transferred to two Ti45 tubes, and ammonium sulfate added to 33% (20.9 g/100 ml). The mixture is incubated for 1 hr at 4° with agitation, then centrifuged in a Ti45 rotor at $100,000g$ for 30 min at 4°. The supernatant is removed and the pellets resuspended in dialysis buffer (20 mM Tris, pH 8, 1 mM MgCl$_2$, 5 μM GDP, 1 mM DTT, 0.4 mM PMSF), and dialyzed against the same buffer in three steps (2 hr, overnight, then 3 hr). PMSF is added just before dialysis to each new batch of buffer. The dialyzed supernatant is then applied to two 5-ml HiTrap Q ion exchange columns (Pharmacia) mounted in series and equilibrated with buffer A (20 mM Tris, pH 8, 1 mM MgCl$_2$, 1 mM DTT). The column is loaded at a rate of 0.5 ml/min, washed with 50 ml of buffer A, then the bound proteins eluted with a 100-ml linear gradient from 0 to 500 mM NaCl in buffer A. Fractions (1 ml) are collected and 10- to 20-μl aliquots of selected fractions analyzed by SDS–PAGE followed by Coomassie blue staining (Fig. 2). The concentration of myr-yARF2 in the peak fraction is approximately 25 μM.

Assay of GTPγS Binding to ARF

Using the procedure outlined above, ARF is purified in its GDP-bound form, and hence exchange activity can be assayed by incubating ARF-GDP with [^{35}S]GTPγS and determining the rate of accumulation of [^{35}S]GTPγS-

[12] M. Franco, P. Chardin, M. Chabre, and S. Paris, *J. Biol. Chem.* **270**, 1337 (1995).

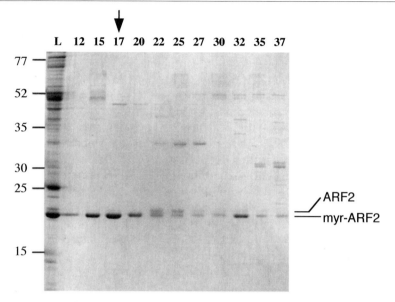

FIG. 2. Coomassie-stained, 15% SDS–polyacrylamide gel (0.375% bisacrylamide) showing the elution profile of the HiTrap Q column. An aliquot of the protein preparation loaded onto the colum is designated by "L," and numbers refer to fractions from the column. Unmyristoylated and myristoylated forms of yeast ARF2 are indicated to the right. The peak fraction, 17, is indicated by an arrow, and contains approximately 25 μM myr-yARF2. The positions of molecular weight standards in kDaltons is indicated on the left-hand side.

ARF. The assay we use is that developed by Paris to monitor [35S]GTPγS binding to recombinant myristoylated bovine ARF1.[12] ARF (in its full-length, myristoylated form) requires lipids in the exchange reaction. We routinely use vesicles made from azolectin, a preparation of soybean lipids containing approximately 20% phosphatidylcholine, prepared by Paris as follows[13]: 20 mg of azolectin (Sigma, St. Louis, MO) is dissolved in 6 ml of diethyl ether, then 1 ml of 50 mM HEPES, pH 7.5, is added. The mixture is sonicated for 2–3 min at 0°. The solvent is removed, and the resulting aqueous suspension of lipids is filtered through a 0.8-μm filter (Millipore, Bedford, MA). The final concentration of this stock suspension is 15 ± 3 mg/ml (dry weight). Aliquots are stored under nitrogen at −80°. In the exchange reaction, the azolectin vesicle stock preparation is diluted to a final concentration of 0.3–1.5 mg/ml. Slow exchange of GDP for GTP on myr-ARF can occur *in vitro* in the presence of lipids and in the absence of an exchange factor. This spontaneous exchange rate on myr-ARF depends on the concentration of lipids used, and can be reduced by decreasing the amount of azolectin vesicles present in the reaction.

[13] M. Franco, P. Chardin, M. Chabre, and S. Paris, *J. Biol. Chem.* **268,** 24531 (1993).

For each reaction, a time course of [^{35}S]GTPγS binding is carried out, with samples of 25 μl (25 pmol myr-yARF2) removed after the desired time of incubation at 30°. Tubes containing phospholipids (0.3–1.5 mg/ml azolectin vesicles), 1× reaction buffer (diluted from a 5× stock consisting of 250 mM HEPES, pH 7.5, 5 mM DTT, 5 mM MgCl$_2$) and H$_2$O to the appropriate volume are preincubated in a 30° water bath. One minute before beginning the time course, one-tenth volume of the GTPγS 10× stock solution [100 μM GTPγS (Boehringer Mannheim), [^{35}S]GTPγS (Amersham) at 10^5 cpm/μl, in H$_2$O] is added to the prewarmed reaction tubes. The GEF protein preparation to be tested (or buffer control) is added 15–30 sec before the reaction is started. The reaction is initiated by addition of myr-yARF2 to a final concentration of 1 pmol/μl. At the appropriate time, a 25-μl sample is removed and mixed with 2 ml of ice-cold stop solution (20 mM HEPES, pH 7.5, 100 mM NaCl, 10 mM MgCl$_2$). The mixture is quickly poured onto a 25-mm-diameter nitrocellulose filter (Schleicher and Schuell, Keene, NH BA85 0.45 μm) mounted on a Millipore filtration unit connected to a vacuum. The filter is washed twice with 2 ml of ice-cold stop solution, then removed to a heat-resistant tray. At the end of the time course, filters are baked for 10 min at 80°, then placed in 5 ml of aqueous scintillation fluid, and protein-bound radioactivity retained on the filter is measured using a scintillation counter. The level of nonspecific binding of [^{35}S]GTPγS to phospholipids and filters and to protein fractions is measured by carrying out parallel reactions in the absence of myr-yARF2.

Analysis of Gea1p and Gea2p Function *in Vivo*

Yeast offers the advantage of combining biochemical analysis with powerful genetic approaches (both classical and molecular) to analyze the function of a given protein both *in vivo* and *in vitro*. Genetic analysis begins with isolation of mutants in the gene encoding the protein of interest. Single disruption strains in which *GEA1* or *GEA2* are deleted do not have any growth or secretion defect that we have noted to date. However, a *gea1Δgea2Δ* double mutant strain is inviable, indicating that these two proteins assure at least one essential function in the cell in a redundant manner. We have generated conditional alleles (thermosensitive or *ts*) by mutagenizing the *GEA1* gene in a strain deleted for *GEA2*, and as expected these *gea1-ts gea2Δ* mutants have defects in protein trafficking in yeast. Most of the techniques used to analyze secretion and endocytosis defects in yeast are standard and are thus not described here. See, for example, Rothblatt *et al.*,[14] Franzusoff *et al.*,[15] Roberts *et al.*,[16] and Dulic

[14] J. Rothblatt and R. Schekman, *Methods Cell Biol.* **32,** 3 (1989).
[15] A. Franzusoff, J. Rothblatt, and R. Schekman, *Methods Enzymol.* **194,** 662 (1991).
[16] C. J. Roberts, C. K. Raymond, C. T. Yamashiro, and T. H. Stevens, *Methods Enzymol.* **194,** 644 (1991).

et al.[17] However, a simple technique not used routinely that monitors total secretion of proteins into the medium[18] has been particularly useful in analysis of mutants compromised in functions related to ARF. Gaynor and Emr discovered that the *sec21-3* mutant (defective in yeast γ-COP, an effector of ARF) continued to secrete a subset of proteins into the medium at a nonpermissive temperature for growth, whereas secretion of other proteins into the medium was completely inhibited.[19] This phenotype has also been reported for yeast *glo3* mutants lacking Glo3p ARF GAP function.[20] The protocol can also be used as a rapid method to monitor general secretion competence under conditions where all secretion is blocked or slowed.[8]

Monitoring Total Secretion of Proteins into Medium

Total cellular protein is first labeled with a mixture of [^{35}S]methionine and [^{35}S]cysteine *in vivo,* then cells are removed and the medium recovered for analysis. Before labelling, cells are grown to an optical density at 600 nm of 0.2–0.3 in SD medium [0.67% (w/v) yeast nitrogen base with ammonium sulfate/without amino acids, 2% (w/v) glucose] supplemented with appropriate amino acids. Cells are resuspended at 5 OD units/ml in SD medium containing amino acids and 350 μg/ml bovine serum albumin (BSA). Cells are preincubated at the appropriate temperature in a water bath, without shaking. Labeling is initiated by adding an appropriate volume of Promix [^{35}S] (Amersham) to obtain 20 μCi/OD unit of cells. After a 10-min incubation, the chase is initiated by adding a 10× chase solution (50 mM methionine, 10 mM cysteine, 4% yeast extract, 20% glucose) to a 1× final concentration. To stop protein transport at the chase points, a 400- to 500-μl aliquot of cells (2–2.5 OD equivalents) is removed and added to 30 μl of stop solution (NaN$_3$ 0.5 M, NaF 0.5 M) preincubated on ice. Cells and media are separated by centrifugation (5 min, 5000g, 4°); 75% of the supernatant volume containing media proteins is then recovered and added to trichloroacetic acid (TCA) on ice so that the final TCA concentration is approximately 7%. After mixing, the tubes are incubated on ice for at least 15 min. The media/TCA mixes are then centrifuged for 5 min at 16,000g, 4°. The supernatants are carefully removed and the pellets are washed twice with 1 ml of ice-cold acetone. After centrifugation (5 min, 16,000g, 4°), the acetone-washed pellets are dried in a Speed-Vac. Proteins are then resuspended in Laemmli sample buffer plus 5% 2-mercaptoethanol, boiled and centrifuged (16,000g, 5 min, 4°). Approximately 0.5 OD equiva-

[17] V. Dulic, M. Egerton, I. Elguindi, S. Raths, B. Singer, and H. Riezman, *Methods Enzymol.* **194,** 697 (1991).

[18] J. S. Robinson, D. J. Klionsky, L. M. Banta, and S. D. Emr, *Mol. Cell Biol.* **8,** 4936 (1988).

[19] E. C. Gaynor and S. D. Emr, *J. Cell Biol.* **136,** 789 (1997).

[20] D. Dogic, B. de Chassey, E. Pick, D. Cassel, Y. Lefkir, S. Hennecke, P. Cosson, and F. Letourneur, *Eur. J. Cell Biol.* **78**(5):305–310.

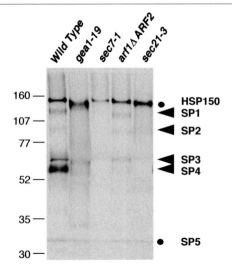

FIG. 3. Total secretion of proteins into the medium in different temperature-sensitive mutants. Strains were incubated at the nonpermissive temperature of 37° for 30 min prior to labeling of total protein with [³⁵S]methionine and [³⁵S]cysteine for 10 min. After a chase period of 30 min, cells were removed and the medium recovered for analysis on an 8% SDS–polyacrylamide gel followed by fluorography. Only one of the proteins secreted into the medium has been identified, HSP150 (see text); the others are labeled SP1–SP5. Filled circles indicate proteins that continue to be secreted into the medium in *sec21-3*, and arrows indicate proteins whose secretion is blocked. The positions of the molecular weight standards are indicated on the left-hand side. Strains used are as follows:

Wild type CJY52-10-2/ *MATα ura3-52 leu2 his3 Δ200 lys2-801 ade2-101 trp1-Δ63 gea1::HIS3 gea2::HIS3*/pCLJ92

gea1-19: CJY52-10-2/ *MATα ura3-52 leu2 his3Δ200 lys2-801 ade2-101 trp1-Δ63 gea1::HIS3 gea2::HIS3*/pNTS19 (CEN-TRP1-*gea1-19.*)

sec7-1: AFY080 *MATα ura3-1 leu2-3, 112 his3-11,15 trp1-1 ade2 sec7-1*

arf1Δ ARF2: CJY045-5-3 *MATa ura3-52 leu2-3,112his3Δ200 lys2-801 trp1-Δ63 arf1::HIS3*

sec21-3: EGY1213 *MATα ura3-52 leu2-3, 112 his3Δ200 trp1-Δ109 suc2-Δ9 sec21::HIS3*/p315sec21-3.

lents are then loaded on a 9% SDS–polyacrylamide gel (Fig. 3). The only protein secreted into the medium identified to date is HSP150, a protein whose rate of secretion is induced by heat shock.[21,22] It has been reported previously that HSP150 continues to be secreted in *sec7* mutants.[21,22]

Analysis of Effects of Brefeldin A on Yeast Cells

The drug brefeldin A (BFA) has profound effects on organelle structure and function in all eukaryotic cells examined including *Saccharomyces*

[21] V. V. Lupashin, S. V. Kononova, N. Ratner Ye, A. B. Tsiomenko, and I. S. Kulaev, *Yeast* **8**, 157 (1992).

[22] P. Russo, N. Kalkkinen, H. Sareneva, J. Paakkola, and M. Makarow, *Proc. Natl. Acad. Sci. U.S.A.* **89**, 3671 (1992).

cerevisiae.[23] Many trafficking pathways are blocked by BFA treatment of cells, including transport out of the ER–Golgi system.[24,25] BFA was shown to specifically inhibit a membrane-associated ARF GEF activity,[26,27] and as predicted by this result, it was recently that Sec7 domain proteins are direct targets of the drug.[8] BFA acts through the relatively rare mechanism of uncompetitive inhibition; that is, it binds to a reaction intermediate in the nucleotide exchange reaction rather than to the Sec7 domain alone.[8] The actual target of BFA is a very short-lived ARF-GDP-Sec7 domain protein reaction intermediate, and BFA acts to stabilize this complex. An interesting question is whether the effects of BFA on cells are mediated exclusively through their effects on sensitive Sec7 domain proteins. This question is difficult if not impossible to resolve in mammalian cells, but in yeast it was possible to demonstrate that the Sec7 domain proteins Gea1p, Gea2p, and Sec7p are the major targets of BFA in the secretory pathway.[8]

Wild-type yeast cells are not sensitive to BFA, so *in vivo* studies of the effects of BFA in yeast are carried out using the mutant *erg6.*[28] The *ERG6* gene encodes *S*-adenosylmethionine Δ^{24}-sterol-*C*-methyltransferase,[29] which catalyzes the final step in the biosynthesis of ergosterol, the equivalent of cholesterol in yeast.[28] The *erg6* mutant strains are hypersensitive to a number of drugs, and because ergosterol is an essential component of the yeast plasma membrane, it was postulated that the *erg6* mutation rendered cells more permeable to drugs. However, it was recently demonstrated that the yeast Pdr5p multidrug resistance pump has greatly reduced activity in an *erg6* mutant.[30] This protein when overexpressed confers resistance to a number of drugs by efficiently transporting the toxic compounds out of the cell; *pdr5* mutants are hypersensitive to such compounds because of a reduced efflux capacity.[30] Hence *erg6* likely confers resistance to drugs such as BFA at least in part because of a decreased capacity to pump these compounds out of the cell.

A difficulty in the use of *erg6* mutants for molecular analysis arises from the fact that these mutants are very difficult to transform using the standard lithium acetate procedure. Electroporation is an alternative method[31] that

[23] R. D. Klausner, J. G. Donaldson, and J. Lippincott-Schwartz, *J. Cell Biol.* **116**, 1071 (1992).
[24] J. Lippincott-Schwartz, L. C. Yuan, J. S. Bonifacino, and R. D. Klausner, *Cell* **56**, 801 (1989).
[25] T. R. Graham, P. A. Scott, and S. D. Emr, *EMBO J.* **12**, 869 (1993).
[26] J. G. Donaldson, D. Finazzi, and R. D. Klausner, *Nature* **360**, 350 (1992).
[27] J. B. Helms and J. E. Rothman, *Nature* **360**, 352 (1992).
[28] N. D. Lees, B. Skaggs, D. R. Kirsch, and M. Bard, *Lipids* **30**, 221 (1995).
[29] K. G. Hardwick and H. R. Pelham, *Yeast* 10, 265 (1994).
[30] R. Kaur and A. K. Bachhawat, *Microbiology* **145**, 809 (1999).
[31] E. Meilhoc, J. M. Masson, and J. Teissie, *Biotechnology (NY)* **8**, 223 (1990).

works very well for DNA transformation of *erg6* mutant strains. Cells are grown to an optical density at 600 nm of 0.6–0.9, centrifuged (5000*g*, 3 min, room temperature), washed twice in one-tenth volume of EB (270 m*M* sucrose, 1 m*M* MgCl$_2$, 10 m*M* Tris-Cl, pH 7.5), and resuspended in EB containing 20 m*M* HEPES, pH 7.5, and 25 m*M* DTT. Cells are incubated 10 min at 30°, then centrifuged as before and resuspended in 1/200 volume of EB. A 50-μl aliquot of the cell suspension is transferred to Eppendorf tubes conaining 0.1–1 μg DNA on ice, mixed, then transferred to a cooled electroporation cuvette with a 2-mm electrode gap (Eppendorf or Bio-Rad, Hercules, CA). Electroporation is carried out using an Eppendorf electroporator model 2510, set at 750 V.

[32] Isolation, Cloning, and Characterization of Brefeldin A-Inhibited Guanine Nucleotide-Exchange Protein for ADP-Ribosylation Factor

By Gustavo Pacheco-Rodriguez, Joel Moss, and Martha Vaughan

Introduction

ADP-ribosylation factors (ARFs) are GTPases that function as molecular switches to regulate intracellular vesicular trafficking pathways.[1] ARFs, which are ubiquitous in eukaryotic cells, have been grouped into class I (ARF1–3), class II (ARF4, 5), and class III (ARF6), based on molecular size, amino acid sequence, and gene structure.[2] All can activate cholera toxin ADP-ribosyltransferase and phospholipase D.[2] Alternation between the inactive GDP-bound and active GTP-bound forms of GTPases is dependent on guanine nucleotide-exchange proteins or GEPs, which accelerate replacement of bound GDP with GTP, and GTPase-activating proteins or GAPs, which enhance GTP hydrolysis.[3] Brefeldin A (BFA), a fungal metabolite that inhibits protein secretion, blocked guanine nucleotide-exchange on ARF,[4,5] although initial attempts to purify the responsible en-

[1] P. Chavrier and B. Goud, *Curr. Opin. Cell Biol.* **11,** 466 (1999).
[2] J. Moss and M. Vaughan, *J. Biol. Chem.* **270,** 12327 (1995).
[3] J. Moss and M. Vaughan, *J. Biol. Chem.* **273,** 21431 (1998).
[4] J. G. Donaldson, D. Finazzi, and R. D. Klausner, *Nature* **360,** 350 (1992).
[5] J. B. Helms and J. E. Rothman, *Nature* **360,** 352 (1992).

zyme yielded only BFA-insensitive preparations.[6,7] When a BFA-sensitive GEP, termed p200[8] or BIG1,[9] was purified from bovine brain cytosol, amino acid sequences of four tryptic peptides were 47–73% identical to the Sec7 protein from *Saccharomyces cerevisiae,* which had been identified during characterization of a temperature-sensitive, secretion-deficient mutant[10] and was later shown to possess GEP activity.[11] Direct evidence that the yeast Sec7 domain[12] and the BIG1 Sec7 domain[13,14] are loci of GEP activity and BFA sensitivity soon appeared.

This chapter describes methods to purify, clone, and characterize BIG1, and compares it briefly with other known ARF GEPs.

Isolation of BIG1

This procedure[8] was devised to purify the BFA-inhibited GEP that had been initially observed associated with an ~670-kDa protein complex.[6] It has been successfully used by several different people. Percentage recovery and fold purification were usually not calculated, because these values can be, at best, only rough estimates.

Fresh bovine brain cortex was homogenized (Polytron; Brinkman) in four volumes of TENDS buffer containing protease inhibitors.[7] To the cytosolic fraction prepared by centrifugation at 100,000g for 1 hr, solid $(NH_4)_2SO_4$ was added to 45% saturation. Precipitated proteins were subjected to three consecutive steps of column chromatography on DEAE-Sephacel (Pharmacia, Piscataway, NJ), hydroxylapatite (Bio-Gel HTPgc), and Mono QHR 10/10 (Pharmacia). BFA-inhibited GEP activity was assayed in eluate fractions. After precipitation at pH 5.8 in 3-morpholinopropanesulfonic acid from pooled active Mono QHR 10/10 fractions and chromatography on Superose-6 HR (Pharmacia), a protein complex of ~670

[6] S.-C. Tsai, R. Adamik, J. Moss, and M. Vaughan, *Proc. Natl. Acad. Sci. U.S.A.* **91,** 3063 (1994).

[7] S.-C. Tsai, R. Adamik, J. Moss, and M. Vaughan, *Proc. Natl. Acad. Sci. U.S.A.* **93,** 1941 (1996).

[8] N. Morinaga, S.-C. Tsai, J. Moss, and M. Vaughan, *Proc. Natl. Acad. Sci. U.S.A.* **93,** 12856 (1996).

[9] A. Togawa, N. Morinaga, M. Ogasawara, J. Moss, and M. Vaughan, *J. Biol. Chem.* **274,** 12308 (1999).

[10] T. Achstetter, A. Franzusoff, C. Field, and R. Scheckman, *J. Biol. Chem.* **263,** 11711 (1988).

[11] M. Sata, J. G. Donaldson, J. Moss, and M. Vaughan, *Proc. Natl. Acad. Sci. U.S.A.* **95,** 4204 (1998).

[12] M. Sata, J. Moss, and M. Vaughan, *Proc. Natl. Acad. Sci. U.S.A.* **96,** 2752 (1999).

[13] N. Morinaga, J. Moss, and M. Vaughan, *Proc. Natl. Acad. Sci. U.S.A.* **94,** 1226 (1997).

[14] N. Morinaga, R. Adamik, J. Moss, and M. Vaughan, *J. Biol. Chem.* **274,** 17417 (1999).

kDa was obtained.[8] This was roughly 12,000-fold purified from the $(NH_4)_2SO_4$ precipitate with ~1% recovery.[8]

The ~670-kDa complex contained (among other proteins), the ~200- and 190-kDa proteins that exhibited BFA-inhibited GEP activity after elution from the electrophoresis gel and renaturation. Samples of eluted protein were transferred to a nitrocellulose membrane for proteolysis and sequencing of tryptic peptides.[8] Four of nine peptides from p200 (BIG1) were 47–73% identical to those of yeast Sec7,[10] and cDNA cloning later showed that all nine peptides are present in the deduced amino acid sequence of BIG1.[13]

Materials

TENDS buffer contains 20 mM Tris-HCl (pH 8.0), 1 mM EDTA, 1 mM NaN₃, 10 mM dithiothreitol (DTT), and 250 mM sucrose unless otherwise indicated. Protease inhibitors present throughout GEP purification and assay were leupeptin, aprotinin, soybean, and lima bean trypsin inhibitors (each 0.1 μg/ml), and 0.5 mM AEBSF [4-(2-aminoethyl)benzenesulfonyl fluoride hydrochloride]. Preparation of native and recombinant ARF proteins is described in Refs. 6 and 15, respectively.

Assays for Brefeldin A-Inhibited Guanine Nucleotide-Exchange Protein (BIG1)

GEP Activation of ARF Assayed by Stimulation of Cholera Toxin A Subunit-Catalyzed ADP-ribosylagmatine Synthesis. In tissue preparations that contain other GTP-binding proteins, guanine nucleotide binding by ARF is not measured directly, but rather by its effect on ARF activity.[6] Stimulation of cholera toxin ADP-ribosyltransferase activity is directly proportional to the amount of active ARF. GTPγS, a poorly hydrolyzed analog, is used instead of GTP to avoid confounding effects of GTP hydrolysis. This assay was used throughout initial steps of BIG1 purification.[8]

For assay of GEP activity in preparations that contain other GTP-binding proteins, ARF stimulation of cholera toxin A subunit (CTA)-catalyzed ADP-ribosylagmatine formation is used as a specific measure of its GTPγS binding.[6] In the first step of the assay, GTPγS binding is carried out with or without 200 μM brefeldin A, in 50 μl of TENDS with the sample to be tested for GEP activity, 1 μM ARF (native or recombinant), 4–20 μM GTPγS, 30–60 μg of bovine serum albumin, (BSA), 20 μg of phosphatidylserine, and 5 mM MgCl₂. Control mixtures lacking GEP or

---

[15] G. Pacheco-Rodriguez, E. Meacci, N. Vitale, J. Moss, and M. Vaughan, *J. Biol. Chem.* **273**, 26543 (1998).

ARF or both are incubated in parallel. After incubation (usually 40 min at 37°), tubes are placed in an ice bath and four additions are made to complete a 300-μl reaction mixture for assay of ARF activation of CTA-catalyzed ADP-ribosylagmatine synthesis: (1) 100 μl of TENDS; (2) 50 μl of solution containing 30 μg of ovalbumin, 120 μM Cibachrome blue, and 20 μg of phosphatidylserine; (3) 50 μl of solution containing 30 mM MgCl$_2$, 3 mM ATP, 120 mM DTT, 60 mM agmatine, 1.2 mM [^{14}C]NAD (10^5 cpm), and 300 mM potassium phosphate buffer, pH 7.5; (4) 2 μg of CTA in 50 μl of TENDS. After incubation for 1 hr at 30°, 150 μl of the reaction mixture is applied to 1 ml of AG 1-X2 (Bio-Rad, Hercules, CA) previously equilibrated with deionized water followed by five washes with 1 ml of water. ^{14}C-Labeled ADP-ribosylagmatine in the effluent was quantified by liquid scintillation counting. After subtraction of ADP-ribosylagmatine synthesized in the presence of ARF, but without GEP, the effect of GEP on ARF activity and its BFA inhibition is calculated.

GEP Activation of ARF Assayed by [^{35}S]GTPγS Binding. This assay is used for characterization of activity of recombinant GEP or other preparations that do not contain GTP-binding proteins. For assay (total volume 100 μl) of the acceleration of GTPγS binding to ARF by BIG1,[7,15] a sample to be tested for GEP activity is incubated in 20 mM Tris-HCl, pH 8.0, with 10–50 pmol (0.2 μg) of native or recombinant ARF1 and/or ARF3, 4 μM [^{35}S]GTPγS (2–4×10^6 cpm), 50 μg of BSA, 20 μg of L-α-phosphatidylserine (PS), 1 mM DTT, 1 mM EDTA, and 5 mM MgCl$_2$, without or with 200 μM BFA (6 μg per assay). Protein with bound GTPγS is collected on a nitrocellulose filter for radioassay. After subtraction of [^{35}S]GTPγS bound to ARF alone and GEP alone, the difference between binding without and with BFA is BFA-inhibited GEP activity.

The temperature and time of incubation should be selected so that GEP activity is constant throughout and is proportional to the amount of enzyme added. Because different GEPs differ greatly in stability, fulfillment of these criteria must be verified whenever a change is made in assay conditions or enzyme preparations.[12] Incubation temperatures between 4°[12] and 37°[14] have been used to arrange incubation times convenient for large numbers of assays.

Guanine nucleotide binding and exchange are dramatically dependent on the concentration of free Mg^{2+},[16] as is GEP activity. This concentration can be influenced by other components of the assay, e.g., phospholipid, and the optimal concentration is not the same for all ARFs. In assays containing 1 mM EDTA (and phosphatidylserine), the activity of a partially purified BFA-insensitive GEP toward native ARF1 or ARF3 was maximal

[16] O. Weiss, J. Holden, C. Rulka, and R. A. Kahn, *J. Biol. Chem.* **264**, 21066 (1989).

with 3–5 mM MgCl$_2$.[7] The optimal concentration for BIG1 has not been established.

Routine assays contained phosphatidylserine because without it GTPγS binding to ARF was very low and GEP activity of BIG1[14] or of a BFA-insensitive GEP[7] was very low. Other phospholipids may be equally or more effective, but have not been systematically studied with BIG1.

BIG1 Cloning and Preparation of Recombinant Protein

Degenerate oligonucleotide primers based on the sequences of BIG1 peptides EVMYAVDQQHDFSGR and PEEYLSAIYNEIAGK, which are present also in the ySec7 protein, were used for PCR amplification of cDNA in a bovine brain λgt11 library.[13] A single plaque was cloned from the same library using two 50-mer oligonucleotides derived from the PCR product. The complete BIG1 cDNA contains 7016 bases with an ORF of 5547 bases encoding 1849 amino acids with a predicted molecular mass of 207.8 kDa.

For synthesis of BIG1 protein with an N-terminal hexahistidine (His$_6$) tag,[13] Sf9 (*Spodoptera frugiperda* ovary) cells are cotransfected with baculovirus containing the BIG1 cDNA construct in pAcHLT-C and Baculogold DNA (PharMingen). To purify the expressed protein, cells from a 75-mm^2 flask are sedimented by centrifugation (~1000 rpm, 5 min), dispersed in 2 ml of 10 mM sodium phosphate, pH 8.0/100 mM NaCl containing protease inhibitors, and lysed by two cycles of freezing and thawing. The homogenate is centrifuged (4°, 15 min, 16,000g) and the supernatant is incubated for 1 hr with nickel nitrilotriacetic acid agarose (2 ml of supernatant/0.5 ml of resin), which is poured into a column and washed extensively with 50 mM sodium phosphate, pH 8.0/300 mM NaCl/10% (v/v) glycerol/0.5 mM AEBSF containing 20 mM imidazole. Bound proteins are eluted with the same phosphate buffer at pH 6.0 containing 100 mM imidazole, and dialyzed against TENDS containing 30 mM NaCl, 5 mM MgCl$_2$, and 0.5 mM AEBSF. The protein is stored in small portions at −20° until use.

GEP activity of the purified protein is assayed by its effect on ARF stimulation of CTA-catalyzed ADP-ribosylagmatine synthesis and by its effect on binding of guanosine 5′-3′ *O*-(thio)triphosphate (GTPγS) by native ARF1/3 purified from bovine brain cytosol.

Structure and Function of BIG1

The deduced amino acid sequence of bovine BIG1[14] contains a central Sec7 domain (residues 697–885) and is 99% identical to human BIG1[9] with most differences located near the N terminus. BIG1 from bovine brain was

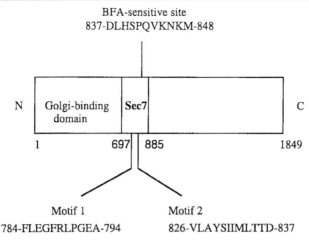

FIG. 1. Structural–functional domains of mammalian brefeldin A-inhibited guanine nucleo-tide-exchange protein 1 (BIG1), p200. Motif 1 (784–794) and motif 2 (826–837), which are critical for GEP activity, are shown within the Sec7 domain. The sequence adjacent to motif 2 contains D837 and M848, corresponding to those that were critical for BFA sensitivity of the yeast Sec7 domain in which they were identified.[11] The N-terminal region that contains Golgi-binding elements[27] and the large segment of unknown function C terminal to the Sec7 domain are also indicated.

the first BFA-inhibited GEP to be purified, cloned, and characterized.[8,13,14] Initial peptide sequencing[8] had revealed the presence of a Sec7 domain, which has since been found in all four groups of ARF GEPs. Two of these are BFA inhibited, represented by mammalian BIG1 and BIG2[9] with the closely related yeast Sec7[10] in one group and yeast Gea1 and Gea2 in another.[17] The BFA-insensitive GEPs include GBF1,[18] a 206-kDa protein that is much more active with ARF5 than ARF1 or ARF3 at a Mg^{2+} concentration of 1 mM, and EFA6,[19] which preferentially activates ARF6 and contains N- and C-terminal coiled-coil regions in addition to Sec7 and pleckstrin-homology domains. The much better characterized cytohesins are ∼50-kDa proteins with an N-terminal coiled-coil segment, a central Sec7 domain, and C-terminal pleckstrin-homology domain.[20] Cytohesin-1 was cloned first by subtractive hybridization as a protein highly expressed

[17] A. Peyroche, S. Paris, and C. L. Jackson, *Nature* **384,** 479 (1996).

[18] A. Claude, B.-P. Zhao, C. E. Kuziemsky, S. Dahan, S. J. Berger, J.-P. Yan, A. D. Arnold, E. M. Sullivan, and P. Melançon, *J. Cell. Biol.* **146,** 71 (1999).

[19] M. Franco, P. J. Peters, J. Boretto, E. Donselaar, A. D. Neri, C. Souza-Schorey, and P. Chavrier, *EMBO J.* **18,** 1480 (1999).

[20] M. Ogasawara, S.-C. Kim, R. Adamik, A. Togawa, V. J. Ferrans, K. Takeda, M. Kirby, J. Moss, and M. Vaughan, *J. Biol. Chem.* **275,** 3221 (2000).

in NK, but not T-helper cells,[21] and then, independently as an intracellular protein that interacted specifically with β_2-integrin and altered cell adhesion,[22] hence the name cytohesin. The presence of a Sec7 domain was noted by both groups, and cytohesin-1 GEP activity was soon reported.[23]

The Sec7 domains of Gea2[24] and cytohesin-1[25] and -2[26] have the same overall conformation, with two domains of five α helices each, and amino acid sequences of all Sec7 domains have a high degree of identity. The active site is in the C-terminal portion and residues critical for catalysis are included in sequences termed motif 1 and motif 2, which are highly conserved among ARF GEPs. In a three-dimensional structure of the Sec7 domain of Gea2 associated with a truncated form of human ARF1, amino acids in the switch I and II regions of ARF participate in specific interactions with the Sec7 domain.[24] All critical residues involved in Gea2 interaction with ARF1 are also present in the BIG1 Sec7 domain.

Some structural elements important for BIG1 function are shown in Fig. 1. Mutation of BIG1 to remove sequence C terminal to the Sec7 domain plus the first 594 amino acids had little effect on GEP activity, but deletion of the next 36 residues (through position 630) decreased it at least 100-fold, to the level of the Sec7 domain itself[14] demonstrating that elements critical for GEP activity are present outside of the Sec7 domain. Formation of a complex of the BIG1 Sec7 domain with Δ13ARF1 (ARF1 lacking the first 13 amino acids) was shown by gel filtration, but it was much less stable than the complex of Δ13ARF1 and the cytohesin-1 Sec7 domain under the same conditions.[14] It was notable also that the Sec7 domain was more efficient than cytohesin-1 in catalyzing guanine nucleotide exchange and less substrate specific,[15] demonstrating again that structure outside of the Sec7 domain can modify its GEP activity and interaction with ARF.

[21] L. Liu and B. Pohajak, *Biochim. Biophys. Acta* **1132,** 75 (1992).
[22] W. Kolanus, W. Nagel, B. Schiller, L. Zeitlmann, S. Godar, H. Stockinger, and B. Seed, B. *Cell* **86,** 233 (1996).
[23] E. Meacci, S.-C. Tsai, R. Adamik, J. Moss, and M. Vaughan, *Proc. Natl. Acad. Sci. U.S.A.* **94,** 1745 (1997).
[24] J. Goldberg, *Cell* **95,** 237 (1998).
[25] S. F. Betz, A. Schnuchel, H. Wang, E. T. Olejniczak, R. P. Meadows, B. P. Lipsky, E. A. S. Harris, D. E. Staunton, and S. W. Fesik, *Proc. Natl. Acad. Sci. U.S.A.* **95,** 7909 (1998).
[26] J. Cherfils, J. Ménétrey, M. Mathieu, G. LeBrass, S. Robineau, S. Béraud-Dufour, B. Antonny, and P. Chardin, *Nature* **392,** 101 (1998).
[27] S. J. Mansour, J. Skaug, X.-H. Zhao, J. Giordano, S. W. Scherer, and P. Melançon, *Proc. Natl. Acad. Sci. U.S.A.* **96,** 7968 (1999).

[33] Expression, Purification, and Properties of ADP-Ribosylation Factor (ARF) GTPase Activating Protein-1

By Irit Huber, Miriam Rotman, Elah Pick, Vardit Makler, Lilah Rothem, Edna Cukierman, and Dan Cassel

Introduction

Small GTPases of the ADP-ribosylation factor (ARF) family act as regulators of vesicular trafficking of proteins in all eukaryotic cells.[1-3] In mammalian cells the family consists of six proteins classified into class I (ARF1, 2, and 3), class II (ARF4 and 5), and class III (ARF6). The most abundant and best characterized ARF protein, ARF1, regulates the budding of COPI-coated vesicles from Golgi stacks[4,5] and of clathrin-coated vesicles from the trans-Golgi network (TGN).[6,7] The conversion of ARF to the GTP-bound form through the action of a guanine nucleotide exchange protein (GEP)[8,9] triggers the recruitment of a cytosolic coat protein onto the membrane, whereas coat protein dissociation depends on the hydrolysis of GTP, a process that requires the action of a GTPase-activating protein (GAP). Thus, regulators of the GTPase cycle of ARF1 play a critical role in membrane traffic.[10]

Our laboratory has reported the purification and cloning of a 45-kDa ARF1-directed GAP from rats (GAP1).[11,12] The "catalytic" domain of GAP1 was found to reside within the first 130–140 amino acids and to contain an essential Cys_4 zinc finger structure. Several lines of evidence

[1] J. D. Donaldson and R. D. Klausner, *Curr. Opin. Cell Biol.* **6,** 527 (1994).

[2] T. E. Kreis, M. Lowe, and R. Pepperkok, *Annu. Rev. Cell Dev. Biol.* **11,** 677 (1995).

[3] J. E. Rothman and F. T. Wieland, *Science* **272,** 227 (1996).

[4] J. G. Donaldson, D. Cassel, R. A. Kahn, and R. D. Klausner, *Proc. Natl. Acad. Sci. U.S.A.* **89,** 6408 (1992).

[5] D. J. Palmer, J. B. Helms, C. J. Becker, L. Orci, and J. E. Rothman, *J. Biol. Chem.* **268,** 12083 (1993).

[6] M. A. Stamnes and J. E. Rothman, *Cell* **73,** 999 (1993).

[7] L. M. Traub, J. A. Ostrom, and S. Kornfeld, *J. Cell. Biol.* **123,** 561 (1993).

[8] J. G. Donaldson, D. Finazzi, and R. D. Klausner, *Nature* **360,** 350 (1992).

[9] J. B. Helms and J. E. Rothman, *Nature* **360,** 352 (1992).

[10] M. G. Roth, *Cell* **16,** 149 (1999).

[11] V. Makler, E. Cukierman, M. Rotman, A. Admon, and D. Cassel, *J. Biol. Chem.* **270,** 5232 (1995).

[12] E. Cukierman, I. Huber, M. Rotman, and D. Cassel, *Science* **270,** 1999 (1995).

Copyright © 2001 by Academic Press
All rights of reproduction in any form reserved.
0076-6879/00 $35.00

suggest that GAP1 functions in the regulation of Golgi traffic. GAP1 cycles between cytosol and Golgi,[12] its overexpression results in an expected phenotype of Golgi disassembly,[13] and the protein interacts with a Golgi-localized receptor that mediates the retrieval of escaped endoplasmic reticulum (ER) glycoproteins bearing a carboxy-terminal KDEL tag.[14–16]

ARF GAPs have been identified in yeast (see [34] in this volume)[17,18] and new mammalian ARF GAPs have been identified and characterized (see [36] and [37] in this volume).[19–21] All of these proteins show high similarity to GAP1 in their catalytic domains, but otherwise mammalian ARF GAPs vary in size, contain different protein motifs, and may have distinct biological functions. In this chapter, we describe the preparation and properties of GAP1.

Purification of GAP1 from Rat Liver

GAP1 is expressed at low level, comprising less than 0.01% of cytosolic proteins in all tissues examined so far.[22] The protein can be obtained from rat liver cytosol as described.[11] Due to the very low yield of pure protein and the current availability of methods for the purification of GAP1 from recombinant expression systems, the original purification procedure will only be outlined. Briefly, a 38.5% ammonium sulfate precipitate obtained from 3–3.5 g cytosol is subjected to chromatography on a DEAE-Toyopearl column (Toyobo, Osaka, Japan), and fractions containing GAP activity are rechromatographed on DEAE-Toyopearl in the presence of 5 M urea, followed by chromatography on a Resource Q column under native conditions and in the presence of 20% (v/v) glycerol. This yields 150–200 μg of

[13] I. Huber, E. Cukierman, M. Rotman, T. Aoe, V. Hsu, and D. Cassel, *J. Biol. Chem.* **273**, 24786 (1998).

[14] T. Aoe, E. Cukierman, A. J. Lee, D. Cassel, P. J. Peters, and V. W. Hsu, *EMBO J.* **16**, 7305 (1997).

[15] T. Aoe, A. J. Lee, E. van Donselaar, P. J. Peters, and V. W. Hsu, *Proc. Natl. Acad. Sci. U.S.A.* **95**, 1624 (1998).

[16] T. Aoe, I. Huber, C. Vasudevan, S. C. Watkins, G. Romero, D. Cassel, and V. W. Hsu, *J. Biol. Chem.* **274**, 20545 (1999).

[17] P. P. Poon, X. Wang, M. Rotman, I. Huber, E. Cukierman, D. Cassel, R. A. Singer, and G. C. Johnston, *Proc. Natl. Acad. Sci. U.S.A.* **93**, 10074 (1996).

[18] P. P. Poon, D. Cassel, A. Spang, M. Rotman, E. Pick, R. A. Singer, and G. C. Johnston, *EMBO J.* **18**, 555 (1999).

[19] M. T. Brown, J. Andrade, H. Radhakrishna, J. G. Donaldson, J. A. Cooper, and P. A. Randazzo, *Mol. Cell Biol.* **18**, 7038 (1998).

[20] R. T. Premont, A. Claing, N. Vitale, J. L. Freeman, J. A. Pitcher, W. A. Patton, J. Moss, M. Vaughan, and R. J. Lefkowitz, *Proc. Natl. Acad. Sci. U.S.A.* **95**, 14082 (1998).

[21] J. Andreev, J. P. Simon, D. D. Sabatini, J. Kam, G. Plowman G, P. A. Randazzo, and J. Schlessinger, *Mol. Cell Biol.* **19**, 2338 (1999).

[22] M. Rotman, and D. Cassel, unpublished results (1997).

preparation containing a major 49-kDa band; this band, however, is a mixture of GAP1 and hnRNP F/H. Pure GAP1 may be obtained by reversed-phase high-performance liquid chromatography (HPLC) on a C_8 column using an acetonitrilotrifluoric acid gradient, yielding a few micrograms of homogeneous protein. This preparation retains partial activity despite the harsh solvent conditions, which decays on extended storage in this solvent. It is therefore preferable to employ the partially purified Resource Q material for biochemical experiments.

Expression and Purification of His_6-GAP1(1–257) from *Escherichia coli*

Bacterial expression systems are the most convenient source for the preparation of large amounts of GAP1. Unfortunately, however, full-length GAP could not be expressed in *E. coli.*[12] This is apparently due to the presence in the carboxy-terminal part of GAP1 of an uncharacterized determinant that is detrimental to the expression of GAP1, as suggested by the finding that fusion of this determinant to glutathione *S*-transferase (GST) abolishes GST expression.[23] Nevertheless, truncated proteins encompassing the entire catalytic domain of GAP1 can be expressed at high levels in *E. coli* and employed in biochemical and structural studies. We obtained the most consistent results in terms of expression level and GAP activity from constructs encoding the first 257 amino acids of GAP1. No expression was detected with constructs encoding more than the first 359 amino acids.

Construction of Expression Vector

Amino-terminal fragments of GAP1 can be expressed from different *E. coli* expression vectors. We routinely employ the T7 RNA polymerase-dependent expression system developed by Studier *et al.*[24] In this system the cDNA is placed under the control of the T7 promoter, and the resulting expression vector is introduced into *E. coli* strain BL21(DE3), which contains the T7 RNA polymerase under the control of the *lacZ* promoter. Induction of the T7 polymerase with isopropyl-β-D-thiogalactopyranoside (IPTG) triggers the expression of the cDNA. GAP1 cDNA can be expressed from vectors encoding a hexahistidine (His_6) tag at either the amino or carboxy terminus with or without a protease cleavage site.

The GAP1 cDNA fragment encoding the first 257 amino acids was prepared by polymerase chain reaction (PCR) amplification of GAP1 cDNA (from clone Z6[12]). Where appropriate, primers with extensions that introduce restriction sites were employed. The initiating methionine of

[23] I. Huber and D. Cassel, unpublished results (1997).
[24] F. W. Studier, A. H. Rosenberg, J. J. Dunn, and J. W. Dubendorff, *Methods Enzymol.* **185**, 60 (1991).

GAP1 is part of an *NcoI* site, which was used for cloning into vectors containing this cloning site. For cloning into pKM260[12] or pET20 expression vectors that introduce a hexahistidine tag at either the amino or the carboxy terminus (respectively), clone Z6 DNA was amplified with the T3 primer and a GAP1 cDNA-derived antisense primer including a *BamHI* site, and was cloned between the *NcoI* and *BamHI* sites of these vectors. This way of cloning results in the inclusion of a TEV protease cleavage site (pKM260) or signal peptide (pET20) in the expressed proteins. To eliminate these motifs, sense primers including an *NheI* site (pKM260) or an *NdeI* site (pET20) were used for PCR amplification, and amplification products were directly cloned into pUC57 and then subcloned into the expression vectors between *NheI* and *BamHI* (pKM260) or *NdeI* and *BamHI* (pET20).

Expression and Purification

Proteins are expressed in *E. coli* strain BL21(DE3) or in this strain bearing the pLysS plasmid encoding T7 lysosyme using standard conditions. Cultures are grown in LB containing the appropriate antibiotics to 0.6–0.8 OD_{600}, and proteins are induced for 2.5 hr at 37° in the presence of 0.4 mM IPTG. Proteins produced under these conditions (as well as under all other incubation conditions tested) are largely insoluble, and are thus prepared either as inclusion bodies,[12] or preferably by purification under denaturing conditions as follows.

Bacteria (400 ml) are centrifuged at 4000 rpm for 10 min, resuspended in 16 ml 6 M guanidine hydrochloride, 0.1 M sodium phosphate, 10 mM Tris, pH 8 (guanidine solution), mixed by inversion for 20 min and centrifuged at 15,000 rpm for 15 min. Supernatants are mixed with Ni^{2+}-NTA beads (2.5-ml packed volume, prewashed in the guanidine solution) by inversion for 45 min. The beads are washed twice with 20 ml of the guanidine solution by centrifugation and resuspension, and are transferred to a disposable plastic column. The column is washed with 10 ml of the guanidine solution, and proteins are eluted with guanidine solution containing 250 mM imidazole hydrochloride, pH 8. Fractions (0.5 ml) are collected and are supplemented with 5 mM dithiothreitol (DTT) immediately following elution. The fractions are tested for total protein using the Bio-Rad (Hercules, CA) protein assay reagent prediluted 1:4 in water. (Predilution of the reagent prevents precipitation of proteins that takes place when fractions in guanidine are diluted with water first.) Fractions containing the protein peak are pooled and are dialyzed overnight against 50 mM NaCl, 25 mM Tris, pH 7.4, 1 mM DTT with one buffer replacement. The dialysate is spun at 15,000 rpm for 10 min, and the supernatant is loaded on a 1-ml Resource Q column (a Mono Q column can be used as well). The column is developed with a

NaCl gradient (50–500 mM in 25 mM Tris, pH 7.4/1 mM DTT) run over 30 min at 1 ml/min. GAP1 elutes as a major, symmetrical peak in a position corresponding to 0.2 M salt, yielding 0.5–1 mg of protein. SDS–PAGE analysis (11% gels) should reveal a major band of 31 kDa; lower bands representing GAP1 degradation products may be present as minor contaminants. Storage of the preparation at 4° may result in some loss of activity, whereas full activity is retained during several weeks of storage in a freezer in the presence of 40% glycerol.

Comments

1. Even though Resource Q chromatography results in only a minor increase in purity, this step results in the removal of denatured forms of GAP1(1–257) and thus improves the homogeneity and stability of the preparation. The denatured forms tend to absorb to hydrophobic surfaces, and therefore the Resource Q step is essential for measurements of specific interaction of GAP1(1–257) with phospholipids (see below).
2. The eluate from the Ni-NTA column (containing 6 M guanidine-HCl) can be prepared as a stock and stored at −70° indefinitely. This material can also be shipped to other laboratories without cooling with high retention of activity.
3. The procedure described above relies on the high efficacy at which GAP1 renatures from guanidine solutions,[11,12] which may be due in part to the tight coordination of a zinc ion in the Cys$_4$ structure, which is not disrupted by guanidine.[23]
4. In the original study[12] we did not detect any soluble pool of GAP1 (1–257) following the lysis of bacteria by a freeze–thaw cycle and DNAse treatment. Subsequently we did observe a small soluble pool in extract from bacteria lysed with lysozyme followed by sonication. GAP1(1–257) could be purified from the soluble pool by Ni^{2+}-NTA chromatography under native conditions followed by Resource Q chromatography. However, this gave a low and variable yield of preparation often containing a significant amount of degradation products, and specific GAP activity was not higher than that of protein purified in the presence of guanidine. Based on this experience the purification under denaturing conditions as described above is the better choice in most applications.

Expression and Purification of Full-Length GAP1 from SF9 Cells

Even though the catalytic part of GAP1 can be expressed in *E. coli*, expression of the noncatalytic carboxy-terminal part cannot be accomplished in this system. However, a full-length GAP1 protein can be prepared by using the baculovirus expression system. The Bac-to-Bac system (Gibco-BRL, Gaithersburg, MD) was employed by closely following the instruction manual (URL: www2.lifetech.com/catalog/techline/molecular_biology/Manuals_PPS/bac.pdf). The entire GAP1 cDNA (from the initating Met through the 3'-UTR) was excised from clone Z6 by digestion with NcoI and KpnI and was cloned into the corresponding sites of plasmid pFASTBAC HTb that encodes a hexahistidine addition at the amino terminus. A recombinant baculovirus shuttle vector (bacmid) was generated by transposition into bacmid bMON14272 in E. coli DH10BAC in which the transposition function is provided by a helper plasmid.

To express the GAP1 protein, the recombinant bacmid is transfected into Sf9 (Spodoptera frugiperda ovary) cells, and viruses harvested from the cells are subjected to two rounds of amplification. For protein production, Sf9 cells are grown in serum-free Bioinsect-1 medium (Biological Industries, Kibbutz Beit-Haemek, Israel). Cells are infected with recombinant viruses, and are harvested 72 hr after infection. Cell are lysed in the presence of Nonidet P-40 and GAP1 is purified by Ni^{2+}-NTA chromatography using 0.6-ml packed volume of Ni^{2+}-NTA per 250 ml of suspension culture of Sf9 cells, yielding 0.5–1 mg purified GAP1 preparation (consult the instruction manual for details). The preparation can be frozen with liquid nitrogen and stored at $-80°$.

Properties of Recombinant GAP

In vitro ARF GAP Activity

A number of ARF GAP activity assays have been described.[11,19,25] All assays are based on measurement of single round hydrolysis of ARF-bound GTP. ARF is first loaded with GTP in the presence of lipids and/or detergents (which are required for the guanine nucleotide exchange reaction, either spontaneous or GEP stimulated). Subsequently a GAP is added and GTP hydrolysis is measured either by following the breakdown of radiolabeled GTP or by a continuous fluorimetric measurement that is based on the large decrease in ARF fluorescence on transition from the GTP- to the GDP-bound state.[25] Here we describe a simple assay of ARF GAP1 activity that is a modification of a previously described procedure.[11]

[25] B. Antonny, I. Huber, S. Paris, M. Chabre, and D. Cassel, *J. Biol. Chem* **272**, 30848 (1997).

Reagents

ARF: The procedure below employs fully myristoylated recombinant ARF1 (myr-ARF1) prepared as described.[26] Partially myristoylated ARF1 preparations[27] can also be employed, but their concentration during loading with GTP must be increased in order to achieve a sufficiently high percentage of ARF-bound GTP.[11] Myristoylated yeast ARF1 and ARF2, prepared with carboxy-terminal hexahistidine tags and purified by Ni-NTA chromatography, have also been employed in this assay.[17,18] (Consult the appropriate references for GTP loading conditions.)

DMPC/cholate: 3–5 mg of dimyristoylphosphatidylcholine (DMPC, Sigma, St. Louis, MO) are weighted in a 1-cm-diameter thin glass test tube and suspended in 25 mM MOPS, pH 7.4, to give a DMPC concentration of 22 mg/ml. The mixture is sonicated in a water bath sonicator for a few minutes and sodium cholate is added from a 10% stock to give a 1% final concentration, whereupon the suspension should become completely clear.

Charcoal suspension: Fine charcoal is weighed and suspended in water in a narrow vessel, and is allowed to settle. Floating material is removed by decantation, and the procedure is repeated once more. Finally the charcoal is suspended in 50 mM NaH_2PO_4 to give a 5% suspension.

GTP Loading

The reaction mixture contains 4 μM myr-ARF1, 1 μM [γ-^{32}P]GTP (a mixture of labeled and unlabeled GTP containing 10,000–20,000 cpm/pmol), 2 mM EDTA, 1 mM $MgCl_2$, 50 mM NaCl, 25 mM MOPS, pH 7.4, and a 1:10 dilution of the DMPC/cholate mixture (added last following a brief prewarming of the reaction). Final volume is usually 100–250 μl. The reaction is incubated for 15 min at 30°, and is terminated by the addition of $MgCl_2$ to give 2 mM. To determine loading efficiency, a 2-μl sample from the reaction is diluted with 2.5 ml of cold 50 mM KCl/25 mM Tris, pH 7.4/10 mM $MgCl_2$, followed by filtration through a 0.45-μm nitrocellulose filter and a wash with 5 ml of the same solution. The filter is counted in a scintillation counter and the results are compared with the radioactivity in a parallel unfiltered sample. Typically 50–75% of the GTP in the system becomes ARF bound. The preparation can be stored in aliquots at −80°.

[26] M. Franco, P. Chardin, M. Chabre, and S. Paris, *J. Biol. Chem.* **270**, 1337 (1995).
[27] P. A. Randazzo, O. Weiss, and R. A. Kahn, *Methods Enzymol.* **257**, 128 (1995).

Assay Procedure

GAP activity is determined at 30° in 100-μl assay systems containing 5 mM MgCl$_2$, 25 mM MOPS, pH 7.4, and different concentrations of GAP. Following a brief warm-up, reactions are initiated by the addition of 10 μl [γ-^{32}P]GTP loaded ARF. At different times 10-μl aliquots are transferred into microfuge tubes containing 0.5 ml of cold charcoal suspension, the tubes are vortexed briefly, and are left on ice. At the end of the assay, all samples are centrifuged for 1 min in the microfuge, and 0.4 ml of the supernatant is taken for radioactivity measurement by scintillation counting either by using the Cerenkov radiation of ^{32}P, or in the presence of a scintillation fluid.

Comments:

1. Hydrolysis of GTP during a 10-min assay reaches half maximal values at GAP1 concentrations ranging between 1 and 10 nM. Different GAP1 preparations from Sf9 cells showed similar to severalfold higher activity as compared with bacterially expressed proteins. Because the assay operates well bellow saturating conditions with respect to ARF1–GTP, meaningful specific activity values cannot be determined.
2. The dilution of ARF into the assay (at least fivefold) is required for activity, apparently because it results in the formation of DMPC/cholate micelles containing bound ARF–GTP. For this reason, the assay is sensitive to excess of detergents such as cholate, Triton X-100, and CHAPS (above 0.05%).
3. In some experiments inhibition was observed at very high GAP concentrations (above 4 μM).
4. GAP1 catalytic activity is insensitive to zinc chelators, but is sensitive to treatment with iodoacetamide (20 mM), apparently due to disruption of the zinc finger structure following the alkylation of its scaffold cysteine residues.[23]
5. The assay can be modified in order to determine the lipid requirements of the reaction by replacing the DMPC/cholate mixture in the ARF loading protocol with defined mixtures of pure phospholipids (0.4–1 mg/ml final concentration). These mixtures are best prepared as small unilamelar liposomes by filter extrusion (see next section). Dilution into the assay of ARF preloaded in the presence of lipids without detergents is not necessary.
6. Because some free radiolabeled GTP is always present in the loaded ARF preparations, activity assays employing [γ-^{32}P]GTP are sensitive to nucleotidases that may be present in crude GAP preparations

as well as in certain preparations of recombinant ARF. When nucleotidase contamination is low, nonspecific GTP hydrolysis may be decreased by the addition of 1–3 mM NaP$_i$. However, whenever crude GAP preparations are employed, it is necessary to employ [α-^{32}P]GTP and a nucleoside triphosphate regeneration system, as previously described.[11]

7. A lipid-free assay using ARF1 lacking the first 17 amino acids was described[28] and was reproduced in our laboratory using both full-length and truncated GAP1 preparations. GAP activity with the truncated ARF1 is low in itself but is strongly stimulated in the presence of coatomer.[28]

Interaction of GAP1(1–257) with Liposomes

GAP1(1–257), as well as the yeast ARF GAP Gcs1p, were found to interact with phospholipid vesicles.[25] This interaction depends on the type of phospholipids employed [phosphatidylethanolamine (PE) > phosphatidylserine (PS) > phosphatidylcholine (PC)] and is strongly stimulated by diacylgycerols (DAGs) bearing one or two unsaturated fatty acids.

Reagents

Lipids: Prepare chloroform stocks of lipids of high purity. We have employed the following from Sigma: phosphatidylcholine, phosphatidylethanolamine, phosphatidylserine, dioleylglycerol (DOG).

Polycarbonate filters: 0.4-μm filters from Poretics (Livermore, CA).

Extrusion device: We employ a homemade, two syringe-based device kindly provided by Dr. Bruno Antonny. A commercial device is also available from Lipex Biomembranes Inc., Vancouver, BC.

Preparation of Liposomes

Solutions of the lipids in chloroform at the desired proportions (2 mg total lipid) are mixed in 50-ml conical evaporation flasks and if necessary are diluted with chloroform to give 0.5 ml. A film is formed in a rotary evaporator and a nitrogen stream is applied to ensure complete removal of chloroform. Lipids are suspended in 0.5 ml of 220 mM sucrose, 20 mM Tris, pH 7.5, by rotation of the flask (without vacuum) for approximately 15 min. The suspension is transferred to a 2-ml plastic microfuge tube that is filled with nitrogen and capped, vortexed for 20 min, and subjected to five cycles of freezing in liquid nitrogen and thawing at 37°. After the last thaw the lipids must be kept at room temperature through all subsequent

[28] J. Goldberg, *Cell* **96,** 893 (1999).

stages. (*Do not chill!*) The suspension is transferred back and forth through a polycarbonate filter until its turbidity decreases (approximately 20 times), diluted fivefold with 120 mM NaCl, 20 mM Tris, pH 7.5, and spun at 40,000 rpm for 30 min at 25°. Finally the liposomes are suspended in 250 μl of the above buffer by pipetting. Yield is quantitative, resulting in an 8 mg/ml suspension.

Binding Assay

Reactions are set at room temperature in microultracentrifuge tubes. Reaction mixtures in a final volume of 80 μl contain 2 mg/ml liposomes, 1 mM MgCl$_2$, 120 mM NaCl, 20 mM Tris, pH 7.5, and GAP1(1–257) (prepared as described above) added last to give 75 μg/ml. The tubes are incubated for 15 min at room temperature and spun at 45,000 rpm for 30 min. Supernatants are transferred to new tubes, and pellets are suspended in 80 μl of 120 mM NaCl/20 mM Tris, pH 7.5. Supernatants and resuspended pellets are treated with SDS sample buffer, and equal volumes from each sample (30–50 μl) are run by SDS–PAGE (11% minigels). The gels are stained with Coomassie blue, destained, and the intensity of the 31-kDa GAP band is determined by scanning densitometry. The percentage of GAP that was bound to the liposomes is calculated by comparing the amounts of GAP in the pellet and supernatant.

Comments

1. As already noted, GAP1(1–257) must be purified through the Re-source Q step, because less purified preparations contain denatured forms that show high absorption to liposomes of all compositions. Preparations that have been stored with glycerol at −18° may be employed.
2. A good correlation was observed between GAP1(1–257) interaction with vesicles of a certain composition and GAP1-dependent hydroly-sis of ARF-bound GTP under the same conditions, suggesting that lipids (such as DAG) that stimulate GAP activity do so by promoting membrane recruitment.[25]

Acknowledgments

The research on which this chapter has been based was supported by grants from the Israel Science Foundation and by the Fund for Promotion of Research at Technion.

[34] Expression, Analysis, and Properties of Yeast ADP-Ribosylation Factor (ARF) GTPase Activating Proteins (GAPs) Gcs1 and Glo3

By Pak Phi Poon, Dan Cassel, Irit Huber, Richard A. Singer, and Gerald C. Johnston

Introduction

Eukaryotic cells use membrane vesicles to transport lumenal and membrane-associated cargo molecules between intracellular membrane compartments. A transport vesicle generated from a donor membrane by a budding process moves to a target membrane, where the vesicle fuses with the acceptor compartment to deliver its cargo. At the Golgi apparatus a transport vesicle destined for the endoplasmic reticulum (ER, retrograde movement) is formed by recruitment of cytoplasmic coat proteins that constitute the COPI protein complex.[1,2] Likewise, vesicles produced from the *trans*-Golgi network associate with clathrin adaptor proteins.[3] COPI and clathrin-related coats bind to the membrane in association with GTP-binding proteins termed ADP-ribosylation factor (ARF) proteins.[4,5] The binding of GTP activates ARF, which then recruits coat proteins to form a transport vesicle.[6] Hydrolysis of the ARF-bound GTP deactivates ARF and allows for the efficient fusion of a transport vesicle with a target membrane. This cycle of GTP binding and hydrolysis by ARF may actually mediate transport-vesicle formation and the selection of cargo.[7–10]

ARF proteins have negligible intrinsic GTPase activity and must rely

[1] F. Letourneur, E. C. Gaynor, S. Hennecke, C. Demolliere, R. Duden, S. D. Emr, H. Reizman, and P. Cosson, *Cell* **79,** 1199 (1994).

[2] P. Cosson, C. Demolliere, S. Hennecke, R. Duden, and F. Letourneur, *EMBO J.* **15,** 1792 (1996).

[3] R. Le Borgne and B. Hoflack, *Curr. Opin. Cell Biol.* **10,** 499 (1998).

[4] A. L. Boman and R. A. Kahn, *Trends Biochem. Sci.* **20,** 147 (1995).

[5] J. G. Donaldson, D. Cassel, R. A. Kahn, and R. D. Klausner, *Proc. Natl. Acad. Sci. U.S.A.* **89,** 6408 (1992).

[6] J. E. Rothman and F. T. Wieland, *Science* **272,** 227 (1996).

[7] J. Goldberg, *Cell* **96,** 893 (1999).

[8] J. Lanoix, J. Ouwendijk, C.-C. Lin, A. Stark, H. D. Love, J. Ostermann, and T. Nilsson, *EMBO J.* **18,** 4935 (1999).

[9] M. G. Roth, *Cell* **97,** 149 (1999).

[10] S. Springer, A. Spang, and R. Schekman, *Cell* **97,** 145 (1999).

on an accessory GTPase-activating protein (GAP) for GTP hydrolysis.[11] Two ARF GTPase activating proteins (GAPs) have been identified in the yeast *Saccharomyces cerevisiae*. These proteins, called Gcs1 and Glo3, were originally identified through genetic analyses and subsequently shown to function as ARF GAPs both *in vivo* and *in vitro*.[12,13] Both proteins exhibit a distinctive $C-X_2-C-X_{16-17}-C-X_2-C$ zinc finger motif (where X is any amino acid). This motif is found in ARF GAPs identified in different organisms and is required for ARF-GAP function.[7,12,14,15] Inspection of the yeast genome reveals that four other yeast proteins also contain this characteristic motif (Ref. 16 and our unpublished observations). The region of highest similarity among these yeast proteins includes a block of approximately 80 amino acids that encompasses the zinc finger motif.

Production of Recombinant Yeast ARF-GAP Proteins

Construction of Expression Vectors Encoding
  Hexahistidine-Tagged Proteins

The demonstration that Gcs1 and Glo3 proteins are ARF GAPs relied in part on an *in vitro* GAP assay using yeast proteins expressed in *Escherichia coli* (below). To facilitate the purification of the Gcs1 and Glo3 proteins from *E. coli*, we created versions of the proteins with a hexahistidine (His_6) tract. For Gcs1 we used an *Aha*II/*Bam*HI fragment that lacks the first 13 codons of the *GCS1* open reading frame (ORF) and includes a 532-bp region downstream of the stop codon of the *GCS1* ORF. The *Aha*II-cleaved site was filled in and the fragment was then ligated into a *Bam*HI/*Bgl*II digested pTrcHisC plasmid (Invitrogen, Carlsbad, CA) after filling in the *Bam*HI end. This strategy places the *GCS1* ORF downstream of the His_6 sequences. The plasmid encoding the His_6–Gcs1 fusion protein was then digested with *Nco*I and *Eco*RI and the fragment was transferred to plasmid pET16b (Novagen, Madison, WI; also digested with *Nco*I and *Eco*RI) to place expression of the His_6–Gcs1 construct under the control of the T7 promoter. For Glo3, we generated the ORF of the *GLO3* gene using PCR to create a *Bam*HI site before the *GLO3* start codon, and a

[11] R. A. Kahn and A. G. Gilman, *J. Biol. Chem.* **261**, 7906 (1986).
[12] P. P. Poon, X. Wang, M. Rotman, I. Huber, E. Cukierman, D. Cassel, R. A. Singer, and G. C. Johnston, *Proc. Natl. Acad. Sci. U.S.A.* **93**, 10074 (1996).
[13] P. P. Poon, D. Cassel, A. Spang, M. Rotman, E. Pick, R. A. Singer, and G. C. Johnston, *EMBO J.* **18**, 555 (1999).
[14] E. Cukierman, I. Huber, M. Rotman, and D. Cassel, *Science* **270**, 1999 (1995).
[15] V. Mandiyan, J. Andreev, J. Schlessinger, and S. R. Hubbard, *EMBO J.* **18**, 6890 (1999).
[16] C.-J. Zhang, M. M. Cavenaugh, and R. A. Kahn, *J. Biol. Chem.* **279**, 19792 (1998).

*Hind*III site immediately before the stop codon. The fragment was then inserted into the *Bam*HI/*Hind*III digested pET21b plasmid that encodes a hexahistidine tract to produce a Glo3 protein with the hexahistidine tract at the C terminus.

A Gcs1 protein that is His$_6$-tagged at its N terminus has significant ARF-GAP activity *in vitro*,[12] whereas a Glo3 protein similarly tagged at its N terminus exhibits only inefficient ARF-GAP activity relative to that of Gcs1. The basis underlying this different tolerance for added N-terminal sequences has not been investigated. Placement of a His$_6$ tag at the C terminus of Glo3 allows effective *in vitro* ARF-GAP activity.[13]

Purification of Recombinant Gcs1 and Glo3 Proteins

Recombinant Gcs1 and Glo3 proteins are expressed at moderate to high levels in *E. coli*. Plasmids were introduced into *E. coli* strain BL21(DE3) and proteins are produced during induction with 0.4 m*M* isopropylthiogalactoside (IPTG) for 2.5 hr at 37°. Both Gcs1 and Glo3 proteins produced under these conditions are largely insoluble and are thus prepared under denaturing conditions by guanidine extraction followed by Ni-NTA purification as described for the purification of recombinant ARF GAP1.[17]

In Vitro ARF-GAP Assays

GAP activity is assayed by a single-round hydrolysis of ARF-bound GTP, either using [α-^{32}P]GTP and measuring the amount of [α-^{32}P]GDP formed following separation by thin-layer chromatography on polyethyleneimine (PEI)-cellulose, or using [γ-^{32}P]GTP and measuring [^{32}P]P$_i$ release following charcoal absorption of the nucleotides.[17] ARF is first loaded with [^{32}P]GTP in the presence of DMPC and cholate, and is diluted into the assay systems. Yeast Arf1 or Arf2 protein, or mammalian ARF1 can be used with similar results. Yeast Arf1 and Arf2 were expressed in *E. coli* as His$_6$-tagged versions and purified by Ni-NTA chromatography.[12,13]

Dilution of guanidine solutions of the purified yeast GAPs results in at least partial refolding, as manifested by *in vitro* GAP activity. Under these conditions Gcs1 gives a consistently high GAP activity, causing 50% hydrolysis of ARF-bound GTP at approximately 1 n*M*. Things are less straightforward with Glo3. With Glo3 protein tagged at the C terminus, activity was about 10-fold lower than that of Gcs1.[13] This lower activity could be attributed in part to a lower purity of Glo3 preparations, which contain

[17] I. Huber, M. Rotman, E. Pick, V. Makler, L. Rothem, E. Cukierman, and D. Cassel, *Methods Enzymol.* **329**, [33], 2000 (this volume).

degradation products, but other possible reasons such as specific lipid requirements for Glo3 cannot be ruled out. We recently found that Gcs1 and Glo3 display similar and high activities when assayed in a lipid-free environment and in the presence of coatomer,[7] consistent with the idea that recombinant Glo3 preparations have high catalytic capacity (unpublished observations).

Genetic Analysis of Yeast ARF-GAP Function

Deletions

The genetic facility of yeast allows the exploration of ARF-GAP function *in vivo*. Deletion of either the *GCS1* or the *GLO3* gene does not affect cell viability, whereas deletion of both genes is lethal.[13] Thus, Gcs1 and Glo3 ARF GAPs provide an overlapping essential function, and one such function has been demonstrated to be retrograde vesicular transport from the Golgi to the ER.[13] Although single-deletion mutant cells are viable, these deletion mutations do impair certain activities. The absence of the Gcs1 protein from diploid yeast cells prevents the developmental process of sporulation (unpublished data), and in both diploid and haploid yeast cells the deletion of the *GCS1* gene (*gcs1*Δ) causes a cold sensitivity for cell proliferation at the otherwise normal growth temperature of 15° under certain defined physiologic conditions.[18] In particular, starved (stationary-phase) *gcs1*Δ mutant cells are unable to resume the process of cell proliferation when incubated in nutrient-replete liquid medium at 15°, but are able to resume and maintain cell proliferation like wild-type cells at higher temperatures (for example, 29°). Similarly, *gcs1*Δ mutant cells that have formed a colony on solid medium are unable to grow when transferred (by replica plating) to fresh solid medium and incubated at 15°. (This is so because most cells in a colony on solid medium are in a nutrient-deprived, stationary-phase state.) On the other hand, actively proliferating *gcs1*Δ mutant cells can maintain cell proliferation after transfer to 15° as long as nutrients are sufficient.[18,19]

The effects of *GLO3* gene deletion are different than those of *GCS1* deletion. Although the absence of Glo3 protein also causes a cold sensitivity for cell proliferation, this cold sensitivity is independent of growth status so that both stationary-phase and actively proliferating *glo3*Δ cells are unable to undergo cell proliferation after transfer to 15°. The absence of

[18] L. S. Ireland, G. C. Johnston, M. A. Drebot, N. Dhillon, A. J. DeMaggio, M. F. Hoekstra, and R. A. Singer, *EMBO J.* **13**, 3812 (1994).

[19] M. A. Drebot, G. C. Johnston, and R. A. Singer, *Proc. Natl. Acad. Sci. U.S.A.* **84**, 7948 (1987).

Glo3, unlike the absence of Gcs1, also affects cell growth at 29°, extending generation time by 50%.[13]

These genetic effects were determined by mutating the GCS1 and GLO3 genes in strain W303; however, we have found that for at least one other yeast strain, deletion of the GCS1 gene causes a cold sensitivity for cell proliferation that is independent of growth status (Ref. 19 and unpublished observations). Thus, we recommend assessment of the deletion phenotype before employing the gcs1Δ mutation in any other yeast strain.

A cautionary note: we have found that gcs1 mutants, and to a lesser extent glo3 mutants, display a high rate of genetic suppression, with easily identified cold-resistant derivatives arising after prolonged incubation at 15° or storage in the cold. This effect is due to true genetic suppression rather than simple reversion, for cold-resistant derivatives arise even in the complete absence of GCS1 or GLO3 sequences, and the retesting (by replica plating) of cold-resistant derivatives yields only cold-resistant cells. To avoid this kind of genetic suppression, we recommend that glycerol stocks of gcs1 and glo3 cells be maintained at −70°, and that colonies on solid media not be stored at 4° (these are, after all, "cold-sensitive" strains). We routinely use fresh cultures from a −70° freezer, and always inoculate new cultures from a single colony.

Synthetic Enhancement

Genetic strategies available in yeast allow the identification of proteins that interact functionally—in this case, proteins that affect ARF-GAP function. Often the combination of two functionally related mutations in the same cell causes a more severe phenotype than expected from the effects of each single mutation, a phenomenon referred to as "synthetic enhancement."[20] For example, as already mentioned, deletion of either the GCS1 or GLO3 genes allows such mutant cells to survive, whereas deletion of both is lethal. Likewise, deletion of the ARF1 gene, encoding one of two yeast Arf proteins with overlapping function, has little effect on cell growth, but deletion of both the ARF1 and GCS1 genes, or deletion of the ARF1 and GLO3 genes, severely impairs cell proliferation.[12,13] These instances of synthetic enhancement indicate that the gene products affect a common process.

For this synthetic enhancement analysis, the mutations were first introduced separately, by molecular procedures,[21] into a diploid cell to ensure that in each case one wild-type version of each gene remains to provide function. From the resulting heterozygous diploid strain, double-mutant

[20] L. Guarente, *Trends Genet.* **9**, 362 (1993).
[21] C. Guthrie and G. R. Fink, *Methods Enzymol.* **194**, (1991).

cells were then produced by sporulation, which generates haploid progeny. Such haploid segregants with the appropriate genetic makeup were then identified and examined. We recommend the use of heterozygous diploid strains when dealing with the *GCS1* gene because cells lacking both copies of *GCS1* (homozygous diploid cells) are unable to sporulate, a situation that precludes easy genetic analysis. This sporulation defect is not seen for *glo3Δ* diploid cells homozygous for deletion of the *GLO3* gene.

Expression of ARF-GAP Proteins in Yeast

Regulated Yeast ARF-GAP Expression

To control expression of the Gcs1 and Glo3 ARF-GAP proteins *in vivo,* these genes can be placed under the control of the galactose-inducible *GAL1* promoter from the pEMBLyex vector.[22] This promoter fusion was then transferred to the *CEN*-based vector pRS315[23] to provide function in *gcs1Δ glo3Δ* double-mutant cells. With galactose as the sole carbon source we observed good growth of *gcs1Δ glo3Δ* double-deletion mutant cells that contain a plasmid carrying either a *GAL1*-driven *GCS1* gene or a *GAL1*-driven *GLO3* gene (Ref. 13 and unpublished observations). Use of the *GAL1* promoter allows expression of these ARF-GAP genes to be shut off. However, when *gcs1Δ glo3Δ* cells harboring a *GAL1*-driven *GLO3* gene are transferred to growth medium containing glucose, which strongly represses expression from the *GAL1* promoter, Glo3 activity remains high enough to allow continued cell proliferation. Based on this finding, we assume that cells need very little Glo3 function. We find that a truncated version of the *GAL1*-driven *GLO3* gene, encoding a Glo3 protein missing the C-terminal 33 residues, provides adequate Glo3 function in the presence of galactose but insufficient activity when the transfer of cells to glucose leads to a markedly decreased supply of the (presumably enfeebled) truncated Glo3 protein. Such mutant cells after transfer to glucose medium cease cell proliferation and exhibit several defects in vesicular transport.[13] Similarly, placing the full-length *GCS1* gene under the control of the *GAL1* promoter also fails to decrease Gcs1 activity (on glucose medium) to levels that are inadequate for cell proliferation. We have not attempted to make a truncated (enfeebled) form of the Gcs1 protein analogous to the Glo3 situation, but presume that such an approach may also lead to insufficient Gcs1 activity under conditions that repress expression of the *GAL1*-driven *GCS1* gene.

[22] G. Cesarini and J. Murray, *Genet. Eng.* **9,** 135 (1987).
[23] R. Sikorski and P. Hieter, *Genetics* **122,** 19 (1989).

Rat ARF-GAP Expression in Yeast Cells

An ARF-GAP protein purified from rat resembles the yeast Gcs1 and Glo3 proteins structurally.[14] To test the ability of this rat ARF GAP to function in yeast cells, the nucleotide sequences[14] encoding rat ARF-GAP residues 14–415 were cloned downstream of the *Aha*II site within the *GCS1* ORF.[13] The resulting plasmid (pPPL25) encodes a chimeric protein consisting of the first 12 residues of Gcs1 protein fused to residues 14–415 of the rat ARF GAP, under the transcriptional control of a 337-bp fragment from the 5′ noncoding region of the *GCS1* gene.

As a test for ARF-GAP function *in vivo* we determined whether the rat ARF-GAP protein could support the proliferation of *gcs1*Δ *glo3*Δ double-deletion cells. Plasmid pPPL25 was transformed into *gcs1*Δ *glo3*Δ cells that were sustained by the presence of the *GAL1-GLO3* truncation plasmid described above. The transformed cells were grown as patches on galactose-containing rich solid medium and then transferred by replica plating to solid medium with glucose as the carbon source, conditions that prevent adequate activity of the *GAL1-GLO3* truncation gene. Under these conditions the rat ARF-GAP protein, expressed from the *GCS1* promoter, can supply activity for cell proliferation.[13] Curiously, however, plasmid pPPL25 carrying the rat ARF-GAP gene is unable to alleviate the cold sensitivity of *gcs1*Δ and *glo3*Δ single-mutant cells (unpublished observations), suggesting that the rat ARF-GAP protein may not function adequately in the cold, or that the cold sensitivity of *gcs1*Δ and/or *glo3*Δ single-mutant cells may not reflect the overlapping function of these yeast proteins for retrograde Golgi-to-ER transport (i.e., these ARF GAPs may function in compartments other than the ER-Golgi).

New ARF-GAP Mutations by PCR Mutagenesis

Temperature-sensitive mutant alleles of the *GCS1* gene were created by a mutagenic polymerase chain reaction (PCR) "gap-repair" procedure.[24] A "gapped" plasmid (pPPL18) was first constructed by cloning the 5′ and 3′ noncoding regions of the *GCS1* gene, separated by a *Hin*dIII restriction site for this *LEU2*-based vector. PCR under mutagenic conditions[24] was then used to amplify the *GCS1* coding sequences and flanking noncoding regions. These PCR products were transformed, along with an equivalent amount of *Hin*dIII-digested pPPL18 plasmid containing the *GCS1* flanking regions, into *gcs1*Δ *glo3*Δ cells that were kept alive by the *GAL1*-driven *GLO3* truncation plasmid. Under these cotransformation conditions the

---

[24] D. Muhlrad, R. Hunter, and R. Parker, *Yeast* **8,** 79 (1992).

high rate of homologous recombination in yeast cells leads to the incorporation of PCR-amplified (and possibly mutated) *GCS1* sequences into the "gapped" plasmid,[21] thus reconstructing an intact plasmid-borne *GCS1* gene. Transformants were selected by growth at 23° on defined solid medium lacking leucine (to select for transformants harboring derivatives of the pPPL18 plasmid) and tryptophan (to maintain selection for the *GAL1-GLO3* plasmid) and with galactose as the carbon source to maintain expression of the truncated Glo3 protein. PCR-generated mutant alleles of the *GCS1* gene that confer temperature sensitivity were then identified by replica-plating to glucose medium, which shuts off expression of the *GAL1-GLO3* truncation plasmid, and incubation at 37° and 23°. Temperature-sensitive function of a newly produced, plasmid-borne *gcs1* mutant gene was confirmed in derivatives that had spontaneously lost the *GAL1-GLO3* truncation plasmid, which were identified as cells that were no longer tryptophan prototrophs.

Acknowledgments

This work was supported by grants from the National Cancer Institute of Canada with funds made available from the Canadian Cancer Society (held jointly by G.C.J. and R.A.S.) and from the Israel Science Foundation (D.C.).

[35] Purification and Properties of ARD1, An ADP-Ribosylation Factor (ARF)-Related Protein with GTPase-Activating Domain

By Nicolas Vitale, Joel Moss, and Martha Vaughan

Introduction

Originally identified by their ability to support the cholera toxin-catalyzed ADP-ribosylation of Gs,[1] ADP-ribosylation factors (ARFs), were later shown to be critical GTPases, associated with specific intracellular organelles such as the Golgi apparatus[2] and endosomes,[3] and regulating the formation and fusion of transport vesicles (for a review see Ref. 4).

[1] R. A. Kahn and A. G. Gilman, *J. Biol. Chem.* **259,** 6228 (1984).
[2] D. Botstein, N. Segev, T. Stearns, M. A. Hoyt, J. Holden, and R. A. Kahn, *Cold Spring Harbor Symp. Quant. Biol.* **53,** 629 (1988).
[3] C. D'Souza-Schorey, G. Li, M. I. Colombo, and P. D. Stahl, *Science* **267,** 1175 (1995).
[4] J. Moss and M. Vaughan, *J. Biol. Chem.* **270,** 12327 (1995).

ARFs have also been shown to be activators of phospholipase D.[5-7] The six known mammalian ARFs are ≈180 amino acid proteins that exhibit a high percentage of sequence identity and are ubiquitous in eukaryotes (for review see Ref. 8). They fall into three classes based on amino acid and DNA sequence similarities and gene structure.[4] Most amino acid differences among ARF proteins are concentrated in the amino- and carboxyl-terminal ends, and sequences involved in guanine nucleotide binding are completely conserved.[10] Crystal structures of ARF1 in GDP- and GTP-bound states revealed those residues that form the nucleotide-binding pocket and showed nicely the conformational changes in the overall structure of ARF1 that result from GTP binding.[11-13]

ARD1, A Unique Member of ARF Family

ADP-ribosylation factor domain protein 1 (ARD1) is a 64-kDa protein that contains an 18-kDa carboxyl-terminal ARF domain and a 46-kDa amino-terminal domain. This distinctive member of the ARF family was originally identified during screening of a HL-60 cell cDNA 1 ZAP library.[14] A single clone hybridized with the ARF2 cDNA and failed to hybridize with oligonucleotide probes specific for the six known mammalian ARFs. The new clone contained a 1660-base-pair (bp) insert, including the sequence of ARD1 (1207–1722) that encodes the ARF domain of 172 amino acids. Clones isolated from human fetal brain cDNA λ ZAP library screened with probes specific for the 1660-bp original insert revealed that about 1200 nucleotides preceded the ARF region without a stop codon in the same

[5] H. A. Brown, S. Gutowski, C. R. Moomaw, C. Slaughter, and P. C. Sternweis, *Cell* **75,** 1137 (1993).

[6] S. Cockroft, G. M. H. Thomas, A. Fensome, B. Geny, E. Cunningham, I. Gout, I. Hiles, N. F. Totty, O. Truong, and J. J. Hsuan, *Science* **263,** 523 (1994).

[7] D. Massenburg, J. S. Han, M. Liyanage, W. A. Patton, S. G. Rhee, J. Moss, and M. Vaughan, *Proc. Natl. Acad. Sci. U.S.A.* **91,** 11718 (1994).

[8] J. Moss and M. Vaughan, "Handbook of Experimental Pharmacology," Vol. 108/I, "GTPases in Biology. Molecular Characterization of ADP-Ribosylation Factors." Springer-Verlag, Berlin, 1993.

[9] C. M. Lee, R. S. Haun, S.-C. Tsai, J. Moss, and M. Vaughan, *J. Biol. Chem.* **267,** 9028 (1992).

[10] S. R. Price, A. Barber, and J. Moss, *in* "ADP-Ribosylating Toxins and G Proteins: Insights into Signal Transduction," p. 397. American Society for Microbiology, Washington, DC, 1990.

[11] J. C. Amor, D. H. Harrison, R. A. Kahn, and D. Ringe, *Nature* **372,** 704 (1994).

[12] S. E. Greasley, H. Jhoti, C. Teahan, R. Solari, A. Fensome, G. M. H. Thomas, S. Cockroft, and B. Bax, *Nature Struct. Biol.* **2,** 797 (1995).

[13] J. Goldberg, *Cell* **95,** 237 (1998).

[14] K. Mishima, M. Tsuchiya, M. S. Nightingale, J. Moss, and M. Vaughan, *J. Biol. Chem.* **268,** 8801 (1993).

reading frame. The putative open reading frame of ARD1 consists of 1722 nucleotides encoding a protein of 574 amino acids.[14]

The nucleotide sequence of the ARF domain from ARD1 is 60–66% identical to those of mammalian ARFs. At the amino acid level, the ARF domain of ARD1 is 55–60% identical and 69–72% similar to other ARFs. Human and rat ARD1 are 92 and 98% identical at the nucleotide and amino acid levels, respectively.[14] Regions common to ARFs that are believed to be involved in guanine nucleotide binding and GTP hydrolysis are also conserved in ARD1.[14] Recombinant proteins were used to demonstrate that ARD1 and its ARF domain specifically bind GDP and GTP in the presence of millimolar concentrations of magnesium.[15] As observed for ARFs, binding of GTPγS to ARD1 is considerably stimulated by certain phospholipids, especially cardiolipin. Accordingly, ARD1 stimulated cholera toxin A subunit (CTA) ADP-ribosyltransferase activity in a GTP- and phospholipid-dependent manner.[15]

Most of the differences between the ARF sequence in ARD1 and other ARFs are in the amino- and carboxyl-terminal regions. In ARD1, the 15 amino acids corresponding to the N-terminal α helix of ARF influence GDP dissociation,[16] in a manner similar to that of the analogous N-terminus of ARF1.[17–19] Site-specific mutagenesis suggested that hydrophobic residues in this region have a critical effect on GDP dissociation from ARD1.[16] Because no guanine nucleotide-dissociation inhibitory (GDI) protein for ARFs has been identified, it was postulated that ARFs and ARD1 have an intrinsic regulatory GDI domain that accounts for the slow GDP dissociation from ARFs and ARD1.[16]

Under physiological conditions, hydrolysis of bound GTP by ARFs is essentially undetectable and is markedly accelerated by a GTPase-activating protein (GAP), several of which are now known (for references see Conclusion). Although the ARF domain of ARD1 has no detectable intrinsic GTPase activity, ARD1 hydrolyzes bound GTP at a significantly higher rate.[21] The 46-kDa N-terminal extension, synthesized separately from the ARF domain in *Escherichia coli,* was demonstrated to be responsible for this GAP activity.[21] The smallest GTPase-activating domain identified in

[15] N. Vitale, J. Moss, and M. Vaughan, *J. Biol. Chem.* **272,** 3897 (1997).

[16] N. Vitale, J. Moss, and M. Vaughan, *J. Biol. Chem.* **272,** 250077 (1997).

[17] J. X. Hong, R. S. Haun, S.-C. Tsai, J. Moss, and M. Vaughan, *J. Biol. Chem.* **269,** 9743 (1994).

[18] J.-X. Hong, X. Zhang, J. Moss, and M. Vaughan, *Proc. Natl. Acad. Sci. U.S.A.* **92,** 3056 (1995).

[19] P. A. Randazzo, T. Terui, S. Sturch, H. M. Fales, A. G. Ferrige, and R. A. Kahn, *J. Biol. Chem.* **270,** 14809 (1995).

[20] J. Moss and M. Vaughan, *J. Biol. Chem.* **273,** 21431 (1998).

[21] N. Vitale, J. Moss, and M. Vaughan, *Proc. Natl. Acad. Sci. U.S.A.* **93,** 1941 (1996).

ARD1 was between amino acids 101 and 333,[22] a region similar in size to the minimal catalytic domains of the Ras GAPs.[23] The GAP site of ARD1 can be divided into a segment important for physical association with the ARF domain (residues 200–333) and a domain directly involved in acceleration of GTP hydrolysis. Therefore, ARD1, unlike other ARFs, possesses its own domain regulating GTP hydrolysis.

Site-specific mutagenesis in the latter region revealed that a zinc finger, a motif that resembles a consensus sequence in Rho/Rac GAPs, and two arginine residues are required for GAP activity, but not for physical interaction with the ARF domain.[22] Mutations that abolished physical interaction of the GAP and the ARF domains of ARD1 also prevented stimulation of the hydrolysis of bound GTP, suggesting that a stable association of the two domains is required for GTP hydrolysis.[22]

The GAP activity of the amino terminus of ARD1 is specific for its ARF domain because other ARFs are not substrates.[24] A small region of seven amino acids in the effector region of the ARF domain is responsible for this specificity.[15] Single or double amino acid substitutions in this segment demonstrated that a proline contributes to the physical interaction with the GAP domain, presumably by creating a curve in the β-sheet structure, which could place important charged residues (Asp-427 and Glu-428) in correct position for interaction with the GAP domain.[15] These negatively charged residues are believed to form salt bridges with positively charged Arg-249 and Lys-250 in the GAP region.[22] Similarly, the Ras/GAP association is based on interactions between positively charged residues in GAPs and negatively charged residues in the effector region of Ras.[25] Hydrophobic interactions have also been postulated to be important for the interaction of the two functional domains of ARD1.[22]

Northern blot analyses revealed 4.2-kb and 3.7-kb ARD1 mRNAs present in every tissue tested from rat, mouse, rabbit, and human, suggesting that ARD1 is ubiquitous and highly conserved in eukaryotic cells.[14] RT-PCR (reverse transcription–polymerase chain reaction) experiments confirmed that ARD1 mRNAs were present in numerous tissues from human, bovine, and mouse (Vitale, unpublished observations). A protein of 64 kDa, potentially an analog of ARD1, was also identified in *Plasmodium falciparum* (Goldman *et al.,* personal communication, ASCB meeting 1996, abstract 3485). A gene resembling that of ARD1 was detected in *Caeno-*

[22] N. Vitale, J. Moss, and M. Vaughan, *J. Biol. Chem.* **273,** 2553 (1998).
[23] M. A. Ahmadian, L. Wiesmüller, A. Lautwein, F. R. Bischoff, and A. Wittinghofer, *J. Biol. Chem.* **271,** 16409 (1996).
[24] M. Ding, N. Vitale, S.-C. Tsai, J. Moss, and M. Vaughan, *J. Biol. Chem.* **271,** 24005 (1996).
[25] W. Miao, L. Eichelberger, L. Baker, and M. S. Marshall, *J. Biol. Chem.* **271,** 15322 (1996).

rhabditis elegans, but not in yeast (Vitale, unpublished observations). Native ARD1 was associated with purified lysosomal and Golgi membranes from human liver.[26] Accordingly, when overexpressed in NIH 3T3, COS 7, and HeLa cells, ARD1 had a subcellular localization typical of the Golgi apparatus and lysosomes, which is different from that of other ARFs.[26] ARD1, synthesized as a green fluorescent fusion protein, was initially associated with the Golgi network and subsequently appeared in lysosomes, suggesting that ARD1 might undergo vectoral transport between the two organelles.[26] Thus, by analogy to ARFs, a role for ARD1 in transport between the Golgi apparatus and lysosomes might be postulated.

Methods

Immunodetection of Native ARD1 in Lysosomal and Golgi Membranes

Membranes from lysosomes and Golgi are purified from a crude membrane fraction from human liver (Brain and Tissue Bank, University of Miami, FL) by immunoadsorption with Dynabeads M-500 (Dynal). 500 μg of rat anti-mouse Fc antibody is bound to 8×10^8 Dynabeads, to serve as linker for either 100 μg of antibody against lysosomal LAMP1 or 80 μg of antibody against Golgi AP1. Crude membranes (20 mg protein), prepared by centrifugation ($4°$, $100,000g$, 30 min) of a postnuclear supernatant from human liver homogenate, are incubated (12 hr, $4°$) with 2×10^8 Dynabeads. Beads are then washed five times each with 1 ml of ice-cold phosphate-buffered saline (PBS) containing 5% (w/v) bovine serum albumin (BSA), PBS with 0.1% (w/v) BSA, and PBS. Molecules bound to beads are eluted in 1 ml of 0.2 M glycine, pH 2.7/10% (v/v) ethylene glycol (which disrupts antigen–antibody interactions) and the pH is immediately adjusted to 7.0 with 1 M Tris. Free antibodies are bound to protein A/G Sepharose (Pierce, Rockford, IL) and the purified membrane fraction is dialyzed against PBS and concentrated (Centricon 10, Amicon, Danvers, MA). Protein is estimated by a dye-binding assay. Membrane proteins (20 μg) are separated by SDS–PAGE in 4–20% gel and transferred to nitrocellulose membrane for Western blotting with affinity-purified antibodies raised against recombinant ARD1 or undecapeptides corresponding to the N- or C-terminal sequences of ARD1.[26] A single band corresponding in size to recombinant ARD1 is detected, consistent with the association of native ARD1 with Golgi and lysosomal membranes from human liver.[26]

[26] N. Vitale, K. Horiba, V. J. Ferrans, J. Moss, and M. Vaughan, *Proc. Natl. Acad. Sci. U.S.A.* **98**, 8613 (1998).

*Synthesis and Purification of Recombinant Full-Length ARD1 and Its
ARF and GAP Domains*

Open reading frames (ORF) encoding the entire human ARD1 protein, the ARF domain, or the GAP domain are prepared from human fetal brain λ ZAP library (Stratagene, La Jolla, CA) using a ligation-independent cloning method in the expression vector pGEX-5G/LIC.[27] Ultracompetent cells (Stratagene) are transformed with the plasmids. Sequences of constructs are confirmed by automatic sequencing (Applied Biosystems, 373 DNA Sequencer). Individual colonies of transformed BL21(DE3) bacteria (Novagen) are tested for expression of recombinant glutathione *S*-transferase (GST)-fusion protein after 2 hr of induction with 0.5 mM isopropyl-β-D-thiogalactopyranoside (IPTG). Colonies with good expression of the GST-fusion protein without expression of GST are selected. For large-scale production, 10 ml of overnight bacterial culture is added to 1 liter of Luria–Bertani (LB) broth with ampicillin, 100 μg/ml, followed by incubation at 37° with shaking. When the culture reaches an A_{600} of 0.6, 500 μl of 1 M IPTG is added (0.5 mM IPTG final concentration). After incubation for 3 hr more, bacteria are collected by centrifugation (Sorvall GSA, 6000 rpm, 4°, 10 min), and stored at −20° until protein purification.

Bacterial pellets are dispersed in 10 ml of cold PBS with trypsin inhibitor, 20 μg/ml, leupeptin and aprotinin, each 5 μg/ml, and 0.5 mM phenylmethylsulfonyl fluoride (PMSF). Lysozyme (20 mg in 10 ml) and Tween 20 (0.1% final) are added. After 30 min at 4°, cells are disrupted by sonication and centrifuged (Sorvall SS34, 16,000 rpm, 4°, 20 min).

At this stage, ARD1 and the GAP domain fusion proteins remain insoluble in inclusion bodies, whereas the ARF domain fusion protein is soluble. Insoluble proteins are denatured by incubation for 1 hr at 4° in 30 ml of buffer A (7 M urea, 50 mM Tris, pH 8, 1 mM EDTA, 2 mM MgCl$_2$ and with trypsin inhibitor, 20 μg/ml, leupeptin and aprotinin, each 5 μg/ml, and 0.5 mM PMSF). Proteins are sonified 20 times with short pulses. Samples are then dialyzed three times for 8–12 hr against 500 ml of buffer B [buffer A without urea, and with 2.5 mM dithiothreitol (DTT)]. To allow protein renaturation, DTT is removed stepwise by successive dialyses against 500 ml of buffer B containing 1.5 mM, 0.75 mM, and 0.1 mM DTT, followed by dialysis twice against 500 ml of buffer B without DTT. The solution is clarified by centrifugation (Sorvall SS34, 16,000 rpm, 4°, 20 min).

GST-fusion proteins, purified on glutathione-Sepharose (Pharmacia, Piscataway, NJ) in 3 ml of 10 mM glutathione and 50 mM Tris, pH 8, are ~90% pure as estimated by silver staining after SDS–PAGE. After cleavage

---

[27] R. S. Haun and J. Moss, *Gene* **112**, 37 (1992).

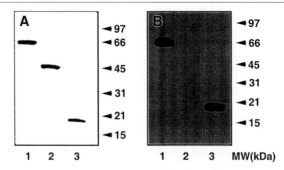

Fig. 1. The ARF domain of ARD1 binds GTP. 22 pmol of purified recombinant ARD1 (lane 1), N-terminal domain (lane 2), or C-terminal ARF domain (lane 3) was subjected to SDS–PAGE in 4–20 gels. One gel was silver-stained to estimate purity (A). Proteins from an equivalent gel were transferred to nitrocellulose membrane, which was used to detect [α-^{32}P]GTP (B). Positions of protein standards are given at the right of each gel.

by bovine thrombin, GST is removed with glutathione-Sepharose beads and thrombin with benzamidine-Sepharose 6B.[28] Proteins are further purified by successive gel filtration through Ultrogel AcA 54 (1.5 × 45 cm) and Ultrogel AcA 34 (1.5 × 30 cm) before storage in small portions at $-20°$. Purity estimated by silver staining after SDS–PAGE is >98% (Fig. 1A). Based on the nucleotide sequence the expected sizes of ARD1, the C-terminal ARF domain and the N-terminal domain are, respectively, 64, 18, and 46 kDa. Amounts of purified proteins are estimated by a dye-binding assay[29] and by SDS–PAGE using bovine serum albumin as standard.

Assays Used for Functional Characterization of ARD1

GTP-Binding Assay. To visualize GTP-binding to purified recombinant ARD1 (Fig. 1), the N-terminal GAP domain or C-terminal ARF domain (Fig. 1B), 22 pmol of each (respectively, 1.4, 1, and 0.4 μg) is subjected to electrophoresis in 4–20% polyacrylamide gel with SDS (Fig. 1A) and transferred to nitrocellulose. The membrane is incubated in 50 mM Tris-HCl, pH 7.5, 150 mM NaCl, 2 mM DTT, 2.5 mM EDTA, soybean trypsin inhibitor (10 μg/ml), 0.5 mM PMSF with BSA (0.3 mg/ml), and cardiolipin (1 mg/ml) at room temperature for 3 hr, transferred to fresh buffer containing 10 mM MgCl$_2$ and [α-^{32}P]GTP (3000 Ci/mmol), 2 μCi/ml, for 2 hr, washed 10 times for 5 min with buffer minus MgCl$_2$ and [α-^{32}P]GTP, briefly dried, and exposed to Kodak (Rochester, NY) XAR film at $-80°$ overnight with intensifying screen.

[28] A. J. Ridley, H. F. Paterson, C. L. Johnson, D. Diekmann, and A. Hall, *Cell* **70**, 401 (1992).
[29] M. M. Bradford, *Anal. Biochem.* **72**, 248 (1976).

For quantification of GTP binding to purified recombinant proteins, 30-pmol samples are incubated for 30 min at 30° in 20 mM Tris, pH 8.0, 10 mM DTT, 2.5 mM EDTA with BSA (0.3 mg/ml), and cardiolipin (1 mg/ml), then for 40 min at 30° in the same medium plus 10 mM MgCl$_2$ and 3 μM [^{35}S]GTPγS (~10^6 cpm). From the 150-μl mixture, duplicate samples (70 μl) are transferred to nitrocellulose filters in a manifold (Millipore, Bedford, MA) for rapid filtration,[30] followed by washing five times, each with 1 ml of ice-cold buffer (25 mM Tris-HCl, pH 8.0, 100 mM NaCl, 1 mM DTT, 1 mM EDTA, 5 mM MgCl$_2$). Dried filters are dissolved in scintillation fluid for radioassay.[31] No GTP binding to the N-terminal GAP domain is detected. In a typical experiment, 0.95 pmol of GTPγS is bound to 80 pmol of recombinant ARD1 or the ARF domain.

GDP Release Assay. ARD1 or related proteins (300 pmol) are incubated for 30 min at 30° in 450 μl of 20 mM Tris, pH 8.0, 10 mM DTT, 2.5 mM EDTA, with BSA (0.3 mg/ml), and cardiolipin (1 mg/ml), and then for 40 min in the same medium plus 10 mM MgCl$_2$ and 3 μM [^{35}S]GDPβS (2 × 10^7 cpm) or [^{3}H]GDP (1.5 × 10^7 cpm) (total volume 600 μl). At zero time, 100-μl samples (50 pmol) are transferred to nitrocellulose filters, which are washed five times with 1 ml of 25 mM Tris, pH 8.0, and 5 mM MgCl$_2$ in 100 mM NaCl before radioassay by liquid scintillation counting. The remainders of the mixtures are immediately diluted with an equal volume of reaction buffer (500 μl) containing 2 mM GDPβS or GDP. 200-μl samples (50 pmol) are taken after 5, 15, 30, 45, and 60 min at 30° for quantification of bound radioactivity, as described for the zero-time samples. To calculate values for dissociation curves, radioactivity bound to filters in the absence of protein is subtracted. Identical results are obtained with [^{35}S]GDPβS and with [^{3}H]GDP. The rate of GDP dissociation from the ARF domain of ARD1 is faster than from ARD1 or NΔ387ARD1, which represents the ARF domain plus 15 amino acids preceding (Fig. 2). The 15 amino acid region is, therefore, responsible for slowing GDP dissociation from ARD1.

Assay of GTPase Activity. To demonstrate GTPase-activating action of the non-ARF domain of ARD1, samples of ARD1 or ARF domain are incubated for 30 min at 30° in 20 mM Tris, pH 8.0, 10 mM DTT, 2.5 mM EDTA with BSA (0.3 mg/ml), and cardiolipin (1 mg/ml), then for 40 min at 30° in the same medium with 0.5 μM [α-^{32}P]GTP (3000 Ci/mmol) and 10 mM MgCl$_2$ (total volume 120 μl). After addition of the N-terminal GAP domain of ARD1 or vehicle (40 μl), incubation at room temperature is continued for 30 min (final volume 160 μl), before proteins with bound nucleotides are collected on nitrocellulose. Bound nucleotides are eluted

[30] S.-C. Tsai, R. Adamik, J. Moss, and M. Vaughan, *J. Biol. Chem.* **268**, 10820 (1993).
[31] S.-C. Tsai, R. Adamik, J. Moss, and M. Vaughan, *Proc. Natl. Acad. Sci. U.S.A.* **91**, 3063 (1994).

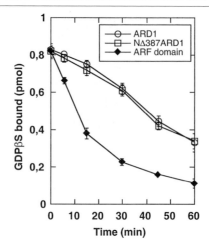

Fig. 2. Effect of N-terminal deletion on dissociation of GDP from ARD1. 300 pmol of ARD1, NΔ387 ARD1, and ARF domain proteins with similar amounts of radiolabeled GDP bound were incubated in the presence of unlabeled GDP, and protein-bound labeled GDP was quantified at the indicated times. Preparation of recombinant NΔ387ARD1, which contains the ARF domain plus the 15 preceding amino acids of ARD1, is described in Ref 16.

in 250 μl of 2 M formic acid, of which samples (3–4 μl) are analyzed by TLC on polythyleneimine-cellulose plates[32] and 240 μl is used for radioassay to quantify total nucleotide. TLC plates are subjected to autoradiography at −80° for 15–30 hr. The intrinsic rate of GTP hydrolysis by ARD1 is much higher than that by the ARF domain itself (Fig. 3, lane 2 versus lane 1). Addition of the N-terminal GAP domain to the ARF domain completely restores the GTP hydrolysis (Fig. 3, lane 3). Total amounts of labeled nucleotides (GTP + GDP) bound, whether quantified by radioassay of the formic acid solution, by counting total radioactivity on the filter, or by phosphorimaging (Molecular Dynamics) after TLC, are not significantly different under any conditions.

    Assay of Cholera Toxin-Catalyzed ADP-Ribosylagmatine Formation. To evaluate ARF activity (stimulation of cholera toxin ADP-ribosyltransferase activity) samples of ARD1 or related proteins are incubated for 30 min at 30° in 40 μl of 20 mM Tris pH 8.0, 10 mM DTT, 2.5 mM EDTA with BSA (0.3 mg/ml), and cardiolipin (1 mg/ml), before addition of 20 μl of a solution to yield final concentrations of 100 μM GTPγS or GTP and 10 mM MgCl₂. Samples are further incubated for 20 min at 30° with the amino-terminal domain of ARD1 or water (total volume 80 μl). Components needed to

[32] P. A. Randazzo and R. A. Kahn, *J. Biol. Chem.* **269**, 10758 (1994).

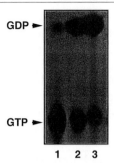

FIG. 3. Effect of the N-terminal domain on the GTPase activity of ARD1. 50 pmol of the ARF domain of ARD1 with [α-^{32}P]GTP bound was incubated with water (lane 1) or 50 pmol of the amino-terminal domain of ARD1 for 30 min. 50 pmol of ARD1 with [α-^{32}P]GTP bound was incubated with water for 30 min (lane 2). Bound nucleotides were separated by TLC. Positions of standard GDP and GTP are indicated at left.

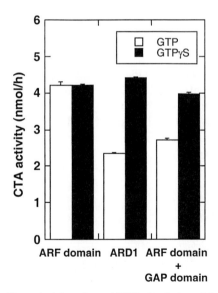

FIG. 4. Effect of the N-terminal domain on ARD1 stimulation of the ADP-ribosyltransferase activity of CTA. 900 pmol of the ARF domain of ARD1 with GTP or GTPγS bound was incubated with 900 pmol of the N-terminal GAP domain or with water for 20 min at 30°. 200 pmol of ARD1 with GTP or GTPγS bound was incubated with water for 20 min at 30°. Components needed to quantify ARF stimulation of CTA-catalyzed ADP-ribosylagmatine formation were then added, and incubation continued for 1 hr at 30°. Basal activity in the absence of nucleotide in each condition has been subtracted.

quantify ARF stimulation of cholera toxin-catalyzed ADP-ribosylagmatine formation are then added in 70 μl to yield final concentrations of 50 mM potassium phosphate, pH 7.5, 6 mM MgCl$_2$, 20 mM DTT, ovalbumin (0.3 mg/ml), 0.2 mM [adenine-^{14}C]NAD (0.05 μCi), 20 mM agmatine, cardiolipin (1 mg/ml), and 100 μM GTPγS or GTP with 0.5 μg of cholera toxin.[33] After incubation at 30° for 1 hr, samples (70 μl) are transferred to columns of AG1-X2 equilibrated with water and eluted with five 1-ml volumes of water. The eluate, containing [^{14}C]ADP-ribosylagmatine, is collected for radioassay.

As shown in Fig. 4, the ARF domain of ARD1 stimulates ADP-ribosylagmatine formation to the same extent in the presence of GTP or GTPγS, whereas ARD1 is much less active with GTP than with GTPγS (Fig. 4). Addition of the N-terminal GAP domain to the ARF domain reduces stimulation with GTP but not with GTPγS (Fig. 4), consistent with GTPase-activating function of the non-ARF domain of ARD1.

Conclusion

The Gα subunits of heterotrimeric G proteins contain a GTP-binding core and an inserted domain involved in GTP hydrolysis.[34] ARFs, like other Ras-related GTPases, also contain GTP-binding domains but lack the latter region. As a consequence, the ARF GTPases are inactivated by interaction with specific GTPase-activating proteins or GAPs.[35–40] The covalent attachment of a GAP-like domain to the GTP-binding core of an ARF protein seen in ARD1 is unique and is probably important in the regulation of its activity. Localization of ARD1 to the Golgi and lysosomal subcompartments suggests that it may play a role in the formation and function of these organelles.

[33] S.-C. Tsai, M. Noda, R. Adamik, P. P. Chang, H.-C. Chen, J. Moss, and M. Vaughan, *J. Biol. Chem.* **263**, 1768 (1988).

[34] H. R. Bourne, *Nature* **366**, 628 (1993).

[35] E. Cukierman, I. Huber, M. Rotman, and D. Cassel, *Science* **270**, 1999 (1995).

[36] M. Ding, N. Vitale, S.-C. Tsai, R. Adamik, J. Moss, and M. Vaughan, *J. Biol. Chem.* **271**, 24005.

[37] P. Randazzo, *Biochem J.* **324**, 413 (1997).

[38] R. T. Premont, A. Claing, N. Vitale, J. L. R. Freeman, J. A. Pitcher, W. A., Patton, J. Moss, M. Vaughan, and R. J. Lefkowitz, *Proc. Natl. Acad. Sci. U.S.A.* **95**, 14082 (1998).

[39] M. T. Brown, J. Andrade, H. Radhakrishna, J. G. Donaldson, J. A. Cooper, and P. A. Randazzo, *Mol. Cell. Biol.* **12**, 7038 (1998).

[40] J. Andreev, J. P. Simon, D. D. Sabatini, J. Kam, G. Plowman, P. A. Randazzo, and J. Schlessinger, *Mol. Cell. Biol.* **3**, 2338 (1999).

[36] Purification and Characterization of GIT Family of ADP-Ribosylation Factor (ARF) GTPase-Activating Proteins

By RICHARD T. PREMONT and NICOLAS VITALE

Introduction

The ADP-ribosylation factor (ARF) family of small GTP-binding proteins is known to be an important component of the regulatory apparatus controlling vesicular formation and trafficking throughout the cell. Six mammalian ARF proteins have been described, and all share a remarkable degree of sequence similarity.[1] However, these have been categorized into three classes based on primary sequence conservation and gene organization: class I (ARF1, 2, and 3), class II (ARF4 and 5), and class III (ARF6). ARF1 function has been most extensively studied in the Golgi, where it is required for clathrin-coated and non-clathrin-coated vesicle formation.[1,2]

Like other small GTP-binding proteins, ARF family members are inactive when bound to GDP and active when bound to GTP.[1,2] The precise effectors that bind the activated ARF proteins are generally ill defined, but certainly include the phosphatidyl choline-specific phospholipase D and vesicular coat proteins such as the AP1 clathrin adaptor complex and coatomer.[1,3] The activation and deactivation of ARF proteins appear to require accessory proteins that stimulate GDP–GTP exchange and accelerate GTPase activity, respectively.[3] Many such ARF guanine nucleotide exchange proteins (GEPs) and GTPase-activating proteins (GAPs) have been identified recently. ARF GEPs share a common sec7-homology domain, while GAPs share a conserved GATA-like $CX_2CX_{16}CX_2C$ zinc finger domain.[3] The structures of both types of domain, in complex with ARF1, have been solved.[4,5]

The first ARF GAP sequence to be cloned and extensively characterized, called ARF-GAP1, contains such a GATA-like $CX_2CX_{16}CX_2C$ zinc finger domain followed by several ankyrin repeats.[6] Mutagenesis to delete the amino-terminal zinc finger-like sequence or to eliminate individual

[1] J. Moss and M. Vaughan, *J. Biol. Chem.* **270,** 12327 (1995).
[2] A. Boman and R. A. Kahn, *Trends Biol. Sci.* **20,** 147 (1995).
[3] J. Moss and M. Vaughan, *J. Biol. Chem.* **273,** 21431 (1998).
[4] J. Goldberg, *Cell* **95,** 237 (1998).
[5] J. Goldberg, *Cell* **96,** 893 (1999).
[6] E. Cukierman, I. Huber, M. Rotman, and D. Cassel, *Science* **270,** 1999 (1995).

cysteine residues within this sequence yielded protein with greatly attenuated GAP activity, suggesting that this zinc finger region is critical to catalysis. The X-ray crystal structure of the complex of ARF1 and the amino-terminal half of ARF-GAP1 shows that this zinc finger does coordinate zinc and does indeed bind to ARF.[5] However, no part of the GAP appeared to interact directly with the nucleotide, suggesting that the GAP may function indirectly rather than catalytically.

Based on this conserved ARF GAP zinc finger domain, several distinct ARF GAPs and putative ARF GAPs have been identified recently. These include the GIT family, the ASAP/PAP family, centaurin/PIP$_3$-binding protein (PIPBP), the *Saccharomyces cerevisiae* SAT proteins, and many uncharacterized sequences of unknown function.[7–13] All share a conserved 60 amino acid domain containing a $CX_2CX_{16}CX_2C$ zinc finger, and many also contain several ankyrin repeat units. Many of these GAPs are significantly larger than ARF-GAP1 and appear to serve as multidomain adaptor proteins in addition to being GAPs for ARF proteins.

The GIT family of ARF GAPs contains at least three members. GIT family members have been identified independently by several distinct strategies. We identified GIT1 as a G-protein receptor kinase (GRK)-interacting protein, and demonstrated that overexpression of GIT1 in HEK293 cells alters the function and regulation of the G-protein-coupled β_2-adrenergic receptor.[7] Using a two-hybrid strategy, we identified GIT1 interaction with the PIX/COOL family of rac1/cdc42 guanine nucleotide exchange factors, and through them the interaction with the p21-activated (PAK) kinases. We also identified a related GIT2 protein, which shares many properties (including ARF GAP activity) with GIT1, but which undergoes extensive alternative splicing. One splice variant of GIT2 has also been identified as the KIAA0148 gene product.[14] Further, the CAT2 protein, another splice variant of GIT2, was identified by interaction with the

[7] R. T. Premont, A. Claing, N. Vitale, J. L. R. Freeman, J. A. Pitcher, W. A. Patton, J. Moss, M. Vaughan, and R. J. Lefkowitz, *Proc. Natl. Acad. Sci. U.S.A.* **95,** 14082 (1998).

[8] M. T. Brown, J. Andrade, H. Radhakrishna, J. G. Donaldson, J. A. Cooper, and P. A. Randazzo, *Mol. Cell. Biol.* **18,** 7038 (1998).

[9] F. J. King, E. Hu, D. F. Harris, P. Sarraf, B. M. Speigelman, and T. M. Roberts, *Mol. Cell. Biol.* **19,** 2330 (1999).

[10] J. Andreev, J.-P. Simon, D. D. Sabatini, J. Kam, G. Plowman, P. A. Randazzo, and J. Schlessinger, *Mol. Cell. Biol.* **19,** 2338 (1999).

[11] L. P. Hammonds-Odie, T. R. Jackson, A. A. Profit, I. J. Blader, C. W. Turck, G. D. Prestwich, and A. B. Thiebert, *J. Biol. Chem.* **271,** 18859 (1996).

[12] K. Venkateswarlu, P. B. Oatey, J. M. Tavare, T. R. Jackson, and P. J. Cullen, *Biochem. J.* **340,** 359 (1999).

[13] C. Zhang, M. M. Cavenagh, and R. A. Kahn, *J. Biol. Chem.* **273,** 19792 (1998).

[14] T. Nagase, N. Seki, A. Tanaka, K.-I. Ishikawa, and N. Nomura, *DNA Res.* **2,** 167 (1995).

PIX/COOL proteins.[15] The third GIT family member, called the paxillin-kinase linker (PKL), was identified as a paxillin binding protein responsible for the association of the PIX/PAK complex with paxillin.[16] Thus GIT family members, in addition to being ARF GAPs, also have been reported to interact with PIX proteins (and through them with PAK kinases) and with paxillin. All of these proteins appear to be enriched at focal adhesions.

We have shown previously that purified recombinant GIT1 is capable of acting as a GAP for ARF1.[7] In this chapter, we describe the purification of recombinant GIT1 and GIT2, and the assay of these proteins as ARF GAPs. At this time, we are specifically interested in understanding the specificity of these newly described GAPs for members of the ARF family.

Methods

Preparation of Baculoviruses Encoding ARF GAPs

Preparation of GIT1/His$_6$. The rat GIT1 cDNA (GenBank AF085693) is modified to add an *Eco*RI site in place of the native stop codon by amplification. A carboxyl-terminal His$_6$ tag followed by a stop codon is then added to the carboxyl-terminal end of the GIT1 sequence by insertion of two annealed oligonucleotides at the *Eco*RI site. The modified cDNA is sequenced to verify the desired alterations and the absence of mutations. The modified GIT1 insert is released by digestion with *Not*I followed by partial digestion by *Bam*HI, and ligated into the *Bam*HI and *Not*I sites of the pVL1393 baculovirus shuttle vector (Pharmingen, San Diego, CA).

Preparation of GIT2/His$_6$. The shortest splice variant of the human GIT2 (KIAA0148, CAT2) cDNA (GenBank D63482) is obtained from Dr. Takahiro Nagase (Kazusa DNA Research Institute, Chiba, Japan). The GIT2 sequence is amplified using a 3′ oligonucleotide containing a His$_6$ tag immediately prior to the stop codon, subcloned into the *Xho*I and *Xba*I sites of pBK-CMV, and resequenced. The His$_6$-tagged GIT2 short insert is released by digestion with *Xho*I, which is blunted with T4 DNA polymerase, and subsequent digestion with *Xba*I. This fragment is then ligated into the pVL1393 baculovirus shuttle vector using the *Sma*I and *Xba*I sites.

Preparation of ARF-GAP1/His$_6$. The rat ARF-GAP1 cDNA (GenBank U35776) is obtained by amplification of a rat liver cDNA library using oligonucleotide primers specific to the 5′ and 3′ ends of the rat ARF-GAP1

[15] S. Bagrodia, D. Bailey, Z. Lenard, M. Hart, J. L. Guan, R. T. Premont, S. J. Taylor, and R. A. Cerione, *J. Biol. Chem.* **274,** 22393 (1999).
[16] C. E. Turner, M. C. Brown, J. A. Perrotta, M. C. Riedy, S. N. Nikolopoulos, A. R. McDonald, S. Bagrodia, S. Thomas, and P. S. Leventhal, *J. Cell Biol.* **145,** 851 (1999).

sequence.[6] The amplified product is subcloned into pBK-CMV using the *Eco*RI and *Xho*I sites, and sequenced to verify the absence of mutations. The ARF-GAP1 sequence is then reamplified using a 3' oligonucleotide containing a His_6 tag immediately prior to the stop codon, and resequenced. The His_6-tagged ARF-GAP1 insert is ligated into the pVL1393 baculovirus shuttle vector using the *Eco*RI and *Xho*I sites.

Recombinant pVL1393 vectors bearing GAP cDNA inserts are cotransfected with BaculoGold baculovirus DNA (Pharmingen) into Sf9 (*Spodoptera frugiperda* ovary) cells, according to the manufacturer's protocol, and recombined baculoviruses are amplified by two sequential infections of Sf9 cells. High-titer baculovirus stocks are prepared by infection of 500 ml of 1×10^6 cell/ml with 3 ml of amplified parental virus stock. After 4 days, the cells are pelleted by centrifugation at 2000g for 15 minutes at 4° in a Sorvall RT6000 centrifuge, and the supernatant sterile-filtered through 0.2-μm low-binding filter (Nalgene). The filtrate, containing the active viral particles, is stored in the dark at 4°.

Expression and Purification of GAP Proteins

The His_6-tagged GAP proteins are expressed in Sf9 cells infected with the appropriate recombinant baculovirus. Sf9 cells are grown at 25° using the ambient air in Grace's insect media (Life Technologies) supplemented with 10% heat-inactivated fetal bovine serum and 25 μg/ml gentamicin. Cells are seeded at 0.7×10^6 cell/ml and allowed to grow overnight to 1.5×10^6/ml for infection. Cells are infected by addition of the appropriate high-titer baculovirus stock [1 : 8 by volume, approximately 5–10 MOI (multiplicity of infection)]. Infected cells are allowed to grow an additional 48–72 hr before harvest.

Infected Sf9 cells are harvested by centrifugation in a Sorvall RC3C Plus centrifuge (HBB6 rotor) at 2000g for 15 min at 4°. Media is decanted and the cell pellets resuspended at 30 ml/liter of orginal culture in lysis buffer (20 mM Tris-HCl, pH 7.4, 1 mM EDTA, 100 mM NaCl, and a cocktail of protease inhibitors: 1 μg/ml aprotinin, 10 μg/ml benzamidine, 1 μg/ml leupeptin, 1 μg/ml pepstatin, 10 mM phenylmethylsulfonyl fluoride). Resuspended cells are immediately frozen in liquid nitrogen and stored at −80°. Lysates prepared in this way appear stable at −80° for at least 6 months.

All three GAP proteins are purified to near homogeneity by the same basic method. Frozen Sf9 cells (in lysis buffer) are thawed gently in a beaker of warm water. Cell lysates are prepared using 10 strokes of a tight Dounce homogenizer on ice, NaCl is added to 250 mM, and followed by another 10 strokes. Lysates are spun in a Sorvall RC5C centrifuge (SS-34 rotor) at 20,000g for 30 min at 4°. The 20,000g supernatant (approximately 20 ml/

liter original culture) is transferred to a screw cap 50-cm^3 tube, and 0.5 ml
of a 50% slurry of ProBond resin (Invitrogen) is added. The lysates are
rotated with the beads for 2 hr at 4°. Beads are washed batchwise three
times with 40 ml of lysis buffer containing 250 mM NaCl, using a table-top
Sorvall RT6000 centrifuge (2000g) to pellet the beads. The His$_6$-tagged
proteins are eluted from the ProBond beads with 20 ml (per ml resin) 20
mM Tris (pH 7.4) and 100 mM imidazole.

The ProBond eluate is passed through a 0.2-μm Millex GV syringe filter
(Millipore) and applied manually to a 5-ml HiTrap-Q desalting column
(Pharmacia). The column is attached to a Pharmacia Model 500 FPLC unit
and washed with 20 ml of lysis buffer (100 mM NaCl) at 1 ml/min. The
column is eluted using a linear gradient of NaCl (100–1500 mM) in 20 mM
Tris (pH 7.4), 1 mM EDTA, at 1 ml/min over 90 min. One-ml fractions
are collected, and the eluted protein peaks identified by separating 10-
μl aliquots of each fraction on 10% SDS–PAGE gels and staining with
Coomassie blue. Representative gels are shown for the purification of each
protein in Fig. 1. Pooled eluate fractions are further desalted and concentrated using three cycles of concentration in CentriPrep-30 devices (Amicon) in lysis buffer (100 mM NaCl), and stored frozen at −80° after supplementation to 30% glycerol.

Purification of the His$_6$-tagged GAP proteins by this method yields
substantially pure (>95%) preparations that are suitable for use in *in vitro*
GAP assays. Purified GIT1 appears to elute from the HiTrap-Q column
as two distinct peaks, one requiring significantly higher NaCl for elution
(Fig. 1). Initial experiments with the two isolated peaks indicated equivalent
GAP activity for ARF1 (data not shown), and in further experiments we
have used a pool of the two peaks. The ARF-GAP1 protein appears as a

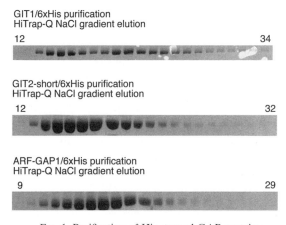

FIG. 1. Purification of His$_6$-tagged GAP proteins.

closely spaced doublet, as assessed by Coomassie blue staining (Fig. 1). This doublet pattern is also present in the starting Sf9 cell homogenates, as assessed by immunoblotting using a His_6 probe antiserum (data not shown). In either case, the cause of heterogeneity remains unknown, but does not appear to be due to proteolysis.

Expression and Purification of ARF Proteins

The ARF family of GTP-binding proteins consists of six distinct ARF proteins, as well as several ARF-like (ARL) proteins and the ARF domain 1 (ARD1) protein. Individual recombinant ARF proteins are expressed in *E. coli* using a T7 RNA polymerase system and purified using ion exchange chromatography on DEAE-Sephacel and gel filtration chromotography using UltraGel AcA 54, as described in detail in Vol. 257 of this series.[17]

Measurement of ARF GAP Activity

Principle of GAP Assay. The ability of GAPs to alter the guanine nucleotides bound to ARF proteins is assessed using a single-round GTP hydrolysis assay. ARF proteins are incubated with $[\alpha\text{-}^{32}P]GTP$ to create the substrate for the GAP proteins (GTP-bound ARF) prior to addition of any GAP proteins. After incubation with the GAP, free $[\alpha\text{-}^{32}P]GTP$ is separated from ARF-bound radioisotope (GTP and GDP) by rapid filtration of the reaction through nitrocellulose filters. The radioactivity retained with the ARF protein on the filters is eluted with formic acid, and a fraction of the eluate is spotted onto a thin-layer chromatography (TLC) plate while the majority is counted in scintillation fluid. The TLC plate is developed to separate the GTP from any GDP formed, and the ratio of radioactivity in the GTP and GDP spots determined using a phosphorimager. The total mass of GTP plus GDP bound to the ARF is determined from the total counts. Results can be expressed as moles of GTP hydrolyzed per unit time per mg of GAP protein, or as a percentage of the originally bound GTP that is lost or that is found to be converted to GDP.

Limitations of GAP Assay. The ARF GTP-binding proteins have little or no intrinsic GTPase activity. However, like all GTP binding proteins, ARF proteins bind GDP with somewhat higher affinity than GTP. Therefore, if the $[\alpha\text{-}^{32}P]GTP$ preparation contains GDP as a contaminant, this GDP will preferentially bind to the ARF protein during the loading step. This GDP-bound ARF protein is not a substrate for the GAP, but will be seen as a background of apparent ARF GTPase activity. Therefore, a

---

[17] P. A. Randazzo, O. Weiss, and R. A. Kahn, *Methods Enzymol.* **257,** 128 (1995).

control sample consisting of only the loaded ARF protein (no GAP added) is essential. In commercial [α-^{32}P]GTP preparations, an apparent GDP background of 5–10% is not uncommon. Therefore, GAP assays must consume somewhat greater than 10% of the ARF–GTP substrate to be detectable above this background. This presents a relatively narrow window of GAP activity in a kinetically favorable range, and assays may well deviate from simple kinetics as 50% or more of the available substrate is consumed. This may be somewhat alleviated by repurifying the [α-^{32}P]GTP immediately before assay. As ARF proteins release bound GDP poorly, the exchange of unlabeled guanine nucleotides for the radiolabeled GTP during the loading step is very inefficient (2–5% loading). Therefore, significantly more ARF protein, most of which is invisible in the assay as it is not bound to the radiolabel, must be used. To help load the GTP onto the ARF protein, nucleotide destabilizers such as phosphatidylserine are added to the loading reaction. This inefficient loading could be offset by the use of a specific ARF guanine nucleotide exchange factor to facilitate the loading step, but care must be taken to completely remove the exchange factor prior to the GAP assay.

Preparation of GTP-Bound ARF Substrate. To measure GAP activity, the purified ARF proteins must first be loaded with [α-^{32}P]GTP to create the GTP-bound ARF protein substrate for the GAP proteins. A master loading mix sufficient for all assay points (10 μl per assay point) is prepared by mixing the appropriate ARF protein (0.5 μg) and 15 μCi [α-^{32}P]GTP (3000 Ci/mmol) in 20 mM Tris-HCl (pH 8.0), 2 mM DTT, 1 mM MgCl$_2$, 500 μg/ml bovine serum albumin, and phosphatidylserine. This loading reaction is incubated for 20 min at 30°, and then held on ice.

ARF GAP Assay. GAP activity of test proteins is determined by adding the proteins to the loaded ARF mixture. The ARF loading mixture (10 μl) is added to the appropriate number of tubes on ice, each containing all of the other assay components in 40 μl. This 40 μl contains the appropriate dilution of GAP protein (or GAP protein buffer) in assay buffer [20 mM Tris-HCl (pH 8.0), 2 mM DTT, 1 mM MgCl$_2$, 500 μg/ml bovine serum albumin, and 10 μg/ml phosphatidylserine] that has been supplemented with an additional 3.2 mM MgCl$_2$ (yielding a 3.6 mM final concentration). The assays are then incubated for an additional 10 min at 30°. Reactions are terminated by addition of 1 ml of ice-cold wash buffer [20 mM Tris-HCl (pH 8.0), 100 mM NaCl, 5 mM MgCl$_2$, and 2 mM DTT]. The diluted reaction is filtered through a 0.45-μm nitrocellulose filter (25-mm circle, Millipore) under vacuum, and washed six times with 2 ml of ice-cold wash buffer. The filter is allowed to dry for 5 min under the vacuum, and placed into a tube containing 250 μl of 2 M formic acid. The tubes are capped and vortexed at room temperature to elute the bound nucleotides.

The ratio of GTP and GDP in the eluates is determined by applying 3-μl spots to polyethyleneimine (PEI)-cellulose TLC plates. After the spots dry (10 min), the plate is developed in 1 M LiCl/ 1 M formic acid to separate the GDP product from the GTP. Authentic GTP and GDP spots are identified by separating ^{32}P-labeled nucleotides in adjacent lanes. After the plate dries, it is exposed to a phosphorimager screen overnight, and radioactivity is detected as accumulated phosphorescence in a Molecular Dynamics PhosphorImager. The ratio of GTP and GDP is determined for each assay point by quantifying the "volume" or accumulated counts of the GTP and GDP spots using the ImageQuant program (Molecular Dynamics). Total ARF-bound counts (GTP plus GDP) is determined by scintillation counting after adding the remaining 250 μl of eluted nucleotides into 15 ml of cocktail. Data are expressed as % GTP hydrolysis, calculated as the volume of GDP bound divided by the sum of the GTP and GDP volumes for each point. Absolute GAP activity can be determined, knowing the total mass of guanine nucleotide bound, by calculating the moles of GTP converted to GDP per unit time per unit GAP protein.

ARF Family Specificity of GIT Proteins

To further understand the potential roles for the GIT proteins as regulators of ARF GTP-binding proteins, we have examined the ability of GIT1 to act as a GAP for distinct members of the ARF protein family, in comparison with the ARF-GAP1 protein. Several distinct members of the ARF protein superfamily are loaded with [α-^{32}P]GTP and tested for their ability to be GAP substrates for purified GIT1 and ARF-GAP1 proteins (Fig. 2).

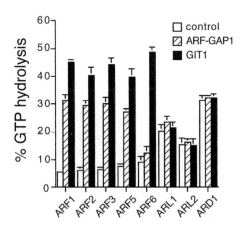

FIG. 2. ARF protein superfamily as GAP substrates for purified GIT1 and ARF–GAP1 proteins.

The GIT1 protein is capable of increasing GTP hydrolysis of all the ARF proteins tested, and notably ARF6. However, GIT1 does not recognize as substrates the more distant ARF superfamily members, GTP-bound ARLs or ARD1. The ARF-GAP1 protein, by contrast, recognizes as substrate the class I (ARF1, 2, and 3) and class II (ARF4 and 5) ARF proteins, but does not activate the GTPase activity of ARF6, a class III ARF protein. ARF-GAP1 is also unable to stimulate GTP hydrolysis by ARL or ARD1 proteins. Note that ARD1 contains an intrinsic GAP domain and thus elevated GAP activity in the absence of added GAP proteins.[18] Examination of other GAP proteins has shown that the ability to GAP the class I and class II ARF proteins is generally common to all characterized GAPs,[8,10,19] but the GIT1 protein is the first GAP known to have high activity for the ARF6 protein. While GIT1 does not appear to be selective for ARF6, this ability to GAP ARF6 sets GIT1 apart from other known ARF GAP proteins. Further studies will address whether this is a property common to other GIT family members, and whether GIT proteins regulate ARF6 in the cell.

[18] N. Vitale, J. Moss, and M. Vaughan, *Methods Enzymol.* **329**, [35], (2001) (this volume).
[19] P. A. Randazzo, *Biochem. J.* **324**, 413 (1997).

[37] Assay and Purification of Phosphoinositide-Dependent ADP-Ribosylation Factor (ARF) GTPase Activating Proteins

By Paul A. Randazzo, Koichi Miura, and Trevor R. Jackson

Introduction

The ADP-ribosylation factor (ARF) family of GTP-binding proteins regulates membrane trafficking events[1,2] and, as more recently described, actin cytoskeleton remodeling.[3–5] The ARF proteins have also been found

[1] J. G. Donaldson and R. D. Klausner, *Curr. Opin. Cell Biol.* **6**, 527 (1994).
[2] J. Moss and M. Vaughan, *J. Biol. Chem.* **273**, 21431 (1998).
[3] H. Radhakrishna, O. Al-Awar, Z. Khachikian, and J. G. Donaldson, *J. Cell Sci.* **112**, 855 (1999).
[4] S. R. Frank, J. C. Hatfield, and J. E. Casanova, *Mol. Biol. Cell* **9**, 3133 (1998).
[5] M. Franco, P. J. Peters, J. Boretto, E. Van Donselaar, A. Neri, C. D'Souza-Schorey, and P. Chavrier, *EMBO J.* **18**, 1480 (1999).

to activate phospholipase D[6,7] and phosphatidylinositol kinases.[8-10] ARF affects membrane traffic by controlling the assembly of vesicle coat proteins, but details of the molecular mechanisms for ARF function in membrane traffic, the mechanism by which actin cytoskeleton is affected, and the role of ARF-induced changes in phospholipid metabolism are not known. Identification of proteins that interact with ARFs and a characterization of the biological activity of these proteins are anticipated to contribute to understanding the molecular basis of ARF action. Among these proteins are the ARF GTPase-activating proteins (GAPs).[11-13] Six mammalian proteins are known to have ARF GAP activity[11-14] and more than a dozen proteins, based on sequence homology, are predicted to be ARF GAPs. Many of these proteins contain multiple domains and interact with signaling molecules in addition to ARF, including nonreceptor tyrosine kinases and phosphoinositides.[12,13] These properties of the ARF GAPs, together with the finding that multiple ARF GAPs have similar ARF specificities, have raised the possibility that the GAPs themselves may contribute to the action of an ARF at a particular site. The availability of purified ARF GAPs will enable examination of the role of ARF GAP in the cellular functions of ARF and characterization of the signals that regulate ARF.

The ARF GAPs can be divided into two groups based on phospholipid dependence: diacylglycerol activated[15] and phosphoinositide dependent.[12,13,16,17] This latter group forms at least one subfamily within the centaurin family of proteins, characterized by PH, zinc finger of the ARF GAP type and ANK repeat domains. Our laboratories have been focusing

[6] H. A. Brown, S. Gutowski, C. R. Moomaw, C. Slaughter, and P. C. Sternweis, *Cell* **75**, 1137 (1993).

[7] S. Cockcroft, G. M. H. Thomas, A. Fensome, B. Geny, E. Cunningham, I. Gout, I. Hiles, N. F. Totty, O. Truong, and J. J. Hsuan, *Science* **263**, 523 (1994).

[8] A. Martin, F. D. Brown, M. N. Hodgkin, A. J. Bradwell, S. J. Cook, M. Hart, and M. J. O. Wakelam, *J. Biol. Chem.* **271**, 17397 (1996).

[9] A. Fensome, I. Cunningham, S. Prosser, S. K. Tan, P. Swigart, G. Thomas, J. Hsuan, and S. Cockcroft, *Curr. Biol.* **6**, 730 (1996).

[10] A. Godi, P. Pertile, R. Meyers, P. Marra, G. DiTullio, C. Iurisci, A. Luini, A. Corda, and M. A. DeMatteis, *Nature Cell Biol.* **1**, 280 (1999).

[11] E. Cukierman, I. Huber, M. Rotman, and D. Cassel, *Science* **270**, 1999 (1995).

[12] M. T. Brown, J. Andrade, H. Radhakrishna, J. G. Donaldson, J. A. Cooper, and P. A. Randazzo, *Mol. Cell. Biol.* **18**, 7038 (1998).

[13] J. Andreev, J. P. Simon, D. Sabatini, J. Kam, G. Plowman, P. A. Randazzo, and J. Schlessinger, *Mol. Cell. Biol.* **19**, 2338 (1999).

[14] R. T. Premont, A. Claing, N. Vitale, J. L. Freeman, J. A. Pitcher, W. A. Patton, J. Moss, M. Vaughan, and R. J. Lefkowitz, *Proc. Natl. Acad. Sci. U.S.A.* **95**, 14082 (1998).

[15] B. Antonny, I. Huber, S. Paris, M. Chabre, and D. Cassel, *J. Biol. Chem.* **272**, 30848 (1997).

[16] P. A. Randazzo and R. A. Kahn, *J. Biol. Chem.* **269**, 10758 (1994).

[17] P. A. Randazzo, *Biochem. J.* **324**, 413 (1997).

on the centaurin β's (Fig. 1), proteins that contain coiled coil, pleckstrin homology (PH), ARF GAP, and at least three ANK repeat domains. The PH domains are most similar to those of the ARF GEFs and other type I PH domains. The members of the family differ most in the amino-terminal and carboxy-terminal domains. Centaurins β_3 and β_4 contain proline-rich domains that bind src family proteins and an SH3 domain which centaurins β_1 and β_2 lack. We have found that each of the centaurin β's is a phosphoinositide-dependent ARF GAP. Here, we describe the assay and purification of this group of ARF GAPs. Two types of preparations will be described. Truncated recombinant proteins expressed in bacteria have been useful for studying interaction with ARF–GTP. Partially purified recombinant centaurin β's expressed in mammalian cells have been useful for applications where the full-length protein is required and for centaurins that we have not been able to express in bacteria.

ARF GAP Assay

In this assay, the hydrolysis of GTP by ARF is measured. The assay involves two parts. First, radionucleotide-labeled GTP is bound to ARF, forming the "substrate," ARF–GTP, for the GAP. In the second part of the assay, AFR–GTP and a source of GAP are incubated under conditions that minimize interference by nucleotide exchange or contaminating nucleotidases.

In studying GAP activity, two decisions concerning the substrate must be made. First is the choice of ARF gene product. The centaurin β's are active against class I, II, and III ARFs (ARF1, ARF3, ARF5, and ARF6).[12,13] No activity has been detected using ARL2,[12,13] and other ARF and ARL gene products have not been tested. The particular gene product used will depend on the specific questions being addressed. The second decision is whether to use an acylated ARF. Although native ARF is myristoylated, the centaurin β's use nonmodified ARF as well as myris-

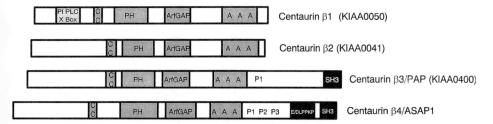

FIG. 1. Schematic of known members of the centaurin β family of ARF GAPs. GenBank accession numbers: centaurin β_1/KIAA0050, D30758; centaurin β_2/KIAA0041, D26069; centaurin β_3/PAP/KIAA0400, AB007860; centaurin β_4/ASAP1, AF075462.

toylated ARF. The use of a myristolylated or nonmodified ARF depends on whether the substrate will be used for routine assays or for saturation kinetics. Nonmodified ARFs are simple to prepare following protocols described in this series[18,19] and are used in our laboratory for routine assays. However, nonmyristoylated Arfs are difficult to load to high concentrations with GTP. Therefore, when examining saturation kinetics, we use myristoylated ARF. By following the protocol of Randazzo,[17] we have been able to routinely prepare ARF1 that is 100% myristoylated and will bind more than 0.7 mol GTP/mol ARF. We obtain similar results with ARF6 by following the protocol in Brown et al.[12]

Despite ARF3's high similarity to ARF1 (96% identical), we have not been able to purify the myristoylated form by the same methods used for ARF1. Furthermore, nonmyristoylated ARF3 binds GTP very poorly and has been difficult to prepare as a substrate. We have found that recombinant ARF3 is best prepared as described in Kanoh et al.[20] It is about 50% myristoylated, and GTP-binding stoichiometries of 25% can be achieved. Alternatively, a mixture of 90% ARF3, 10% ARF1 can be prepared from bovine brain as described by Claude et al.[21] ARF5 can be prepared as described in Liang et al.[22] or in Claude et al.[21]

In the assay, the conversion of $[\alpha\text{-}^{32}\text{P}]$GTP to $[\alpha\text{-}^{32}\text{P}]$GTP is measured. Both myristoylated ARF and nonmodified ARF are purified with GDP. Exceptions to this include the truncated ARFs, but these are poor substrates, with 100- to 1000-fold lower affinity than full-length ARF for the centaurin β's[23,24] and, therefore, are not routinely used to examine interaction of ARFs with centaurins. To prepare either myristoylated or nonmodified ARF to be used as a substrate for ARF GAP, GDP is exchanged for $[\alpha\text{-}^{32}\text{P}]$GTP. Exchange is most efficient in the presence of a hydrophobic surface. For nonmodified ARF, 0.1% Triton X-100 works more efficiently than mixed micelles or liposomes. For myristoylated ARF, phospholipid micelles composed of dimyristoylphosphatidylcholine and cholate (described in this series[19]) or phospholipid vesicles work well. For routine assays, nonmyristoylated ARF at 1–10 μM is incubated in either buffer A (25 mM HEPES, pH 7.4, 100 mM NaCl, 3.5 mM MgCl$_2$, 1 mM EDTA, 1 mM ATP, 1 μM $[\alpha\text{-}^{32}\text{P}]$GTP (specific activity = 50,000–250,000 cpm/

[18] P. A. Randazzo and R. A. Kahn, *Methods Enzymol.* **250,** 394 (1995).

[19] P. A. Randazzo, O. Weiss, and R. A. Kahn, *Methods Enzymol.* **219,** 362 (1992).

[20] H. Kanoh, B.-T. Williger, and J. H. Exton, *J. Biol. Chem.* **272,** 5421 (1997).

[21] A. Claude, B.-P. Zhao, C. E. Kuziemsky, S. Dahan, S. J. Berger, J. P. Yan, A. D. Armold, L. M. Sullivan, and P. Melancon, *J. Cell Biol.* **146,** 71 (1999).

[22] J. O. Liang and S. Kornfeld, *J. Biol. Chem.* **272,** 4141 (1997).

[23] J. X. Hong, X. Zhang, J. Moss, and M. Vaughan, *Proc. Natl. Acad. Sci. U.S.A.* **92,** 3056 (1995).

[24] P. A. Randazzo, T. Terui, S. Sturch, and R. A. Kahn, *J. Biol. Chem.* **269,** 29490 (1994).

pmol), 25 mM KCl, 1.25 U/ml pyruvate kinase, and 3 mM phosphoenolpyruvate) or buffer B [25 mM HEPES, pH 7.4, 100 mM NaCl, 0.5 mM MgCl$_2$, 1 mM EDTA, 1 mM ATP, 1–100 μM [α-^{32}P]GTP (specific activity = 10,000–50,000 cpm/pmol), and 1 mM dithiothreitol (DTT)] containing 0.1% Triton X-100 for 1 hr at 30°. In buffer A, a regeneration system is included that converts GDP, produced by any contaminating nucleotidases in the protein preparations, back to GTP. In buffer B, high concentrations of ATP prevent the nonspecific hydrolysis of GTP by nucleotidases. Buffer B, with low Mg^{2+} concentrations, maximizes the rate of loading and is used, with either 3 mM DMPC/0.1% sodium cholate or vesicles composed of 700 μM bovine brain PC and 300 μM PA,[25] to load myristoylated ARF. Myristoylated ARF at concentrations up to 200 μM and 200–400 μM GTP can be used. Myristoylated ARF has a higher affinity for GTP than GDP[26] and, therefore, large excesses of GTP over ARF–GDP are not required to load preferentially with GTP. The most critical point in this loading reaction is to minimize the activity of nonspecific nucleotidases than may copurify with ARF. The binding stoichiometry is determined as described in this series[19] and the ARF–GTP in the binding buffer is added directly to the GAP assay. Note that nonmyristoylated ARF–GTP is not stable at 4° and should be used on the day it is prepared. Myristoylated ARF–GTP is more stable and we have stored it at 4° for up to 1 week prior to use.

GAP activity is measured using a single round hydrolysis assay. This assay was chosen because of the exchange properties of ARF. The assay has the benefit of little interference from contaminating nucleotidases. GTP labeled in the α position is used to both minimize and detect any interference from increased rates of nucleotide exchange. The GAP assay cocktail (25 mM HEPES, pH 7.4, 100 mM NaCl, 2 mM MgCl$_2$, 1 mM GTP, 1 mM DTT) contains a high concentration of GTP to dilute the radiolabeled GTP carried over from the loading reaction or that is released from ARF. At the much lower specific activity, any GTP that might be hydrolyzed while free in solution and bind either to ARF or another contaminating GTP-binding protein makes little or no contribution to the signal. Phosphoinositides and phosphatidic acid (PA) stimulate the centaurin β's 10,000-fold,[25] and are included in the assay to detect activity. For routine assays, a convenient source of these lipids is a crude mixture of acid phospholipids sold by Sigma Chemical Co. (St. Louis, MO) as "phosphoinositides." Purified phospholipids can also be used. The lipids can be added to the reaction as

[25] J. L. Kam, K. Miura, T. R. Jackson, J. Gruschus, P. Roller, S. Stauffer, J. Clark, R. Aneja, and P. A. Randazzo, *J. Biol. Chem.* **275**, 9653 (2000).
[26] P. A. Randazzo, T. Terui, S. Sturch, H. M. Fales, A. G. Ferrige, and R. A. Kahn, *J. Biol. Chem.* **270**, 14809 (1995).

mixed micelles with Triton X-100 or as vesicles, prepared either by sonication[27] or extrusion.[28]

Maximal activity is achieved with 40–100 μM phosphatidylinositol 4,5-bisphosphate (PtdIns4,5-P$_2$) and 100–500 μM phosphatidic acid. Synthetic Ptd4,5-P$_2$ with saturated acyl groups is less effective than PtdInsP$_2$ purified from bovine brain.[25] We suggest titrating each lipid preparation into the reaction to determine optimal conditions. There is variation among preparations and some preparations may be inhibitory at high concentrations. Mg^{2+} is required for activity and needs to be in excess of the nucleotide, but at free concentrations of greater than 1 mM will inhibit activity. Typically, the GAP assay cocktail is prepared as a 1.43-fold concentrate. Seven parts of this buffer (17.5 μl for a 25-μl reaction) is added to 1 part (2.5 μl) of a 10-fold lipid concentrate and 1 part of a 10-fold concentrate of GAP.

The reaction is initiated by the addition of one part ARF–GTP, i.e., for a 25-μl reaction, 2.5 μl of the exchange reaction, and the temperature is immediately shifted to 30°. Incubations are continued for 1–15 min, and stopped by dilution into 20 mM Tris-HCl, pH 8.0, 100 mM NaCl, 10 mM MgCl$_2$ and 1 mM DTT at 4°. Typically, for a 10–100 μl reaction, we dilute into 2 ml of the stop buffer. The protein-bound guanine nucleotide is separated from nucleotide free in solution by binding the protein to nitrocellulose filters (BA85, pore size 0.45 μM, size 25 mm, from Schleicher and Schuell, Keene, NH) as described in this series.[19] The nucleotides are released from the protein by extraction into 0.75–1 ml 1 M formic acid, conveniently accomplished by placing the nitrocellulose filter containing the protein into a 20-ml glass scintillation vial containing the formic acid solution. Fifty μl of the extract, 10–12.5 μl at a time, is applied to polyethyleneimine (PEI) cellulose thin-layer chromatography (TLC) plates (Fischer Scientific, Pittsburg, PA, or Macherey-Nagel, Düren, Germany) and GTP and GDP are separated by developing the chromatogram in 1 M formic acid, 1 M lithium chloride (see inset of Fig. 2A for example chromatogram).

The relative amounts of GTP and GDP from the incubations are quantified, either by a phosphorimager or by scintillation counting. When examining saturation kinetics, the reaction must be examined at several time points to allow the determination of an initial rate and we strongly suggest that progress curves be examined to validate the assay even if it is not being used for kinetic analysis. GTP hydrolysis should fit a first-order rate equation (Fig. 2A). Once first-order rates are established, single time points can

[27] L. Stephens, K. Anderson, D. Stokoe, H. Erdjument-Bromage, G. F. Painter, A. B. Holmes, P. R. J. Gafney, C. B. Reese, F. McCormick, P. Tempst, J. Coadwell, and P. T. Hawkins, *Science* **279**, 710 (1998).

[28] S. Paris, S. Beraud-Dufour, S. Robineau, J. Bigay, B. Antonny, M. Chabre, and P. Chardin, *J. Biol. Chem.* **272**, 22221 (1997).

be used for routine assay. In this case, the linear range of the assay can be extended by keeping concentration of ARF–GTP lower than the K_m (which is approximately 5 μM[17]) and using a simplified integrated rate equation, i.e.,

$$\text{Rate} = V_{\max}/K_m = \ln(S_0/S)/t$$

to calculate relative activity. The linear range of the assay is illustrated in Fig. 2B.

We do not recommend use of GTP labeled in the γ position because there is no means of detecting a number of artifacts that can be a particular problem with nonmyristoylated ARF.

Purification

A purified native protein or at least a full-length protein would be ideal for *in vitro* studies. Unfortunately, only one native protein has been purified and in that case, the homogeneous protein requires very specific conditions for stability.[12] We have been unsuccessful in expressing and purifying the full-length centaurin β's in bacteria. However, we have found that truncated forms of centaurin β_3 and β_4 can be expressed as soluble proteins in bacteria. The truncated proteins are similar to native proteins in GAP activity and have been useful for studying the interaction with ARF. For other applications, we have been able to express the full-length protein in cultured mammalian cells and, by two chromatographic steps, obtain a stable preparation enriched in activity.

Recombinant Proteins from Bacteria

The soluble, active forms of centaurin β_3 (AB007860) and centaurin β_4 (AF075462) include the PH domain, ARF GAP, and ANK repeats (residues 325–724 for centaurin β_4, residues 290–694 for centaurin β_3).[12,13] If the PH domain is deleted, the protein is soluble but inactive. If the ANK repeats are deleted, the protein is insoluble and we have not been able to solubilize and refold it into an active form. *Nde*I sites and *Xho*I sites are used to ligate the open reading frame into either pET19, for a histidine-tagged protein, or pET21 (Novagen), for a protein lacking the tag. BL21(DE3) strain of *Escherichia coli* is used to express the protein.

Transformed bacteria are streaked on Luria broth (LB) agar plates containing 50 μg/ml ampicillin. A single colony is used to inoculate 100 ml of LB containing 100 μg/ml ampicillin. The culture is grown to an OD_{600} of 0.6 and refrigerated overnight. The next morning, the bacteria from the culture are collected by centrifugation and used to inoculate 1–2 liters of LB containing 100 μg of ampicillin. When the OD_{600} is 0.6, protein expression is induced by the addition of 1 mM isopropyl-β-D-thiogalactylpyranoside. The

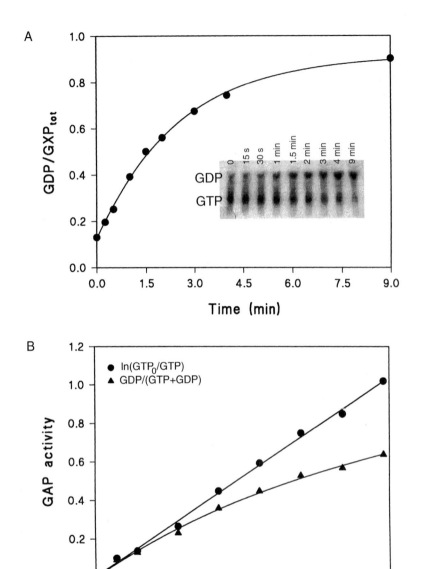

Fig. 2. Quantitation of ARF GAP activity. (A) Establishing first-order hydrolysis rate. ARF–GTP was incubated at 30° with 0.7 nM bacterially expressed [325–724]ASAP1 and GTP hydrolysis was determined at the indicated times. *Inset:* Autoradiogram of the chromatographic separation of protein-bound guanine nucleotide. (B) Use of the integrated rate equation to extend the linear range of the assay. The fraction of starting GTP on ARF hydrolyzed in 2 min in the presence of the indicated concentration of [325–724]ASAP1 is represented by circles. The transformed data (lnGTP_0/GTP) are shown by the triangles.

incubation is continued for 3–4 hr and the bacteria from 250-ml aliquots are collected by centrifugation and stored at −70°.

The bacterial pellet from 250 ml of culture is thawed and resuspended in 10 ml of 20 mM Tris, pH 8.0, 25 mM NaCl, and 10% glycerol, and one tablet of Complete Protease Inhibitor Cocktail Tablets (Roche, Indianapolis, IN) is added. The bacteria are lysed using a French press and the lysate is cleared by centrifugation at 100,000g for 60 min at 4°. The supernatant is applied to a 5-ml HiTrap-Q column (Pharmacia, Piscataway, NJ). The proteins are eluted using a gradient of 25–500 mM NaCl in 20 mM Tris, pH 8.0, 10% glycerol over 30 ml. The protein can be detected by assaying for GAP activity, but expression is generally high enough that the purification can be followed by SDS–PAGE of the appropriate fractions followed by Coomassie blue staining. The protein elutes at approximately 200 mM NaCl and is more than 95% pure (see Fig. 3A). Nontagged protein can be further purified by chromatography on HAP (see description below for the full-length protein) and the histidine-tagged protein can be further purified by metal chelate chromatography. In this latter case, the fractions from the HiTrap-Q column containing the protein are applied to a 1-ml HiTrap chelating column (Pharmacia) charged with NiCl$_2$ and equilibrated with 20 mM Tris, pH 8.0, 10 mM imidazole, 500 mM NaCl, and 10% glycerol. The protein is eluted with an ascending gradient of 10–500 mM imidazole in 15 ml (Fig. 3B). The protein elutes at 200–300 mM imidazole. The imidazole can be removed by dialysis, but usually this step is not necessary for assays. The ARF GAP is typically assayed at concentrations of 0.5–2 nM, and these preparations are consequently diluted over 10,000-fold into GAP reactions. The proteins are snap frozen in aliquots and stored at −70°.

We have not been successful in expressing similar fragments of centaurin β_1 and β_2 in bacteria. The proteins are insoluble and we have not been able to extract the protein from the inclusion bodies and refold the protein into an active form.

Purification of Full-Length Recombinant Protein from Mammalian Cells

We have purified centaurin β's from cells overexpressing recombinant proteins. Centaurin β's have been successfully expressed by introducing cDNA into cells by either viral infection[13] or transient transfection.[12,13] Here, we describe the latter. We usually express epitope-tagged proteins for ease in following and troubleshooting purifications but the epitope tag is not necessary. The open reading frames from centaurin β_1 (accession D30758) and β_2 (accession D26069) were integrated into the *Not*I–*Kpn*I sites of pFLAG-CMV2 (Sigma Chemical Company).[29] The open reading

[29] T. R. Jackson, F. D. Brown, Z. Nie, K. Miura, L. Foroni, J. Sun, V. Hsu, J. G. Donaldson, and P. A. Randazzo, *J. Cell Biol.*, in press.

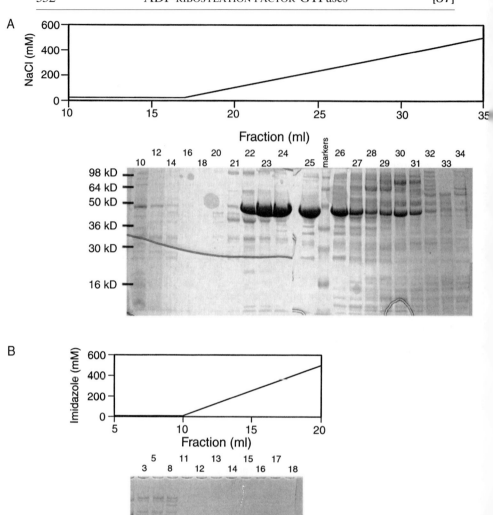

Fig. 3. Purification of [325–724]ASAP1. (A) Fractionation of bacterially expressed proteins on HiTrap-Q. The NaCl gradient is indicated. (B) Further purification of histidine-tagged [325–724]ASAP1 on a HiTrap metal chelating column.

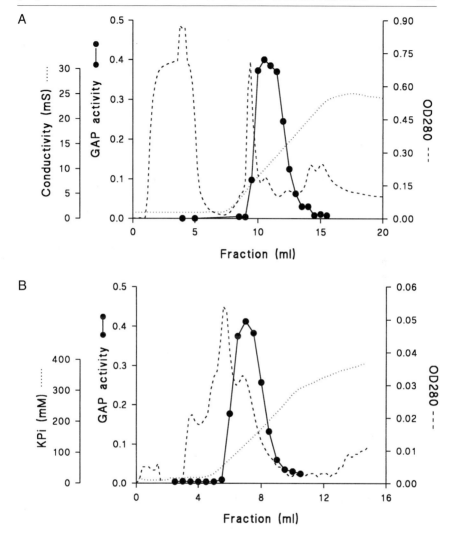

FIG. 4. Purification of ectopically expressed centaurin β_1 from COS7 cell lysates. (A) Fractionation on HiTrap-Q. (B) Fractionation on Econo-Pac CHT-II.

frame of centaurin β_4 was amplified from a cDNA clone using primers that incorporated an *Eco*RI site at the 5' end and a *Not*I site at the 3' end. Sequence encoding a FLAG epitope tag was incorporated into the primers used for PCR. The product was ligated into the *Eco*RI and *Not*I sites of PCI (Promega, Madison, WI) or pcDNA3 (Invitrogen).

We have been able to obtain adequate quantities of protein for biochemical analysis with three to four 100-mm plates of COS7, 293, or NIH 3T3 cells. Transfections are performed using Fugene-6 (Roche). Fifteen μg of DNA and 30 μl reagent are used for one 100-mm plate. Following transfections, cells are incubated 24–48 hr. Cells from one 100-mm plate are then lysed in 1 ml of 20 mM Tris, pH 8.0, 25 mM NaCl, and 10% (v/v) glycerol containing 1% Triton X-100 and Complete Protease Inhibitor (Roche). The lysate is cleared by centrifugation at 100,000g for 60 min at 4° and material from three or four plates is applied to a 1-ml HiTrap-Q column (Pharmacia) (Fig. 4A). The column is developed in a 20-ml gradient of 25–500 mM NaCl. Fractions containing activity are then applied to a 1-ml Econo-Pac CHT-II cartridge (Bio-Rad, Hercules, CA) (Fig. 4B). This column is developed in a 10–400 mM KP$_i$ (pH 7.0) gradient in 15 ml of a buffer that also contains 100 mM KCl and 20% (v/v) glycerol. Activity elutes at 150–250 mM KP$_i$.

Other means of purifying the proteins have been attempted without success. All centaurin β's we have examined irreversibly adhere to nonderivatized matrices of agarose cross-linked with acrylamide and similar matrices typically used for gel permeation columns. Thus, we have not been able to recover even the truncated forms of the protein from Superose (Pharmacia), Superdex (Pharmacia), or AcA (Biosepra, Life Technologies)-based sizing columns. The interaction is likely occurring through the PH domain, since the isolated PH domain does adhere to the columns but the inactive form of the protein consisting of the ARF GAP and ANK repeat domains does elute. We have not been able to recover activity by immunoprecipitation. The full-length form of centaurin β_4 rapidly and irreversibly loses activity when highly purified if removed from millimolar levels of phosphate and 20% (v/v) glycerol.

Note Added in Proof. The name of this family of Arf GAPs has been changed to the ASAP/ACAP family because of potential confusion arising from the name "centaurin." The prototypical centaurin, centaurin α has no GAP activity and binds phosphatidylinositol 3,4,5 trisphosphate (PIP3). In contrast, the ASAP/ACAP proteins do have Arf GAP activity and, where examined, bind and are activated by phosphatidylinositol 4,5-bisphosphate in preference to PIP3. Furthermore, although the centaurins and Arf GAPs have common domains, identity between the domains is limited and the domains occur within the proteins in different arrangements. Given these differences, the nomenclature should clearly distinguish the Arf GAPs from centaurin α. ACAP1 and ACAP2, for Arf GAP with coiled-coil, ANK repeat and PH domains, the new names for centaurin β1 and β2, better reflect the enzymology and structure of the proteins. PAP and ASAP1 will not be called centaurins but, instead, will retain their original names.

[38] Biological Properties and Measurement of Phospholipase D Activation by ADP-Ribosylation Factor (ARF)

By Shamshad Cockcroft, Kun Bi, Nicholas T. Ktistakis,
and Michael G. Roth

Introduction

Two laboratories independently discovered that ADP-ribosylation factor 1 (ARF1) is a potent activator of a phospholipase D(PLD), now known to be PLD1.[1,2] Because the known biological functions of ARF1 are to control aspects of membrane traffic, there is considerable interest in determining whether PLD1 is an effector for ARF in any of these activities, or serves other purposes. There is good evidence that PLD1 functions in regulated secretion and more controversial evidence that PLD1 may function in constitutive secretory processes.

PLD1 converts phosphatidylcholine (PC) to phosphatidic acid (PA) but in the presence of sufficient concentrations of primary alcohols will produce phosphatidyl alcohol (PEt) instead. Secondary alcohols function very poorly in this transphosphatidylation reaction. Because PA is actively metabolized and phosphatidyl alcohols are turned over slowly, a convenient assay for PLD activity in intact or permeabilized cells is to measure the production of the latter in the presence of ethanol or butanol. This chapter describes methods for investigating the role of PLD1 in regulated and constitutive secretory processes, as well as assays of PLD activity.

Role of Phospholipase D in Regulated Secretion

Because alcohols interfere with the production of PLD-catalyzed PA production, they can be used to investigate functional responses in cells.[3] Mast cells and neutrophils contain secretory granules that are lysosomal in origin and therefore contain many lysosomal enzymes including β-hexosaminidase and β-glucuronidase. The release of these granules is regulated by cell surface receptors. The physiologic stimulus to secretion in mast cells

[1] H. A. Brown, S. Gutowski, C. R. Moomaw, C. Slaughter, and P. C. Sternweis, *Cell* **75,** 1137 (1993).
[2] S. Cockcroft, G. M. Thomas, A. Fensome, B. Geny, E. Cunningham, I. Gout, I. Hiles, N. F. Totty, O. Truong, and J. J. Hsuan, *Science* **263,** 523 (1994).
[3] J. Stutchfield and S. Cockcroft, *Biochem. J.* **293,** 649 (1993).

is the cross-linking of the high affinity receptor, FcεR1, with antigen. On stimulation, multiple signaling pathways are activated, including the activation of phosphoinositide 3-kinase, phospholipase Cγ, and phospholipase D, which are upstream to secretion.[4–6] For these studies, cultured mast cells (RBL-2H3 cell line) are suited for functional analysis of regulated secretion because they can be labeled with appropriate radioactive lipid precursors in culture. Intact RBL mast cells are first incubated with IgE to sensitize the cells for subsequent challenge with antigen. Antigen binds to the IgE and cross-links the Fcε receptor, thus initiating a signaling cascade that culminates in the release of secretory granules. Secretion from cells is assayed by measuring the release of hexosaminidase in the external medium. When secretion is examined in permeabilized cells, the cells can be stimulated with either antigen or with GTPγS, which can directly activate G proteins. To examine the function of ARF and phospholipase D in the regulated secretory process, one strategy is to permeabilize cells to deplete the cells of endogenous proteins including ARFs. Having confirmed that the cytosol-depleted cells are refractory to stimulation, ARF proteins can be specifically reintroduced to reconstitute secretory function and phospholipase D activity simultaneously.[7]

Assays of PLD Activity and Secretion in Intact and Permeabilized RBL-2H3 Mast Cells

Reagents

RBL-2H3 mast cells can be obtained from the American Type Culture collection (Rockville, MD). GTPγS is purchased from Boehringer Mannheim (Germany) as 100 mM solution and prepared as a 5 mM stock solution in 0.9% NaCl and stored at $-20°$ in small aliquots. Streptolysin O is obtained from Murex (Dartford, UK). IgE [anti-dinitrophenol (DNP)] and antigen, DNP conjugated to human serum albumin (DNP-HSA) are obtained from Sigma (St. Louis, MO). DNP-HSA is prepared as a stock solution in water at 1 mg/ml and stored at 4° for a maximum of 3 months.

[4] R. S. Gruchalla, T. T. Dinh, and D. A. Kennerly, *J. Immunol.* **144,** 2334 (1990).
[5] H. Yano, S. Nakanishi, K. Kimura, N. Hanai, Y. Saitoh, F. Yasuhisa, N. Yoshiaki, and Y. J. Matsuda, *Biol. Chem.* **268,** 25846 (1993).
[6] S. A. Barker, K. K. Caldwell, J. R. Pfeiffer, and B. S. Wilson, *Mol. Biol. Cell* **9,** 483 (1998).
[7] G. Way, N. O'Luanaigh, and S. Cockcroft, *Biochem. J.* **345,** 63 (2000).

Procedure

Studies with Intact RBL Cells. RBL-2H3 mast cells are grown in Dulbecco's modified Eagle's medium (DMEM) supplemented with 12.5% fetal calf serum (FCS), 4 mM glutamine, 50 μg/ml penicillin, and 50 IU/ml streptomycin as a monolayer in 175-cm^2 vented flasks, at 37° with 5% (v/v) CO_2 and 100% humidity. Each flask when confluent contains approximately 4×10^7 cells. Prior to stimulation with antigen, the cells are incubated with immunoglobulin E (IgE) (2 μg/ml) overnight. The following day, the cells are washed in a HEPES buffer, pH 7.2 [20 mM HEPES, 137 mM NaCl, 3 mM KCl, 2 mM MgCl$_2$, 1 mM CaCl$_2$, 5.6 mM glucose, and 0.1 mg/ml bovine serum albumin (BSA)], and the cells recovered in suspension by scraping into 5 ml of buffer. The cells are incubated at 37° for 10 min and then 100-μl aliquots of the cells are transferred to Eppendorf tubes that contain antigen (40 ng/ml) and varying amounts of ethanol (1–3% final) all prepared in HEPES buffer. The cells are incubated at 37° for 25 min and the reactions terminated by centrifugation at 4° for 5 min at 1000g. Secretion of hexosaminidase is measured by incubating 50 μl of the supernatant with 50 μl of 1 mM 4-methylumbelliferyl-N-acetyl-β-D-glucosaminide in 0.2 M citrate (pH 4.5), a fluorescent substrate for hexosaminidase, in a black multiwelled dish for 1–2 hr at 37°. The reaction is then quenched with 150 μl of 0.2 M Tris and the resulting fluorescence is measured at 405 nm with a Fluoroscan II fluorescence plate reader. To obtain % secretion, total enzyme content is also determined. An identical aliquot of cells is lysed with 0.2% Triton X-100 to measure total enzyme content. Figure 1 illustrates the results of a typical experiment. The cells are capable of releasing between 30 and 70% of the total hexosaminidase. The percent secretion varies depending on passage number. Low passage number cells secrete higher amounts and this declines as the passage number increases to 20. Near maximal inhibition of secretion is observed with 1% ethanol.

Studies with Permeabilized Cells. Permeabilization is achieved by using streptolysin O, a bacterial (streptococcal) cytolysin that generates large lesions (approximately 15 nm in diameter) in the plasma membrane of cells to allow the efflux of cytosolic proteins. Proteins such as lactate dehydrogenase and ARF leak out of the cells within 5–10 min.[8] To obtain a high level of cytosol depletion, permeabilization is carried out in cells in suspension. Collecting the released material, concentrating it by trichloroacetic acid (TCA) precipitation, and using ARF antibodies, which are commercially

[8] A. Fensome, E. Cunningham, S. Prosser, S. K. Tan, P. Swigart, G. Thomas, J. Hsuan, and S. Cockcroft, *Curr. Biol.* **6**, 730 (1996).

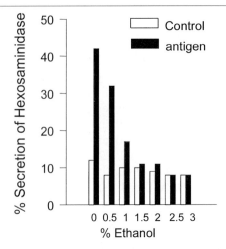

FIG. 1. Inhibition of secretion by ethanol from antigen-stimulated RBL mast cells. Intact RBL mast cells were stimulated for 25 min with antigen in the presence of increasing concentrations of ethanol. At the end of stimulation, the secreted product was measured in the external medium.

available, monitors the release of ARF. The time course of loss of ARF and loss of secretory function and PLD activity can be readily monitored (Fig. 2).

RBL-2H3 mast cells are grown in Dulbecco's modified Eagle's medium (DMEM) as described in 175-cm^2 vented flasks. Each flask when confluent contains approximately 4×10^7 cells and provides sufficient cells for 20 assays. For the simultaneous measurement of PLD activity and secretion from the same cell preparation, release of [^{3}H]choline rather than phosphatidylethanol is used. This avoids the addition of ethanol that would inhibit secretion. RBL-2H3 mast cells are prelabeled with [^{3}H]choline to label the choline-containing lipids, which are mainly PC and some sphingomyelin. The label (0.5 μCi/ml) is added directly to the complete medium and the cells are allowed to proliferate for 48 hr to reach 80% confluence. Cell monolayers are washed twice with permeabilization buffer, pH 7.2 (137 mM NaCl, 3 mM KCl, 2 mM MgCl$_2$, 5.6 mM glucose, and 0.1 mg/ml bovine serum albumin), and the cells recovered in suspension by scraping into 2.5 ml of buffer. The cells are centrifuged at 450g for 5 min at room temperature and resuspended in 4 ml (10^7/ml). The cells are transferred at 37° and 0.5 ml of a permeabilization cocktail is added such that the final concentration of the permeabilization agent, streptolysin O, is 0.4 IU/ml. In addition, MgATP, 1 mM (final) and Ca^{2+}, 100 nM buffered with 3 mM EGTA (final) are also included in the permeabilization cocktail. To monitor loss of the secretory response and PLD activity simultaneously, 100-μl aliquots of cells are transferred immediately (time 0, permeabilization time), and subsequently at timed intervals (see Fig. 2), to Eppendorf tubes containing antigen or GTPγS as a stimulus, 10 μM Ca^{2+}, and 1 mM MgATP. The final assay volume is 200 μl. The cells are incubated for a further 25 min at 37° and the reaction terminated by centrifugation (2000g) of the cells for two minutes at 4°. 50 μl of the supernatant is sampled to assay for released β-hexosaminidase for measurement of secretion as described above, and 50 μl of the supernatant is analyzed for the release of [^{3}H]choline.

[^{3}H]Choline is separated from the rest of the phosphorylated choline metabolites (glycerophosphocholine and phosphorylcholine) by cation chromatography. The supernatant containing the choline metabolites is applied to a 1-ml bed volume of Bio-Rex 70 cation exchange resin (sodium form, mesh size 200–400 purchased from Bio-Rad, Hercules, CA) in a Bio-Rad column. The column is washed with 3 ml of water to elute the phosphorylated choline metabolites. Radiolabeled choline is quantitatively eluted with 3 ml 50 mM glycine containing 500 mM NaCl, pH 3, directly into scintillation vials. The Bio-Rex resin is regenerated by extensively washing the resin with 0.5 M NaOH, pH 9, followed by washing with water. The resin is then washed with 0.1 M sodium phosphate, pH 7, and finally washed with water. The radioactivity is measured after addition of a scintillation cocktail that is able to accommodate high salt. PCS (phase combining system) from Amersham (Pharmacia, Piscataway, NJ) or Ultima Gold from Canberra Packard (Didcot, UK) are both suitable. The results can be expressed as a percent of radiolabeled choline incorporated into the lipids. Thus an aliquot of cells is used to extract the total phospholipids, and the radioactivity determined.

To monitor leakage of ARF from permeabilized cells, a larger amount of cells needs to be used. Four ml (5 × 10^7 cells) is added to 1 ml of permeabilization cocktail, to give final concentrations of 1 mM MgATP, 0.4 IU/ml streptolysin O, and 100 nM Ca^{2+} buffered by 3 mM EGTA. At stated time points (5, 10, and 30 min) 1-ml samples are removed. After centrifugation of the cells, the supernatant is harvested, and treated with 10% TCA at 4° for 30 min to precipitate proteins. The precipitate is recovered by centrifugation, resuspended in 50 μl of 1 M NaOH, and neutralized with 50 μl of 1 M Tris, pH 6.8. Precipitated proteins are separated by SDS–PAGE (14% acrylamide), transferred to polyvinyldifluoride membranes, and Western blotted with ARF antibodies. To monitor how much ARF is retained in the cells after permeabilization, the cell pellet is resuspended in 100 μl RIPA buffer [150 mM NaCl, 1% Triton X-100 (v/v), 0.5% deoxycholate (w/v), 0.1% SDS (w/v) and 50 mM Tris, pH 7.5], vortexed thoroughly, and placed at 4° for 30 min. After centrifugation for 15 min at 4°, the solubilized proteins are concentrated by treatment with TCA as described above and used for Western blot analysis.

Stimulus: GTPγS

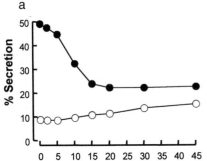

Stimulus:antigen

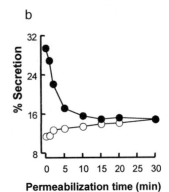

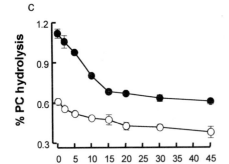

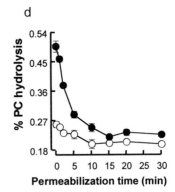

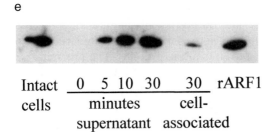

Intact cells   0   5   10   30   30   rARF1
            minutes   cell-
          supernatant   associated

Comments

Zero time permeabilization provides an indicator of the response when the full complement of the cytosolic proteins is still present. In permeabilized cells, antigen or GTPγS stimulation of secretion and PLD activity are dependent on the presence of 10 μM Ca^{2+} and MgATP. Ca^{2+} is buffered with 3 mM EGTA. The loss of responsiveness is dependent on how long the cells are permeabilized for and on the stimulus. With antigen as stimulus, both the secretory response and the PLD activity have declined to control values by 10 min (Figs. 2c and 2d). When the cells are stimulated with 10 μM GTPγS, the loss of responsiveness is slower for both PLD activity and for secretion and is incomplete (Figs. 2a and 2b).

Reconstitution of Secretory Function and Phospholipase D Activity in Cytosol-Depleted RBL Mast Cells

The cytosol-depleted cells can be reconstituted with ARF proteins with respect to PLD activity and secretion. To monitor the relationship between the amount of PA and secretion, ethanol can be used to titrate out the PA formed. To measure PEt production, the lipid moiety of PC needs to be prelabeled and here we describe a method for monitoring PLD activity in permeabilized cells. Two ml of suspended RBL-2H3 mast cells are prelabeled with [^{3}H]alkyl-lyso-PC (5 μCi/ml) for 1 hr in the HEPES buffer. Bovine serum albumin is replaced with fatty acid-free BSA. The cells are washed and finally resuspended in permeabilization buffer in 4 ml (10^7/ml). One ml of a permeabilization cocktail is added such that the final concentration of the permeabilization agent, streptolysin O, is 0.4 IU/ml with 1 mM MgATP, and 100 nM Ca^{2+} (buffered with 3 mM EGTA). The cells are transferred to 37° for permeabilization. For restoration of antigen responses, a 5-min permeabilization is used. Longer permeabilization time is detrimental and the cytosol-depleted cells are no longer responsive to antigen. When GTPγS is used, the cells are permeabilized for 10 min. After

---

FIG. 2. Time-dependent loss of GTPγS- (a, b) or antigen-stimulated (c, d) secretion, PLD activation and cell-associated ARF (e) from permeabilized RBL mast cells. [^{3}H]Choline-labeled RBL-2H3 mast cells were permeabilized with 0.4 IU/ml streptolysin O for the indicated times and assayed in the presence (closed symbols) or absence (open symbols) of either GTPγS (10 μM) *plus* Ca^{2+} (10 μM) (a and b) or antigen (DNP-HSA, 40 ng/ml) plus Ca^{2+} (10 μM) (c and d). The cells were stimulated for 25 min at 37°, the samples quenched and the supernatants analyzed for released β-hexosaminidase (a and c) and [^{3}H]choline release (b and d). (e) Release of ARF in the medium following permeabilization with streptolysin O. (Reproduced from Way *et al.*[7])

permeabilization, the cells are pelleted by centrifugation at $4°$ ($1800g$ for 5 min) and the supernatant discarded. The cells are resuspended in permeabilization buffer at $4°$ and divided into 50-μl aliquots in Eppendorf tubes containing the appropriate reagents for the reconstitution assays. The final assay volume for the reconstitution assay is reduced to 100 μl to conserve proteins. Fifty μl of reaction mixture is prepared at $4°$ in Eppendorf tubes containing twice the concentration of MgATP (1 mM final), 10 μM Ca^{2+} buffered with 3 mM EGTA, antigen (40 ng/ml final), recombinant myristoylated ARF1, and a varying amount of ethanol (0–3%). The assay tubes are incubated for 25 min at $37°$ and the reaction terminated by centrifugation of the cells at $4°$. Fifty μl of the supernatant is sampled to assay for released β-hexosaminidase as a measurement of secretion as described above.

To the remainder of the cells, 375 μl 1 : 2 chloroform : methanol is added, together with 50 μl of water. The samples are vortexed vigorously, and phase separation is obtained by adding 125 μl of chloroform and 125 μl of water. The samples are spiked with phosphatidylethanol and vortexed again. After separation of the phases, the lower chloroform phase is transferred to a clean Eppendorf tube, dried under vacuum and the lipids resuspended in 50 μl chloroform. PEt is separated from the rest of the lipids by chromatography on Merck silica gel 60 TLC plates using the solvent system composed of chloroform : methanol : acetic acid : water (75 : 45 : 5 : 0.5). The lipids are visualized with iodine vapor, and PEt spot is marked using a pencil. The amount of radiolabel incorporated into PC in resting cells is also measured. The silica gel containing the lipids is excised and transferred to scintillation vials. The lipids are extracted with 0.2 ml methanol and counted for radioactivity after addition of 3 ml of scintillation fluid (Canberra Packard). The results are expressed as the percentage of disintegrations per minute (dpm) in PEt/dpm in PC from resting cells.

Figure 3 illustrates the results from a reconstitution assay. In the absence of ethanol, secretion is restored in the cytosol-depleted cells with both antigen (Fig. 3a) and GTPγS (Fig. 3b) with ARF1. Under these conditions no PEt is formed due to the absence of ethanol (Fig. 3b); instead an increase in PA can be observed (data not shown). As the concentration of ethanol is increased in the incubation, the formation of PEt is increased at the expense of PA. The effect of ethanol is to inhibit ARF1 *plus* antigen- or GTPγS-stimulated restoration of secretory function. As the increase in PEt occurs, a decrease in secretion is observed in the ARF-reconstituted cells. Of note is that when GTPγS is the stimulus, the inhibition by ethanol is less marked compared to when antigen is used. This is most likely due to a stronger and sustained PLD activation with GTPγS compared to antigen.

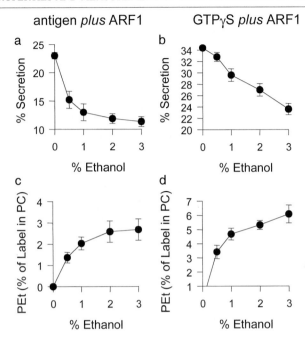

FIG. 3. Ethanol inhibits ARF1-reconstituted secretion and increases phosphatidylethanol production. Ethanol inhibits secretion reconstituted with ARF1 *plus* antigen (a) and ARF1 *plus* GTPγS (b). Ethanol increases PEt production in cells reconstituted with ARF1 *plus* antigen (c) and ARF1 *plus* GTPγS (d). RBL-2H3 mast cells were permeabilized with 0.4 IU/ml SLO, washed, and incubated with either antigen *plus* ARF1 (a and c) or with ARF1 *plus* GTPγS (b and d) in the presence of increasing concentrations of ethanol. For measurement of PEt, the cells were prelabeled with [³H]alkyl-lyso-PC and the amount of PEt made was expressed as the amount of label incorporated into PC. Experiments were conducted in duplicate and were repeated between three to seven occasions. Error bars reflect the S.E.M. from pooled experiments (a) $n = 3$; (b) $n = 7$; (c) $n = 3$; and (d) $n = 4$. (Reproduced from Way *et al.*[7])

Expression of Recombinant Phospholipase D in Continuous Cell Lines

Establishing continuous cell lines that constitutively overexpress PLD1 is quite difficult. After many transfections of Chinese hamster ovary (CHO) cells with a plasmid expressing human PLD1, Bi and Roth succeeded in obtaining only two surviving cell clones. These cells constitutively express PLD activity at approximately the same level as did untransfected CHO cells that were stimulated with PMA, and grew threefold slower than un-transfected cells, suggesting that higher levels of constitutive PLD1 activity

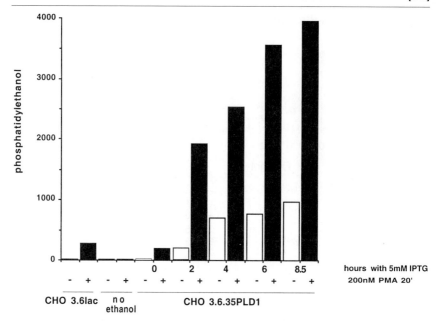

FIG. 4. Time course of PLD activity after induction of PLD1 expression. Parental cells expressing the *lac* repressor, CHO 3.6lac, or cells stable expressing PLD, CHO 3.6.35PLD1, were labeled for 18 hr with [³H]palmitate. For the last part of the labeling period, samples were incubated with 5 m*M* IPTG and 1 m*M* sodium butyrate for the times shown. At the end of the labeling period, the cells were incubated with medium containing 3% ethanol ± 200 nm PMA for 20 min and lipids were extracted and separated on a TLC plate and quantified as described in the text.

are probably toxic. Cell lines expressing PLD1 in an inducible manner were made by using the system devised by Welch *et al.*[9] CHO cells were transfected with a plasmid that constitutively expressed the *lac* repressor protein and these were transfected in turn with a plasmid in which PLD expression was controlled by a cytomegalovirus (CMV) promoter that contained binding sites for the *lac* repressor. We obtained many cell clones that could be induced to express high levels of PLD1 activity by including isopropyl-D-thiogalactoside (IPTG) in the culture medium, and this activity could be increased additionally by treating cells with sodium butyrate (see comments below). A time course of PLD1 activity after induction is shown in Fig. 4. Methods for establishing lines of cells that overexpress PLD1 and for measuring PLD activity are described below.

[9] H. Welch, A. Eguinoa, L. R. Stephens, and P. T. Hawkins, *J. Biol. Chem.* **273,** 11248 (1998).

Establishing Cell Lines with Inducible Phospholipase D Activity

Reagents

CHO K1 cells can be obtained from the American Type Culture Collection. Cells are cultured at 37° in normal culture medium: Dulbecco's modified Eagle's medium (DMEM, Life Science Technologies) containing 0.35 mM proline, 50 U/ml penicillin, 50 μg/ml streptomycin, 10 mM HEPES (pH 7.3), and 5% fetal bovine serum (Hyclone, Logan, UT). Cell lines capable of inducible protein expression are made using plasmid pCMVLac1, in which the *lac* repressor protein is controlled by a CMV1 immediate early promoter, and pCMV3Rluc, which contains a 2.1-kb CMV5 immediate early promoter with triplicate binding sites for the *lac* repressor protein inserted near the TATA binding site.[3] We obtained a cDNA for human PLD1a from Michael Frohman,[10] but PLD1 sequences from several species are currently in the database and available. To control for effects of protein expression that are independent of enzyme activity, we obtained a cDNA encoding a catalytically inactive PLD1a in which the codon for lysine-466 had been converted to alanine[11] from Bill Singer and Paul Sternweis (UT Southwestern). Anti-PLD antiserum was prepared by immunizing rabbits with peptides containing amino- or carboxyl-terminal sequences of PLD1.[12] Antibodies for the *lac* repressor protein can be purchased from Stratagene (La Jolla, CA). We use IPTG from Boehringer Mannheim (Indianapolis, IN), stored at 4° as a stock solution of 1 M in water. Hygromycin (Fisher Scientific) and G418 (GIBCO-BRL, Gaithersburg, MD) are stored as 1000× stocks at 4° in 1-ml aliquots of 400 mg drug/ml in cell culture medium and activity is maintained for several months. Other reagents can be obtained from Sigma.

Procedure

Transfect CHO cells with pCMVLac1 by a convenient method and select clonal cell lines that stably express the plasmid by growth in 400 μg hygromycin/ml culture medium. Isolate individual colonies with glass cloning rings and screen by immunofluorescence with anti-*lac* repressor antibodies to identify cell lines that express high levels of the Lac protein. Choose one or several of these and transfect them with plasmid pCMV3RhPLD1, in which the luciferase cDNA in the plasmid pCMV3Rluc

[10] S. M. Hammond, Y. M. Altshuller, T. C. Sung, S. A. Rudge, K. Rose, J. Engebrecht, A. J. Morris, and M. A. Frohman, *J. Biol. Chem.* **270,** 29640 (1995).
[11] T. C. Sung, R. L. Roper, Y. Zhang, S. A. Rudge, R. Temel, S. M. Hammond, A. J. Morris, B. Moss, J. Engebrecht, and M. A. Frohman, *EMBO J.* **16,** 4519 (1997).
[12] M. Manifava, J. Sugars, and N. T. Ktistakis, *J. Biol. Chem.* **274,** 1072 (1999).

has been replaced with a cDNA for human PLD1a. As a control, DNA encoding an inactive PLD subcloned into pCMV3RLuc can be transfected into cells expressing the *lac* repressor. Select stably transfected clones by growth in the presence of 400 μg G418/ml culture medium until cell colonies form and isolate some with cloning cylinders. After inducing PLD expression with 5 mM IPTG and 7.5 mM butyric acid in normal culture medium for 15 hr, identify positive clones by immunofluorescence using anti-PLD1 antibody. For cells expressing the active PLD, measure the enzyme activity by labeling them with [^{3}H]palmitic acid and measuring the production of phosphatidylethanol in response to phorbol myristate acetate (PMA) at various intervals after induction with IPTG and sodium butyrate (see below). A representative experiment is shown in Fig. 4. Clonal cell lines in which PLD activity is low in the absence of the inducers but which increases rapidly in the presence of inducers can be identified by these two screens.

Comments

It is very important to identify a cell line in which the *lac* repressor protein expression is high and stable. Thus, a sufficient number of independently cloned, hygromycin-resistant cells transfected with pCMVlac should be screened by immunofluorescence using antibodies to the *lac* repressor protein. The ability of candidate cell lines to inducibly express proteins can be tested by transfecting them with pCMV3Rluc and measuring luciferase activity in the presence and absence of IPTG in the cell culture medium. The *lac* repressor cell lines that look promising should be maintained in culture for some time and stained again with anti-*lac* repressor antibodies. We have found that our 3.6lac cell line is relatively stable, maintaining expression of the repressor after 3 or more months in culture. However, it is wise to freeze a large number of samples of a good *lac* repressor cell line as soon as it is identified, so that low passage *lac* repressor cells can be used to make inducible cell lines. In turn, these should be used at relatively low passage to ensure reproducible results.

Clonal cell lines stably expressing a transgene often exhibit heterogeneous levels of expression. Sodium butyrate is a nonspecific inducer of transcription, and we observe more uniform expression of the *lac* repressor protein and of PLD when cells are treated with 1 mM sodium butyrate in addition to IPTG. However, approximately 5% of control CHO 3.6lac show elevated staining with anti-PLD antibodies after treatment with sodium butyrate. This is not a problem for experiments measuring biochemical characteristics of the cell population as a whole, but may influence experiments where activity is measured in single cells by techniques such as immunofluorescence.

Assay of Phospholipase D Activity in Intact Cells

Phorbol 12-myristate 13-acetate (PMA), a stimulator of protein kinase C, stimulates activity of PLD1 but not PLD2.[13] Thus, a convenient assay of the extent of exogenous PLD1 activity is to compare the production of phosphatidylethanol in the presence of PMA in nontransfected cells to that produced by transfected cells in which PLD expression has been induced.

Reagents

CHO cells expressing PLD or the parental *lac* repressor cells as controls are grown on six-well culture plates (Costar) in normal culture medium (see above). Cells are labeled in culture medium containing 25 μg/ml of [^{3}H]palmitic acid (10 μCi/ml, NEN Dupont, Boston, MA) and only 1% FCS. PLD expression is stimulated by adding 1 mM sodium butyrate and 5 mM IPTG to the culture medium; typically we run parallel samples without inducers as controls. PLD activity is stimulated with PMA, which is stored at $-20°$ as a stock solution at a concentration of 200 μM in ethanol. The following solutions are used for lipid extraction: methanol, chloroform, phosphate-buffered saline (PBS, Life Technologies) and 23 mM HCl. A source of nitrogen gas is required for drying samples. Lipid samples are analyzed by thin-layer chromatography (TLC) on LK6DF silica plates (Whatman, Clifton, NJ) in a solvent system of chloroform : methanol : acetic acid : acetone : water at $270 : 54 : 54 : 108 : 27$ (v/v) ratios.[14] The plates are sprayed with a solution of scintillator: 25 ml 2-methylnaphthalene (comes as a solid, melt at 37°), 15 ml toluene, 0.1 g 2,5-diphenyloxazole) and exposed to X-ray film (Konica or Kodak). Exposed film is scanned by densitometer.

Procedure

Cells are grown to 80% confluence on six-well culture plates. This allows duplicate or triplicate samples of cells to be easily handled together. Cells are incubated 15–18 hr with [^{3}H]palmitate-labeling medium. The labeling medium is removed and the cells are incubated for 30 min at 37° in medium containing 1% FCS, 3% ethanol, and $\pm$ 200 nM PMA. At the end of this incubation cells are put on ice and washed three times with ice-cold PBS. Lipids are extracted by adding 0.4 ml of ice-cold PBS to the cells, then 0.4 ml of ice-cold methanol. The plate is swirled to mix the two solutions and the cells are scraped loose with a Teflon cell scraper. An additional 0.4 ml of methanol is added to the well. The contents of each well are transferred

---

[13] J. M. Jenco, A. Rawlingson, B. Daniels, and A. J. Morris, *Biochemistry* **37,** 4901 (1998).
[14] E. G. Bligh and W. J. Dyer, *Can. J. Biochem. Physiol.* **37,** 911 (1959).

to 2-ml polypropylene microfuge tubes, and 150 μl of methanol and 0.4 ml of chloroform are added. The samples are incubated at room temperature for 30 min with vortexing each 10 min. Samples are centrifuged in excess of 12,000g for 5 min to pellet cell debris. The clear supernatant is transferred to a glass tube (Fisher Scientific) and 0.5 ml of chloroform and 0.4 ml of 23 mM HCl are added to produce an aqueous and an organic phase. The phases are mixed by votexing and separated by centrifugation at 3000g at 22° for 5 min. If no clear phase separation is visible, add 100 μl of chloroform, mix, and centrifuge again. The aqueous phase is discarded and the organic phase is dried either under a stream of nitrogen gas or by evaporation under vacuum during centrifugation. Dissolve the dried lipids in 50 μl of chloroform and count 5 μl in a liquid scintillation counter. Load equal counts of each sample onto a TLC plate and resolve using the solvent system listed above, allowing the solvent to reach 1 cm from the top of the plate. To determine the identity of various lipid species, it is useful to run pure lipid species in parallel in the same solvent system and to identify their positions of migration by staining the plate with iodine. Dry the plate in a chemical hood and spray evenly with scintillator. Allow this solution to dry, then expose the plate to X-ray film at −80°. Scan the exposed film with a densitometer and for each sample compare the optical density of the phosphatidylethanol band to the intensities of the other lipid bands to correct for variations in sample loading. A typical image of a TLC plate is shown in Figure 5.

Comments

It is very important to evenly spray the scintillator solution on the TLC plate to avoid uneven fluorography. A good indication on LK6DF plates, in which strips of silica are separated by clear glass, is to obtain an even deposition of scintillator on the glass separators. Dry the plates thoroughly to prevent the silica from sticking to the X-ray film during exposure. Be careful to quantify only bands in which the optical density is within the linear range of the X-ray film (usually 0.2 and 2 OD).

Assay of Phospholipase D Activity in Isolated Membrane Preparations Using Exogenous Substrates

This assay is essentially that published by Brown and Sternweis[15] with the modifications that Golgi membranes provide the source of PLD and a

---

[15] H. A. Brown and P. C. Sternweis, *Methods Enzymol.* **257,** 313 (1995).

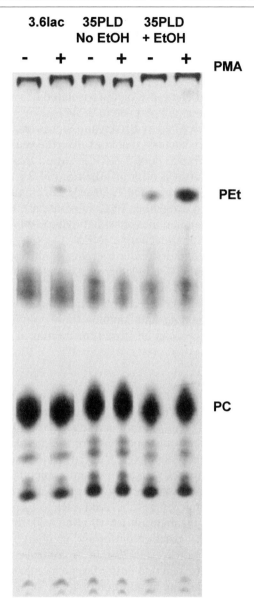

FIG. 5. A representative autoradiograph of a TLC plate resolving phosphatidylethanol (PEt) is shown. Cells were labeled as described for Fig. 4. In the absence of ethanol, no band migrating at the position of PEt is observed. The position of phosphatidylcholine (PC) is shown.

more stable preparation of substrate lipid is prepared by the protocol of Jiang.[16]

Reagents

Light membranes enriched in Golgi are isolated according to Balch *et al.*,[17] and stored as aliquots of 3 μg protein in sucrose at $-80°$. ARF can be purified from tissue as described by Brown and Sternweis,[15] or recombinant ARF can be isolated from bacteria that coexpress the yeast myristoyltransferase as described by Franco *et al.*[18] ARF is stored frozen in aliquots at 10 μM. PLD reaction buffer is 50 mM HEPES, pH 7.5, 3 mM EGTA, 3 mM CaCl$_2$, 3 mM MgCl$_2$, 1 mM DTT, 80 mM KCl, and 10 μM GTPγS. Vesicles to be used as an exogenous PLD substrate are made as follows.

Mix phosphatidylethanolamine, phosphatidylinositol 4,5-bisphosphate, and [^{3}H]dipalmitoylphosphatidycholine (NEN, Dupont) in a molar ratio of 10:1:1 in chloroform and dry under a stream of nitrogen. Add 1× PLD reaction buffer containing 1% octylglucoside and sonicate for 10 min at room temperature. Filter the vesicles through a column (0.7 cm × 30 cm) of AcA34 (Biosepra) that has been equilibrated with 1× PLD reaction buffer. Collect 300-μl fractions and combine those containing the peak of radioactivity. These vesicles can be stored for months at $-80°$.

Procedure

Thaw Golgi membranes for 5 min in a 37° water bath. Dilute membranes to 400 μl with 20 mM HEPES, pH 7.5, containing 0.2 M sucrose and centrifuge 7 min at 14,000g at 4° in a microfuge. Carefully aspirate with a 27-gauge needle along the wall opposite from the membrane pellet. Add 20 mM HEPES, pH 7.5, containing 0.2 M sucrose and resuspend the membranes by pipetting 15 strokes with a 200-μl micropipette tip. Add 5 μl of the resuspended membranes containing 0.5 μg protein to 15 μl of 1.5× PLD reaction buffer in a microfuge tube. Add ARF to 1 μM final concentration, vortex gently, and incubate on ice 10 min. Add 10 μl (40,000 cpm) of substrate vesicles in 1× reaction buffer and incubate for 20 min at 37°. Leave duplicate samples on ice for 20 min as controls. Prepare controls containing ARF, substrate, reaction buffer but no membranes and incubate 20 min at 37°. Stop the reaction by addition of 200 μl of 10% trichloroacetic

---

[16] X. Jiang, Ph.D. Dissertation, Univ. Texas Southwestern Medical Center, Dallas, pp. 159 (1999).

[17] W. E. Balch, E. Fries, W. H. Dunphy, L. J. Urbani, and J. E. Rothman, *Methods Enzymol.* **98**, 37 (1983).

[18] M. Franco, P. Chardin, M. Chabre, and S. Paris, *J. Biol. Chem.* **270**, 1337 (1995).

acid and 100 µl of 10% bovine serum albumin. Centrifuge the samples and measure the radioactivity (released choline) in the supernatant with a scintillation counter.

Assay of Phospholipase D Activity in Immunoprecipitates Using Exogenous Substrates

Ktistakis and colleagues have adapted use of exogenous substrates to measure PLD activity in immunoprecipitates.[19,20] In an important modification, they have used PC labeled in the acyl chain as substrate, so that the production of PA can be measured directly by TLC.

Reagents

Lysis buffer: 50 mM Tris-HCl, pH 8.0, 50 mM KCl, 10 mM EDTA, 1% Nonidet P-40, and protease inhibitors

Wash buffer: 50 mM Tris-HCl, pH 8.0, 150 mM NaCl, 5 mM EDTA, 0.1% Tween 20 and 0.02% sodium azide

Purified ARF as described above

Lipid substrate can either be prepared as described above, or as described by Sugars et al.;[20] however, [14C]dipalmitoylphosphatidylcholine (Amersham Pharmacia Biotech) mixed with nonradioactive PC should be used.

The assay is similar to that described above for isolated membranes, with the exception that the source of PLD is an immunoprecipitate. Cells are incubated with lysis buffer for 15 min at 4°, scraped into centrifuge tubes, and spun at 17,000g for 10 min at 4°. The supernatant from this spin is incubated for 30 min on a rocking platform with 2 µl preimmune serum and 50 µl 10% protein A-Sepharose in wash buffer. The Sepharose is pelleted and the supernatant incubated with 3 µl anti-PLD serum for 1 hr and with 50 µl 10% protein A-Sepharose for 30 min. The immunoprecipitate is washed 4 times with wash buffer and once with PBS. The Sepharose beads are centrifuged and incubated with liposomes, GTPγS, and ARF in the buffers described above for the assay of PLD in membranes. The samples are mixed by gentle vortexing every 5 min during the reaction. Lipids are extracted and analyzed by TLC as described above.

[19] M. Manifava, J. Sugars, and N. T. Ktistakis, J. Biol. Chem. **274,** 1072 (1999).
[20] J. M. Sugars, S. Cellek, M. Manifava, J. Coadwell, and N. T. Ktistakis, J. Biol. Chem. **274,** 30023 (1999).

Comments

The method for labeling the PC substrate can influence experimental outcomes. PC used for measuring PLD activity can labeled with [³H]choline in the head group, ³²P in the phosphate linkage, or ¹⁴C or ³H in the acyl chains. It is simple to quantify released choline by scintillation counting; however, such assays do not distinguish between PLD and PLC activities. ³²P-labeled PC provides a strong signal in PA but light membrane fractions enriched in PLD also contain a very active phosphatidic acid phosphatase that removes the phosphate, leading to underestimates of PA production. The use of PC labeled in the acyl chain reveals this activity as an ARF-dependent production of diacylglycerol and provides information about lipid metabolism downstream of PLD, as monoacylglycerol is also produced when Golgi-enriched membranes are treated with ARF.

[39] Use of Aminoglycoside Antibiotics and Related Compounds to Study ADP-Ribosylation Factor (ARF)/Coatomer Function in Golgi Traffic

By Rockford K. Draper, Robert Tod Hudson, and Tonghuan Hu

Introduction

Coatomer is a soluble macromolecular complex involved in membrane traffic through the Golgi complex and contains seven nonidentical protein subunits (α, β, β', γ, δ, ε, and ζ) termed coat proteins (COPs).[1,2] During the process of binding to membranes, coatomer associates with a small molecular weight GTP-binding protein, ADP-ribosylation factor 1 (ARF1), and the complex of coatomer and ARF1 constitutes coat protein I (COPI).[3] COPI was originally thought to coat Golgi-derived vesicles moving in the anterograde direction (toward the plasma membrane).[4,5] Subsequent work revealed that coatomer bound to peptides displaying the C-terminal amino

[1] V. Malhotra, T. Serafini, L. Orci, J. C. Shepard, and J. E. Rothman, *Cell* **58,** 329 (1989).
[2] M. G. Waters, T. Serafini, and J. E. Rothman, *Nature* **349,** 248 (1991).
[3] T. Serafini, L. Orci, M. Amherdt, M. Brunner, R. A. Kahn, and J. E. Rothman, *Cell* **67,** 239 (1991).
[4] L. Orci, D. J. Palmer, M. Amherdt, and J. E. Rothman, *Nature* **364,** 732 (1993).
[5] J. Ostermann, L. Orci, K. Tani, M. Amherdt, M. Ravazzola, Z. Elazar, and J. E. Rothman, *Cell* **75,** 1015 (1993).

acid sequence KKXX (K is lysine and X is any amino acid)[6,7] that is present in the cytoplasmic domains of many transmembrane proteins that are residents of the endoplasmic reticulum (ER). The KKXX motif, and related sequences, are retrieval signals for returning resident ER proteins to the ER should they escape to the Golgi complex.[8–10] The interaction of coatomer with proteins displaying ER retrieval motifs suggested that COPI directs the budding of vesicles trafficking in the retrograde direction, from Golgi membranes to the ER. There is nevertheless evidence that different types of COPI-coated vesicles may be involved in both anterograde and retrograde traffic[11,12]; however, the precise function and directionality of COPI-coated vesicles remains controversial.

This article describes the use of aminoglycoside antibiotics and related compounds that inhibit the binding of coatomer to Golgi membranes *in vitro*[13] and *in vivo*,[14] apparently by interacting with KKXX binding sites on coatomer. These compounds, referred to for convenience as KKXX mimetics, were discovered after it was observed that the aminoglycoside antibiotic neomycin precipitated coatomer from solution with features implying that it cross-linked coatomer molecules to create large insoluble complexes.[13] Neomycin (Fig. 1A) contains three pairs of amino groups, each pair suggesting resemblance to the two ε-amino groups in the lysine residues of the KKXX motif. Dilysine, 2-deoxystreptamine (a constituent of neomycin, Fig. 1B), and a variety of other compounds containing only one pair of amino groups fail to precipitate coatomer but they nevertheless block the ability of neomycin to precipitate coatomer. This suggested that neomycin cross-links coatomer by binding to two or more dilysine binding sites on coatomer.[13] Moreover, compounds that blocked the precipitation of coatomer by neomycin also blocked the binding of coatomer to Golgi membranes *in vitro*. It would be useful to extend the application of KKXX mimetics to intact cells; however, KKXX mimetics containing hydroxyl groups, such as those present in aminoglycoside antibiotics, are not effective inhibitors of coatomer binding to Golgi membranes with intact cells, pre-

---

[6] F. Letourneur, E. C. Gaynor, S. Hennecke, C. Démolliére, R. Duden, S. D. Emr, H. Riezman, and P. Cosson, *Cell* **79,** 1199 (1994).

[7] P. Cosson and F. Letourneur, *Science* **263,** 1629 (1994).

[8] T. Nilsson, M. Jackson, and P. A. Peterson, *Cell* **58,** 707 (1989).

[9] M. R. Jackson, T. Nilsson, and P. A. Peterson, *EMBO J.* **9,** 3153 (1990).

[10] M. Jackson, T. Nilsson, and P. A. Peterson, *J. Cell Biol.* **121,** 317 (1993).

[11] K. Fiedler, M. Veit, M. A. Stamnes, and J. E. Rothman, *Science* **273,** 1396 (1996).

[12] L. Orci, M. Stamnes, M. Ravazzola, M. Amherdt, A. Perrelet, T. H. Söllner, and J. E. Rothman, *Cell* **90,** 335 (1997).

[13] R. T. Hudson and R. K. Draper, *Mol. Biol. Cell* **8,** 1901 (1997).

[14] T. Hu, C.-Y. Kao, R. T. Hudson, A. Chen, and R. K. Draper, *Mol. Biol. Cell* **10,** 921 (1999).

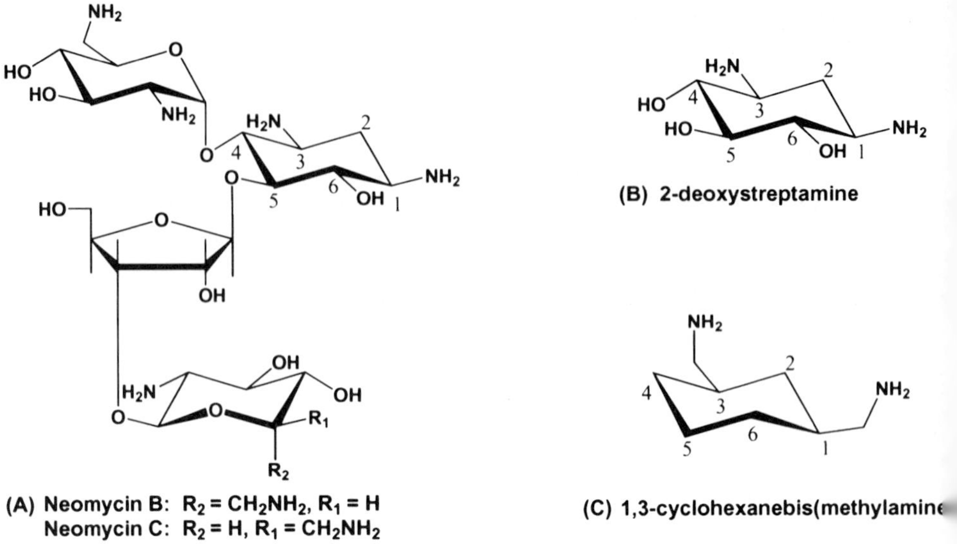

Fig. 1. Structures of (A) neomycin, (B) 2-deoxystreptamine, and (C) 1,3-cyclohexanebis (methylamine).

sumably because they are too hydrophilic to accumulate sufficiently within cells. This led to a search for more hydrophobic KKXX mimetics that would enter cells and resulted in the identification of 1,3-cyclohexanebis (methylamine) (CBM) (Fig. 1C). CBM inhibited coatomer binding to Golgi membranes both *in vitro* and *in vivo* and impaired secretion in live cells.[14] It was also interesting that, unlike brefeldin A, CBM did not block ER-to-Golgi transport nor did it induce Golgi membranes to fuse with the ER.[14] CBM should be useful when it is desired to impair COPI function in cells by a mechanism that is different from that of brefeldin A.

Inhibition by CBM of Neomycin-Induced Precipitation of Coatomer

Neomycin, a multivalent (three pairs of amino groups) KKXX mimetic, precipitates coatomer from solution and the basis of the assay described here is the ability of monovalent (one amino group pair) KKXX mimetics to inhibit this precipitation by competing with neomycin for binding sites on coatomer.[13] Two compounds tested here are dilysine (a positive control) and CBM. The assay can be used to calibrate the activity of commercially available CBM, to screen other compounds for their ability to interact with KKXX binding sites of coatomer, or to study the effect of KKXX mimetics on aspects of membrane traffic amenable to analysis with Golgi membranes *in vitro.*

Procedures

Solutions and Materials

Neomycin (Sigma, St. Louis, MO) stock solution: 100 mM in water adjusted to pH 7.4

1,3-Cyclohexanebis(methylamine) (CBM) (Acros Organics, Pittsburgh, PA) stock solution: 500 mM in water adjusted to pH 7.4

Dilysine (Sigma, St. Louis, MO) solution: 500 mM in water adjusted to pH 7.4

10× HM-S buffer: 250 mM HEPES, pH 7.4, 500 mM KCl, 25 mM magnesium acetate

10× H buffer: 250 mM HEPES, pH 7.4, 250 mM sucrose

Phosphate-buffered saline (PBS): 10 mM Na_2HPO_4/NaH_2PO_4, pH 7.4

Reagents and apparatus for polyacrylamide gel electrophoresis

Reagents and apparatus for electrophoretic transfer of proteins from a polyacrylamide gel to nitrocellulose

Mouse monoclonal antibody M3A5 to β-COP[15]

Horseradish peroxidase-conjugated goat anti-mouse

Supersignal chemiluminescence system for detecting horseradish peroxidase in immunoblots (Pierce Biochemical, Rockford, IL)

Preparation of CHO Cell Cytosol. CHO cells from 20 (15-cm-diameter) plates grown to confluency are trypsinized to detach the cells. Further treatments are all at 4°. The cells are centrifuged, washed once with PBS and once with H buffer, and resuspended in five times the cell pellet volume of H buffer. After 30 min, the cells are homogenized in a 15-ml stainless steel Dounce homogenizer. The homogenate is centrifuged at 100,000g for 1 hr at 4° in a Beckman SW50.1 rotor. The supernatants, which contain soluble coatomer, are pooled and recentrifuged using the same conditions. The final supernatant is desalted over a Sephadex G-25 column equilibrated with HM-S buffer. Protein concentration should be in the range of 5–7.5 mg/ml. The cytosol is quickly frozen with liquid nitrogen in 400-μl aliquots.

Inhibition by CBM of Neomycin-Induced Precipitation of Coatomer from Cytosol. Assays are at 4° in a final volume of 100 μl. To 1.5-ml conical centrifuge tubes are added 10 μl of 10× HM-S, 1 μl of stock neomycin, and volumes of either the dilysine or CBM stock solutions to give the desired concentration ranges. Enough water is added to give a final volume of 100 μl after the cytosol is added and the contents of the tube are mixed.

---

[15] M. Lowe and T. E. Kreis, *J. Biol. Chem.* **270,** 31364 (1995).

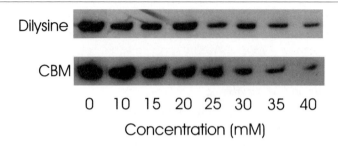

Dilysine

CBM

0 10 15 20 25 30 35 40

Concentration (mM)

Fɪɢ. 2. The inhibition by dilysine and CBM of neomycin-induced precipitation of β-COP from cytosol. Cytosol was treated with neomycin in the presence of dilysine or CBM and precipitated as described in the text. β-COP was detected by immunoblotting.

Prior to addition, the CHO cell cytosol is centrifuged at 14,000 rpm for 30 min in the Eppendorf microfuge (about 16,000g in an F-45-18-11 rotor) to remove coatomer that may have precipitated on freezing and thawing. A volume of cytosol is added to give 1.5 mg/ml of protein, which should bring the final volume to 100 μl, and the contents are mixed again. The samples are incubated for 2 hr at 4° to allow precipitation and then centrifuged again for 30 min at 4° in the microfuge to separate supernatants from precipitates. The supernatant is carefully removed and to the pellets is added 120 μl of Laemmli sample buffer.[16] If desired, the supernatant can also be analyzed for the presence of coatomer, which should decline as coatomer is precipitated. The samples are heated to boiling and electrophoresed in an 8% polyacrylamide gel with sodium dodecyl sulfate. Proteins are transferred to nitrocellulose by electrophoresis and the nitrocellulose is incubated with 0.1% Tween 20 as a blocking agent. Immunodetection of β-COP is with monoclonal antibody M3A5 to β-COP[15] and the secondary antibody is horseradish peroxidase-conjugated goat anti-mouse. The blots are developed with the Pierce Biochemical Supersignal system according to manufacturer instructions. Inhibition by both dilysine and CBM of neomycin-induced precipitation of β-COP is illustrated in Fig. 2.

Comments

The ratio of neomycin concentration (1 mM) to cytosol protein (1.5 mg/ml) used in these assays was empirically determined to give maximal precipitation of coatomer.[13] Changing either the neomycin or protein concentrations will require recalibration of the concentration ratio. The precipitation reaction is sensitive to salt concentration and will need to be recali-

[16] U. K. Laemmli, *Nature* **227,** 680 (1970).

brated if the salt is changed.[13] We have been unable to solubilize neomycin-precipitated coatomer with either high salt concentrations or 0.1% Triton X-100, suggesting that the precipitation reaction is unsuitable for facile purification of coatomer.

Inhibition of Coatomer Binding to Golgi Complex of Intact Cells with CBM

This assay describes treating cells with CBM. For purposes of illustration, the effect of CBM is documented by the loss of δ-COP and the persistence of the KDEL receptor associated with the Golgi complex as monitored by immunofluorescence microscopy. The procedure can be adapted to a variety of biochemical and morphological assays to measure what effects CBM might have on other functions of interest.

Procedures

Solutions and Materials

Vero cells cultured in Dulbecco's modified Eagle's medium (DMEM) with 10% (v/v) dialyzed fetal bovine serum (FBS) according to standard practice

DMEM-T: DMEM lacking sodium bicarbonate, buffered to pH 7.4 with 10 mM TRICINE, supplemented with 10% dialyzed FBS

2 mM CBM in DMEM-T, pH 8.8, prepared fresh each day of use; after addition of CBM, adjust to pH 8.8 with NaOH

4% (w/v) Paraformaldehyde in PBS, pH 7.4

Cold methanol ($-20°$)

PBS–BSA: PBS containing 1% (w/v) bovine serum albumin (BSA)

Polyclonal rabbit anti-δ-COP (kindly provided Drs. C. Harter and F. Wieland, University of Heidelberg, Heidelberg, Germany) diluted 1:100 in PBS–BSA

Monoclonal mouse anti-KDEL (StressGen Biotechnologies, Victoria, BC, Canada) diluted 1:100 in PBS–BSA

Tetramethylrhodamine isothiocyanate (TRITC)-labeled goat anti-rabbit IgG and fluorescein isothiocyanate (FITC)-labeled goat anti-mouse IgG (Sigma, St. Louis, MO)

Light microscope equipped for immunofluorescence microscopy

Assay for CBM-Induced Release of δ-COP from Golgi Complex. Seed Vero cells at a density of 1.5×10^5 cells/well in 24-well culture dishes containing round glass coverslips 24 hr before the experiment. Rinse the cells with PBS and add 0.3 ml of 2 mM CBM in DMEM-T, pH 8.8, to each well. Incubate at 37° for 15–60 min, remove the CBM solution and without washing the cells add 4% paraformaldehyde for 20–30 min. Remove the paraformaldehyde and add cold methanol for 10 min at $-20°$. Rinse the cells three times with PBS and add PBS–BSA for 10 min. Incubate the cells with primary antibodies (mouse anti-KDEL receptor and rabbit anti-δ-COP) for 30 min at room temperature followed by rinsing three times with PBS and incubation for 10 min with PBS–BSA. Incubate the cells with the secondary antibodies for 30 min, followed by the same washing procedure. Remove the coverslips and view with a fluorescence microscope using filters appropriate for TRITC and FITC. In the presence of CBM, δ-COP staining associated with the Golgi complex of Vero cells is greatly reduced, while the KDEL receptor still maintains a Golgi-like distribution (Fig. 3).

Comments

Cultured cells tend to round up during treatment with CBM. The longer the exposure to the drug, the more pronounced the rounding. This problem

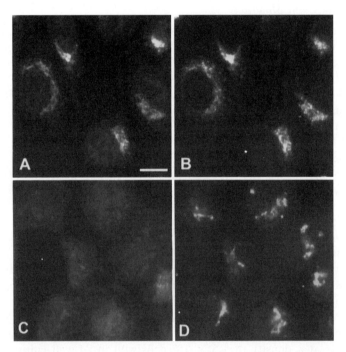

Fɪɢ. 3. The effect of CBM on the distribution of δ-COP and the KDEL receptor in Vero cells. Staining for δ-COP (A and C) and the KDEL receptor (C and D) by double immunofluorescence in the absence (A and B) or the presence (C and D) of 2 mM CBM. Experimental details are given in the text. Bar: 10 μm.

can be reduced by using cells that adhere strongly to slides, such as Vero cells, and by treating the slides with fibronectin or related adherence factors. Another important parameter is the pH of the medium during exposure to CBM. A basic pH is chosen because it improves the uptake of CBM by reducing the charge on the amino groups. In our experience, different cell lines may respond best to a different balance of high pH and CBM concentration. Small changes in pH can be important and it is recommended that a survey of CBM concentrations from 1 to 3 mM at pH values from 8.2 to 8.8 (in 0.2 pH increments) be tested to find the optimum conditions.

Acknowledgments

This work was supported by grants from the National Institutes of Health (GM-34297) and the National Science Foundation (MCB-9513244). We thank Dr. Carole Mikoryak for reading the manuscript.

[40] Adaptor Protein 1-Dependent Clathrin Coat Assembly on Synthetic Liposomes and Golgi Membranes

By Yunxiang Zhu, Matthew T. Drake, and Stuart Kornfeld

Introduction

The assembly of clathrin-coated vesicles on Golgi membranes is initiated by the GTP-binding protein ADP-ribosylation factor (ARF), which generates high-affinity membrane-binding sites for the adaptor protein 1 (AP-1) complex.[1–3] Once bound, the heterotetrameric AP-1 recruits clathrin triskelia, which polymerize to form the coat. The factors and/or properties of the Golgi membrane that act in concert with ARF · GTP to initiate AP-1 binding at the trans-Golgi network (TGN) remain unclear. Putative docking proteins[3,4] or cargo molecules such as the mannose 6-phosphate receptors[5] have been proposed to account for this spatial selectivity.

To understand the AP-1 clathrin coat assembly process more completely, we have developed an *in vitro* assay that uses protein-free liposomes, ARF1, GTP, and cytosol or soluble clathrin coat proteins to reconstitute

[1] M. S. Robinson and T. E. Kreis, *Cell* **69,** 129 (1992).
[2] M. A. Stamnes and J. E. Rothman, *Cell* **73,** 999 (1993).
[3] L. M. Traub, J. A. Ostrom, and S. Kornfeld, *J. Cell Biol.* **123,** 561 (1993).
[4] Y. X. Zhu, L. M. Traub, and S. Kornfeld, *Mol. Biol. Cell* **9,** 1323 (1998).
[5] R. Le Borgne and B. Hoflack, *J. Cell Biol.* **137,** 335 (1997).

AP-1-dependent clathrin-coated vesicle formation.[6] This simple assay allows for the dissection of the contributions of membrane and soluble cytosolic components to coat assembly that is not possible in a reconstituted system using purified Golgi membranes. Specific phospholipid(s), particularly phosphatidylserine, a cytosolic factor(s), and ARF · GTP are the minimal components required for AP-1-mediated clathrin coat assembly.

Preparation of Liposomes

L-α-Phosphatidylcholine from soybeans, containing either 20% phosphatidylcholine (PC) (Sigma, St. Louis, MO) or 40% PC (Sigma) are used to make liposomes designated as 20% or 40% PC liposomes. Liposomes are prepared by dissolving 4 mg of the lipids in 1 ml chloroform in glass tubes. The chloroform is then evaporated under a stream of nitrogen, and the thin film of lipids is hydrated with 1 ml of assay buffer [25 mM HEPES–KOH, pH 7.0, 125 mM potassium acetate, 2.5 mM magnesium acetate, 1 mM dithiothreitol (DTT)].[3] The samples are vortexed several times over a period of 30 min to release the lipids from the wall of the tubes, followed by sonication on ice to opaque translucence, typically by three bursts of 5 sec per burst at a power setting of 3 with a microtip (550 Sonic Dismembrator, Fisher Scientific, Pittsburgh, PA). Care must be taken not to oversonicate the liposomes. If this occurs, the resultant liposomes are still useful for coat binding assays, but are no longer suitable for clathrin-coated vesicle morphology studies by electron microscopy. The final liposomes usually remain in suspension following overnight storage at 4°.

To make 40% PC liposomes supplemented with different phospholipids, phosphatidic acid (PA) or the dioleoyl forms of either phosphatidylserine (PS) or phosphatidylinositol (PI) are mixed with the 40% PC preparation in chloroform at different proportions according to the weight of each lipid. In other reports,[7,8] the techniques used for liposome preparation are slightly different. Liposomes prepared according to other described methods should work as well. It is the lipid composition of the liposome that determines the efficiency of coat assembly.

[6] Y. X. Zhu, M. T. Drake, and S. Kornfeld, *Proc. Natl. Acad. Sci. U.S.A.* **96**, 5013 (1999).
[7] K. Matsuoka, L. Orci, M. Amherdt, S. Y. Bednarek, S. Hamamoto, R. Schekman, and T. Yeung, *Cell* **93**, 263 (1998).
[8] K. Takei, V. Haucke, V. Slepnev, K. Farsad, M. Salazar, H. Chen, and P. De Camilli, *Cell* **94**, 131 (1998).

Preparation of Cytosols

Bovine adrenal glands or rat livers are obtained fresh at a local abattoir or from Sprague-Dawley rats, respectively. Tissues are minced with a razor blade and homogenized at 4° in homogenization buffer (25 mM HEPES–KOH, pH 7.4, 250 mM sucrose, 2 mM EDTA) using a Polytron homogenizer for 3 × 15-sec bursts at a speed setting of 5. A 1 : 2 ratio of tissue weight : buffer volume is used, and protease inhibitors are included at final concentrations of 1 mM phenylmethylsulfonyl fluoride (PMSF), 0.1 U/ml aprotinin, and 5 µg/ml leupeptin. The homogenates are centrifuged at 12,000g for 20 min at 4°. The resulting supernatants are decanted, pooled, and recentrifuged at 100,000g for 1 hr at 4°. The final supernatants are quick frozen on dry ice in small aliqouts and stored at −80°. Prior to each binding assay, cytosols are desalted through PD-10 columns (Pharmacia Biotech, Piscataway, NJ) and recentrifuged at 240,000g for 20 min at 4° in a Beckman tabletop ultracentrifuge (Palo Alto, CA).

Preparation of Clathrin-Coated Vesicles and Clathrin Coat Proteins

Crude clathrin-coated vesicles from fresh rat liver are isolated according to the procedure described by Linder.[9] A final discontinous sucrose gradient step is added to remove contaminating vaults.[10] Briefly, the crude clathrin-coated vesicles are resuspended in buffer A (100 mM 2-[N-morpholino] ethanesulfonic acid (MES), pH 6.5, 0.5 mM MgCl$_2$, 1 mM EGTA, 0.02% NaN$_3$) containing 0.1 mM PMSF using a Dounce homogenizer. Discontinous sucrose gradients are prepared in SW28 tubes by carefully layering 5 ml of 40%, 5 ml 30%, 6 ml 20%, 8.5 ml 10%, and 8.5 ml 5% sucrose solutions prepared in buffer A from bottom to top. The crude clathrin-coated vesicle preparation (4–5 ml) is laid on top of the gradient and centrifuged at 25,000 rpm (100,000g) in an SW28 rotor for 1 hr at 4°. Twenty-six 1.5-ml fractions are collected from the top. Small aliquots from every other fraction are analyzed for clathrin-coated vesicles using SDS–PAGE followed by immunoblotting to detect clathrin and AP-1. The fractions containing the clathrin-coated vesicles (typically fractions 12–21 as numbered from the top of the gradient) are combined, diluted with 3 volumes of buffer A, and collected by centrifugation at 100,000g for 1 hr at 4°. The resultant vesicle pellet is resuspended in buffer A in a Dounce homogenizer, quick frozen on dry ice, and stored in small aliqouts at −80°.

Soluble clathrin coat proteins, mainly clathrin triskelia and adaptor

---

[9] R. Lindner, in "Cell Biology: A Laboratory Handbook" (J. E. Celis, ed.), p. 525. Academic Press, San Diego, 1994.
[10] N. L. Kedersha and L. H. Rome, *J. Cell Biol.* **103,** 699 (1986).

proteins, are released from purified clathrin-coated vesicles with 0.5 M Tris-HCl (pH 7.0) essentially as described by Keen et al.,[11] with a minor modification. The frozen clathrin-coated vesicles in buffer A are thawed and mixed with 0.5 M Tris-HCl (pH 7.0) in 0.5× assay buffer [by mixing equal volumes of 1.0 M Tris-HCl (pH 7.0) and assay buffer] on ice for 10 min, then diluted with 20 volumes of assay buffer. The soluble coat proteins are separated from the residual clathrin-coated vesicle membranes by centrifugation at 240,000g for 20 min at 4°. The coat proteins are stored at a concentration of 15–30 μg/ml.

Preparation of Golgi Membranes

Golgi membranes are prepared according to Leelevathi et al.[12] with slight modifications. Fresh livers excised from Sprague-Dawley rats previously fasted overnight are minced in homogenization buffer (10 mM Tris-HCl, pH 7.0, 5 mM EDTA, 0.5 M sucrose) at a tissue weight-to-buffer ratio of 1:4 (g:ml). The resultant mixture is homogenized using a motor-driven Potter–Elvehjem homogenizer, with 1 mM PMSF, 10 μg/ml leupeptin, and 30 U/ml aprotinin added as protease inhibitors. Approximately five strokes at a speed setting of 5 are used to obtain a homogeneous solution devoid of tissue remnants. The homogenates are centrifuged at 1000g for 10 min at 4°. The supernatants are collected and layered (up to 30 ml) on top of 28 ml of 1.3 M sucrose (all sucrose buffers used for Golgi membrane isolation are prepared in 10 mM Tris HCl, pH 7.0, 5 mM EDTA) in Ti45 tubes and centrifuged at 31,500 rpm (78,000g) for 1 hr at 4°. Membranes at the 0.5/1.3 M sucrose interface are collected, adjusted to 1.2 M sucrose, and approximately 15-ml aliquots transferred to SW28 tubes. Fourteen ml of 1.0 M sucrose and 9 ml of 0.5 M sucrose are carefully overlaid sequentially above each aliquot, and the resultant tripartite preparation is centrifuged at 25,000 rpm (100,000g) for 90 min at 4°. The white, fluffy membranes at the 0.5 M/1.0 M sucrose interface are carefully harvested, and this Golgi membrane-enriched fraction is quickly frozen in small aliquots on dry ice and stored at −80°.

Coat Recruitment Assays and Morphologic Studies by Electron Microscopy

One-stage coat recruitment assays are performed in presiliconized microcentrifuge tubes by mixing either Golgi membranes (50 μg protein/ml)

[11] J. H. Keen, M. C. Willingham, and I. H. Pastan, *Cell* **16**, 303 (1979).
[12] O. E. Leelevathi, L. W. Esters, D. C. Feingold, and B. Lombardi, *Biochem. Biophys. Acta* **211**, 124 (1970).

or 20% PC liposomes (200 μg/ml) with 5 mg/ml cytosol, 4 μM recombinant myristoylated ARF1,[13] and 100 μM GTPγS (all concentrations given are final) and adjusting reactions to a final volume of 200 μl with assay buffer, either with or without soluble clathrin coat protein supplementation to a final concentration of 15 μg/ml. Following incubation at 37° for 15 min, Golgi membranes and liposomes are recovered by centrifugation at 20,000g for 15 min at 4°, and the pellets resuspended by boiling in 50 μl of 1× SDS sample buffer for 5 min. After separation by 7–15% gradient SDS–PAGE and transfer to nitrocellulose membranes, coat proteins are detected with antibodies against specific coat proteins and labeled bands visualized by enhanced chemiluminescence (ECL).

For morphologic studies by electron microscopy (EM), membrane pellets are fixed with 1% glutaraldehyde in 0.1 M sodium cacodylate buffer, pH 7.0, for 1 hr on ice, postfixed with 1% osmium tetroxide (v/v), and contrasted with tannic acid according to Orci et al.[14] Membranes are then embedded in Epon and thin sectioned. The thin sections are further contrasted with uranyl acetate and lead citrate followed by analysis in a Zeiss 902 electron microscope.

The ability of liposomes made from a commercial soybean lipid preparation containing 20% PC to recruit AP-1 and clathrin relative to purified Golgi membranes is seen in Fig. 1. AP-1 and clathrin binding do not occur in the absence of GTPγS (Fig. 1A, lanes 1, 3, 5, 7) with either Golgi membranes or liposomes. On addition of GTPγS, however, the liposomes recruit AP-1 and clathrin almost as efficiently as Golgi membranes (Fig. 1A, lanes 2, 4, 6, 8), indicating that integral membrane proteins are not essential for coat recruitment. When 100 μg/ml brefeldin A, a fungal metabolite known to inhibit the Golgi-associated guanine nucleotide exchange factor (GEF) activity[15] as well as some cytosolic GEFs,[16] is included in the coat recruitment assay with liposomes, AP-1 binding is greatly inhibited, indicating that cytosolic GEF activity is required for maximal ARF–GTP exchange onto liposomes.[16] The residual ARF · GTP binding results from the spontaneous nucleotide exchange that occurs on the liposome surface.[17,18] By preincubating ARF and GTPγS with the liposomes, it is possible to preload the membranes with ARF · GTP.

Thin section EM analysis of liposomes incubated with cytosol supple-

[13] J. O. Liang, T.-C. Sung, A. J. Morris, M. A. Frohman, and S. Kornfeld, J. Biol. Chem. 272, 33001 (1997).
[14] L. Orci, B. S. Glick, and J. E. Rothman, Cell 46, 171 (1986).
[15] P. A. Randazzo, Y. C. Yang, C. Rulka, and R. A. Kahn, J. Biol. Chem. 268, 9555 (1993).
[16] N. Morinaga, J. Moss, and M. Vaughan, Proc. Natl. Acad. Sci. U.S.A. 94, 12926 (1997).
[17] T. Terui, R. A. Kahn, and P. A. Randazzo, J. Biol. Chem. 269, 28130 (1994).
[18] M. Franco, P. Chardin, M. Chabre, and S. Paris, J. Biol. Chem. 270, 1337 (1995).

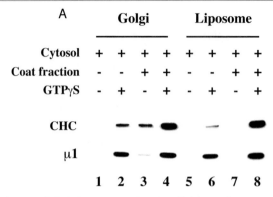

Fig. 1. Recruitment of clathrin coat proteins onto Golgi membranes and liposomes. (A) Golgi membranes and soybean 20% PC liposomes were incubated with 5 mg/ml bovine adrenal cytosol supplemented with 15 μg/ml soluble clathrin coat proteins, 4 μM mARF1, and 100 μM GTPγS as indicated. The recruitment of clathrin (CHC) and AP-1 (μ1) was determined by immunoblotting. (B) Membrane pellets recovered from Golgi membranes and liposomes that had been incubated as above were fixed and processed for electron microscopy as described in the text. Golgi membranes incubated with cytosol and coat fraction in the absence (a) and presence (b) of GTPγS. Liposomes incubated with cytosol and coat fraction in the absence (d) or presence (e and f) of GTPγS. (c) Purified clathrin-coated vesicles from rat liver. Clathrin-coat assembly occurred only in the presence of GTPγS (b, e, and f). Bar: 100 nm.

mented with soluble clathrin coat proteins and ARF · GTPγS reveals that coat assembly proceeds to the complete formation of clathrin-coated buds and vesicles (Fig. 1B, panels e and f), similar to those seen with Golgi membranes (panel b) or purified clathrin-coated vesicles (panel c). This demonstrates that clathrin-coated vesicles can form in the absence of cargo molecules, although the finding does not exclude the possibility that cargo molecules might make the process more efficient. Importantly, vesicle formation occurs to a much greater extent when cytosol is supplemented with the soluble clathrin-coat proteins. Without supplementation, almost no clathrin-coated buds or vesicles are found on either Golgi membranes or liposomes, consistent with the greatly reduced clathrin heavy chain detected in Fig. 1A (lanes 2 vs. 4 and 6 vs. 8). This indicates that the concentration of clathrin, and perhaps AP-1 as well, in the cytosol is limiting under standard assay conditions (5 mg cytosol/ml).

Incubation of the clathrin coat proteins with liposomes plus ARF · GTP in the absence of cytosol does not result in AP-1 or clathrin recruitment, indicating the requirement for a cytosolic factor in addition to ARF.[6] The identity of this cytosolic factor(s) is currently unknown.

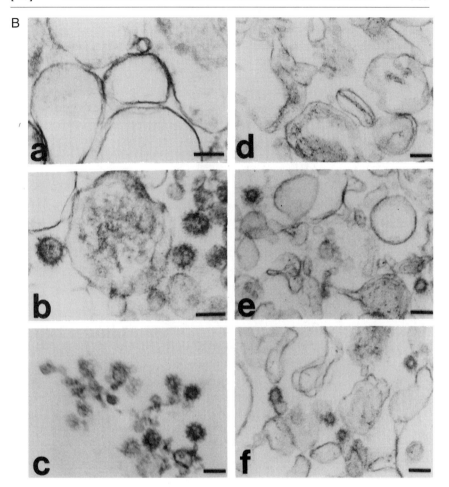

FIG. 1. (*continued*)

Lipid Composition Modulating ARF and AP-1 Binding

The lipid requirements for ARF and AP-1 recruitment have been studied in standard assays using liposomes made from two soybean lipid extracts, one containing 20% PC and the other, 40% PC. Liposomes prepared from the 40% PC material are unable to recruit ARF and AP-1 efficiently compared to liposomes made from the 20% PC source (Fig. 2A), indicating that the purification process has removed or destroyed the specific lipid(s) required for recruitment of this coat material. Analysis of the phospholipid

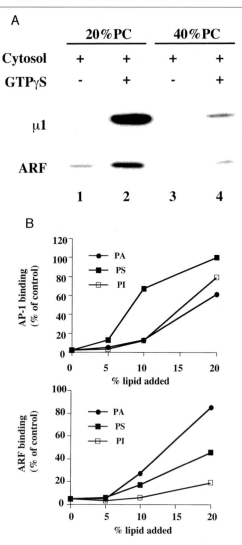

FIG. 2. Lipid-dependent AP-1 recruitment. (A) Liposomes made from soybean lipid fractions containing 20 and 40% PC were incubated with 5 mg/ml gel-filtered rat liver cytosol in the presence or absence of 100 μM GTPγS. The binding of AP-1 and ARF was detected by immunoblotting. (B) Liposomes were prepared from 40% PC soybean material supplemented with 5–20% of PA, PS, or PI and incubated as described in the text. The binding of AP-1 and ARF was determined by immunoblotting and densitometry and was plotted as percentage of binding relative to the 20% PC liposomes.

composition of the two preparations by thin-layer chromatography, reveals that the 40% PC soybean preparation contains significantly less PS and PA as well as lower levels of phosphatidylinositides (data not shown).

To determine whether loss of these lipids accounts for the decreased ARF and AP-1 binding to liposomes made from the 40% PC preparation, various amounts of PS, PI, and PA were added to this material and liposomes were generated. As seen in Fig. 2B, addition of 10% PS increases AP-1 binding sevenfold whereas 10% PA or PI have little effect on AP-1 recruitment. Under these conditions, ARF1 recruitment is only moderately increased. When the lipids are added to final concentrations of 20%, PS restores AP-1 binding to the level obtained with the 20% PC soybean preparation. PA and PI also enhance AP-1 binding, but to a lesser extent. This indicates that PS in the 20% PC soybean liposomes most likely accounts for the efficient recruitment of ARF and AP-1, with PA and PI having more minor roles. Because the 10% PS enhances AP-1 recruitment two- to threefold more than ARF recruitment, this lipid appears to facilitate AP-1 binding beyond the step of ARF recruitment.

Conclusions

It is apparent from these data that AP-1 and clathrin can be recruited onto protein-free liposomes with an efficiency that is comparable to that obtained with purified Golgi-enriched membranes. In both instances, the recruitment of these coat components is absolutely dependent on ARF · GTP and clathrin-coated vesicle formation is observed. The phospholipid composition of the liposomes is critical for efficient AP-1 recruitment, with phosphatidylserine being the most critical component. When soluble clathrin coat proteins are used as a source of AP-1, it can be shown that AP-1 recruitment requires a cytosolic factor(s) in addition to ARF · GTP.[6]

These findings, together with that of Spang et al.,[19] establish that the three coat proteins known to require ARF · GTP for membrane binding, namely, AP-1, AP-3,[20] and COPI, can be recruited onto protein-free liposomes in an ARF · GTP-dependent manner. This system should prove useful for characterizing the various requirements for the recruitment of these different coats.

[19] A. Spang, K. Matsuoka, S. Hamamoto, R. Schekman, and L. Orci, *Proc. Natl. Acad. Sci. U.S.A.* **95,** 11199 (1998).
[20] M. T. Drake, Y. Zhu, and S. Kornfeld, *Mol. Biol. Cell* **11,** (2000).

[41] Receptor-Dependent Formation of COPI-Coated Vesicles from Chemically Defined Donor Liposomes

By WALTER NICKEL and FELIX T. WIELAND

Introduction

COPI-coated vesicles are thought to mediate bidirectional transport in the early secretory pathway (for recent reviews see Refs. 1–3). Much focus has been placed on the minimal requirements for vesicle budding resulting in the identification of the cytosolic factors [the GTPase ADP-ribosylation factor (ARF)[4] and coatomer,[5] a heterooligomeric complex consisting of seven subunits termed α- to ζ-COP] essential for COPI vesicle biogenesis.[6,7] More recently, various nucleotide exchange factors have been discovered that catalyze the conversion of ARF proteins from their inactive GDP-bound form into the active GTP-bound form.[8–12] However, because native Golgi membranes (i.e., consisting of a complex mixture of membrane proteins) are used for the *in vitro* generation of COPI vesicles, it was technically difficult to assess specific requirements for membrane factors in the overall process of vesicle budding.

The p24 family of integral membrane proteins implicated in coated vesicle formation has been discovered recently[13–15] whose members are

[1] M. Lowe and T. E. Kreis, *Biochim. Biophys. Acta* **1404**, 53 (1998).

[2] W. Nickel, B. Brügger, and F. T. Wieland, *Semin. Cell Dev. Biol.* **9**, 493 (1998).

[3] G. Warren and V. Malhotra, *Curr. Opin. Cell Biol.* **10**, 493 (1998).

[4] T. Serafini, L. Orci, M. Amherdt, M. Brunner, R. A. Kahn, and J. E. Rothman, *Cell* **67**, 239 (1991).

[5] M. G. Waters, T. Serafini, and J. E. Rothman, *Nature* **349**, 248 (1991).

[6] L. Orci, D. J. Palmer, M. Amherdt, and J. E. Rothman, *Nature* **364**, 732 (1993).

[7] L. Orci, D. J. Palmer, M. Ravazzola, A. Perrelet, M. Amherdt, and J. E. Rothman, *Nature* **362**, 648 (1993).

[8] P. Chardin, S. Paris, B. Antonny, S. Robineau, S. Beraud-Dufour, C. L. Jackson, and M. Chabre, *Nature* **384**, 481 (1996).

[9] M. Franco, J. Boretto, S. Robineau, S. Monier, B. Goud, P. Chardin, and P. Chavrier, *Proc. Natl. Acad. Sci. U.S.A.* **95**, 9926 (1998).

[10] A. Claude, B. P. Zhao, C. E. Kuziemsky, S. Dahan, S. J. Berger, J. P. Yan, A. D. Armold, E. M. Sullivan, and P. Melancon, *J. Cell Biol.* **146**, 71 (1999).

[11] S. J. Mansour, J. Skaug, X. H. Zhao, J. Giordano, S. W. Scherer, and P. Melancon, *Proc. Natl. Acad. Sci. U.S.A.* **96**, 7968 (1999).

[12] A. Togawa, N. Morinaga, M. Ogasawara, J. Moss, and M. Vaughan, *J. Biol. Chem.* **274**, 12308 (1999).

[13] F. Schimmöller, B. Singer-Krüger, S. Schröder, U. Krüger, C. Barlowe, and H. Riezman, *EMBO J.* **14**, 1329 (1995).

localized to the early secretory pathway.[13–18] The cytoplasmic domains of these proteins were shown to recruit coat proteins,[15,17,19] implying that the corresponding cellular proteins may act as coat receptors directly involved in coated vesicle formation. *In vivo* experiments and the experimental data obtained with subcellular fractions partially enriched in Golgi membranes had suggested a model of COPI vesicle budding, in which ARF binds to the membrane in its GTP-loaded state followed by the recruitment of coatomer. Coatomer binds to ARF–GTP via its β and γ subunits[20,21] and to p24 family proteins via its γ subunit.[22] However, functional proof for a direct role of p24 proteins in COPI vesicle biogenesis was missing until recently.

To analyze the function of membrane factors in a direct way, purified components need to be reconstituted in a synthetic and chemically defined membrane followed by the establishment of the function in question. For example, isolated v- and t-SNARE proteins[23] have been reconstituted in separate populations of synthetic liposomes and shown to catalyze membrane fusion.[24–27] Although several groups have reported lipid compositions of liposomes that promote binding of various kinds of coat proteins resulting in vesicle budding and fission,[28–31] we have recently shown that coated

[14] M. A. Stamnes, M. W. Craighead, M. H. Hoe, N. Lampen, S. Geromanos, P. Tempst, and J. E. Rothman, *Proc. Natl. Acad. Sci. U.S.A.* **92**, 8011 (1995).

[15] K. Sohn *et al., J. Cell Biol.* **135**, 1239 (1996).

[16] M. Rojo, R. Pepperkok, G. Emery, R. Kellner, E. Stang, R. G. Parton, and J. Gruenberg, *J. Cell Biol.* **139**, 1119 (1997).

[17] M. Dominguez, K. Dejgaard, J. Füllekrug, S. Dahan, A. Fazel, J. P. Paccaud, D. Y. Thomas, J. J. Bergeron, and T. Nilsson, *J. Cell Biol.* **140**, 751 (1998).

[18] J. Füllekrug, T. Suganuma, B. L. Tang, W. Hong, B. Storrie, and T. Nilsson, *Mol. Biol. Cell* **10**, 1939 (1999).

[19] K. Fiedler, M. Veit, M. A. Stamnes, and J. E. Rothman, *Science* **273**, 1396 (1996).

[20] L. Zhao, J. B. Helms, B. Brügger, C. Harter, B. Martoglio, R. Graf, J. Brunner, and F. T. Wieland, *Proc. Natl. Acad. Sci. U.S.A.* **94**, 4418 (1997).

[21] L. Zhao, J. B. Helms, J. Brunner, and F. T. Wieland, *J. Biol. Chem.* **274**, 14198 (1999).

[22] C. Harter and F. T. Wieland, *Proc. Natl. Acad. Sci. U.S.A.* **95**, 11649 (1998).

[23] T. Söllner, S. W. Whiteheart, M. Brunner, H. Erdjument-Bromage, S. Geromanos, P. Tempst, and J. E. Rothman, *Nature* **362**, 318 (1993).

[24] T. Weber, B. V. Zemelman, J. A. McNew, B. Westermann, M. Gmachl, F. Parlati, T. H. Sollner, and J. E. Rothman, *Cell* **92**, 759 (1998).

[25] J. A. McNew, T. Weber, D. M. Engelman, T. H. Söllner, and J. E. Rothman, *Mol. Cell* **4**, 415 (1999).

[26] W. Nickel, T. Weber, J. A. McNew, F. Parlati, T. H. Söllner, and J. E. Rothman, *Proc. Natl. Acad. Sci. U.S.A.* **96**, 12571 (1999).

[27] F. Parlati, T. Weber, J. A. McNew, B. Westermann, T. H. Söllner, and J. E. Rothman, *Proc. Natl. Acad. Sci. U.S.A.* **96**, 12565 (1999).

[28] K. Matsuoka, L. Orci, M. Amherdt, S. Y. Bednarek, S. Hamamoto, R. Schekman, and T. Yeung, *Cell* **93**, 263 (1998).

vesicle formation from liposomes is completely independent of any specific lipid requirement given the presence of appropriate coat receptor domains.[32] Here we describe how COPI-coated vesicle budding can be reconstituted employing purified coat proteins and chemically defined liposomes that contain many copies of the cytoplasmic domains of p24 proteins projecting from the membrane surface.

Preparation of Various Components Used in Liposome-Based COPI Budding Assay

As outlined above, cytosolic factors are involved in COPI-coated vesicle formation and are used here as purified proteins. Moreover, the cytoplasmic domains of integral membrane proteins are employed as lipid-anchored peptides and are integrated into liposomes of a defined lipid composition. In the following we describe how these reagents can be prepared and purified.

Preparation of Recombinant Human ARF1

The preparation of recombinant human ARF1 described here is a modification of the procedure described by Franco et al.[33,34]

Solutions

100× Myristate solution: 5 mM myristic acid (sodium salt [Sigma, St. Louis, MO]), 0.6 mM bovine albumin (fatty acid free [Sigma]); sodium myristate is first dissolved in water at 50° followed by its addition to a prewarmed (37°) solution of albumin.

Buffer A: 25 mM Tris-HCl (pH 8.0 at 4°), 50 mM NaCl, 1 mM dithiothreitol (DTT), 2 mM MgCl$_2$, protease inhibitor cocktail tablets (complete) (Roche Diagnostics, Mannheim, Germany, 1 tablet per 50 ml buffer as suggested by the instructions of the manufacturer), 200 μM GDP

Buffer B: 25 mM Tris-HCl (pH 8.0 at 4°), 2.5 mM MgCl$_2$, 1 mM DTT, protease inhibitor cocktail tablets (complete) (1 tablet per 50 ml buffer)

[29] A. Spang, K. Matsuoka, S. Hamamoto, R. Schekman, and L. Orci, Proc. Natl. Acad. Sci. U.S.A. **95**, 11199 (1998).

[30] K. Takei, V. Haucke, V. Slepnev, K. Farsad, M. Salazar, H. Chen, and P. De Camilli, Cell **94**, 131 (1998).

[31] Y. Zhu, M. T. Drake, and S. Kornfeld, Proc. Natl. Acad. Sci. U.S.A. **96**, 5013 (1999).

[32] M. Bremser, W. Nickel, M. Schweikert, M. Ravazzola, M. Amherdt, C. A. Hughes, T. H. Söllner, J. E. Rothman, and F. T. Wieland, Cell **96**, 495 (1999).

[33] M. Franco, P. Chardin, M. Chabre, and S. Paris, J. Biol. Chem. **270**, 1337 (1995).

[34] M. Franco, P. Chardin, M. Chabre, and S. Paris, J. Biol. Chem. **271**, 1573 (1996).

Buffer C: 25 mM Tris-HCl (pH 8.0 at 4°), 2.5 mM MgCl$_2$, 1 mM DTT, protease inhibitor cocktail tablets (complete) (1 tablet per 50 ml buffer), 1 M NaCl

Method. For each preparation, competent cells [*Escherichia coli* strain BL21(DE3)] are cotransformed with bacterial expression vectors containing the full-length human ARF1 cDNA (pET11d[35]) and a cDNA encoding yeast N-myristoyltransferase (pBB131[36]), respectively. Double transformants are obtained by both ampicillin and kanamycin selection. Cells are grown at 30° in NZCYM broth (Sigma) to an optical density of OD$_{600}$ = 0.6 at which time myristate solution (kept at 37°) is added at a final concentration of 1×. After 10 min, protein expression is induced by the addition of isopropylthiogalactoside (IPTG) (1 mM final concentration). Following 3 hr of incubation at 30°, cells are collected by centrifugation at 7000g_{av} for 10 min and the pellet is resuspended in prechilled buffer A (typically cells of a 10-liter culture are resuspended in 200 ml buffer A). All subsequent steps are performed at 4°.

Cells are broken using a cell disrupter (Avestin EmulsiFlex-C5, Ottawa, Canada) at a pressure of 10,000–15,000 psi. The homogenate is centrifuged at 10,000g_{av} for 15 min; the supernatant is carefully recovered and subjected to ultracentrifugation at 100,000g_{av} for 60 min. The resulting supernatant is supplemented with additional GDP at a final concentration of 200 μM.

This protein solution is subjected to ammonium sulfate precipitation at a saturation of 45% (26.2 g/100 ml). After gradual addition of pulverized ammonium sulfate over a period of about 15 min, the suspension is stirred for 60 min. Precipitates are collected by centrifugation at 8000g_{av} for 20 min. The supernatant is carefully removed and the pellet is resuspended in buffer B employing a tight-fitting glass–glass Dounce homogenizer. The suspension is then dialyzed (M_r cutoff: 6000–8000) against buffer B (once against 5 liter for 2 hr, once against 5 liter overnight, and once against 5 liter for additional 2 hr).

The dialyzate is recovered and any remaining unsoluble material is removed by ultracentrifugation (100,000g_{av} for 60 min). The resulting supernatant is loaded onto a HiPrep DEAE agarose column (20-ml bed volume; Pharmacia, Piscataway) equilibrated in buffer B at a flow rate of 2 ml/min. Fractionation (2 ml per fraction) is started on loading. After collecting 40 ml of flow-through (2 bed volumes), a linear gradient (200 ml) from 0 to 1 M NaCl is applied using buffers B and C. Typically, the late fractions of the flow-through contain ARF at a purity of about 75% and a protein

[35] G. Tanigawa, L. Orci, M. Amherdt, M. Ravazzola, J. B. Helms, and J. E. Rothman, *J. Cell Biol.* **123,** 1365 (1993).
[36] R. J. Duronio *et al., Proc. Natl. Acad. Sci. U.S.A.* **87,** 1506 (1990).

concentration of about 0.1–0.2 mg/ml. For further purification, the ARF-containing fractions can be subjected to cation-exchange (Mono S, Amersham-Pharmacia, Uppsala, Sweden) chromatography as described by Franco *et al.*[33] Alternatively, the DEAE fractions are pooled and supplemented with bovine serum albumin (BSA) to prevent precipitation during the following concentration step. A mild procedure to concentrate this fraction is solid polyethylene glycol (PEG) dialysis. The pooled fractions are transferred into a dialysis tubing (M_r cutoff: 6000–8000) and placed into a beaker containing solid PEG (average M_r 35,000). Within approximately 1–2 hr, the fraction can be concentrated up to fivefold (0.5–1 mg/ml; Fig. 1A, lane 2). Small aliquots are flash-frozen employing liquid nitrogen and should be stored at −80°.

Preparation of Recombinant His₆-Tagged ARNO

Preparation of Recombinant His₆-Tagged ARNO

The preparation of the ARF-specific nucleotide exchange factor ARNO[8] is based on a common affinity purification procedure employing a Ni-NTA resin (Qiagen, Hilden, Germany).

Solutions

Solutions

Buffer D: 50 mM Tris-HCl (pH 8.0 at 4°), 300 mM KCl, 5 mM 2-Mercaptoethanol, 10% (w/v) Glycerol, EDTA-free protease inhibitor cocktail tablets (1 tablet per 50 ml buffer)

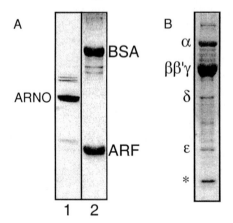

FIG. 1. Protein preparations used for COPI-coated vesicle formation. (A) Lane 1, 1 μl affinity-purified ARNO (2.5 mg/ml); lane 2, 5 μl ARF (substituted with BSA, total protein concentration: 1 mg/ml). (B) 5 μl coatomer, peak fraction of ResQ 10 ml column. (For details see text, 2 mg/ml.) The asterisk indicates a contamination running at the putative position of ζ-COP. In (A), a 12% SDS polyacrylamide gel was used whereas in (B), an 8–16% gradient SDS polyacrylamide gel was used.

Buffer E: 50 mM Tris-HCl (pH 8.0 at 4°), 300 mM KCl, 5 mM 2-mercaptoethanol, 10% (w/v) glycerol, EDTA-free protease inhibitor cocktail tablets (1 tablet per 50 ml buffer), 1 M imidazole

Method. For each preparation, competent cells (*E. coli* strain BL21 [DE3]) are transformed with the bacterial expression vector pET15b containing a cDNA encoding an N-terminally hexahistidine (His$_6$)-tagged version of human ARNO.[37] Cells are grown in superbroth (Q-Biogene, Illkirch, France) at 30° to an optical density of OD$_{600}$ = 0.6. Protein expression is induced by adding IPTG at a final concentration of 1 mM. Cultures are further incubated for 3 hr at 30°. Cells are collected by centrifugation and the sediment is resuspended in buffer D (typically cells of a 10-liter culture are resuspended in a volume of 200 ml). All subsequent steps are performed at 4°.

Cells are broken using a cell disrupter (Avestin EmulsiFlex-C5) at a pressure of 10,000–15,000 psi. The homogenate is centrifuged at 10,000g_{av} for 15 min; the supernatant is carefully recovered and subjected to ultracentrifugation at 100,000g_{av} for 60 min. The resulting supernatant is supplemented with Ni-NTA beads (4 ml packed beads for a 10-liter culture) equilibrated in buffer D and the suspension is incubated overnight on a rotating wheel.

Ni-NTA beads are collected by a low-speed centrifugation and then resuspended in fresh buffer D. The suspension is poured into a chromatography column and connected to a low-pressure chromatography system (Pharmacia). The resin is washed with buffer D until a stable baseline (OD$_{280}$) is observed. Nonspecifically bound protein is eluted with 95% buffer D/5% buffer E (imadazole: final concentration 50 mM). Protein specifically bound to the Ni-NTA resin is eluted applying a steep gradient (6 bed volumes) ranging from 50 to 1000 mM imidazole. One-ml fractions are collected and analyzed by SDS–PAGE/Coomassie blue staining. The peak fractions (His$_6$-tagged ARNO typically elutes at about 150 mM imidazole) are pooled and divided into small aliquots. Following flash-freezing with liquid nitrogen, samples are stored at −80°. Typically, this procedure yields about 20 mg recombinant ARNO (from a 10-liter culture) with a purity greater than 90% (Fig. 1A, lane 1) and a total protein concentration of about 2.5 mg/ml.

Preparation of Coatomer from Rabbit Liver

The preparation of mammalian coatomer from rabbit liver tissue described here is a modification of the procedure described earlier.[38]

[37] E. Mossessova, J. M. Gulbis, and J. Goldberg, *Cell* **92,** 415 (1998).
[38] J. Pavel, C. Harter, and F. T. Wieland, *Proc. Natl. Acad. Sci. U.S.A.* **95,** 2140 (1998).

Solutions

Buffer F: 25 mM Tris-HCl (pH 7.4 at 4°), 500 mM KCl, 250 mM sucrose, 2 mM EGTA, 1 mM DTT, protease inhibitor cocktail tablets (complete) (1 tablet per 50 ml buffer)

Buffer G: 25 mM Tris-HCl (pH 7.4 at 4°), 200 mM KCl, 1 mM DTT, 10% (w/v) glycerol, protease inhibitor cocktail tablets (complete), (1 tablet per 50 ml buffer)

Buffer H: 25 mM Tris-HCl (pH 7.4 at 4°), 500 mM KCl, 1 mM DTT, 10% (w/v) glycerol

Buffer J: 25 mM Tris-HCl (pH 7.4 at 4°), 1 mM DTT, 10% (w/v) glycerol

Buffer K: 25 mM Tris-HCl (pH 7.4 at 4°), 1000 mM KCl, 1 mM DTT, 10% (w/v) glycerol

Method. Typically, four rabbit livers are needed for a coatomer preparation and should be kept in ice-cold buffer F until further processed. All steps are carried out at 4°. The liver tissue (30–40 g) is cut into small pieces using a scalpel or a razor blade. After addition of 200 ml buffer F the tissue is homogenized using a Waring blender (3 × 20 sec, in between keep sample on ice for 2 min).

The homogenate is centrifuged at $10,000g_{av}$ for 1 hr followed by filtration through cheesecloth. The sample is then subjected to two rounds of ultracentrifugation at $100,000g_{av}$ for 1 hr. The supernatant is carefully recovered and filtrated through cheesecloth.

The resulting preparation of cytosolic proteins is diluted with buffer F to a final volume of 500 ml and then fractionated by ammonium sulfate precipitation. Pulverized ammonium sulfate is added gradually up to 35% saturation (20.9 g/100 ml) and the suspension is stirred for 60 min. Precipitated protein is collected by centrifugation at $8000g_{av}$ for 30 min. The sediment is resuspended with 100 ml buffer G employing a tight-fitting glass–glass Dounce homogenizer. After transferring the sample to a dialysis tubing (M_r cutoff: 12,000–14,000), it is dialyzed against 5 liters buffer G lacking protease inhibitors. The buffer is changed twice with one interval proceeding overnight.

The dialyzate is recovered and subjected to ultracentrifugation at $100,000g_{av}$ for 60 min. The resulting supernatant is loaded onto a chromatography column containing DEAE Sepharose Fast Flow (Pharmacia; bed volume 500 ml) equilibrated in buffer G (without protease inhibitors). The column is operated at a flow rate of 10 ml/min. Following the loading procedure, the resin is washed with buffer G (without protease inhibitors) until the OD_{280} value (measured against buffer F) is below 0.2 (requires approximately 1.5 liters of buffer F). Bound protein is eluted with buffer H at a flow rate of 10 ml/min and 10-ml fractions are collected. Fractions characterized by an OD_{280} value >0.25 are pooled (typically 500 ml) and diluted with buffer J to lower the salt concentration to 200 mM NaCl. This procedure is monitored by use of a conductivity meter comparing 1/100 dilutions of the sample and buffer G (conductivity ≈280 μS/cm).

Using a high-performance pump (P-500, Pharmacia), the sample is loaded onto a FPLC (fast protein liquid chromatography) chromatography column containing 10 ml Resource Q resin (Pharmacia) equilibrated in buffer G. The column is operated at a flow rate of 2 ml/min. Following a wash procedure applying 100 ml buffer G, protein is eluted with 200 ml salt gradient ranging from 200 to 1000 mM NaCl using buffers G and K. Fractions of 2 ml are collected. Coatomer elutes at a NaCl concentration of about 350–400 mM. The corresponding fractions are analyzed by SDS–PAGE/Coomassie blue staining and Western blotting employing sub-unit-specific antibodies (e.g., anti-β'-COP C1BL[39] or anti-β-COP M3A5[40]) directed against coatomer. Typically, the peak fractions contain coatomer with a purity greater than 80% (Fig. 1B). However, to improve the total yield, all fractions containing significant amounts of coatomer are pooled and diluted with buffer J to give a NaCl concentration of about 200 mM (again monitored by conductivity measurements, see above).

Following ultracentrifugation at 100,000g_{av} for 1 hr, the sample is loaded onto a FPLC chromatography column containing 1 ml Resource Q resin (Pharmacia) equilibrated in buffer G. The column is operated at a flow rate of 0.5 ml/min. After washing with 5 ml buffer G, protein is eluted applying a steep salt gradient ranging from 200 to 1000 mM NaCl in a total volume of 5 ml. Fractions of 250 μl are collected and analyzed by SDS–PAGE and Coomassie blue staining. The peak fractions are pooled and divided into small aliquots. Following flash-freezing using liquid nitrogen, samples are stored at $-80°$. The protocol described here yields coatomer with a purity of 50–70% at a total protein concentration of about 5 mg/ml.

Preparation of p23 Lipopeptide

To generate a molecule containing a cysteine-modified peptide cova-lently linked to the headgroup of a lipid, the commercially available phos-phatidylethanolamine (PE) derivative 1,2-dioleoyl-sn-glycero-3-phospho-ethanolamine-n-[4-(p-maleimidophenyl)butyramide] (MPB-PE; Avanti Polar Lipids, Birmingham, AL) is used that contains a functional maleimido group. Under suited conditions in homogeneous solution, the maleimido group spontaneously reacts (based on a nucleophilic attack) with a free

[39] G. Stenbeck, C. Harter, A. Brecht, D. Herrmann, F. Lottspeich, L. Orci, and F. T. Wieland, *EMBO J.* **12,** 2841 (1993).
[40] V. J. Allan and T. E. Kreis, *J. Cell Biol.* **103,** 2229 (1986).

thiol group to form a stable thioether (Fig. 2A). The lipopeptide can be purified and incorporated into liposomes that present the peptide on both surfaces of the membrane. The protocol described here contains both improvements and simplifications as compared to our original procedure.[32]

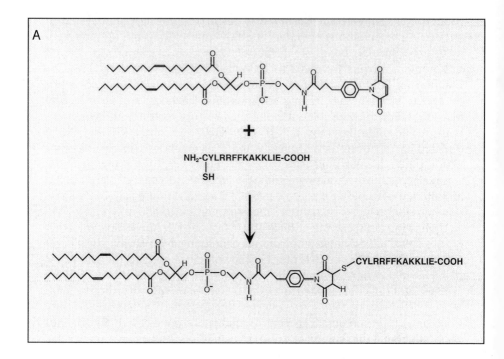

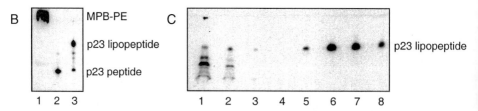

Fig. 2. Synthesis of p23 lipopeptide. (A) Structure of MPB-PE, p23 peptide, and p23 lipopeptide (for further details see text). (B) Analysis of the coupling reaction employing thin-layer chromotography (TLC). Lane 1, 40 nmol MPB-PE; lane 2, 40 nmol p23 peptide; lane 3, 5% of a 500-nmol (based on MPB-PE) incubation. (C) Purification of p23 lipopeptide employing reversed-phase chromatography on C18. Lane 1, flow through; lane 2, 30% AN wash; lane 3, 40% AN wash; lane 4, 60% AN wash; lane 5, 80% AN wash; lanes 6 and 7, 90% AN wash; lane 8, 100% AN wash. Five percent of each fraction was subjected to TLC analysis. Samples were visualized by iodine.

Solutions

MPB-PE stock solution in chloroform: 25 mg/ml (24.75 mM)

p23 Peptide stock solution in dimethylformamide (DMF): 4 mg/ml (2.2 mM) (prepare freshly prior to use)

Method. The cysteine-containing p23 peptide[32] (corresponding to the cytoplasmic domain of the COPI coat receptor p23) is first analyzed with respect to a potential oxidation of the thiol group needed for coupling to MPB-PE. This can be achieved employing Ellman's reagent (Pierce, Rockford, IL) according to the instructions of the manufacturer. Typically, 60–80% (dependent on storage time and conditions) of peptide were found to have a functional thiol group.

The coupling reaction of MPB-PE and cysteine-containing p23 peptide is performed in glass reaction vials (Pierce) because organic solvents are used. The glass vial should be rinsed prior to use with chloroform in order to remove any contaminations.

500 nmol of MPB-PE (~20 μl of the MPB-PE stock solution) is added to the glass vial (using Drummond pipettes equipped with glass capillaries) and solvent is removed applying a gentle stream of nitrogen. The glass vial is then kept under vacuum for 15 min to remove any residual chloroform. Depending on the degree of oxidation of the p23 peptide, an appropriate volume (e.g., 333 μl of peptide stock solution with 75% functional thiol groups) of p23 peptide stock solution is added to the lipid film. This amount corresponds to a 10% excess of peptide (based on functional thiol groups) over MPB-PE. Typically, the mixture clears up rapidly and is further incubated for at least 4 hr at room temperature (the reaction may proceed overnight) on a rotating wheel. Any remaining free maleimido groups are quenched by adding 10 μl of 2-mercaptoethanol (100 mM). Solvent is removed by evaporation and the sample can be stored under argon at $-80°$.

The lipopeptide is purified employing Sep-Pak C_{18} cartridges (1-ml bed volume; Waters, Milford, MA) equilibrated in 30% acetonitrile (AN)/0.1% trifluoroacetic acid (TFA). The cartridge should be first treated with 100% AN followed by the addition of 30 bed volumes of 30% AN/0.1% TFA. The dried sample is dissolved in 1 ml 30% AN/0.1% TFA and loaded onto the resin using a syringe. Bound material is eluted by applying 2 ml 30% AN/0.1% TFA, 2 ml 40% AN/0.1% TFA, 2 ml 60% AN/0.1% TFA, 2 ml 80% AN/0.1% TFA, 4 ml 90% AN/0.1% TFA, and 2 ml 100% AN/0.1% TFA. Two-ml fractions are collected and solvent is removed by evaporation. Dried material of each fraction is dissolved in 200 μl 30% AN/0.1 TFA (i.e., 10× concentrated) and 10 μl are analyzed by thin-layer chromatography (Fig. 2C). A mixture of butanol, pyridine, acetic acid (glacial), and water (9.7:7.5:1.5:6; v:v:v:v) is used as solvent system with standard silica gel 60 plates (20 × 20 cm) equipped with a concentration zone (Merck, Darmstadt, Germany). Plates are heated (10 min at 180°) prior to use and the chamber containing the solvent system is equilibrated for at least 30 min. Silica plates are developed for approximately 3 hr. Remaining solvent is removed by keeping the plate in a fume hood for at least 1 hr. Plates are placed in a chamber containing solid iodine that stains MPB-PE, p23 peptide, and lipopeptide.

Most of the lipopeptide is typically eluted at an acetonitrile concentration of 90% (see Fig. 2C). These fractions are pooled and an aliquot (10%) is subjected to phosphate determination as described.[41] Potassium dihydrogen phosphate is used to generate a standard curve that should range from 0 to 80 nmol phosphate. According to the amounts of lipopeptide determined (typically, the yield of lipopeptide based on MPB-PE input is >50%), the sample is divided into aliquots containing 15 nmol lipopeptide and solvent is removed by evaporation. The dried lipopeptide should be stored under argon at −80°.

The procedure described here has been successfully applied for use with other peptides. However, depending on the chemical properties of the peptide, it may be necessary to make solvent adjustments in order to ensure a homogeneous solution during coupling of MPB-PE and the peptide in question.

Preparation of p23 Lipopeptide-Containing Liposomes

Here we describe the preparation of liposomes consisting of a lipid composition similar to that of Golgi membranes.[42] However, as shown previously, COPI-coated vesicle formation from liposomes does not depend on specific lipid requirements provided the presence of appropriate COPI coat receptor domains.[32]

Solutions. Golgi lipid mix stock solution (all lipids are purchased from Avanti Polar Lipids) final lipid concentration: 3 mM in CHCl$_3$ (150 μmol/ 50 ml)

> Phosphatidylcholine (PC) (from bovine liver, molecular mass 786.12 g/mol): 43 mol% (64.5 μmol): 50.7 mg/50 ml
> Phosphatidylethanolamine (PE) (from bovine liver, molecular mass 768.06 g/mol): 19 mol% (28.5 μmol): 21.9 mg/50 ml
> Phosphatidylserine (PS) (from bovine brain, molecular mass 810.03 g/ mol): 5 mol% (7.5 μmol): 6.1 mg/50 ml
> Phosphatidylinositol (PI) (from bovine liver, molecular mass 909.12 g/mol): 10 mol% (15 μmol): 13.6 mg/50 ml

[41] G. Rouser, S. Fleischer, and A. Yamamoto, *Lipids* **5**, 494 (1970).
[42] G. van Meer, *Trends Cell Biol.* **8**, 29 (1998).

Sphingomyelin (SM) (from bovine brain, molecular mass 703.44 g/ mol): 7 mol% (10.5 μmol): 7.4 mg/50 ml

Cholesterol (from wool grease, molecular mass 386.66 g/mol): 16 mol% (24 μmol): 9.3 mg/50 ml

Buffer L: 50 mM HEPES-KOH (pH 7.4), 100 mM KCl

Method. Liposomes are prepared in glass tubes rinsed with chloroform prior to use. To prepare the lipid film, 100 μl of lipid stock mix (300 nmol total lipid) is added to 15 nmol of dried lipopeptide followed by transfer to the glass tube. The solvent is removed by applying a gentle stream of nitrogen. The glass tube is kept under a vacuum for at least 15 min in order to remove residual solvent. Liposomes are formed by adding 300 μl buffer L and the sample is vortexed for 15 min at room temperature. The liposome suspension is then subjected to five cycles of freezing (use dry ice in 2-propanol) and thawing. In case of visible aggregates, the sample can be subjected to sonication.

To generate a liposome preparation consisting of vesicles characterized by a homogeneous size distribution, the suspension is extruded through polycarbonate membranes of a defined pore size (Avestin extruder). In case of the liposome budding assay, filters with a pore diameter of 400 nm are used, which typically results in liposomes with an average diameter of about 300 nm as determined by EM.[32] The standard procedure involves 21 passages through two membranes. The liposome suspension is recovered and subjected to centrifugation for 5 min at 15,000g_{av}. The supernatant is transferred into a new test tube and the liposome preparation is stored at 4°. Lipid recovery is determined by either including a tritiated lipid tracer ([^{3}H]PC) in the lipid stock solution or by phosphate determination.[41]

p23 Lipopeptide-Dependent Formation of Liposome-Derived COPI Vesicles[32]

Solutions

Buffer M: 250 mM HEPES–KOH (pH 7.4), 25 mM MgCl$_2$, 1000 mM KCl, 10 mM ATP, 10 mM DTT

Buffer N: 25 mM HEPES–KOH (pH 7.4), 100 mM KCl, 2.5 mM MgCl$_2$

Buffer O: 100 mM K$_2$HPO$_4$/KH$_2$PO$_4$ (pH 7.4)

Buffer P: 100 mM sodium cacodylate, (pH 7.2, pH adjusted with HCl)

GTP or GTPγS stock solution: 8 mM in 25 mM HEPES–KOH, pH 7.5

Liposomes: 1 mM lipid concentration including 5 mol% lipopeptide

hARF1: 0.5 mg/ml

Rabbit liver coatomer: 5 mg/ml

His$_6$-ARNO: 0.25 mg/ml

Incubation mixture: 4 μl buffer M, 1 μl GTP or GTPγS (final concentra-

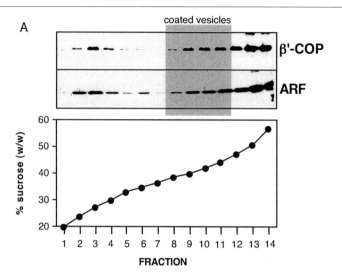

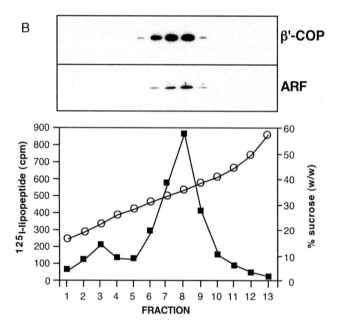

tion 200 μM), 20 μl liposomes (final lipid concentration 500 μM), 5 μl ARF (final concentration 0.0625 mg/ml), 5 μl coatomer (final concentration 0.625 mg/ml), 2 μl ARNO (final concentration 0.0125 mg/ml), 3 μl water (Millipore, Bedford, MA)

Method. The incubation mixture is prepared in the order of addition indicated above. The sample is incubated for 30 min at 37° to allow budding to occur followed by incubation on ice for 5 min. The sample is adjusted to 55% sucrose (w/w) in a final volume of 500 μl and placed on the bottom of a Beckman SW55 centrifugation tube. The gradient is prepared by carefully adding sucrose solutions of the following concentrations: 500 μl 50% (w/w), 500 μl 45% (w/w), 500 μl 40% (w/w), 1 ml 35% (w/w), 1 ml 30% (w/w), and 1 ml 20% (w/w). All sucrose solutions are prepared in buffer N.

Flotation of membranes is achieved by ultracentrifugation at 150,000g_{av} for 18 hr at 4°. Under these conditions, large protein-free liposomes float to the top part of the gradient whereas soluble protein or protein precipitates stay at the bottom of the tube. The gradient is designed in a way that allows for separation of coated vesicles from donor liposomes and protein not associated with membranes. Fractionation is performed from the top into five fractions of 500 μl, eight fractions of 250 μl, and one fraction of 500 μl, which represents the original load.

For a second round of gradient centrifugation, fractions 8 to 11 (containing liposome-derived COPI vesicles; see Figs. 3 and 4) are pooled and subjected to sedimentation analysis. This procedure allows us to verify the purity of the coated vesicle fraction as well as to analyze the buoyant density of liposome-derived COPI vesicles. Gradients are prepared in Beckman SW41 centrifugation tubes using the following solutions: 1 ml of 65% (w/w) sucrose, 1 ml 50% (w/w) sucrose, 1 ml 45% (w/w) sucrose, 1 ml 40% (w/w) sucrose, 2 ml 35% (w/w) sucrose, 2 ml 30% (w/w) sucrose, and 2 ml of sample adjusted to 20% (w/w) sucrose. All sucrose solutions are prepared in buffer N. After ultracentrifugation at 150,000g_{av} for 18 hr at 4°, 13 fractions of 750 μl are collected from the top of the gradient. To characterize gradient fractions (Fig. 3), aliquots are TCA-precipitated and subjected to

---

FIG. 3. Biochemical analysis of budding incubations employing gradient centrifugation. (A) Analysis of the flotation gradient. Fifty percent of each fraction was TCA-precipitated and subjected to 12% SDS–PAGE/Western blot analysis. Coated vesicles migrate in fractions 8 to 11. (B) Analysis of the sedimentation gradient. Fractions 8 to 11 (highlighted by the transparent box in part A) of the flotation gradient were pooled and subjected to sedimentation analysis. Seventy-five percent of each fraction was TCA-precipitated and subjected to 12% SDS–PAGE/Western blot analysis. Coated vesicles migrate in fractions 7 to 9 with an apparent buoyant density of about 40% sucrose. As outlined in detail in Ref. 32, for these kinds of experiments lipopeptide modified with ^{125}I can be used as membrane marker.

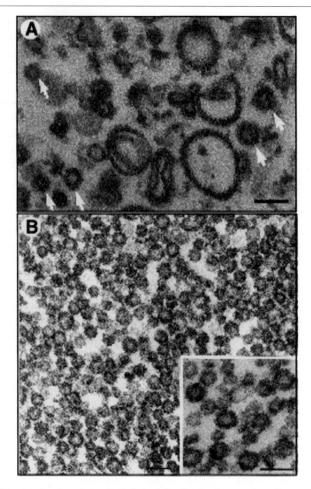

Fig. 4. Ultrastructural analysis of whole budding incubations and gradient-purified coated vesicles. (A) Whole budding incubation with p23-lipopeptide-containing liposomes, coatomer, and ARF–GTPγS incubated at 37°. (B) Gradient-purified liposome-derived COPI vesicles (fractions 8 to 11 of the flotation gradient; see Fig. 3). Electron micrographs were generated as described in the text. Bar: 100 nm.

13% SDS–PAGE/Western blot analysis employing polyclonal antibodies against β'-COP (C1BL[39]) and ARF (2048[43]).

For ultrastructural analysis of whole incubations or gradient fractions (Fig. 4) samples are diluted in buffer O followed by the addition of glutaral-

[43] J. B. Helms, D. J. Palmer, and J. E. Rothman, *J. Cell Biol.* **121,** 751 (1993).

dehyde at a final concentration of 1% (w/v). Fixed material is collected by ultracentrifugation at 100,000g_{av} for 60 min at 4°. Sediments are washed with buffer P and then incubated with 0.5% tannic acid for 60 min at 4°. After washing with buffer P, postfixation with 2% osmium tetroxide is carried out at room temperature. For dehydration samples are treated with increasing concentrations of ethanol (30, 50, 70, 90% and 2 × 100%) followed by incubation with 100% acetone for 15 min. Fixed material is embedded in Spurr's resin (Sigma) and incubated for 48 hr at 60° to allow for polymerization. Ultrathin sections are prepared and stained with 1% uranyl acetate and 1% lead citrate. Samples are analyzed with a transmission electron microscope.

As shown in our original report,[32] formation of COPI-coated vesicles from liposomes depends on elevated temperature, ARF–GTP, and p23 or p24 lipopeptide with functional coatomer binding motifs. The ARF-specific nucleotide exchange factor ARNO was not found to be a requirement consistent with the observation that, under the conditions used, spontaneous nucleotide exchange onto ARF occurs to a significant extent.[32] Once the assay has been further developed in terms of quantitative analysis, it might be worth to study a potential stimulatory effect of ARF-specific nucleotide exchange factors more closely.

Liposome-derived COPI-coated vesicles have a buoyant density corresponding to approximately 40% (w/w) sucrose (fractions 8–11 in Fig. 3A and fractions 7–9 in Fig. 3B). Ultrastructural analysis of their morphology (Figs. 4A and 4B) reveals that they are strikingly similar when compared to their native counterparts[44,45] (e.g., Golgi-derived COPI-coated vesicles).

Concluding Remarks

The basic machinery needed for the formation of COPI-coated vesicles has been identified as employing a reconstituted system that is entirely based on purified components. Only the coat proteins ARF (in its GTP-bound form) and coatomer as soluble molecules as well as the cytoplasmic domains of p23 or p24 presented on the membrane surface of a liposome are required to promote the formation of a synthetic COPI vesicle.

The basic protocol reported here now allows the development of many variations of this system that can help to solve important questions concerning COPI vesicle function. Because two populations of COPI vesicles implicated in anterograde and retrograde transport, respectively, have been

[44] L. Orci, B. S. Glick, and J. E. Rothman, *Cell* **46,** 171 (1986).
[45] V. Malhotra, T. Serafini, L. Orci, J. C. Shepherd, and J. E. Rothman, *Cell* **58,** 329 (1989).

demonstrated,[46] it is now a major challenge to identify the machinery that governs this segregation. Because p23 and p24 have been found to contain opposite sorting signals (i.e., retrograde and anterograde, respectively),[17,19,47] they are obvious candidates. Moreover, other machinery molecules implicated in COPI-mediated transport such as the KDEL receptor,[46] ERGIC 53,[48] and v-SNAREs[23] can be studied in this system. For the COPII system, it has already been shown that coat assembly alone can direct incorporation of v-SNAREs into liposome-derived COPII vesicles.[49] A most important aspect now is to study the role of appropriate v-SNAREs such as GOS28[50] in the COPI system, because they may directly be involved in the segregation process of functionally distinct COPI vesicle populations.

Another aspect of COPI vesicle biogenesis is uptake into these carriers of anterograde and retrograde cargo, which has been shown to be affected by reagents interfering with ARF-catalyzed GTP hydrolysis.[51–53] Because ARF has almost no detectable intrinsic GTPase activity,[54,55] ARF–GAP[55,56] activity appears to be needed to promote cargo incorporation.[52] However, it is not yet known whether other cytosolic factors are required for this process as well. It would, therefore, be interesting to reconstitute cargo uptake employing the liposome system in order to define the minimal machinery required for this process.

Acknowledgments

We thank Anja Zeigerer (Cellular Biochemistry and Biophysics Program, Memorial Sloan-Kettering Cancer Center) for providing data used in Fig. 2.

[46] L. Orci, M. Stamnes, M. Ravazzola, M. Amherdt, A. Perrelet, T. H. Söllner, and J. E. Rothman, *Cell* **90**, 335 (1997).

[47] W. Nickel, K. Sohn, C. Bunning, and F. T. Wieland, *Proc. Natl. Acad. Sci. U.S.A.* **94**, 11393 (1997).

[48] E. J. Tisdale, H. Plutner, J. Matteson, and W. E. Balch, *J. Cell Biol.* **137**, 581 (1997).

[49] K. Matsuoka, Y. Morimitsu, K. Uchida, and R. Schekman, *Mol. Cell* **2**, 703 (1998).

[50] M. Nagahama, L. Orci, M. Ravazzola, M. Amherdt, L. Lacomis, P. Tempst, J. E. Rothman, and T. H. Söllner, *J. Cell Biol.* **133**, 507 (1996).

[51] W. Nickel, J. Malsam, K. Gorgas, M. Ravazzola, N. Jenne, J. B. Helms, and F. T. Wieland, *J. Cell Sci.* **111**, 3081 (1998).

[52] J. Malsam, D. Gommel, F. T. Wieland, and W. Nickel, *FEBS Lett.* **462**, 267 (1999).

[53] R. Pepperkok, J. A. Whitney, M. Gomez, and T. E. Kreis, *J. Cell Sci.,* in press (1999).

[54] R. A. Kahn and A. G. Gilman, *J. Biol. Chem.* **261**, 7906 (1986).

[55] P. A. Randazzo and R. A. Kahn, *J. Biol. Chem.* **269**, 10758 (1994).

[56] E. Cukierman, I. Huber, M. Rotman, and D. Cassel, *Science* **270**, 1999 (1995).

[42] ADP-Ribosylation Factor (ARF) as Regulator of Spectrin Assembly at Golgi Complex

By MARIA ANTONIETTA DE MATTEIS and JON S. MORROW

Introduction

First identified in association with the erythrocyte plasma membrane, spectrin is now appreciated as the central component in an ubiquitous and complex system linking membrane proteins, membrane lipids, and cytosolic factors with the major cytoskeletal filament systems of the cell. Its overall role is to provide both order and support to the membrane. Recent studies identify homologs of β-spectrin in association with the Golgi and other organelles of the secretory and endosomal/lysosomal pathways (for reviews, see Refs. 1–3). With these observations has come an understanding that the Golgi spectrin skeleton (like other components of the secretory pathway) is actively regulated by ADP-ribosylation factor (ARF), a small G protein controlling the architecture and dynamics of the Golgi. Understanding the molecular details of how ARF regulates the binding of spectrin to the Golgi has yielded novel insights into both spectrin function and ARF.[4,5]

A methodological approach to the study of ARF and its regulation of spectrin follows three experimental strategies: (1) the use of expressed spectrin peptides in cultured cell lines, observed by indirect immunofluorescent microscopy, to identify the spectrin sequence motifs responsible for targeting it to the Golgi or other organelles *in vivo;* (2) an immunofluorescence analysis of changes in the intracellular distribution of spectrin in permeabilized cultured cells in response to activators (GTPγS) or inhibitors (brefeldin A, BFA) of ARF; and (3) *in vitro* assays of the binding of spectrin or recombinant spectrin peptides to isolated Golgi membranes. Typically, these approaches utilize recombinant βI- or βIII-spectrin peptides representing different functional domains of spectrin [epitope tagged or green fluorescent protein (GFP) labeled], or antibodies specific for βI- or βIII-spectrin. Also required are methods for introducing these peptides into cells, methods for preparing isolated Golgi membranes, and methods for manipulating the activity or abundance of ARF.

[1] K. A. Beck and W. J. Nelson, *Biochim. Biophys. Acta* **1404,** 153 (1998).
[2] M. A. De Matteis and J. S. Morrow, *Curr. Opin. Cell Biol.* **10,** 542 (1998).
[3] M. A. De Matteis and J. S. Morrow, *J. Cell Sci.* **113,** 2331 (2000).
[4] A. Godi *et al., Proc. Natl. Acad. Sci. U.S.A.* **95,** 8607 (1998).
[5] A. Godi *et al., Nat. Cell Biol.* **1,** 280 (1999).

Construction of Plasmids for Expression of Spectrin Fusion Proteins

The overall structure of βIII-spectrin and the sites responsible for some of its functions are depicted in Fig. 1. The residue boundaries chosen for the preparation of spectrin structural repeats should respect the phasing

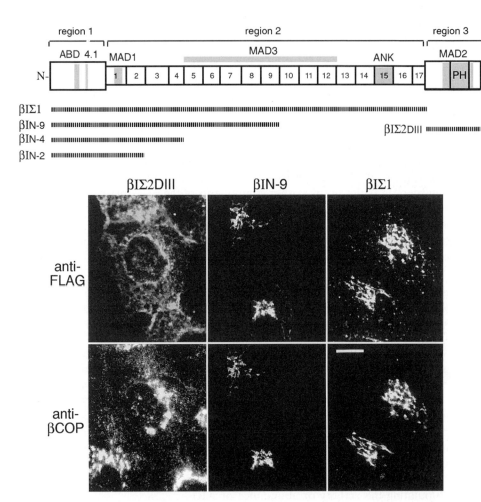

FIG. 1. *Top:* Structure of β-spectrin. Depicted are the two nonhomologous regions at either end of the molecule, along with region 2 composed of 17 homologous ≈106 residue repeat units. The sites of some of the major functional domains are indicated, along with a diagram of the peptides generated from the primers presented in Table I. (MAD1–3, membrane association domains 1–3; ABD, actin binding domain; 4.1, protein 4.1 binding domain; PH, pleckstrin homology domain; ANK, ankyrin binding domain). *Bottom:* FLAG-epitope-tagged spectrin peptides containing different functional domains localize to different regions within these stably transfected MDCK cells. (Adapted with permission from Devarajan *et al.*[17])

of *Drosophila* α-spectrin,[6] a strategy that preserves the basic triple helical folding motif of each ≈106 residue spectrin repeat unit.[7,8] Recombinant glutathione *S*-transferase (GST)-labeled fusion peptides are prepared for prokaryotic expression using the vector pGEX (usually pGEX-2T, Pharmacia, Piscataway, NJ).[9] FLAG- or GFP-labeled fusion peptides for eukaryotic cell expression are respectively prepared with the vectors pcDNAIII 1-3AB or pEGFP-N2 (Clontech, Palo Alto, CA). The vectors pcDNAIII-1-3AB incorporate the FLAG epitope in each of the reading frames at the *Hin*dIII site of the parent vector (pcDNAIII, Invitrogen, Carlsbad, CA), thereby enabling convenient insertion of the FLAG epitope onto any spectrin peptide.[10] Two spectrins most closely associated with the Golgi are βIΣ2-spectrin and βIII-spectrin. Plasmids are therefore typically prepared based on the published sequences of these molecules.[11–13] The desired segment of cDNA is excised by *Bam*HI or other suitable restriction enzymes from pGEM 7z(−), purified from agarose gels by electroelution, and subcloned into the appropriate vector for expression either in *Escherichia coli* or in cultured cells (see below).

Alternatively (and more easily), specific spectrin domains can be generated by PCR (polymerase chain reaction) using oligonucleotide primers that contain the *Eco*RI restriction site plus two extra CG base pairs at the 5′ end (for examples, see Refs. 14–16). These extra bases facilitate efficient digestion by *Eco*RI after amplification. It may be useful if the PCR primers are engineered to have *Bam*HI sites 5′ and *Eco*RI sites 3′ to ease directional subcloning into pGEX-2T; however, because many βI-spectrin peptides contain internal *Bam*HI sites, an alternative and more general strategy is to prepare primers with *Eco*RI sites only, and then just screen for the correct orientation using restriction digests with asymmetric endonucleases.

The oligonucleotides used to generate some of the more interesting spectrin peptides are listed in Table I. With these primer sets, PCR is

[6] E. Winograd, D. Hume, and D. Branton, *Proc. Natl. Acad. Sci. U.S.A.* **88,** 10788 (1991).

[7] Y. Yan *et al., Science* **262,** 2027 (1993).

[8] V. L. Grum, D. Li, R. I. MacDonald, and A. Mondragon, *Cell* **98,** 523 (1999).

[9] D. B. Smith and K. S. Johnson, *Gene* **67,** 31 (1988).

[10] S. A. Weed, P. R. Stabach, C. E. Oyer, P. G. Gallagher, and J. S. Morrow, *Lab. Invest.* **74,** 1117 (1996).

[11] J. C. Winkelmann, F. F. Costa, B. L. Linzie, and B. G. Forget, *J. Biol. Chem.* **265,** 20449 (1990).

[12] J. C. Winkelmann *et al., J. Biol. Chem.* **265,** 11827 (1990).

[13] M. C. Stankewich *et al., Proc. Natl. Acad. Sci. U.S.A.* **95,** 14158 (1998).

[14] S. P. Kennedy, S. L. Warren, B. G. Forget, and J. S. Morrow, *J. Cell Biol.* **115,** 267 (1991).

[15] S. P. Kennedy, S. A. Weed, B. G. Forget, and J. S. Morrow, *J. Biol. Chem.* **269,** 11400 (1994).

[16] C. R. Lombardo, S. A. Weed, S. P. Kennedy, B. G. Forget, and J. S. Morrow, *J. Biol. Chem.* **269,** 29212 (1994).

TABLE I
OLIGONUCLEOTIDE PRIMERS USED TO GENERATE SPECTRIN PEPTIDES

Peptide	Oligo 5'-3'	Template	Region	Strand
βI N-2	cggaattcATGACATCGGCCACAGAG	βI Spectrin	Start region I	Sense
	cggaattcGGTCTCGAGCCTCTGGCG	βI Spectrin	End repeat 2	Antisense
βI N-4	cggaattcATGACATCGGCCACAGAG	βI Spectrin	Start region I	Sense
	cggaattcATCCTGGAGGTTCTTCTT	βI Spectrin	End repeat 4	Antisense
βI N-9	cggaattcATGACATCGGCCACAGAG	βI Spectrin	Start region I	Sense
	cggaattcGTCTCTCAGTAGGACAGA	βI Spectrin	end repeat 9	Antisense
βIE2 Domain III	cggaattcAGACCCGCAGAGGAGACT	βI Spectrin	Start region III	Sense
	cggaattcCTTCTTTTTGGGGAAGAA	βI Spectrin	End region III	Antisense
βIII N-2	cggaattcATGAGCAGCACGCTGTCA	βIII Spectrin	Start region I	Sense
	cggaattcGAGGAGGAGCCGCTCCCG	βIII Spectrin	End repeat 2	Antisense
βIII N-4	cggaattcATGAGCAGCACGCTGTCA	βIII Spectrin	Start region I	Sense
	cggaattcTTGGGCCAGCCGCTGGGC	βIII Spectrin	End repeat 4	Antisense
βIII N-9	cggaattcATGAGCAGCACGCTGTCA	βIII Spectrin	Start region I	Sense
	cggaattcGTCCCGAAGACGGCCCAG	βIII Spectrin	End repeat 9	Antisense
βIII Domain III	cggaattcGAGGAGCGGCGGAAAACAG	βIII Spectrin	Start region III	Sense
	cggaattcCTTGTTCTTCTTAAAGAA	βIII Spectrin	End domain III	Antisense

performed with Platinum *Taq* high fidelity DNA polymerase (Life Technologies, Rockville, MD). The procedure typically utilizes 10 ng of template plasmid in a 50-μl reaction, containing 60 mM Tris–SO$_4$ (pH 8.9), 18 mM ammonium sulfate, 2 mM MgSO$_4$, 0.2 mM dNTP, 0.2 μM each primer, and 2 units of *Taq* polymerase. The PCR conditions are generally 94° denaturation, 50° annealing, and 72° extension for 28 cycles.

After amplification, 30 μl of the PCR reaction is loaded onto a 1% NuSieve GTG low melt agarose gel (BioWhittaker, Rockland, ME) and amplimers of the appropriate size are excised from the gel using a clean razor blade. The DNA is gel purified using a GFX PCR Gel Band Purification Kit from Pharmacia Biotech (Piscataway, NJ). In addition to *Eco*RI digestion, the plasmid vector used in the cloning is dephosphorylated with 1 unit of calf intestinal alkaline phosphatase (New England BioLabs, Beverly, MA), and gel purified as described above. Standard ligation procedures utilize a 3 : 1 molar ratio of insert to vector, using 1 unit of T4 DNA ligase from New England BioLabs. Ligations are designed to allow in-frame fusion at the amino terminus with either the pGEX-2T vector for prokaryotic expression, or with the FLAG epitope tag of vector pcDNAIII 2AB (Invitrogen).[10] Insertion into pEGFP-N2 from Clontech can also be utilized to express spectrin peptides with carboxy-terminal enhanced green fluorescent protein (EGFP). Ligations are screened for insert orientation using restriction mapping, and the sequence of favorable constructs verified to rule out amplification artifacts.

Protein Expression in *Escherichia coli* and in Cultured Cells

Recombinant GST-fusion peptides are expressed in *E. coli* strains HB101, DH5α, or the protease-deficient strain CAG-456, and purified by affinity chromatography on glutathione Sepharose.[9] With many of the spectrin peptides, induction of the GST fusion peptides in bacteria may render the fusion peptide insoluble. This problem can often be alleviated by induction at a lower temperature (30°) and for a shorter time (1 hr), using less isopropylthiogalactoside (IPTG) (0.1 mM). When required, peptides can be further purified by gel filtration on a 1× 100-cm column of Sephacryl S-200 HR in the presence of 1 M NaBr, 10 mM Na$_2$HPO$_4$, 0.5 mM EGTA, 0.5 mM EDTA, 15 mM sodium pyrophosphate, 1 mM NaN$_3$, 0.5 mM dithiothreitol (DTT), pH 8.0.[16] Purified proteins are dialyzed into 20 mM Tris, 150 mM NaCl, 1 mM EDTA, 1 mM NaN$_3$, 1 mM DTT, 0.2 mM phenylmethylsulfonyl fluoride (PMSF), 25% glycerol, pH 7.5, and stored at −80°.

To express spectrin peptides in cultured cells, Madin-Darby canine kidney (MDCK) epithelial cells, NRK cells, or African Green Monkey *Cercopithecus aethiops* (COS7) cells are grown to partial confluence (40–

60%) on either chamber wells from Nunc (Nalge Nunc International, Rochester, NY) or p60 tissue culture dishes. After washing three times with phosphate-buffered saline (PBS) to remove serum, the cells are transfected with 2 μg plasmid DNA using 20 μl of Lipofectamine reagent from Life Technologies and 200 μl serum-free medium (Opti MEM, Life Technologies). After 16 hr at 5% (v/v) CO_2 and 37°, the medium is replaced with normal Dulbecco's modified Eagle's medium (DMEM), 10% fetal bovine serum (FBS), for 24–48 hr. Cells are either used in detection assays directly at this point as transient transfectants, of selected for stable transfectants with G-418 sulfate (geneticin, Life Technologies). For stable selection with G-418 sulfate the cells are released from the p60 dishes with 0.25% trypsin–EDTA, washed once with PBS, and the cell pellet resuspended with 2 ml of media. The cell suspension is then diluted 1:10 and 1:100 into p100 tissue culture dishes containing 40 μg/ml Geneticin. Cells are incubated for 2–3 weeks at 37° and 5% CO_2, changing the media containing G-418 every 3 or 4 days. Untransfected cells will perish; surviving cells are resistant to G-418 sulfate. Individual colonies are expanded by removing all media from the dish and by adding trypsin/EDTA inside a 10 × 10-mm cloning cylinder (Bellco Glass, Inc., Vineland, NJ) prepared with a small amount of autoclaved vacuum grease placed around the perimeter of the colony. In general, it is necessary to capture up to 100 or more distinct colonies to identify cells lines that are viably expressing the fusion gene of interest. Many of the spectrin peptides are potentially toxic to the cells, and therefore stable lines are often difficult to sustain except at very low expression levels.

Antibodies and Immunodetection Procedures

Cell lines selected for G-418 sulfate resistance are assayed for the production of fusion peptide by Western blot analysis. Constructs epitope labeled by the 8-residue FLAG marker (IBI Scientific, Shelton, CT) are detected with monoclonal antibody (MAb) M5 (IBI Scientific). Cell lines expressing EGFP can also be conveniently viewed vitally by epifluorescent microscopy using 480/525-nm excitation/emission filters. Typically, cells are grown on glass coverslips or culture dishes, washed three times with PBS, fixed 15 min in acetone, blocked 30 min in goat serum, and incubated with primary (1°) antibody in 2% BSA in PBS + 10% goat serum at room temperature for 60 min. The bound primary antibody is detected with the appropriate secondary antibodies conjugated to CY3 or CY2 (Vector, Burlingame, CA).

The immunofluorescent distribution pattern of spectrin or different spectrin peptides is quite variable (Fig. 1).[13,17] Some of this variability relates

---

[17] P. Devarajan, P. R. Stabach, M. A. De Matteis, and J. S. Morrow, *Proc. Natl. Acad. Sci. U.S.A.* **94**, 10711 (1997).

to the specific targeting functions of different sequences in the molecule. Additional variability arises from the fact that spectrin is only dynamically associated with the Golgi and other organelles (see below). Thus, unlike resident Golgi proteins, spectrin sometimes appears as concentrated on cytosolic tubular–vesicular structures as it does over the Golgi. Different antibodies to ostensibly the same spectrin also emphasize to some degree different intracellular compartments. The origins of this variability are not yet fully understood, and it remains possible that additional undiscovered forms of spectrin exist.[3] Other factors affecting the intracellular distribution of spectrin relate to the state of cell confluence and possibly the rate of protein flux through the secretory pathway.[18] Finally, when spectrin is expressed at high levels in cultured cells, it is generally toxic. In cells that survive, massive expansions of what appears to be the intermediate compartment can occur.[18] Even when expressed at high levels, however, spectrin does not appear to accumulate in aggresomes,[19] since its concentrations are neither vimentin nor ubiquitin rich.[18] Beyond this, further work is needed to fully understand the nature of the complexes seen at high expression levels.

Immunofluorescence Analysis In Permeabilized Cells

NRK cells are grown in DMEM supplemented with 10% fetal calf serum and plated at 10^5 cells/well on Nunc Chamber slides on the day before the experiment. Cells are incubated with SLO buffer containing 1 U/ml streptolysin O (SLO, Biomerieux, France), 25 mM HEPES–KOH, pH 6.95, 125 mM potassium acetate, 2.5 mM magnesium acetate, 10 mM glucose, 1 mM DTT, for 8 min at 0°. The treatment at 0°, a nonpermissive temperature for the permeabilizing activity of SLO, allows SLO to bind to the plasma membrane but prevents its entry into the cell and any effect on intracellular membranes. The cells are then washed three times in the same buffer (devoid of SLO) to remove the excess of unbound SLO, and then exposed to permeabilization buffer containing 25 mM HEPES–KOH, pH 6.95, 125 mM potassium acetate, 2.5 mM magnesium acetate, 10 mM glucose, 1 mM DTT, 1 mM ATP, 2 mM creatine phosphate, 7.3 IU/ml creatine phosphokinase, and 2 mg/ml rat brain or NRK cell cytosol, at 37° for 15 min. Under these conditions 100% of the cells become permeable to Trypan blue (molecular mass 960 Da) and 80% of the intracellular lactate dehydrogenase (LDH) activity (taken as a marker of cytosolic proteins) is lost in the supernatant. The cells are processed for immunofluorescence by fixation in 1.9% paraformaldehyde at 4° for 1 hr, permeabilized with 0.01% saponin,

[18] Y. Ch'ng, M. C. Stankewich, and J. S. Morrow, submitted (2000).
[19] J. A. Johnston, C. L. Ward, and R. R. Kopito, *J. Cell Biol.* **143**, 1883 (1998).

blocked with 10% normal goat serum for 1 hr, treated with primary anti-spectrin antibodies for 16 hr at 4°, and visualized with the appropriate secondary antibodies as above.

Under these permeabilization conditions the intracellular distribution of membrane Golgi proteins, such as giantin or mannosidase II, is unaffected compared to intact cells, whereas certain peripheral Golgi proteins, such as the coat protein I (COPI) complex (namely, β-COP) and spectrin itself, whose association with the Golgi complex is under the control of the small GTPase ARF, are partially redistributed into the cytosol and eventually lost in the supernatant.[20] The redistribution is likely to be due to an impairment of ARF activation (GDP–GTP exchange) and/or to an acceleration of its inactivation (GTP hydrolysis) since the cytosolic redistribution of both COPI and spectrin is prevented by the irreversible activator of ARF, GTPγS. Indeed, when cell permeabilization is carried out in the presence of 50 μM GTPγS, both spectrin and COPI remain associated with the Golgi complex. The effect of GTPγS on spectrin is mediated by ARF, and not by other G proteins activated by GTPγS, because it is specifically prevented by a pretreatment of the cells with BFA. BFA is a fungal toxin that selectively interferes with nucleotide exchange on ARF but does not affect other G proteins.

In Vitro Binding Assays on Isolated Golgi Membranes

Cytosol and Golgi Preparation

Golgi membrane-enriched fractions are prepared from NRK cells or from rat liver following either of two protocols.[21,22] Subcellular fractions are characterized by Western blot analysis, using specific antibodies to characterize different cellular organelles. The Golgi-rich fractions contain over 90% of the mannosidase II (a Golgi marker), 3% of CaBP1 [a calcium-binding protein that marks the endoplasmic reticulum (ER)] about 10% of the Na$^+$,K$^+$-ATPase in NRK cells (a plasma membrane marker), or in rat liver preparations about 4% of ecto-ATPase HA4 (another plasma membrane marker). Cytosol is obtained from rat brain or from NRK cells.[21]

[20] M. A. De Matteis, G. Santini, R. A. Kahn, G. Di Tullio, and A. Luini, *Nature* **364**, 818 (1993).

[21] V. Malhotra, T. Serafini, L. Orci, J. C. Shepherd, and J. E. Rothman, *Cell* **58**, 329 (1989).

[22] D. E. Leelavathi, L. W. Estes, D. S. Feingold, and B. Lombardi, *Biochim. Biophys. Acta* **211**, 124 (1970).

ARF- and COPI-Depleted Cytosol

The role of ARF in regulating the association of spectrin with the Golgi membranes can be more directly assessed by using an ARF-depleted cytosol, studying the effect of the readdition of purified ARF to the depleted cytosol. ARF depleted cytosol is prepared as published.[23] ARF depletion (loss of the ARF band at $\approx 20,000$ M_r) is verified by SDS–PAGE and Western blot. To achieve a complete depletion of ARF during the binding assay, Golgi membranes must also be washed with 1 M KCl before the incubation, so as to remove the residual ARF that copurifies with the Golgi fractions.

COPI-depleted cytosol is prepared by immunodepletion using the CM1A10 anti-coatomer antibody,[24] or alternatively by high-speed centrifugation to lower the coatomer content to 10% of its original value.[25] A COPI-depleted and spectrin-containing cytosol can also be prepared by taking advantage of the different ATP requirements for COPI and spectrin recruitment to the Golgi (see below). If rat brain cytosol is incubated with excess Golgi membranes and 100 μM GTPγS in the absence of ATP and an ATP-regenerating system for 20 min at 37°, COPI, but not spectrin, is massively recruited to the membranes and removed from the cytosol. Residual COPI is assessed in cleared cytosol using anti-β-COP antibodies; typically reductions to less than 20% of the original value are achieved.

Binding Assays

The role of ARF in regulating the association of spectrin with Golgi membranes can also be studied by *in vitro* binding assays. A single-step protocol involves the incubation of Golgi membranes with complete cytosol or with an ARF-depleted cytosol, with or without GTPγS or BFA. Multistep incubation protocols can also be used to selectively activate ARF.

Single-Step Incubation

Golgi membranes and cytosol are incubated in binding buffer (25 mM HEPES, pH 7.2, 25 mM NaCl, 2.5 mM MgCl$_2$, 0.2 M sucrose, 1 mM DTT, 1 mM ATP, 5 mM creatine phosphate, 10 IU/ml creatine phosphokinase, and GTPγS for at last 10 min) for 20 min at 37° with cytosol with or without 20 μM GTPγS. After incubation, membranes are pelleted by centrifuging at 15,000g for 10 min at 0°. The membrane pellet is washed, solubilized in Laemmli sample buffer, and its protein composition analyzed by SDS–

[23] T. C. Taylor, M. Kanstein, P. Weidman, and P. Melancon, *Mol. Biol. Cell* **5**, 237 (1994).
[24] L. Orci *et al.*, *Nature* **362**, 648 (1993).
[25] F. Aniento, F. Gu, R. G. Parton, and J. Gruenberg, *J. Cell Biol.* **133**, 29 (1996).

PAGE (4–15%) and Western blotting. Alternatively, at the end of the incubation, samples (100 μl) are brought to 1.24 M sucrose (by the addition of 140 μl of 2 M sucrose), overlaid with 300 μl 1 M sucrose, 150 μl 0.5 M sucrose, 100 μl 10 mM Tris-HCl, pH 7.4, and centrifuged at 90,000g for 120 min at 4°. Material floating to the interface between 0.5 and 1 M sucrose is analyzed by SDS–PAGE (4–15%) and Western blotting.

Notably, GTPγS recruits to the Golgi at least six cytosolic proteins, The effect of GTPγS is largely prevented by pretreatment (5 min) with 40 μg/ml BFA (a selective inhibitor of ARF). In the presence of an ARF-depleted cytosol, GTPγS loses its ability to recruit β-spectrin, β-COP, or actin to Golgi membranes. When purified ARF (1 μM) is returned to the incubation mixture, GTPγS once again becomes effective. These *in vitro* results mirror the data obtained in permeabilized cells (see above) and are consistent with the involvement of ARF in the recruitment of spectrin to the Golgi membranes.[4]

Two-Step Incubation

The purpose of this experimental scheme is to achieve the selective activation of ARF. The first step, i.e., incubation of pure ARF, Golgi membranes, and GTPγS, allows for the activation of ARF and the performance of the subsequent incubation step of ARF-primed membranes with cytosol in the absence of GTPγS, thus preventing the activation of cytosolic GTPases by the nucleotide. Typical results with this protocol are presented in Fig. 2.

Golgi fractions, washed with 1 M KCl to ensure the removal of residual ARF, are first incubated in binding buffer (25 mM HEPES, pH 7.2, 25 mM NaCl, 2.5 mM MgCl$_2$, 0.2 M sucrose, 1 mM DTT, 1 mM ATP, 5 mM creatine phosphate, 10 IU/ml creatine phosphokinase) with 20 μM GTPγS or 1 μM purified bovine ARF, or with both, for 15 min at 37° (first step). Membranes are then pelleted (at 15,000g for 10 min at 4°), rinsed, and incubated in the same buffer in the presence or absence of 1 mg/ml cytosol (second step) for an additional 15 min at 37°. At the end of the second incubation, membranes are pelleted and rinsed and proteins are analyzed by SDS–PAGE (4–15%) and Western blot. Only membranes preincubated with ARF and GTPγS are able to recruit spectrin, β-COP, Ank$_{G119}$, and actin, indicating that the GTPγS-induced binding of spectrin (and a subset of other proteins) is strictly and specifically ARF dependent. The recruitment efficiency of the ARF-preloading protocol vs. the single-step addition of GTPγS to the cytosol is similar for spectrin, β-COP, and Ank$_{G119}$, but lower for actin, consistent with the possibility that cytosolic G proteins other than ARF may also promote actin binding and/or polymerization.

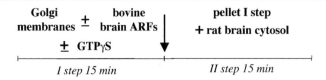

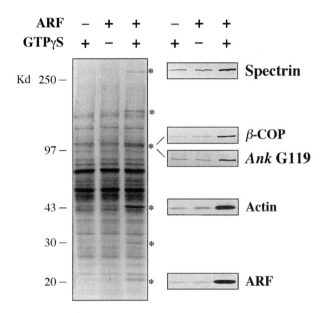

FIG. 2. Two-step binding protocol illustrating that ARF regulates the association of spectrin with Golgi fractions. Golgi membranes are incubated first (step 1) with 1 μM purified bovine ARF and/or 50 μM GTPγS for 15 min at 37°. Membranes are pelleted, rinsed, and then incubated with rat brain cytosol (1.5 mg/ml) for an additional 15 min at 37° (step 2), After this second incubation, membranes are again pelleted, rinsed, and analyzed by SDS–PAGE (4–15% acrylamide–bisacrylamide) and Western blotted with anti βIΣ2-spectrin polyclonal antibody (e.g., mus1) and anti-β-COP, anti-ARF, and anti-actin MAb.[4] When ARF and GTPγS are present in the first step of the incubation, a set of at least six cytosolic proteins (asterisks), including spectrin, β-COP, Ank$_{G119}$, and actin, is recruited from the cytosol during the second incubation. (Adapted with permission from Godi *et al.*[4])

Features of ARF-Dependent COPI- and Spectrin-Complex Recruitment to Golgi Membranes

ARF regulates the assembly of spectrin and coatomer complexes on Golgi by distinct mechanisms. It is possible to dissociate the recruitment activity of ARF on the COPI complex from its effects on spectrin. If ATP and the ATP-regeneration system are omitted in the second step of the incubation, only COPI, but not spectrin (or ankyrin or actin), will associate with the ARF-primed Golgi membranes. Conversely, if a COPI-depleted cytosol (see above) is used in the second incubation step on ARF-primed membranes, only the spectrin complex, but not the COPI complex, will be recruited onto the Golgi membranes.[4]

Direct Analysis of Spectrin Domains Involved in Golgi Membrane Binding

For both βI- and βII-spectrin, *in vitro* studies have identified two direct membrane association domains, MAD1 and MAD2 (Fig. 1).[16] It is likely that βIII-spectrin shares the same functionality, based on sequence homologies[13] and other data. MAD1 is confined to spectrin repeat unit 1, whereas MAD2 encompasses most of spectrin's region 3 and includes its PH domain. The role of MAD1 and MAD2 in the ARF-dependent recruitment of spectrin to the Golgi is studied by assessing the ability of recombinant GST-fused polypeptides spanning MAD1 or MAD2 of βIΣ2-spectrin (see above) to block the binding of endogenous NRK cell spectrin to isolated Golgi fractions. To this end Golgi membranes (10 μg/sample) are incubated for 15 min at 37° with cytosol (75 μg/sample), GTPγS (20 μM), and with GST or GST-fused polypeptides (i.e., βI$_{N-4}$ or βI$_{N-9}$ from βI-spectrin, encompassing MAD1, or βIΣ2$_{DIII}$, encompassing MAD2). Both MAD1- and MAD2-containing polypeptides, but not GST alone, are able to inhibit the ARF-dependent binding of the endogenous spectrin (as well as the binding of Ank$_{G119}$ and actin) to Golgi fractions. The IC$_{50}$ measured for βI$_{N-9}$ and βIΣ2$_{DIII}$ are 0.2 and 1 μM, respectively. In contrast, none of the spectrin fusion polypeptides affect the binding of β-COP or ARF to the Golgi, confirming the conclusion that spectrin binding is independent of COPI coat assembly.

Acknowledgments

The many contributions of Drs. Anna Godi, Michael Stankewich, Prasad Devarajan, and Mr. Paul R. Stabach to the development and refinement of the methods described in this report are gratefully acknowledged. The work described in this article was supported by grants from Telethon (E732), the Human Frontier Science Program, the Italian National Research Council (99.00127.PF49) and the National Institutes of Health.

[43] Purification, Properties, and Analysis of yARL3

By Fang-Jen S. Lee and Chun-Fang Huang

Introduction

ADP-ribosylation factors (ARFs) are members of the Ras superfamily of low molecular weight GTP-binding proteins that includes both ARFs and ARF-like proteins (ARLs).[1,2] Although ARFs were identified as cofactors required for the cholera toxin-catalyzed ADP ribosylation of heterotrimeric G proteins, they were subsequently shown to play an important role in numerous vesicular transport pathways involving endoplasmic reticulum, Golgi, endosomes, peroxisome, or nuclear membranes. Best characterized is ARF1, which is required for association of coat proteins with Golgi membranes[3,4] and adaptor-related protein complex-1 (AP-1) and AP-3 in the trans-Golgi network.[5–7] ARFs 1, 3, and 5 differed in their binding to Golgi[8] and in their dependence on accessory proteins for interaction with Golgi and, perhaps, other cellular membranes.[9] ARFs also stimulate the activity of phospholipase D, an enzyme found in Golgi membranes, suggesting that they may exert their effects, at least in part, by altering membrane phospholipid metabolism.[10,11] In addition, hARF6 was localized to the plasma membrane and endosomes[12] and was shown to regulate plasma membrane architecture and participate in endocytosis by mediating cytoskeletal reorganization.[13]

[1] J. Moss and M. Vaughan, *J. Biol. Chem.* **273**, 21431 (1998).
[2] A. L. Boman and R. A. Kahn, *Trends Biochem. Sci.* **20**, 147 (1995).
[3] J. G. Donaldson, D. Cassel, R. A. Kahn, and R. D. Klausner, *Proc. Natl. Acad. Sci. U.S.A.* **89**, 6408 (1992).
[4] D. J. Palmer, J. B. Helms, C. J. Becker, L. Orci, and J. E. Rothman, *J. Biol. Chem.* **268**, 12083 (1993).
[5] M. A. Stamnes and J. E. Rothamn, *Cell* **73**, 999 (1993).
[6] L. M. Traub, J. A. Ostrom, and S. Kornfeld, *J. Cell Biol.* **123**, 561 (1993).
[7] C. E. Ooi, C. E. Dell'Angelica, and J. S. Bonifacino, *J. Cell Biol.* **142**, 391 (1998).
[8] S. C. Tsai, R. Adamik, R. S. Haun, J. Moss, and M. Vaughan, *Proc. Natl. Acad. Sci. U.S.A.* **89**, 9272 (1992).
[9] S.-C. Tsai, R. Adamik, R. S. Haun, J. Moss, and M. Vaughan, *J. Biol. Chem.* **268**, 10820 (1993).
[10] H. A. Brown, S. Gutowski, C. R. Moomaw, C. Slaughter, and P. C. Sternweis, *Cell* **75**, 1137 (1993).
[11] S. Cockroft, G. M. H. Thomas, A. Fensome, B. Geny, E. Cunningham, I. Gout, I. Hiles, N. F. Totty, O. Truong, and J. J. Hsuan, *Science* **263**, 523 (1994).
[12] J. A. Whitney, M. Gomez, D. Sheff, T. E. Kreis, and I. Mellman, *Cell* **83**, 703 (1995).
[13] H. Radhakrishna, R. D. Klausner, and J. G. Donaldson, *J. Cell Biol.* **134**, 935 (1996).

ARF function depends on its regulated alternation between inactive GDP- and active GTP-bound conformations.[1,2] ARF has no detectable GTPase activity and exchanges bound nucleotide very slowly at physiologic concentration of Mg^{2+}. Interconversion of ARF1 between its inactive and active states requires a guanine nucleotide exchange factor (GEF) to accelerate the replacement of bound GDP with GTP, and a GTPase-activating protein (GAP) that accelerates the hydrolysis of GTP to GDP.

ARL cDNAs encode ARF-like proteins structurally similar to the ARF isoforms, and have been cloned from several species, including human, rat, mouse, *Drosophila,* and yeast.[1,2] The products of these genes appear to lack ARF activity in cholera toxin or phospholipase D assays, and differ in GTP-binding requirements and GTPase activity from ARF isoforms. The biological functions of ARLs remain unclear, although some exhibit tissue- and/or differentiation-specific expression.

Five members of the ARF family have been reported in the yeast *Saccharomyces cerevisiae,* including three ARFs, yARF1, yARF2, and yARF3, and two ARLs, yARL and yARL3.[14–18] yARF1 and yARF2 are believed to act in endoplasmic reticulum (ER)-to-Golgi secretory protein sorting. yARF3 is most similar in amino acid sequence to human ARF6, which has been implicated in the regulation of early endocytic transport. yARL1 in part resides in the Golgi and has a function distinct from those of yARF1 and yARF2. Unlike the lethal phenotype of double null alleles of *arf1* and *arf2,* however, knockout of the yeast *ARL1* gene was not lethal. yARL3, which is most similar to a mammalian ARF-like protein, ARP, was recently identified and characterized. Like yARL1 and yARF3, yARL3 is not essential for cell viability, however, *arl3* mutants appear to be cold sensitive. yARL3 is required for vacuolar protein transport at nonpermissive temperatures. Unlike mammalian ARP, which was detected exclusively on the plasma membrane, yARL3 was found in the soluble fraction and, in part, associated with ER–nuclear envelope structures, suggesting it may function in a novel ER-associated vesicular trafficking pathway.

This chapter describes a method to prepare recombinant yARL3 from *Escherichia coli* using a pET-based vector. Recombinant yARL3 protein is isolated in a functional form from the soluble fraction of the bacterial lysate using a rapid two-step procedure involving Ni^{2+}-agarose chromatog-

[14] J. L. Sewell and R. A. Kahn, *Proc. Natl. Acad. Sci. U.S.A.* **85,** 4620 (1988).

[15] T. Stearns, R. A. Kahn, D. Botstein, and M. A. Hoyt, *Mol. Cell. Biol.* **10,** 6690 (1990).

[16] F.-J. S. Lee, L. A. Stevens, Y. L. Kao, J. Moss, and M. Vaughan, *J. Biol. Chem.* **269,** 20931 (1994).

[17] F.-J. S. Lee, C.-F. Huang, W.-L. Yu, L.-M. Buu, C.-Y. Lin, M.-C. Huang, J. Moss, and M. Vaughan, *J. Biol. Chem.* **272,** 30998 (1997).

[18] C.-F. Huang, W.-L. Yu, L.-M. Buu, and F.-J. S. Lee, *J. Biol. Chem.* **274,** 3819 (1999).

raphy followed by gel filtration. This chapter also describes assays to quantify guanine nucleotide binding and GTP hydrolysis.

Purification of Bacterially Expressed yARL3

Polymerase Chain Reaction

The protocol used for polymerase chain reaction (PCR) amplification is carried out for 30 cycles of 1 min at 95°, and 1 min at 52°, 1 min at 72°, followed by extension at 72° for 10 min, in a solution containing 50 mM KCl, 10 mM Tris-Cl (pH 8.3), 1.5 mM MgCl$_2$, 0.01% gelatin, 20 mM of each dNTP, 0.1% Tween, 25 pmol of each amplification primer, and 2.5 units of *Taq* polymerase (total volume 100 μl). Samples of reaction mixtures are subjected to electrophoresis in a 1.2% agarose gel. All PCR products are purified, subcloned, and sequenced by the dideoxy chain-termination method.[19]

Construction of yARL3 Expression Vectors

The yeast *ARL3* gene is cloned by PCR using yeast genomic DNA as template and primers complementary to sequences upstream or downstream of the *Lpe21p* gene (Yeast Genome Project, accession no. U39205). For producing a His-tagged yARL3 fusion protein, a DNA fragment containing the *yARL3* coding region is generated by amplifying yeast genomic DNA with sequence-specific primers that incorporated unique *Nde*I and *Bam*HI sites beginning at A of the initiating methionine and 6 bp 3′ to the stop codon, respectively. The PCR product is digested with *Nde*I and *Bam*HI, purified and ligated to the expression vector pET15b (Novagen, Madison, WI), yielding pET15byL3. For the nonfusion protein (NFP), PCR products are digested with *Nde*I and *Bam*HI, purified, and ligated to expression vector pT7/Nde,[20] yielding pT7yARL3.

Bacterial Expression of yARL3

BL21 (DE3) *Escherichia coli* (Novagen), transformed with pET15byL3 or pT7yARL3 (to produce His-tagged yARL3 and yARL3-NFP proteins, respectively), are grown on Luria–Bertani (LB)-agar plates containing ampicillin (100 μg/ml). To identify colonies with the highest expression of recombinant protein, several cultures (2 ml) are grown in six-well plates and protein expression is induced for 2 hr with isopropyl-1-thio-β-D-galacto-

[19] F. Sanger, S. Nicklen, and S. Coulson, *Proc. Natl. Acad. Sci. U.S.A.* **74,** 5463 (1977).
[20] J.-X. Hong, R. S. Haun, S.-C. Tsai, J. Moss, and M. Vaughan, *J. Biol. Chem.* **269,** 9754 (1994).

pyranoside (IPTG) after an initial 3-hr growth period. The cell pellet from a 100-μl sample of each culture is subjected to SDS–PAGE under reducing conditions. For large-scale protein production, BL21(DE3) cells containing expression plasmids are grown to a density of $A_{600} = 1.0$ at which time IPTG is added to a final concentration of 1 mM. After 3 hr, cells are harvested by centrifugation, washed once in 20 mM Tris, pH 7.4/1 mM EDTA, and stored at $-80°$ until used. To maximize the yield of soluble yARL3, cultures are grown at 30° with gentle shaking overnight (12–16 hr) in the presence of a low concentration (0.1 mM) of IPTG. After growth overnight, cultures are chilled on ice and cells are pelleted by centrifugation (4000 rpm, 10 min) in a Beckman (Fullerton, CA) Model J2-MC, and lysed on the same day or frozen in liquid nitrogen and stored at $-80°$.

Preparation of Bacterial Lysates and Purification of yARL3

All procedures are performed at 4° unless otherwise stated. For the preparation of His-tagged yARL3, a cell pellet (from a 1-liter culture) is dispersed in 40 ml of ice-cold buffer A (5 mM imidazole, 0.5 M NaCl, 20 mM Tris-HCl, pH 7.4), transferred to a cell disruption bomb (Parr Instruments, Moline, IL), and incubated at 1500 psi with nitrogen gas for 30 min. Because yARL3 behaves like a very hydrophobic protein (although it contains no known lipidic modification, even in yeast), use of a detergent is essential for effective purification. After nitrogen disruption, Triton X-100 is added to 1%, the cell lysate is centrifuged at 22,000g for 20 min, and the clarified supernatant is stored at $-80°$ or used immediately. If frozen, the clarified lysate is thawed and centrifuged at 39,000g for 30 min to remove any precipitated material.

To each 40 ml of clarified lysate is added 4 ml of a 50% slurry of Ni^{2+}-nitrilotriacetic acid-agarose (Ni-NTA-agarose; Qiagen, Chatsworth, CA) that has been equilibrated with buffer A. The His-tagged yARL3 protein is allowed to bind to the Ni-NTA-agarose for 1–2 hr with gentle rocking before the mixture is poured into a 1- × 10-cm column (Bio-Rad, Hercules, CA). The Ni-NTA-agarose settles to the bottom of the column and is then washed successively with 50 ml of buffer A, buffer B (0.5 M NaCl, 50 mM Tris-HCl, pH 7.9) with 10 mM imidazole, and, finally, buffer B with 100 mM imidazole. Each of these buffers is loaded carefully on top of the Ni-NTA-agarose column using a Miniplus 3 pump (Gilson, French) so that the resin is not disturbed (flow rate, 0.5 ml/min).

After the final wash, His-tagged yARL3 is eluted with 50 ml of buffer B containing 0.5 M imidazole (flow rate, 0.5 ml/min), and 2-ml fractions are collected for measurement of A_{280} of each fractions. yARL3 is identified by SDS–PAGE. Fractions containing yARL3 (a single protein band at 22 kDa) are pooled, concentrated to ~1.0 ml using Amicon (Danvers, MA) Centricon-10 (or Centriprep-10) concentrators, dialyzed against two changes of 4 liters of dialysis buffer [20 mM HEPES, pH 7.4, 1 mM EDTA, 2 mM MgCl$_2$, 100 mM NaCl, 1 mM dithiothreitol (DTT)], and stored at $-80°$. The His-tagged yARL3 protein is 80–90% pure as judged by SDS–PAGE and staining with Coomassie blue. Protein is quantified by Coomassie blue or silver stain assays (Bio-Rad).

For the preparation of yARL3-nonfusion protein (NFP), cells are incubated with lysozyme (5 mg/ml) in 50 mM Tris (pH 8.0), 1 mM EDTA for 30 min on ice and disrupted by the cell disruption bomb apparatus described earlier. After centrifugation (6,000g, at 4°, 20 min) the supernatant of the lysate is applied to a column (2.5 × 120 cm) of Ultrogel AcA 54 (BioSepra, Marlborough, MA), which is eluted with 20 mM Tris (pH 8.0) containing 250 mM sucrose, 100 mM NaCl, 5 mM Na$_3$N, 5 mM MgCl$_2$, 1 mM DTT, and 1 mM EDTA. Samples of fractions (4 ml) are subjected to SDS–PAGE under reducing conditions followed by Coomassie blue staining to assess the presence and purity of yARL3-NFP. Fractions containing yARL3-NFP of the highest purity are pooled, concentrated, and centrifuged (14,000 rpm in a microcentrifuge) to remove insoluble aggregates. The supernatant is stored in 100-μl portions at $-80°$. The yARL3 preparations appear to be ~90% pure on SDS–PAGE.

Western Blot Analysis

After SDS–PAGE, proteins are transferred electrophoretically to Immobilon-P membranes (Millipore, Bedford, MA), which are incubated with antibodies in phosphate-buffered saline (pH 7.4) containing 0.1% Tween 20 and 5% dried skim milk at room temperature for 60 min. The anti-yARL3 antibody and horseradish peroxidase-conjugated goat anti-mouse IgG + IgM (H + L) are each diluted 1 : 1000. Bound antibodies are detected with the ECL (enhanced chemiluminescence) system (Amersham) according to the manufacturer's instructions.

Comments

1. A significant fraction of yARL3 (as much as 80%) is found in inclusion bodies. This is purified in the presence of 6 M urea, and results in nucleotide-free yARL3. Because the yield of soluble yARL3 per liter of culture is low, four 1-liter cultures are usually grown for each preparation of recombinant protein.
2. Figure 1 shows that the antibody against yARL3, which is generated by using this purified yARL3 as antigen, reacts specifically with yARL3, but not with other ARF or ARL proteins.

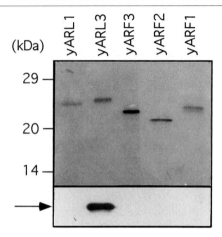

FiG. 1. Immunoblot analysis of purified His-tagged yARF3. Samples (∼30–40 ng) of purified His-tagged yARL1, yARL3, yARL3, yARF2, and yARF1 were separated by SDS–PAGE and visualized with silver stain (*Top*). After transfer to polyvinylidene fluoride (PVDF) filter, His-tagged yARL3 was detected with antibodies against yARL3 using the ECL system, and 10-min exposure of Hyper-film-MP (*bottom,* arrow). Positions of protein standards (kDa) are given at left.

Assay of yARL3

GTPγS Binding Assay

The ability of purified recombinant yARL3 to bind GTPγS is determined by a filter trapping method.[21] Unless otherwise specified, 1 μg of purified recombinant yARL3 is incubated at 30° in 20 mM HEPES (pH 7.5), 100 mM NaCl, 1 mM DTT, 1 mM EDTA, 0.5 mM MgCl$_2$, bovine serum albumin (BSA) (20 μg/ml), 10 μM [^{35}S]GTPγS (Amersham, >1000 Ci/mmol), without or with 3 mM sonified DL-α-dimyristoylphosphatidylcholine (DMPC), and 2.5 mM (0.1%) sodium cholate, in a final volume of 50 μl. Duplicate or triplicate samples are transferred to 2 ml of ice-cold TNMD buffer [20 mM Tris-Cl (pH 7.4), 100 mM NaCl, 10 mM MgCl$_2$, and 1 mM DTT], followed by rapid filtration on 0.45-μm HA filters (Millipore, Bedford), which are then washed four times with 2 ml of ice-cold TNMD buffer. Radiolabeled nucleotide bound to the yARL3 protein retained on the filter is quantified by liquid scintillation counting.

GTPase Assay

GTP hydrolysis is determined by binding 0.5 μM [α-^{32}P]GTP (3000 Ci/ mmol) to 5.0 μM recombinant yARL3 protein, as described by Randazzo

[21] J. K. Northup, P. C. Sternweis, and A. G. Gilman, *J. Biol. Chem.* **258,** 11361 (1983).

and Kahn,[22] followed by dilution with nine volumes of 25 mM HEPES (pH 7.4), 100 mM NaCl, 2.5 mM MgCl$_2$, 0.1% Triton X-100, 1 mM DTT, 1 mM GTP, and bovine brain phosphoinositides (1 mg/ml) and incubation at 30°. Every 5 min, samples (50 μl) are transferred to 2 ml of ice-cold 20 mM Tris-Cl (pH 7.4), 100 mM NaCl, 10 mM MgCl$_2$, and 1 mM DTT. Proteins with bound nucleotides are collected on nitrocellulose filter. Bound GTP and GDP are extracted in 250 μl of 2 M formic acid, of which 4.0-μl samples are analyzed by thin-layer chromatography (TLC) on polyethyleneimine (PEI)-cellulose plates, and 240 μl is used for radioassay to quantify total ^{32}P-labeled nucleotide. TLC plates are subjected to quantification by densitometry after radioautography at $-80°$ for 24 hr. Background binding to sample without protein is subtracted before calculating GTPase activity. Similar results can be obtained by analyzing the reaction products by measuring [^{32}P]P$_i$ released from [γ-^{32}P] GTP, using charcoal to absorb GTP and GDP but not P$_i$.[23]

Comments

GTPγS binding to recombinant His-tagged yARL3 was concentration dependent and was maximal in 60 min at 30°. With and without DMPC/cholate, recombinant yARL3 bound, respectively, 3.2 ± 0.3 and 1.5 ± 0.2 pmol of GTPγS/μg protein. Thus, GTPγS binding to yARL3, like that to ARF, was influenced by the presence of phospholipid/detergent. In contrast, binding of GTPγS to hARL2 and hARL3 was affected very little by added lipid or detergent.[24,25] Purified mammalian ARF1 and ARF3 lack detectable GTPase activity (below 0.0015 min^{-1}). Maximal rates of GTP hydrolysis by hARL2 and hARL3 were 0.0074 and 0.005 min^{-1}, respectively.[24,25] Recombinant His-tagged yARL3 had a rate of $\sim$0.01 min^{-1}, considerably less than that of the mammalian homolog ARP, which was 0.093 min^{-1}.[26]

Acknowledgments

We thank Drs. Randy Haun, Joel Moss, and Martha Vaughan for critical reading of this manuscript. This work was supported by grants from the National Science Council, R.O.C. (NSC-88-2314-B-002-136) to F.-J.S.L.

[22] P. A. Randazzo and R. A. Kahn, *J. Biol. Chem.* **269,** 10758 (1994).

[23] K. Kimura, T. Oka, and A. Nakano, *Methods Enzymol.* **257,** 41 (1995).

[24] J. Clark, L. Moore, A. Krasinskas, J. Way, J. Batttey, J. Tamkun, and R. A. Kahn, *Proc. Natl. Acad. Sci. U.S.A.* **90,** 8952 (1993).

[25] M. M. Cavenagh, M. Breiner, A. Schurmann, A. G. Rosewald, T. Terui, C.-J. Zhang, P. A. Randazzo, M. Adams, H.-G. Joost, and R. A. Kahn, *J. Biol. Chem.* **269,** 18937 (1994).

[26] A. Schurmann, S. Mabmann, and H.-G. Joost, *J. Biol. Chem.* **270,** 30657 (1995).

[44] Preparation and Assay of Recombinant ADP-Ribosylation Factor-Like Protein-1 (ARL1)

By Gustavo Pacheco-Rodriguez, Joel Moss,
and Martha Vaughan

Introduction

ADP-ribosylation factors (ARFs) are ~20-kDa guanine nucleotide-binding proteins that are critical components of intracellular vesicular trafficking pathways.[1,2] Mammalian ARFs are grouped into class I (ARF1, ARF2, ARF3), class II (ARF4, ARF5), and class III (ARF6), based on protein size and structure and nucleotide sequence.[2] ARFs, which lack detectable GTPase activity, are activators of cholera toxin ADP-ribosyltransferase and of a mammalian phospholipase D when GTP is bound. Other guanine nucleotide-binding proteins with sequences similar to ARFs include ARF-related proteins (ARPs),[3] ARF-domain protein (ARD1),[4] and ARF-like protein (ARL1), which was found while investigating the locus of a transactivator of homeotic gene expression in *Drosophila melanogaster*.[5] The encoded 20-kDa protein was very similar in sequence to yeast and mammalian ARFs. Unlike ARFs, however, it was able to hydrolyze GTP and failed to activate cholera toxin.[5] Multiple ARLs have now been identified in many organisms, and localization of ARL1 in the Golgi system of *Saccharomyces cerevisiae*[6] and rat kidney cells[7] is consistent with a role in vesicular transport. Both ARF and ARL molecules have a glycine in position 2 (which is myristoylated in ARFs and probably is also in ARLs), and both have the same consensus sequences for GTP-binding and hydrolysis.

Four human ARLs, in addition to hARL1,[8] are now known. Their

[1] P. Chavrier and B. Goud, *Curr. Opin. Cell Biol.* **11**, 466 (1999).

[2] J. Moss and M. Vaughan, *J. Biol. Chem.* **273**, 21431 (1998).

[3] A. Schurmann, S. Maßmann, and H.-G. Joost, *J. Biol. Chem.* **270**, 30657 (1995).

[4] K. Mishima, M. Tsuchiya, M. S. Nightingale, J. Moss, and M. Vaughan, *J. Biol. Chem.* **268**, 8801 (1993).

[5] J. W. Tamkun, R. A. Kahn, M. Kissinger, B. J. Brizuela, C. Rulka, M. P. Scott, and J. A. Kennison, *Proc. Natl. Acad. Sci. U.S.A.* **88**, 3120 (1991).

[6] F.-J. S. Lee, C.-F. Huang, W.-L. Yu, L.-M. Buu, C.-Y. Lin, M.-C. Huang, J. Moss, and M. Vaughan, *J. Biol. Chem.* **272**, 30998 (1997).

[7] S. L. Lowe, S. H. Wong, and W. Hong, *J. Cell. Sci.* **109**, 209 (1996).

[8] G.-F. Zhang, W. A. Patton, F.-J. S. Lee, M. Liyanage, J.-S. Han, S. G. Rhee, J. Moss, and M. Vaughan, *J. Biol. Chem.* **270**, 21 (1995).

0076-6879/00 $35.00

deduced amino acid sequences appear to fall into three classes, with hARL2[9] and hARL3[10] in class II, hARL4[11] and hARL5[12] in class III, and hARL1,[8] which is more similar to hARF1 than it is to any other hARLs alone, in class I.[13] It was found that in the presence of certain lipids, hARL1 (albeit not as effective as hARF1) was able to activate cholera toxin ADP-ribosyltransferase. It also, to a lesser extent than hARF1, activated mammalian phospholipase D, leading to the suggestion that ARLs and ARFs may be part of a continuum of proteins, rather than members of two separate families.[13] Because hARL1 is 57% identical in sequence to hARF1 (both 181 residues), recombinant proteins in which segments of hARL1 and hARF1 were switched were constructed to probe structure–function relationships. Chimeric proteins in which the N-terminal 73 amino acids were switched demonstrated that different domains of ARF1 are responsible for the activation of cholera toxin and phospholipase D.[8] Other hARL1/hARF1 chimeric proteins are described in Ref. 14.

Information on the physiologic role of hARL1 is very limited. Only recombinant proteins have apparently been used for all *in vitro* studies, because there are no reports of purification of native ARLs. It seems likely for a number of reasons that ARL functions resemble those of ARFs, in part at least, but virtually nothing is known about any specific protein interactions. There are surely regulators of ARL activation analogous to the guanine nucleotide exchange proteins (GEPs) that promote ARF activation by accelerating GTP binding. One of the many ARF GEPs, cytohesin-1 and its Sec7 domain, which is much less substrate specific than the intact protein, failed, however, to activate hARL1.[14] GTPase-activating proteins or GAPs like those that inactivate ARFs presumably exist also for ARLs. Although a partially purified ARF GAP was able to act on hARL1,[15] the biological relevance of this observation remains to be estab-

[9] J. Clark, L. Moore, A. Kraninskas, J. Wang, J. Battey, J. Tamkun, and R. A. Kahn, *Proc. Natl. Acad. Sci. U.S.A.* **90**, 8952 (1993).

[10] M. Cavenagh, M. Breiner, A. Schurmann, H.-G. Joost, and R. A. Kahn, *J. Biol. Chem.* **269**, 18937 (1994).

[11] J. Kim, Y. Lee, I. Lee, B. Kang, Y. Han, H. Kang, and I. Choe, Ph.D. Thesis, Korean Research Institute of Bioscience and Biotechnology, Immune Regulation Research Unit, Yoosung, Taijon, Korea (1996).

[12] S. A. Smith, P. R. Holik, J. Stevens, R. Melis, R. White, and H. Albertsen, *Genomics* **28**, 113 (1995).

[13] J.-H. Hong, F.-J. S. Lee, W. A. Patton, C.-Y. Lin, J. Moss, and M. Vaughan, *J. Biol. Chem.* **273**, 15872 (1998).

[14] G. Pacheco-Rodriguez, W. A. Patton, R. Adamik, H.-S. Yoo, F.-J. S. Lee, J. Moss, and M. Vaughan, *J. Biol. Chem.* **274**, 12438 (1999).

[15] M. Ding, N. Vitale, S.-C. Tsai, R. Adamik, J. Moss, and M. Vaughan, *J. Biol. Chem.* **271**, 24005 (1996).

lished. In the absence of new clues to ARL effectors, functional experiments must focus on molecules and conditions that can accelerate activation (GTP binding or GDP release) or inactivation (GTP hydrolysis) and thereby regulate ARL action.

Preparation of Recombinant hARL1

This procedure for preparation of recombinant hARL1 is published.[8] Briefly, human ARL1 cDNA in the pT7/Nde expression vector is used to transform BL21 (DE3) *Escherichia coli*. Single colonies expressing recombinant hARL1 are grown overnight at 37° in 25 ml of Luria–Bertani medium containing ampicillin (100 μg/ml) before transfer to 500 ml of the same medium and further incubation until the culture reaches an A_{600} of 0.6. Isopropyl-1-thio-β-D-galactopyranoside (final concentration 1 mM) is added, and 2 hr later cells are pelleted by centrifugation (6000g, 10 min, 4°). Cells are washed in ice-cold 50 mM Tris-HCl, pH 8.0, containing 1 mM EDTA (30 μl/ml of culture) and then dispersed in 10 mM Tris-HCl, pH 8.0, with 1 mM EDTA, and lysozyme, 1 mg/ml (30 μl/ml of culture). After 30 min on ice, sonification, and centrifugation (100,000g, 35 min), the supernatant is applied to a column (2.5 × 200 cm) of Ultrogel AcA54, followed by elution with TENDS buffer [Tris-HCl, pH 8.0, 1 mM EDTA, 1 mM NaN$_3$, 10 mM dithiothreitol (DTT), and 200 mM sucrose] containing 100 mM NaCl, 5 mM MgCl$_2$, and 0.5 mM 4-(2-aminoethyl)benzenesulfonyl fluoride hydrochloride. Fractions containing hARL1, as determined by SDS–PAGE, are pooled and stored at −20° in small portions (to avoid repeated freezing and thawing before use).

This procedure has been used successfully several times by different people in the last five years. Preparation of other hARL proteins, with and without His tags, have also been described.[13]

Recombinant ARL proteins (like recombinant and native ARFs) are isolated with bound guanine nucleotide, which is usually GDP, but recombinant proteins are sometimes found with GTP bound. This can be suspected when there is little or no requirement for the addition of GTP to achieve activation of cholera toxin or phospholipase D. Both GDP and GTP were identified bound to *Drosophila* ARL.[5] Extensive dialysis of hARL1 against 7 M urea with EDTA is required before an effect of GTPγS on its activation of cholera toxin could be demonstrated.[13] Nucleotide binding requires prior release of bound nucleotide, which can be favored by a low concentration of free Mg^{2+}. Because binding affinity is greater at higher concentrations of Mg^{2+}, binding assays are terminated with the addition of MgCl$_2$ and the proteins with nucleotide bound are washed with buffer containing 5 mM MgCl$_2$ in the assay described here.

Assays for ARL1

Published assays for ARF-activated phospholipase D[8,13] and cholera toxin ADP-ribosyltransferase activity[8,13] are used to assess the effects of ARL proteins, which have only rarely been reported.[13] It should be emphasized that only activated ARL, i.e., ARL with GTP bound, will be effective. GTP-binding by individual ARF proteins is dramatically influenced by incubation conditions such as the concentration of free Mg^{2+}, the presence of specific phospholipids, and probably additional unrecognized factors. This is likely to be equally true for ARLs. Although systematic studies have not been published, effects of Mg^{2+} concentration and lipid were reported in the first description of *Drosophila* ARL,[5] which also noted that like ARF, it had a higher affinity for GDP than for GTP or GTPγS. The effects of GTPγS concentration on activity of a chimeric protein with the first 73 amino acids of ARL1 and the remainder of ARF1 are markedly altered by different phospholipids and detergents.[8] Similarly, different phospholipids had quite different effects on GTPγS binding to hARF1 and hARL1.[13]

Complications in assessing ARL function that might arise from its GTPase activity[5] (which is lacking in ARFs) can be avoided by using the poorly hydrolyzable analog GTPγS in place of GTP. Nevertheless, it is critical to know the amount of active (GTP or GTPγS bound) ARL (or ARF) in all studies before drawing conclusions about function.

GTPγS-Binding Assay

The procedure described here has been used to assay GTPγS binding to hARL and to ARFs.[14] Briefly, assays (total volume 50 μl) contain 1 μM hARL1, 4 μM [³⁵S]GTPγS in 20 mM Tris-HCl, pH 8.0, 1 mM DTT, 1 mM EDTA, 3 mM $MgCl_2$, 1 mM NaN_3, 200 mM sucrose, 100 mM NaCl, 10 μg of phosphatidylserine, 40 μg of bovine serum albumin, 0.1 μg/ml each of aprotinin, leupeptin, soybean and lima bean trypsin inhibitors, and 0.5 mM 4-(2-aminoethyl)benzenesulfonyl fluoride hydrochloride. After incubation, proteins with [³⁵S]GTPγS bound are collected on nitrocellulose filters and washed with a solution of 25 mM Tris-HCl, pH 8.0, 5 mM $MgCl_2$, 1 mM DTT, and 100 mM NaCl before radioassay of dried filters.[8]

This assay contains phosphatidylserine because it was more effective than cardiolipin or dimyristoylphosphatidylcholine plus sodium cholate in promoting GTPγS binding by hARL1.[13] Protease inhibitors and albumin as a potential stabilizer are included because this assay is often used to assess effects of impure protein preparations that might accelerate guanine nucleotide binding to ARL (or ARF). Under these conditions, nucleotide binding by ARL1 at 37° is slow and essentially constant for at least 100 min.

For most experiments, it is critical to verify that a rate of nucleotide binding is being measured, i.e., that the rate of binding is constant throughout the incubation and that it is proportional to the amount of ARL. Time, temperature, and other conditions of incubation should be adjusted to accomplish this. For other purposes, it is maximal binding and not the rate that is of interest. For example, to evaluate functionality of an ARL preparation, it is important to know how much of the protein is capable of high-affinity nucleotide binding. It is also desirable to achieve maximal binding when preparing ARL with nucleotide bound for quantification of rates of nucleotide release or of GTPase activity by the method described by Vitale for ARD1 elsewhere in this volume.[16]

A similar method is used to assay stimulation of ARF GTPase activity during purification of an ARF GAP that also accelerated GTP hydrolysis by hARL1.[15] No description of a simple assay of steady-state GTP hydrolysis by ARL has been published, although a "derived rate" of 0.05 min^{-1} was reported for *Drosophila* ARL.[5] Better characterization of ARL GTPase activity is clearly needed. A method used to assay activity of other ~20-kDa GTPases, e.g., Ras, can probably be modified to optimize conditions for ARL.

---

[16] N. Vitale, J. Moss, and M. Vaughan, *Methods Enzymol.* **329,** [35], (2001) (this volume).

Section III

Sar GTPases

[45] Purification and Properties of Rat Liver Sec23–Sec24 Complex

By JACQUES T. WEISSMAN, MEIR ARIDOR, and WILLIAM E. BALCH

Introduction

Sec23 and Sec24 are part of a larger coat complex that mediates endoplasmic reticulum (ER) vesicle budding termed COPII (coat protein complex II). COPII is composed of five cytosolic components: the small GTPase Sar1, Sec23–24 complex, and the Sec13–Sec31 complex. Sec23 stimulates the rate of Sar1 GTP hydrolysis to GDP.[1,2] Sec23 and Sec24 form a 300-kDa complex.[3] Although Sec23 association with Sec24 is not required for Sec23 GTPase-activating protein (GAP) activity, both proteins are required for vesicle formation.[2,3] Sec23–24 is recruited to ER membranes on activation of Sar1[4] into a detergent-soluble prebudding protein complex that contains the type 1 transmembrane cargo protein vesicular stomatitis virus glycoprotein (VSV-G).[2] Two mammalian homologs of Sec23 have been identified,[2,5] and at least four mammalian homologs of Sec24 have been identified.[6] This chapter describes the purification of functional endogenous Sec23–24 complex from rat liver. Sec23–24 is isolated from a crude rat liver cytosol preparation using a three-step procedure involving gel filtration and ion-exchange chromatography. In addition, this chapter demonstrates two assays for purified Sec23–24 function. Sec23–24 complex stimulates the intrinsic GTPase activity of Sar1 and also reconstitutes COPII ER vesicle formation from normal rat kidney (NRK) microsomes *in vitro*.

Methods

Preparation of Rat Liver Cytosol

Three adult male Sprague-Dawley rats are anesthetized in a halothane chamber and the livers are excised. Livers are rapidly transferred to ice-

[1] T. Yoshihisa, C. Barlowe, and R. W. Schekman, *Science* **259**, 1466 (1993).
[2] M. A. Aridor, J. T. Weissman, S. Bannykh, C. Nuoffer, and W. E. Balch, *J. Cell Biol.* **6**, 61 (1998).
[3] L. Hicke, T. Yoshihisa, and R. Schekman, *Mol. Biol. Cell.* **3**, 667 (1992).
[4] M. Aridor, S. I. Bannykh, T. Rowe, and W. E. Balch, *J. Cell Biol.* **131**, 875 (1995).
[5] J.-P. Paccaud, W. Reith, J.-L. Carpentier, M. Ravazzola, M. Amherdt, R. W. Schekman, and L. Orci, *Mol. Biol. Cell.* **7**, 1535 (1996).
[6] A. Pagano, F. Letourneur, D. Garcia-Estefania, J. L. Carpentier, L. Orci, and J. P. Paccaud, *J. Biol. Chem.* **19**, 7833 (1999).

cold phosphate-buffered saline (PBS) and the remaining steps are carried out in a 4° cold room. Excess connective tissue is removed using surgical scissors and the livers are washed 2–3 times with PBS to remove excess blood. Livers are weighed (typically 12.5 g per liver) and then transferred to 100 ml of buffer A [25 mM HEPES, pH 7.2 (with KOH), 150 mM potassium acetate, 0.25 M sorbitol, 1 mM dithiothreitol (DTT), 1 mM EDTA (pH 8), 1 mM EGTA (pH 8), complete protease inhibitors (from tablets, Roche, Germany; use one tablet per 50 ml buffer), 1 mM phenyl-methylsulfonyl fluoride (PMSF)]. Livers are each cut into 20 smaller pieces while still in buffer A and then transferred to an ice-cold Waring blender. An additional 100 ml of buffer A is added and the livers are homogenized with four 15-sec pulses on high with 1-min rests. The homogenate is then centrifuged at 960g for 10 min to remove nuclei and unbroken cells. The resulting supernatant is centrifuged for 12,500g for 20 min and the supernatant is recovered. This supernatant is centrifuged at 186,000g for 1 hr to pellet microsomal membranes. The small white lipid layer at the top of the supernatant is removed using a vacuum. This resulting supernatant (rat liver cytosol) is brought to 30% saturation with dropwise addition of 100% saturated ammonium sulfate and is then stirred further on ice for 2–3 hr. The precipitated material containing Sec23–24 complex is then pelleted by centrifugation at 16,000g for 20 min. This 30% ammonium sulfate pellet is then stored at −80°. This procedure is then repeated three more times to collect crude Sec23–24 complex from a total of 12 rat livers.

Notes. The cytosolic Sec23–24 complex is highly susceptible to proteolysis. To minimize this problem, the number of rats sacrificed during a preparation of rat liver cytosol was reduced from 12 to 3. It was also found that during the chromatographic steps, consistent storage of column fractions on ice reduces the amount of Sec23–24 degradation. The addition of 0.4 μg/mL calpain inhibitor I also helps to reduce degradation, particularly in the later chromatographic steps.

Gel Filtration Chromatography

Four 30% ammonium sulfate pellets are thawed on ice and resuspended in buffer B (buffer A containing 1 μg/ml aprotinin, 0.5 μg/ml leupeptin, 1 μg/ml pepstatin, and 0.4 μg/ml calpain inhibitor I). Pellets are resuspended in 24 ml of ice-cold buffer B using 10 strokes in a 40-ml Dounce homogenizer. The resuspended pellets are centrifuged at 10,000g for 15 min and the supernatant is applied to a 440-ml S-300 column (Pharmacia-Amersham, Piscataway, NJ) that has been preequilibrated with buffer B lacking protease inhibitors. One column volume of buffer is then run through the S-300 at 0.4 ml/min and 4-ml fractions are collected (Fig. 1).

Protein-containing fractions are stored on ice and 20-μl aliquots are filtered onto nitrocellulose using a multiwell dot-blot apparatus. Dot blots are then probed with Sec23- and Sec24-specific antisera to locate the immunoreactive fractions.

DEAE Ion-Exchange Chromatography

A 50-ml DEAE column is washed with 2 column volumes of buffer B containing 1 M potassium acetate and then equilibrated with 3 column volumes of buffer B. The Sec23–24 immunoreactive fractions (typically 115 ml) are loaded directly onto the column at a flow rate of 1 ml/min and then the DEAE column is washed with 150 ml of buffer B. The Sec23–24 complex is then eluted in a nonlinear 500-ml salt gradient from 150 mM to 1 M potassium acetate in buffer B using an Econo system (Bio-Rad, Hercules, CA). The first 75 ml of the gradient runs from 150 to 450 mM potassium acetate in order to elute the protein in a sharper peak, and then the DEAE column is washed with 425 ml from 450 mM to 1 M potassium acetate (Fig. 1A). Eight-ml fractions are collected and stored on ice. Twenty-μl fractions are again filtered directly onto nitrocellulose and probed for Sec23 and Sec24 immunoreactivity.

Hydroxyapatite Ion-Exchange Chromatography

A 10-ml hydroxyapatite (HAP) column is washed with 3 column volumes of buffer B containing 500 mM KH_2PO_4 and is then equilibrated with 4 column volumes of buffer B containing 25 mM KH_2PO_4 at 0.8 ml/min. Sec23–24 immunoreactive fractions are pooled, brought to 25 mM KH_2PO_4, and loaded directly onto the HAP column at a flow rate of 0.8 ml/min. The column is washed with 4 volumes of buffer B containing 25 mM KH_2PO_4, and the Sec23–24 complex is eluted with a 100-ml gradient from 25 to 500 mM KH_2PO_4 in buffer B (Fig. 1B). Fractions (2.5 ml) are collected and stored on ice. Aliquots (20 μl) are again screened by dot blot for Sec23 and Sec24 immunoreactivity. Pure Sec23–24 fractions, as judged by silver-stained SDS–PAGE gel (Fig. 1C), are pooled and dialyzed against a buffer containing 25 mM HEPES, pH 7.2, KOH, and 125 mM potassium acetate. Sec23–24 complex is then concentrated to 0.1 mg/ml, frozen in aliquots in liquid nitrogen, and stored at $-80°$ for subsequent use. Typically this procedure yields 2 mg of functional complex from 12 rat livers.

Sec23 GTPase Activation Assay

The Sec23 GTPase activation assay (GAP assay)[2] requires purified mammalian Sar1 that may be isolated after overexpression in *Escherichia*

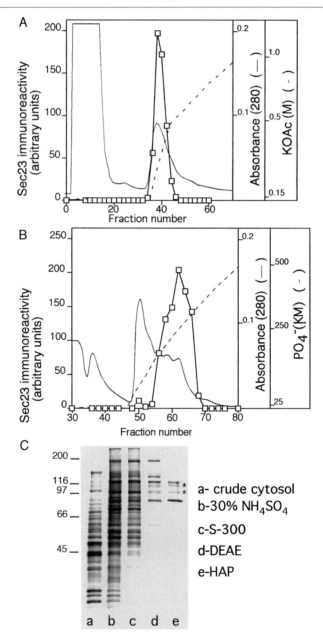

coli.[7] The GAP assay measures the conversion of $[\alpha\text{-}^{32}P]GTP$ to $[\alpha\text{-}^{32}P]GDP$ by the small GTPase Sar1. The two labeled nucleotide forms are separated by thin-layer chromatography (TLC) and quantitated by phosphorimaging (Molecular Dynamics, Sunnyvale, CA).

Sec23–24 complex (1 μM) is incubated in 20 μl with Sar1 (0.5 μM) in buffer C (25 mM HEPES, pH 7.2, KOH; 1 mM DTT; 0.25 mM BSA; 0.25 mM magnesium acetate, 1 mM ATP, and 50 nM $[\alpha\text{-}^{32}P]GTP$) at 32°. At various times 1 μl is removed from this incubation and spotted directly onto a TLC plate (Selecto Scientific, Norcross, GA). The spot is rapidly dried for 20 sec using a hair dryer. Radiolabeled GTP and GDP are separated by placing the bottom of the TLC plate in a shallow reservoir of separation buffer (0.5 M $LiCl_2$, 1 M HCOOH). The separation buffer is allowed to migrate up the TLC plate until it reaches approximately 1 cm from the top of the TLC plate. The TLC plate is then removed from the separation buffer and rapidly dried using a hair dryer. Radiolableled nucleotide is quantitated using a phosphorimager. Under these conditions the purified Sec23–24 complex accelerates GTP hydrolysis of Sar1 up to 100-fold over the intrinsic rate (Fig. 2).

Endoplasmic Reticulum Budding Assay

The ER budding assay[8] requires Sar1, Sec23–24 (as described above), Sec13–Sec31 complex partially purified from rat liver[2] (Fig. 3A), and ER microsomes purified from NRK cells.[8] Microsomes are purified from NRK cells that have been infected with vesicular stomatitis virus (VSV) containing the temperature sensitive form of vesicular stomatitis virus glycoprotein (ts045 VSV-G). When NRK cells are infected at the restrictive temperature (40.5°) ts045 VSV-G misfolds and cannot exit the ER. Transfer of cells to the permissive temperature (32°) results in folding and transport of ts045

[7] T. Rowe and W. E. Balch, *Meth. Enzymol.* **257**, 49 (1995).
[8] T. Rowe, M. A. Aridor, J. M. McCaffery, H. Plutner, and W. E. Balch, *J. Cell. Biol.* **135**, 895 (1996).

FIG. 1. Purification of Sec23–24 from rat liver cytosol. The Sec23–24 complex was purified from rat liver cytosol as described in the Methods section. (A and B) Typical elution profiles from the DEAE and HAP chromatography steps, respectively. (C) A silver-stained gel of representative pooled fractions: lane a, crude cytosol (3.3 μg); lane b, ammonium sulfate precipitate (3 μg); lane c, S-300-Sepharose (2.8 μg), lane d, DEAE (0.7 mg); and lane, HAP (0.2 μg) columns. The asterisks in lane e represent partial proteolytic breakdown products of Sec24 based on immunoblotting. Molecular weight markers are indicated in the left margin (kDa). (Reprinted with permission from *The Journal of Cell Biology*.)

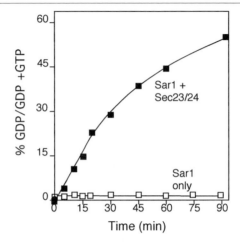

FIG. 2. The Sec23–24 complex is a functional GAP protein for the mammalian Sar1 protein. Recombinant Sar1 was incubated in the presence or absence of purified Sec23–24 and GAP activity was monitored as described in the Methods section. (Reprinted with permission from *The Journal of Cell Biology.*)

VSV-G.[9] To follow ts045 VSV-G export from the ER, 15–20 μg of a washed postnuclear microsomal fraction is incubated in the presence of 1 μg Sar1, 1.5 μg Sec23–24, 20 μg Sec13–Sec31, an ATP regenerating system, and 1–2 mM GTP in export buffer [250 mM sorbitol, 70 mM potassium acetate, 36 mM HEPES, pH 7.2 (KOH), 5 mM EGTA, 1.8 mM calcium acetate, and 2.5 mM magnesium acetate] at 32°. This incubation is then placed on ice, centrifuged at 20,000g for 10 min to pellet membranes, and resuspended in HEPES/sorbitol buffer (25 mM HEPES, pH 7.2, KOH, 250 mM sorbitol). Resuspended membranes are then centrifuged for 16000g for 2 min to separate more rapidly sedimenting ER and Golgi compartments that are recovered in a medium speed pellet (MSP) from slowly sedimenting ER-derived vesicular intermediates released into the medium speed supernatant (MSS). The MSS is then centrifuged at high speed (100,000g for 20 min) to pellet VSV-G containing vesicular intermediates (HSP). The MSP and HSP are then resolved by SDS–PAGE, transferred to nitrocellulose, immunoblotted, and quantitated for VSV-G. VSV-G export is expressed as a percentage of VSV-G in the HSP. Under these conditions, 20–25% export of VSV-G is observed (Fig. 3B).

[9] R. W. Doms, D. S. Keller, A. Helenius, and W. E. Balch, *J. Cell Biol.* **105,** 1957 (1987).

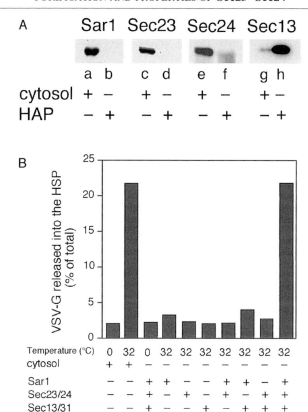

FIG. 3. Vesicle budding from the ER can be reconstituted with purified Sec23–24 in the presence of Sar1 and Sec13–Sec31. (A) The mammalian Sec13–Sec31 complex was partially purified from rat liver cytosol as described.[2] The presence of Sar1, Sec23, Sec24, and Sec13 in crude cytosol (lanes a, c, e, and g) and the HAP pool (lanes b, d, f, and h), respectively, was determined by Western blotting with specific antibodies. Note the absence of Sar1 or Sec23–24 in the HAP fraction highly enriched in the Sec13–Sec31 complex. (B) ER microsomes were incubated in the presence of ATP, GTP, and purified components for 30 min at 32° and then the amount of VSV-G released into the HSP was determined as described in the Methods section. The reactions were carried out in a 40-μl volume supplemented with either 400 μg of rat liver cytosol, 1 μg of His₆-Sar1, 1 μg of Sec23–24 complex, or 24 μg of the Sec13–Sec31 containing fractions as indicated. The typical results of three independent experiments are shown. (Reprinted with permission from *The Journal of Cell Biology*.)

Summary

We have demonstrated a protocol for purifying functional Sec23–24 complex from rat liver cytosol. Because the rat liver Sec23–24 complex is highly susceptible to proteolysis, we have noted several modifications which have allowed us to overcome the problem of degradation. If care is taken to prevent proteolysis, this procedure typically yields 2 mg of functional complex. The Sec23–24 complex can then be used to study Sar1 GTP hydrolysis in the Sec23 GTPase activation assay. Additionally, Sec23–24 can reconstitute ER vesicle formation in the presence of Sar1 and Sec13–31, allowing for the identification of novel proteins or compounds that affect cargo export.

[46] Purification of Functional Sec13p–Sec31p Complex, A Subunit of COPII Coat

By William J. Belden and Charles Barlowe

Introduction

SEC13 was identified in a genetic screen for temperature-sensitive mutants that are defective in the secretory pathway of the yeast *Saccharomyces cerevisiae*.[1] Genetic studies suggested that *SEC13* is involved in vesicle formation at the endoplasmic reticulum (ER).[2] During the development of a cell-free transport assay that measures budding from the ER, fractionation of the required cytosolic components indicated that functional Sec13p, a 33-kDa protein, is found in a complex with a 150-kDa protein.[3,4] Further analysis revealed that the 150-kDa protein is encoded by *SEC31,* another gene essential for vesicle formation from the ER.[5] Both Sec13p and Sec31p contain WD40 motifs that are implicated in protein–protein interactions. Together these two proteins form a 700-kDa heteromeric complex of unknown stoichiometry.[3–5] The Sec13p–Sec31p complex, in addition to the Sec23p–Sec24p complex, and Sar1p, comprise the cytosolic components that are necessary and sufficient to drive COPII vesicle formation.[6,7] The purification protocols for the Sec23p–Sec24p complex and Sar1p are re-

[1] P. J. Novick, C. Field, and R. Schekman, *Cell* **21,** 205 (1980).
[2] C. Kaiser and R. Schekman, *Cell* **61,** 723 (1990).
[3] N. K. Pryer, N. R. Salama, R. Schekman, and C. Kaiser, *J. Cell Biol.* **120,** 865 (1993).
[4] N. R. Salama, T. Yeung, and R. Schekman, *EMBO J.* **12,** 4073 (1993).
[5] N. R. Salama, J. S. Chuang, and R. Schekman, *Mol. Biol. Cell* **8,** 205 (1997).
[6] C. Barlowe, L. Orci, T. Yeung, M. Hosobuchi, S. Hamamoto, N. Salama, M. F. Rexach, M. Ravazzola, M. Amherdt, and R. Schekman, *Cell* **77,** 895 (1994).
[7] K. Matsuoka, L. Orci, M. Amherdt, S. Y. Bednarek, S. Hamamoto, R. Schekman, and T. Yeung, *Cell* **93,** 263 (1998).

ported elsewhere.[8,9] In this chapter we describe the purification of the Sec13p–Sec31p complex that employs a hexahistidine (His$_6$)-tagged version of Sec31p.[5] This procedure requires two purification steps and yields approximately 3 mg of pure Sec13p–Sec31p from a 15-liter culture of yeast cells.

Strains, Materials, and Definitions

The strain RSY1113 is used for overproduction of the Sec13p–Sec31p complex.[5] In this strain, the *SEC31* gene is replaced with *TRP1* (*Matα ura3 trp1 ade3 leu2 his3 sec31::TRP1*) and contains a His$_6$-tagged version of *SEC31* plus *SEC13* in YEp352 (2 μm *SEC13 SEC31-His$_6$ URA3*). Cells are cultured in minimal media (YMD: 0.67% yeast nitrogen base, 2% dextrose, 3 mg/liter leucine, lysine, adenine, histidine, and tryptophan) or rich media (YPD: 2% Bacto-peptone, 1% yeast extract, and 2% dextrose). The protease inhibitors (PI), phenylmethylsulfonyl fluoride (PMSF), pepstatin A, and leupeptin are from Sigma Chemical Co. (St. Louis, MO). Ni-NTA-agarose chromatography resin is obtained from Qiagen Co. (Chatsworth, CA). Antibodies for Sec13p and Sec31p are described in previous reports.[3,4] The Mono Q 5/5 column, concanavalin A (Con-A) Sepharose, secondary antibodies conjugated with horseradish peroxidase (HPR), and chemiluminescent reagents for immunoblot analysis are from Amersham Pharmacia (Piscataway, NJ). Protein concentrations are estimated according to the dye-binding Bradford assay using bovine serum albumin (BSA) as a standard with reagents from Bio-Rad (Hercules, CA). A unit of activity is defined as the amount of Sec13p–Sec31p required for 50% maximal release of protease protected ^{35}S-labeled glyco-pro-α-factor starting with 250 μg of urea-extracted membranes in a 60-μl reaction supplemented with excess Sar1p (12 μg/ml) and Sec23p–Sec24p (12 μg/ml).

Growth, Lysis, and Preparation of Soluble Protein Extract

The strain containing the *SEC13, SEC31-His$_6$* plasmid is grown to late logarithmic phase growth in 1.5 liters of YMD lacking uracil, and this culture is used to innoculate 15 liters of YPD media. Cells are grown for approximately 15 hr after inoculation, and harvested in mid-logarithmic phase growth (OD$_{600}$ 1.4–1.6) by centrifugation at 4000g in a Sorvall GSA rotor. Cells (~90 g) are washed with 600 ml of buffer 88 (20 mM HEPES, pH 7.0, 150 mM potassium acetate, 250 mM sorbitol, 5 mM magnesium

[8] K. Kimura, T. Oka, and A. Nakano, *Methods Enzymol.* **257**, 41 (1995).
[9] T. Yeung, T. Yoshihisa, and R. Schekman, *Methods Enzymol.* **257**, 145 (1995).

acetate) then resuspended in 15 ml of buffer 88. Lysis is preformed by cell disruption in liquid nitrogen as previously described.[10] Briefly, cells are frozen dropwise in a bath of liquid nitrogen forming approximately 0.5-cm-diameter spheres and then lysed by grinding the frozen cells in a Waring blender. Five 2-min intervals of high-speed agitation in the presence of excess liquid nitrogen results in approximately 60% cell lysis. The frozen yeast powder is thawed at 4° and adjusted to a final volume of 300 ml with ice-cold buffer 88 containing PI (5 μg/ml pepstatin A, 1 μg/ml leupeptin, 0.5 mM PMSF). All remaining steps are done at 4°.

A clarified cell extract, containing soluble Sec13p–Sec31p complex, is generated by two consecutive centrifugation steps. First, the lysate is centrifuged at 10,000g for 10 min in a Beckman SS34 rotor to remove unlysed cells and rapidly sedimenting membranes. The medium-speed supernatant is recovered, transferred to Beckman Ti45 rotor tubes, and spun at 100,000g for 1 hr. This sediments the remaining membranes and generates a supernatant (~260 ml) that contains the Sec13p–Sec31p complex.

Purification of Sec13p–Sec31p Complex

The scheme for purifying the Sec13p–Sec31p complex from the clarified cell extract consists of two steps. First the supernatant is loaded onto an 8-ml Ni-NTA-agarose column that is equilibrated with buffer A (150 mM potassium acetate, 20 mM HEPES, pH 7.0, with PI). Sec31p–His$_6$ selectively binds to Ni-NTA-agarose, and the column is washed with greater than 10 column volumes of buffer A. Nonspecifically bound proteins are washed from the column with 50 ml of buffer A containing 15 mM imidazole (pH 7.0). Fractions (5 ml) containing the Sec13p–Sec31p complex are then eluted with buffer A containing 200 mM imidazole (pH 7.0). The pH of the buffers used in the Ni-NTA chromatography step is essential for efficient binding and elution. The protein concentration of each fraction is determined by a Bradford protein assay. The majority of protein elutes in the second and third fractions, which are pooled, and the volume adjusted to 15 ml with buffer A.

The pooled Ni-NTA eluate is clarified by centrifuging at 10,000g for 10 min in a Beckman SS34 rotor prior to loading on a Mono Q HR 5/5 column equilibrated with 10 ml of buffer C (0.3 M potassium acetate, 20 mM HEPES, pH 7.0, 1 mM magnesium acetate). The column is washed with 3 ml of buffer C, and then a 1-ml linear gradient brings the salt concentration to 0.44 M in buffer C. Fractions (1 ml) are collected, and the column is washed with 5 ml of 0.44 M potassium acetate. After the 0.44 M potassium

[10] P. K. Sorger and H. R. B. Pelham, *EMBO J.* **6**, 3035 (1987).

acetate wash, a 5-ml linear gradient raises the salt concentration to 0.72 M potassium acetate in buffer C and this is the salt concentration at which the Sec13p–Sec31p complex elutes from the Mono Q.

Analysis of these purification steps by SDS–PAGE (Fig. 1) is used to indicate the protein content (Fig. 1A) and immunoreactivity of Sec13p and Sec31p (Fig. 1B) after each procedure. As observed in lanes 1 and 2, most of the Sec31p–His$_6$ binds to the Ni-NTA-agarose column, however, a significant amount of Sec13p did not bind to this matrix. Presumably the

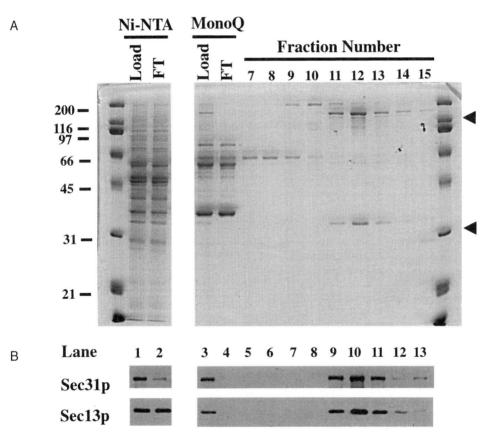

FIG. 1. Analysis of Sec13p–Sec31p complex purification. (A) Colloidal blue stained 12.5% SDS–PAGE and (B) corresponding immunoblots using antibodies against Sec13p and Sec31p. Lane 1 is the starting clarified cell extract (Ni-NTA load); lane 2 is the flow-through from the Ni-NTA column (Ni-NTA FT); lane 3 is the pooled Ni-NTA eluate (Mono Q load); lane 4 is the flow-through from the Mono Q column (Mono Q FT); and lanes 5–13 correspond to the eluted fractions, 7–15, from the Mono Q column. Arrowheads indicate the Sec31p (150-kDa) and Sec13p (33-kDa) subunits.

cell lysate contains an excess of Sec13p over Sec31p, and the flow-through fraction represents a monomeric form of Sec13p. Lanes 3–13 indicate that the Sec13p–Sec31p complex binds efficiently to the Mono Q column and that a majority of pure complex elutes in fractions 11, 12, and 13 (lanes 9–11). Fraction 11 corresponds to a salt concentration of 0.72 M potassium acetate. The volume, protein concentration, and total protein that is contained in the starting cytosol, the pooled Ni-NTA eluate, and the pooled Mono Q eluate are summarized in Table I. A total of 3 mg of purified Sec13p–Sec31p is recovered from 1300 mg of total crude cytosol. The previous method using a Sec13p–DHFR fusion construct yielded 1.4 mg of purified Sec13p–Sec31p complex from 8580 mg of protein in the starting cytosol.[4]

Assay of Functional Sec13p–Sec31p Complex

To determine the activity of purified Sec13p–Sec31p and monitor the yield through each purification step, a cell-free assay that depends on this protein complex is employed. The assay is based on an *in vitro* budding reaction that measures the release of radiolabeled yeast secretory protein, ^{35}S-labeled glyco-pro-α-factor, into ER-derived COPII vesicles.[4,6] Briefly, *in vitro* translated ^{35}S-labeled prepro-α-factor is posttranslationally translocated into purified ER microsomes where core oligosaccharides are added to form ^{35}S-labeled glyco-pro-α-factor. After translocation, the membranes are spun at 13,000g for 3 min, then washed with buffer 88 containing 2.5 M urea, to remove untranslocated ^{35}S-labeled prepro-α-factor and Sec31p–Sec31p complex that is peripherally associated with the microsomes. The membranes are then washed two times with buffer 88 alone to remove urea, and the volume is adjusted to achieve a concentration of 30 mg/ml of membrane protein. For each budding reaction, 250 μg of membranes is incubated with excess Sar1p, and Sec23p–Sec24p, while the concentration

TABLE I
PURIFICATION OF Sec13p–Sec31p COMPLEX

Fraction	Volume (ml)	Concentration (mg/ml)	Weight (mg)	Units[a]	Specific activity (U/mg)	Recovery (%)	Purification (-fold)
Cytosol	260	5	1300	15,000	11.5	100	1
Ni-NTA eluate	15	1.7	25.5	12,400	490	83	43
Mono Q eluate	3	1	3	10,000	3,300	67	287

[a] See unit definitions in Strains, Materials, and Definitions section.

of either cytosol, Ni-NTA eluate or purified Sec13p–Sec31p is varied. The reaction is performed in a 60-μl volume that contains 100 μM GTP and an ATP regeneration system.[3] The reaction is allowed to proceed for 15 min at 20°. After COPII vesicle formation, the ER-derived vesicles are separated from starting microsomes by centrifugation at 13,000g for 5 min. Vesicles contained in the supernatant fluid are treated with 250 μg of trypsin for 10 min at 4°, followed by 250 μg of trypsin inhibitor. Vesicles are solubilized with 1.0% SDS and the amount of ^{35}S-labeled glyco-pro-α-factor is quantified by precipitation with Con-A Sepharose. The percent budding is determined by dividing the amount of protease protected ^{35}S-labeled glyco-pro-α-factor precipitated from the 13,000g fraction by the total ^{35}S-labeled glyco-pro-α-factor from the starting microsomes.[6]

Using this assay, the specific activity of Sec13p–Sec31p complex was determined at each stage of the purification (Table I). In this preparation, a 287-fold purification is obtained and 67% of the activity contained in the starting cytosol is recovered. This compares favorably with a previous method that used a Sec13–DHFR fusion protein where a 3.5% recovery was reported.[4]

Section IV

Dynamin GTPases

[47] Expression, Purification, and Functional Assays for Self-Association of Dynamin-1

By Hanna Damke, Amy B. Muhlberg, Sanja Sever, Steven Sholly, Dale E. Warnock, and Sandra L. Schmid

Introduction

Dynamin is a 100-kDa GTPase required for receptor-mediated endocytosis. It is specifically targeted to clathrin-coated pits where it self-assembles into helical rings and is believed to be required for late stages of clathrin-coated vesicle formation.[1–3] Self-assembly of dynamin tetramers into higher order structures stimulates its GTPase activity,[4,5] indicating that dynamin–dynamin interactions are critical for regulating its cellular function. Dynamin has been shown to contain six distinct domains through both sequence homology and limited proteolysis analysis. These domains include the N-terminal GTPase domain that shares homology with other GTPases, a "middle domain" of uncertain function, a pleckstrin homology (PH) domain, the GTPase effector domain (GED) that performs GTPase-activating protein (GAP) functions, and a C-terminal proline/arginine-rich domain (PRD). It has been demonstrated that the PH domain and the PRD act as negative and positive regulators of self-assembly and GTP hydrolysis, respectively.[6] More importantly, it has been shown that the α-helical GED domain can stimulate GTPase activity directly by interaction with the N-terminal GTPase domain of dynamin.[7] A variety of SH3 domain-containing molecules interact with dynamin through its C-terminal PRD, but the exact role of these partners remains unknown (reviewed in Ref. 8). The fact that dynamin has the intrinsic ability to regulate its own GTPase activity through both intramolecular and intermolecular interactions positions dynamin itself as its most important effector molecule. Here, we describe procedures used for the biochemical analysis and functional consequences of dynamin-1 self-association.

[1] P. De Camilli, K. Takei, and P. S. McPherson, *Curr. Opin. Neurobiol.* **5,** 559 (1995).
[2] D. F. Warnock and S. L. Schmid, *Bioessays* **18,** 885 (1996).
[3] R. Urrutia, J. R. Henley, T. Cook, and M. A. McNiven, *Proc. Natl. Acad. Sci. U.S.A.* **94,** 377 (1997).
[4] P. L. Tuma and C. A. Collins, *J. Biol. Chem.* **269,** 30842 (1994).
[5] D. E. Warnock, J. E. Hinshaw, and S. L. Schmid, *J. Biol. Chem.* **271,** 22310 (1996).
[6] A. B. Muhlberg, D. E. Warnock, and S. L. Schmid, *EMBO J.* **16,** 6676 (1997).
[7] S. Sever, A. B. Muhlberg, and S. L. Schmid, *Nature* **398,** 481 (1999).
[8] S. L. Schmid, M. A. McNiven, and P. De Camilli, *Curr. Opin. Cell Biol.* **10,** 504 (1998).

Expression and Purification of Recombinant Dynamin-1

High-level expression of recombinant dynamin in baculovirus-infected insect cells provides a reproducible technique for the production of biologically active dynamin-1. With this expression system, we are able to apply a relatively simple purification procedure, which includes batch anion-exchange and hydroxyapatite chromatography, that routinely yields >95% pure preparation of dynamin in >10-mg quantities.

Expression of Dynamin-1 in Baculovirus-Infected Insect Cells

Cells and Media. Sf9 (*Spodoptera frugiperda*, GIBCO-BRL, Gaithersburg, MD) insect cells adapted to serum-free suspension culture are grown in Sf-900 II SFM (GIBCO-BRL). TN5 insect cells (comparable to High Five cells, Invitrogen, Carlsbad, CA) adapted to serum-free suspension are grown in EX-CELL 401 (JRH, Lenexa, KS) supplemented with 2 mM L-glutamine and 100 U/ml penicillin/streptomycin. Cells are cultured in non-baffled (Sf9 cells) or baffled (TN5 cells) shaker flasks at 150 rpm and 28°C.

Procedure. For high-level expression of recombinant dynamin-1, we routinely use the Bac-To-Bac baculovirus expression system (GIBCO-BRL). A full-length cDNA encoding human neuronal dynamin-1 is subcloned into the baculovirus vector pFastBac1 and used for the generation of a bacmid through site-specific transposition in *Escherichia coli*. The advantage of this system is that it eliminates the time-consuming requirement for subsequent plaque purification of the virus. Following the manufacturer's instructions (GIBCO-BRL), Sf9 cells are transfected with the bacmid DNA and the transfection reagent CellFECTIN. The virus obtained from the initial transfection supernatant is amplified in exponentially growing Sf9 cells to a high titer virus stock of $\geq 1 \times 10^8$ plaque-forming units (pfu)/ml. This high-titer virus stock is then used to infect 1 liter of TN5 cells, grown to 2×10^6 cell/ml in suspension, with 5 pfu of virus per cell. Under these conditions, TN5 cells can synthesize up to 45 mg/liter of wild-type dynamin-1 (average expression levels are 10–30 mg/liter). The cells are harvested 48 hr after infection while allowing no more than 20% of the cells to lyse (as determined by Trypan-blue exclusion) to minimize proteolysis and release of recombinant protein into the media.

Purification of Dynamin-1

Buffers and Reagents

HEPES column buffer (HCB): 20 mM HEPES, pH 7, 2 mM EGTA, 1 mM MgCl$_2$, and 1 mM dithiothreitol (DTT). Dynamin activity is affected by oxidation (basal rates of GTP hydrolysis can increase, stimulated rates can decrease); therefore, DTT should be maintained in all buffers. HCB50, HCB100, and HCB250 are supplemented with 50, 100, and 250 mM NaCl, respectively.

Phosphate buffers (K-PO$_4$), 200 mM, pH 7.2, or 400 mM, pH 7.2, containing 1 mM DTT.

Protease inhibitors: Dynamin is extremely protease sensitive, therefore protease inhibitors need to be present throughout the purification. Protease inhibitor cocktail tablets (Complete, Boehringer Mannheim, Indianapolis, IN, 1 tablet/50 ml) or a mixture of separate inhibitors [10 μM pepstatin, 5 μM aprotinin, 1 μg/ml tosyl-L-lysine chloromethyl ketone (TLCK), 10 μM leupeptin, 1 mM phenylmethylsulfonyl fluoride (PMSF), or 2 μg/ml 4-(2-aminoethyl)-benzenesulfonyl-fluoride (AEBSF)] can be used. Calpain inhibitor is used at 0.5 μg/ml (Calbiochem, La Jolla, CA).

Q-Sepharose Fast Flow (Amersham Pharmacia Biotech, Piscataway, NJ).

Macro-Prep Ceramic Hydroxyapatite Type I (Bio-Rad, Hercules, CA).

Procedure. Cells are harvested 48 hr after infection (see above) by centrifugation at 1000g for 10 min. If desired, the cell pellets can be frozen and stored for up to 6 months at $-80°$ for subsequent processing. For freezing, the cell pellet from 1 liter of cell suspension is equally divided between two 50-ml Falcon tubes to ensure rapid freezing and thawing. Cells are frozen by submerging the tubes into liquid nitrogen with constant swirling. All of the following steps are performed at 4°.

Cell pellets are resuspended in no more than 50 ml of HCB100 containing protease inhibitors and, if frozen, are quickly thawed by swirling in a 37° water bath. For homogenization, N$_2$-cavitation at 500 psi for 25 min prior to slow release has been shown to be the most reproducible, thorough, and gentle method, although cells can also be broken by sonication. The homogenate is diluted twofold with HCB (no NaCl added) to a final concentration of HCB50 and then centrifuged for 1 hr at 50,000 rpm in a Beckman Ti 60 rotor. The supernatant from this step is referred to as cytosol. Precipitation of the cytosol with 30% ammonium sulfate significantly concentrates and enriches for dynamin in the preparation (Fig. 1). In a glass container, (NH$_4$)$_2$SO$_4$ is slowly added to 30% saturation (16.4 g for every 100 ml of cytosol). The solution is stirred for 30 min at 4°, transferred to screw-cap tubes and spun for 12 min in a Beckman JA-20 rotor at 10,000g. The pellets are gently resuspended in not more than 50 ml total volume of HCB50, containing protease inhibitors, using a small Dounce homogenizer and centrifuged at 10,000g for 8 min to pellet aggregated protein.

The resuspended and clarified ammonium sulfate pellet is loaded onto

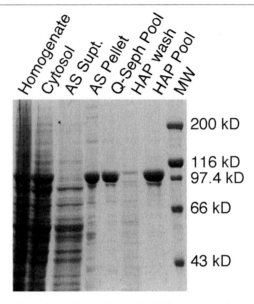

FIG. 1. Coomassie blue-stained 7.5% gel showing SDS–PAGE of fractions obtained during purification of dynamin-1 from baculovirus-infected insect cells.

a Q-Sepharose Fast Flow column (~30-ml volume) preequilibrated with HCB50. The column is washed extensively with HCB100 before elution of dynamin with HCB250. Five-ml fractions are collected at 2 ml/min and the peak fractions of dynamin are combined. The protein concentration of the combined fractions is then determined to calculate the size of the column needed for the subsequent purification step on Macro-Prep Ceramic Hydroxyapatite (HAP) Type I. HAP has a capacity of ~5 mg Q-Sepharose pool/ml resin and average preparations use a 5- to 10-ml column. The HAP column is equilibrated at 0.8 ml/min in HCB250, AEBSF, and calpain inhibitor. Calpain inhibitor is essential at this point because Ca^{2+} proteases seem to be activated during fractionation on HAP. Calpain inhibitor (2 μM), AEBSF (1 mM), and DTT (0.1 mM) are then added to the pooled dynamin fractions before loading onto the HAP column. Washing the HAP column with at least 4 column volumes of 200 mM K-PO$_4$, 1 mM DTT serves as a major purification step for dynamin as substantial amounts of contaminating protein elutes in the wash (Fig. 1). Dynamin is then eluted with 400 mM K-PO$_4$, 0.1 mM DTT and collected in 1-ml fractions (1.25 min/fraction). Protein assays and SDS–PAGE are performed to assess protein yield and purity, which is routinely greater than 95% as judged by Coomassie blue staining. Aliquots are stored at $-80°$ in 400 mM K-PO$_4$ containing 1 mM DTT and 1 mM PMSF. This protocol results in 3–7 mg/ml of pure dynamin and takes about 12 hr when initiated with frozen pellets of insect cells.[6,9] The HAP column can be loaded or washed slowly overnight and eluted the following morning.

Velocity Sedimentation as Measure of Dynamin Assembly

Dynamin has the propensity to self-assemble, in conditions of low ionic strength[10] or under physiologic salt conditions in the presence of GDP and AlF_4^-.[11] The resulting rings and helical stacks of rings are readily sedimentable by ultracentrifugation, providing a simple means for quantitatively assaying self-assembly by velocity sedimentation. The column chromatography purification described in detail in the previous section yields unassembled dynamin tetramers in high-salt buffers. The velocity sedimentation assay can be used to rapidly determine buffer conditions under which both wild-type and mutant forms of dynamin can assemble *in vitro*. Furthermore, the assay can be used to optimize novel buffer conditions or to evaluate the effect of site-directed mutations on the overall assembly process of dynamin. Dynamin self-assembly can also be measured by light scattering at A_{320nm}[12]; however, this assay does not provide quantitative information on the extent of self-assembly and requires significantly more material.

Reagents. All assays are performed with purified protein in the presence of HEPES column buffer (HCB) (20 mM HEPES, pH 7, 2 mM EGTA, 1 mM $MgCl_2$, 0.1 mM DTT) with varying concentrations of NaCl.

Procedure. Following column purification, the dynamin, which is stored in 400 mM K-PO_4 must first be equilibrated in high-salt assay buffer (HCB 150) by dialysis. Routinely, 100-μl samples of purified dynamin are dialyzed against 1 liter of HCB150 at 4° through low molecular weight cutoff dialysis tubing [12,000–15,000 MWCO; Spectra/Por 2 dialysis membrane tubing (Spectrum Medical Industries, Houston, TX)]. The samples are retrieved, centrifuged at 14,000 rpm for 1 min in a microfuge to sediment aggregated material, and the supernatant transferred to new microcentrifuge tubes. The dialyzed dynamin samples are then assayed for total protein concentration using standard microassay techniques (Coomassie Protein Assay Reagent, Pierce, Rockford, IL). Standard conditions for the velocity sedimentation assay are 5 μg dynamin per assay condition in 100 μl total reaction volume. The final salt concentration is then varied from 150 to 10 mM NaCl to determine which best supports dynamin self-assembly. In practice,

[9] D. E. Warnock, L. J. Terlecky, and S. L. Schmid, *EMBO J.* **14,** 1322 (1995).
[10] J. E. Hinshaw and S. L. Schmid, *Nature* **374,** 190 (1995).
[11] J. F. Carr and J. E. Hinshaw, *J. Biol. Chem.* **272,** 28030 (1997).
[12] D. E. Warnock, T. Baba, and S. L. Schmid, *Mol. Biol. Cell* **8,** 2553 (1997).

the precise ionic strength necessary for self-assembly is determined empirically for each new buffer condition or dynamin mutant. To maintain appropriate buffer conditions, HCB150 is diluted with HCB (containing no NaCl) to achieve the desired final salt concentration in the assay.

A range of ionic strength conditions is then prepared in individual ultracentrifuge tubes (Beckman, Fullerton, CA), followed by the addition of 5 μg of dynamin to each ultracentrifuge tube. The dynamin-containing ultracentrifuge tubes are then incubated for 10 min at 20°. Following the incubation, the tubes are transferred to a TLA100 rotor (Beckman) and centrifuged in a tabletop ultracentrifuge (Optima, Beckman) at 100,000g for 10 min.

The supernatant (containing unassembled dynamin) is then carefully collected and transferred to 1.5-ml microcentrifuge tubes. The ultracentrifuge tubes with the pellet (containing the assembled dynamin oligomers) are placed on ice. The protein in the supernatant fraction is then trichloroacetic acid (TCA) precipitated with the addition of 10 μl 100% TCA. (1 μl of 1% Triton X-100 can be included prior to the addition of the TCA to aid the visualization of the protein pellet in subsequent steps.) The tubes are vortexed and incubated at 20° for 30 min. At the end of the incubation, the tubes are centrifuged at 14,000 rpm for 15 min at 4° to pellet the precipitated dynamin. The supernatant is carefully aspirated and 300 μl ice-cold acetone added; the tubes are vortexed and centrifuged as before. Following the final centrifugation of the precipitated supernatant material, the acetone is aspirated and the pellets allowed to air dry. Then, 25 μl of urea solubilization buffer (4 M urea; 0.5% SDS; 100 mM Tris, pH 7) diluted with Laemmli sample loading buffer is added to the pellets representing both the assembled and unassembled dynamin fractions. The samples are then separated by SDS–PAGE (7.5%). Quantitation of dynamin is performed by scanning Coomassie Brilliant Blue stained gels on a Molecular Dynamics (Sunnyvale, CA) densitometer and using Molecular Dynamics ImageQuant software to calculate relative intensities of dynamin in supernatant and pellet fractions.

GTPase Assay for Dynamin Function

As a member of the GTPase family, dynamin has the enzymatic activity to hydrolyze GTP into GDP and inorganic phosphate. Dynamin GTPase activity can be stimulated by a number of functionally diverse effectors, including microtubules (MT), oligomeric SH3 domain-containing proteins (e.g., grb2), and acidic phospholipids, each of which promotes dynamin self-assembly. Unlike most other GTPases, however, dynamin also exhibits a high endogenous GTPase activity (~2 min^{-1}) and relatively low affinity for substrate $(10-100 \ \mu M)$.[13,14] A hallmark of dynamin's intrinsic and stimulated GTPase activities is that they are cooperative (i.e., the specific GTPase activity of dynamin increases with its concentration when assayed under low-salt conditions that favor self-assembly). Numerous mutations at specific amino acids have assisted in delineating important residues in dynamin-1 necessary for GTP binding and hydrolysis and for self-assembly.[7,15,16] Thus, a rapid assay that measures the rate of hydrolysis and affinity for nucleotides provides a meaningful tool to assess the contributions of rationally designed site-directed modifications in the molecule. We describe a simple, 2-hr protocol that can be used to identify dynamin domains that stimulate or inhibit dynamin's GTPase activity or to screen dynamin and its mutants for GTPase activity in a variety of experimental conditions (e.g., salt, interacting molecules).

Reagents. GTPase assays are performed in HCB with the final salt concentration adjusted appropriately. Assays are conveniently performed in 0.5- or 1.5-ml microcentrifuge tubes. $[\alpha\text{-}^{32}P]GTP$ (specific activity 400 Ci/mmol; 10 mCi/ml) is purchased from Amersham Pharmacia Biotech (Arlington Heights, IL). GTP is acquired from Sigma Chemical Co (St. Louis, MO) and stored for up to 4 months at $-80°$ as a 50 mM stock solution in 10 mM Tris, pH 7.4, 50 mM MgCl$_2$. For assays, 3.2 μl of this stock is diluted to 40 μl in water and 2 μl of $[\alpha\text{-}^{32}P]GTP$ is added to generate a 4 mM assay stock of $[\alpha^{32}P]GTP$.

Thin-layer chromatography (TLC) is performed on flexible cellulose polyethyleneimine TLC plates (5 × 20 cm, Selecto Scientific, Norcross, GA). The TLC plates are predeveloped with distilled water overnight, air-dried, and stored at 4°C until needed. Pencil marks are placed at 0.5 cm intervals along a line that is 1 cm from the bottom of the plate to mark the location that will be spotted at the end of each time point.

Procedure. Dynamin's intrinsic GTPase activity is directly related to its degree of self-assembly and the assembly of dynamin is dependent on dynamin concentration and ionic strength. Dynamin's interactions with many of its effectors are also salt sensitive. These variables may account, in part, for the widely divergent rates of intrinsic and stimulated GTP hydrolysis for dynamin reported in the literature. Assays for assembly-stimulated GTPase activity are performed under low-salt ($\leq$50 mM NaCl) conditions. While GED-stimulated GTPase activity is maximal at 50 mM

[13] H. S. Shpetner and R. B. Vallee, *Nature* **355,** 733 (1992).
[14] K. Maeda, T. Nakata, Y. Noda, R. Sato-Yoshitake, and N. Hirokawa, *Mol. Biol. Cell* **3,** 1181 (1992).
[15] J. S. Herskovits, C. C. Burgess, R. A. Obar, and R. B. Vallee, *J. Cell Biol.* **122,** 565 (1993).
[16] H. Damke, T. Baba, D. E. Warnock, and S. L. Schmid, *J. Cell Biol.* **127,** 915 (1994).

NaCl, in our hands, MT-stimulated GTPase activity is greatest when assayed at 20 mM NaCl. To accurately determine dynamin's assembly-independent, basal rate of GTP hydrolysis, assays should be performed under high-salt (100–150 mM NaCl) conditions. Careful attention must also be paid to the order of addition of reaction components so that assembly/disassembly equilibria can be reached before addition of GTP and determination of initial rates of hydrolysis.

The total volume of a typical assay is 20 μl, which is sufficient to provide samples for TLC (the volumes can be adjusted to reflect protein concentration problems if needed). Initially, 10.8 μl of HCB, containing 0.1% BSA is added to each microcentrifuge tube on ice. This is followed by addition of dynamin and, if desired, the effector molecule to be tested in a total volume of 6.7 μl HCB150. Final concentrations of dynamin typically assayed are in the range of 0.5–3 μM for intrinsic, 0.5–1.0 μM for MT-stimulated, or 0.1–0.6 μM for GED-stimulated GTPase activity, respectively. The effector molecules (and final concentrations) used in our lab are taxol-stabilized MT (100 μg/ml), grb2 (2 μM), or isolated GED (5–10 μM; see below). Dynamin GTPase activity is also stimulated by acidic phospholipid vesicles.[17–19] The reaction, containing dynamin and its effector molecules is gently mixed and incubated on ice for 15–30 min. GTP (2.5 μl of the 4 mM [α-^{32}P]GTP assay stock) is then added to each tube. An aliquot (typically 1.5–2 μl) is spotted onto a TLC strip as the "zero-minute" time point. An important negative control to determine backgrounds (MT, for example, themselves have GTPase activity) is to incubate all reaction components except dynamin. In this case, aliquots are spotted at time zero and again at the end of the incubation. The tubes are then transferred to and incubated in a 37° water bath. Aliquots from the reaction mixtures are removed and spotted onto the TLC plates at appropriate time intervals (typically 5 min, assays are linear for 15–30 min). When the incubation is complete, the TLC plates are dried with a hair dryer set to "cool" and then placed in the developing solution (a 1:1 mixture of 1 M LiCl and 2 M formic acid) that is prepared immediately before use. The plates are allowed to develop, then removed from the liquid and dried with a hair dryer. The plates are then wrapped in plastic wrap and placed in a PhosphorImager cassette for approximately 1 hr. The results are then imaged and quantitated using image analysis software for the PhosphorImager (Molecular Dynamics, Sunnyvale, CA). Rates of GTP hydrolysis are calculated from a minimum of five time points and expressed

[17] P. L. Tuma, M. C. Stachniak, and C. A. Collins, *J. Biol. Chem.* **268,** 17240 (1993).
[18] H. C. Lin, B. Barylko, M. Achiriloaie, and J. P. Albanesi, *J. Biol. Chem.* **272,** 25999 (1997).
[19] M. H. B. Stowell, B. Marks, P. Wigge, and H. T. McMahon, *Nature Cell Biol.* **1,** 27 (1999).

as the percentage of GDP/GTP + GDP after subtracting the "zero-minute" background. A representative chromatogram and the corresponding quantitation are shown in Fig. 2.

GED-Stimulated GTPase Activity

Among all the interacting molecules for dynamin identified *in vitro,* dynamin itself is the most important binding partner. Cross-linking studies have established that the GED domain (GTPase effector domain, located between the PH domain and the PRD) of dynamin is closely linked with the N-terminal GTPase domain, and removal of GED by limited proteolysis results in a truncated dynamin with impaired GTPase activity.[6] To establish that the GED domain can act as an intramolecular GTPase-activating protein (GAP), this 14-kDa protein was expressed as a fusion protein with glutathione *S*-transferase (GST) and analyzed in the GTPase assay described in the previous section.

Reagents

pGEX-4T-3 vector (Amersham Pharmacia Biotech)
Glutathione-Sepharose 4B (Amersham Pharmacia Biotech)
Thrombin (Calbiochem)
HEPES column buffer (HCB): 20 mM HEPES, pH 7, 2 mM EGTA, 1 mM MgCl$_2$, and 1 mM DTT; HCB50, HCB100, and HCB250 are supplemented with 50, 100, and 250 mM NaCl, respectively.

Procedure. The vector for GED expression is designed by cloning a 0.4-kb cDNA fragment encoding the amino acids 618–752 of the neuronal isoform dynamin-1 into the pGEX-4T-3 vector. Smaller constructs affecting either the N or C terminus were less active in stimulating dynamin's GTPase activity. GST–GED is expressed by *Escherichia coli* BL21 after induction with 1 mM isopropyl-β-D-thiogalacto-pyranoside (IPTG) for no more than 3 hr at 30°. Cell pellets are resuspended in dPBS with 1 mM DTT and protease inhibitors and treated with 0.5 mg/ml lysozyme for 15 min on ice. The DNA is then degraded by the addition of DNAse I (0.05 mg/ml) in 10 mM MgCl$_2$ and incubation for 15 min on ice. Then, Triton X-100 is added to a final concentration of 1% and the suspension is left for 10 min on ice followed by brief sonication with three short pulses. The suspension is centrifuged in a Beckman JA-20 rotor for 20 min at 4° at 12,000g. At least 2 ml (depending on the expected yield) of a 50% slurry of glutathione-Sepharose (prewashed in dPBS) are added to the supernatant and incubated with rocking for 1 hr at 4°. The suspension is poured into an empty column for subsequent washing with 3 column volumes of dPBS. GED is cleaved from GST while bound to the column. The glutathione-Sepharose bead

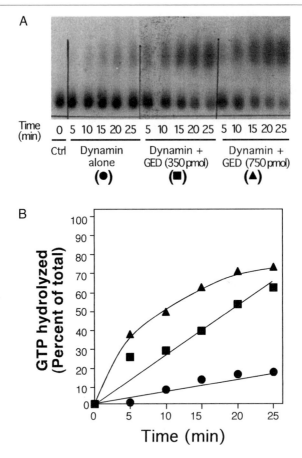

FIG. 2. A thin-layer chromatogram from a typical GED-stimulated GTPase assay (A) and the corresponding quantitation after PhosphorImager analysis (B). Initial rates are determined from the slope of the curves. The specific GTPase activity of dynamin is calculated by converting %GTP hydrolyzed/min to pmol/min and then dividing by the pmol dynamin used in the assay. Note that the curve with high GED concentration has reached saturation kinetics and it is therefore difficult to determine initial rates. This can be corrected by taking shorter time points or using less dynamin. Note also the variability in a single experiment; hence, all rate determinations are calculated from three independent experiments.

volume is diluted 1 : 1 with dPBS and thrombin is added to a final concentration of 25 U/ml, followed by digestion at 4° for 6 hr. The thrombin digest is stopped by addition of PMSF to 0.4 mM. After spinning the beads for 5 min at 500g, the GED-containing supernatant is carefully collected. The thrombin digest can also be performed after eluting GST–GED from the column with 50 mM Tris, pH 8.5, 10 mM glutathione. For elution, the column needs to be incubated for at least 20 min to 1 hr. GED is subsequently separated from GST by chromatography on Q Sepharose and eluted with HCB500 (500 mM NaCl). Yield and purity are assessed by SDS–PAGE and Coomassie blue staining. Isolated GED is dialyzed overnight at 4° into HCB150, 1 mM DTT (GED activity is also sensitive to oxidation) and aggregates formed during dialysis are removed by centrifugation for 5 min at 8000g.

[48] Analysis of Phosphoinositide Binding by Pleckstrin Homology Domain from Dynamin

By ANTHONY LEE *and* MARK A. LEMMON

Introduction

Pleckstrin Homology Domains

Pleckstrin homology (PH) domains are small modules of approximately 120 amino acids that are found in many proteins involved in cell signaling and other membrane-associated processes.[1-3] Harlan *et al.*[4] reported that one of the two PH domains from pleckstrin binds to phosphatidylinositol 4,5-bisphosphate [PtdIns(4,5)P$_2$], albeit with relatively low affinity (K_D in the range of 30 μM). Similar phosphoinositide binding has since been reported for many PH domains,[1-3] and a subset binds with both high affinity and specificity to PtdIns(4,5)P$_2$[5] or products of phosphatidylinositol 3-kinases (PI3Ks).[6] For the PH domains from phospholipase C-δ_1 (PLC-δ_1)[7] and Bruton's tyrosine kinase (Btk),[8] high-resolution X-ray crystal structures

[1] M. A. Lemmon and K. M. Ferguson, *Curr. Top. Microbiol. Immunol.* **228,** 39 (1998).

[2] M. J. Rebecchi and S. Scarlata, *Annu. Rev. Biophys. Biomol. Struct.* **27,** 503 (1998).

[3] N. Blomberg, E. Baraldi, M. Nilges, and M. Saraste, *Trends Biochem. Sci.* **24,** 441 (1999).

[4] J. E. Harlan, P. J. Hajduk, H. S. Yoon, and S. W. Fesik, *Nature* **371,** 168 (1994).

[5] M. A. Lemmon, K. M. Ferguson, R. O'Brien, P. B. Sigler, and J. Schlessinger, *Proc. Natl. Acad. Sci. U.S.A.* **92,** 10472 (1995); P. Garcia, R. Gupta, S. Shah, A. J. Morris, S. A. Rudge, S. Scarlata, V. Petrova, S. McLaughlin, and M. J. Rebecchi, *Biochemistry* **34,** 16228 (1995); H. Yagisawa, M. Hirata, T. Kanematsu, Y. Watanabe, S. Ozaki, K. Sakuma, H. Tanaka, N. Yabuta, H. Kamata, H. Hirata, and H. Nojima, *J. Biol. Chem.* **269,** 20179 (1994).

[6] D. A. Fruman, L. E. Rameh, and L. C. Cantley, *Cell* **97,** 817 (1999); J. K. Klarlund, A. Guilherme, J. J. Holik, A. Virbasius, and M. P. Czech, *Science* **275,** 1927 (1997).

[7] K. M. Ferguson, M. A. Lemmon, J. Schlessinger, and P. B. Sigler, *Cell* **83,** 1037 (1995).

[8] E. Baraldi, K. D. Carugo, M. Hyvonen, P. L. Surdo, A. M. Riley, B. V. Potter, R. O'Brien, J. E. Ladbury, and M. Saraste, *Structure* **7,** 449 (1999).

of the complex formed with the headgroups of PtdIns(4,5)P$_2$ and PtdIns(3,4,5)P$_3$, respectively, have been determined. These structures demonstrate that the high-affinity binding sites are closely related and provide explanations for the apparent specificity. PH domains fall into two broad classes.[9] Members of one class (including those from PLC-δ_1 and Btk) bind strongly and specifically to one or two phosphoinositides, while members of the second class are promiscuous in their phosphoinositide binding and bind with relatively low affinities. Analysis of PH domain binding to PtdIns(4,5)P$_2$ and PI3K products in live yeast[10] suggests a consensus sequence representing the essential components of the high-affinity binding sites seen in crystal structures.[7,8] PH domains that bind weakly and promiscuously to phosphoinositides lack this binding site or have only part of it. These PH domains, which include that from dynamin, are likely to interact with negatively charged phospholipids through a positively charged face seen in all PH domains.[1]

Dynamin

Dynamin is a 100-kDa GTPase with a key role in endocytosis, particularly the scission or "pinching-off" event that leads to endocytic vesicle formation.[11] The precise nature of dynamin's role in this process, and whether it is directly or indirectly responsible for the pinching-off event, is still a subject of debate.[11] As well as its GTPase domain, dynamin contains a PH domain, a GTPase effector domain (GED), and a C-terminal proline/arginine-rich domain that participates in recruitment of the protein to clathrin-coated pits.[12] There is substantial evidence that phosphoinositides, particularly PtdIns(4,5)P$_2$, play a critical role in controlling endocytic vesicle formation. The clathrin adaptor protein AP2 requires interaction with phosphoinositides for its recruitment into clathrin coated pits.[13] A comparison of the ability of different PH domains to inhibit receptor-mediated endocytosis has also suggested an important role for PtdIns(4,5)P$_2$ at a step in

---

[9] J. M. Kavran, D. E. Klein, A. Lee, M. Falasca, S. J. Isakoff, E. Y. Skolnik, and M. A. Lemmon, *J. Biol. Chem.* **273,** 30497 (1998); L. E. Rameh, A.-K. Arvidsson, K. L. Carraway, A. D. Couvillon, G. Rathbun, A. Crompton, B. VanRenterghem, M. P. Czech, K. S. Ravichandran, S. J. Burakoff, D.-S. Wang, C.-S. Chen, and L. C. Cantley, *J. Biol. Chem.* **272,** 22059 (1997).

[10] S. J. Isakoff, T. Cardozo, J. Andreev, Z. Li, K. M. Ferguson, R. Abagyan, M. A. Lemmon, A. Aronheim, and E. Y. Skolnik, *EMBO J.* **17,** 5374 (1998).

[11] W. Yang and R. A. Cerione, *Curr. Biol.* **9,** R511 (1999); S. Sever, A. B. Muhlberg, and S. L. Schmid, *Nature* **398,** 481; M. H. Stowell, B. Marks, P. Wigge, and H. T. McMahon, *Nature Cell Biol.* **1,** 27 (1999); S. M. Sweitzer and J. E. Hinshaw, *Cell* **93,** 1021 (1998).

[12] D. E. Warnock and S. L. Schmid, *BioEssays* **18,** 885 (1996); A. B. Muhlberg, D. E. Warnock, and S. L. Schmid, *EMBO J.* **16,** 6676 (1997).

[13] I. Gaidarov and J. H. Keen, *J. Cell Biol.* **146,** 755 (1999).

endocytosis that follows AP2 recruitment.[14] Targeted disruption of synaptojanin, an inositol-5-phosphatase concentrated in nerve terminals, causes clathrin-coated vesicles to accumulate.[15]

Finally, as discussed in more detail below, phosphoinositide binding by the PH domain of dynamin has been shown to be critical for participation of dynamin in receptor-mediated endocytosis.[16,17] The PH domain from dynamin binds to phosphoinositides and their soluble headgroups with very low affinities. The K_D value for inositol 1,4,5-trisphosphate [Ins(1,4,5)P$_3$] binding by the dynamin-1 PH domain (Dyn1-PH) is between 1.2 and 4.3 mM.[18,19] Using approaches described below we have shown that, unlike some PH domains but in common with many, Dyn1-PH requires oligomerization in order to interact strongly with phosphoinositides in membranes.[20] This is the result of a simple avidity effect. Isolated dynamin forms tetramers and higher order oligomers.[12] According to some models of dynamin function, its oligomerization can be regulated by interactions with other components of the endocytic machinery.[21] If there is such a step in which dynamin oligomerization is tightly regulated, the avidity of its PH domain-mediated membrane targeting will be regulated similarly tightly. Such "avidity regulation" could be important in controlling a step in endocytosis at which dynamin is targeted to the necks of invaginated coated pits prior to vesicle scission.

Materials and Reagents

Reagents

Synthetic dipalmitoyl-PtdIns-3-P, PtdIns(3,4)P$_2$, and PtdIns(3,4,5)P$_3$ are purchased from Matreya, Inc. (Pleasant Gap, PA). Phosphatidylserine

[14] M. Jost, F. Simpson, J. M. Kavran, M. A. Lemmon, and S. L. Schmid, *Curr. Biol.* **8,** 1399 (1998).

[15] O. Cremona, G. Di Paolo, M. R. Wenk, A. Lüthi, W. T. Kim, K. Takei, L. Daniell, Y. Nemoto, S. B. Shears, R. A. Flavell, D. A. McCormick, and P. De Camilli, *Cell* **99,** 179 (1999).

[16] M. Achiriloaie, B. Barylko, and J. P. Albanesi, *Mol. Cell. Biol.* **19,** 1410 (1999); Y. Vallis, P. Wigge, B. Marks, P. R. Evans, and H. T. McMahon, *Curr. Biol.* **9,** 257 (1999).

[17] A. Lee, D. W. Frank, M. S. Marks, and M. A. Lemmon, *Curr. Biol.* **9,** 261 (1999).

[18] J. Zheng, S. M. Cahill, M. A. Lemmon, D. Fushman, J. Schlessinger, and D. Cowburn, *J. Mol. Biol.* **255,** 14 (1996).

[19] K. Salim, M. J. Bottomley, E. Querfurth, M. J. Zvelebil, I. Gout, R. Scaife, R. L. Margolis, R. Gigg, C. I. Smith, P. C. Driscoll, M. D. Waterfield, and G. Panayotou, *EMBO J.* **15,** 6241 (1996).

[20] D. E. Klein, A. Lee, D. W. Frank, M. S. Marks, and M. A. Lemmon, *J. Biol. Chem.* **273,** 27725 (1998).

[21] D. J. Owen, P. Wigge, Y. Vallis, J. D. A. Moore, P. R. Evans, and H. T. McMahon, *EMBO J.* **18,** 5273 (1998).

(PtdSer), PtdIns-4-P, and PtdIns(4,5)P$_2$, from bovine brain, are purchased from Sigma (St. Louis, MO). PtdIns-5-P and PtdIns(3,5)P$_2$ are purchased from Echelon Inc. (Salt Lake City, UT), who now supply all of the other phosphoinositides. Phosphatidylinositol (PtdIns), phosphatidylcholine (PtdCho), and di(dibromostearoyl)-PtdCho are from Avanti Polar Lipids (Birmingham, AL). Protein kinase A (PKA) (catalytic subunit: P-2645) and glutathione agarose are from Sigma (St. Louis, MO), [γ-^{32}P]ATP (unpurified) is obtained from NEN (Boston, MA). Glutathione and isopropylthiogalactoside (IPTG) are from Alexis Inc. (San Diego, CA).

Buffer Solutions

TBS: 50 mM Tris-HCl, pH 8.0, 150 mM NaCl, containing 1 mM phenylmethylsulfonyl fluoride (PMSF), 1 mM dithiothreitol (DTT), and 1 mM EDTA

MBS6: 25 mM 2-[N-morpholino]ethanesulfonic acid (MES), pH 6.0, 100 mM NaCl, containing 5 mM DTT and 1 mM PMSF

MB6: 25 mM MES, pH 6.0, containing 1 mM DTT

MBSG6: 50 mM MES, pH 6.0, 150 mM NaCl, 10% (w/v) glycerol, containing 1 mM DTT

DK buffer: 50 mM potassium phosphate, pH 7.15, containing 10 mM MgCl$_2$, 5 mM NaF, and 4.5 mM DTT

HBS: 25 mM HEPES, pH 7.2, 100 mM NaCl

GE buffer: 50 mM Tris-HCl, pH 8.0, 5 mM NaF, 1 mM EDTA, 0.5 mg/ml bovine serum albumin (BSA), 1 mM DTT, 1 mM PMSF

Methods

Expression and Purification of Dyn1-PH

cDNA fragments corresponding to the region encoding residues 510–633 of human dynamin-1 or residues 509–629 of rat dynamin-2 are subcloned into several *Escherichia coli* expression vectors. For expression of glutathione *S*-transferase (GST) fusion proteins, pGEX-2TK and pGEX-2T (Pharmacia, Piscataway, NJ) are used. The former has a PKA site between GST and the PH domain, allowing labeling by phosphorylation. For expression of the PH domain alone, the same DNA fragments are cloned between the *Nde*I and *Bam*HI sites of pET11a (Novagen, Madison, WI). Proteins are expressed in Luria broth by IPTG induction (1 mM final concentration) for 3 hr at 37° of a culture (OD$_{600}$ 0.6–0.8) of *E. coli* BL-21(DE3) harboring the relevant plasmid. Following lysis of pelleted *E. coli* by sonication in TBS or MBS6, essentially all of the protein is found in the supernatant after a 14,000 rpm spin at 4° for 20 min in an SS34 rotor (Sorvall, Newtown, CT). With some other PH domains, it is necessary to induce expression at 25° or lower, or to alter the domain boundaries in order to achieve soluble expression.

GST fusion proteins are purified by incubating the lysis supernatant (in TBS) with glutathione agarose beads for 15 min at 4°. The beads are then washed 4 times in TBS, once in TBS containing 1 M NaCl, and finally the protein is eluted in TBS containing 15 mM reduced glutathione (ensuring the pH of the elution buffer is 8.0). When protein is to be used for dot-blots analysis (see below), labeling with ^{32}P precedes elution.

Dyn1-PH that is not fused to GST is purified as described in detail by Ferguson et al.[22] Briefly, the clarified lysate (in MBS6) is passed through DEAE-cellulose (Sigma) equilibrated in MBS6. The DEAE flow-through is diluted threefold (not necessary for all PH domains) with MB6 buffer, and loaded onto a EMD-650S$^-$ (Merck, Dahmstadt, Germany) tentacle-type cation-exchange column on a GradiFrac chromatography system (Pharmacia), preequilibrated with buffer MB6. Protein is then eluted with a gradient over 5 column volumes in NaCl from 0 to 250 mM NaCl (in MB6). Dyn1-PH elutes at approximately 150 mM NaCl. Fractions containing Dyn1-PH are identified by OD$_{280}$, pooled, and protein is precipitated by gradual addition of solid $(NH_4)_2SO_4$ (ICN, Aurora, OH) to 75% saturation on ice. Precipitated protein is collected by centrifugation at 12,000 rpm at 4° for 20 min in an SS-34 rotor (Sorvall), and dissolved gently in the minimum volume of MBSG6. After filtration (0.22 μm; Costar, Cambridge, MA), 0.5-ml aliquots of the protein are gel-filtered using a Superose 12 or Superdex 75 HR 10/30 column (Pharmacia, Piscataway, NJ). Dyn1-PH elutes as a 14-kDa monomer, and is at least 99% pure judging from over-loaded silver-stained SDS gels. Purified protein is stored in 50% (w/v) glycerol at −20° or −80°. This basic purification procedure is suitable for all PH domains that we have expressed solubly from pET vectors to date. The different proteins differ only in the NaCl concentration at which they elute from the cation-exchange column.

Labeling of GST-Dyn1PH with ^{32}P

For use in dot blots to investigate lipid binding, a Dyn1-PH GST-fusion protcin is phosphorylated while attached to glutathione-agarose beads following the method described by Margolis and Young.[23] The protein used

[22] K. M. Ferguson, M. A. Lemmon, J. Schlessinger, and P. B. Sigler, *Cell* **79**, 199 (1994).
[23] B. Margolis and R. A. Young, *in* "DNA Cloning 2" (D. M. Glover and B. D. Hames, eds.), p. 1. IRL Press, Oxford, UK, 1995; D. Ron and H. Dressler, *Biotechniques* **13**, 866 (1992).

is produced using pGEX-2TK, in which a PKA site is encoded by the sequence that precedes the cloning site. Approximately 10 μg of the GST-PH protein, bound to glutathione-agarose beads in a slurry of approximately 60 μl, is washed three times with DK buffer. Ten units of PKA (which has been dissolved in 6 mg/ml DTT 5 min earlier) and 0.75 mCi of [γ^{32}-P]ATP (6000 Ci/mmol, 150 mCi/ml) are added. The 75-μl reaction is then nutated at room temperature for 30 min. After labeling, the beads are spun at 1000 rpm for 2 min at room temperature, and the supernatant discarded. Beads are then washed extensively with phosphate-buffered saline (PBS) containing 1 mM DTT and 1 mM PMSF, and the ^{32}P-labeled GST–PH fusion protein is eluted twice for 30 min at room temperature with 400 μl of 15 mM reduced glutathione in GE buffer. Following elution, ^{32}P-labeled protein is filtered using a 0.22-μm Spin-X (Corning) centrifugal filter, prior to its use in the dot-blot assay. Using glutathione agarose, it is important for maximum yield that the beads (which sediment very easily) are not spun at speeds in excess of 1000 rpm before elution with glutathione. As an alternative to this procedure, ^{32}P-labeled PH domain can be removed from the glutathione agarose by thrombin cleavage.[23] It is important to determine whether these methods are effective for any new binding domain under study, and to establish which approach is most efficient before proceeding with labeling.

Dot Blots for Qualitative Analysis of Phosphoinositide Binding by PH Domains

Phosphoinositides and other lipids are dissolved to a final concentration of 2 mg/ml in a 1 : 1 mixture of chloroform and methanol (containing 0.1% HCl). Using a glass positive-displacement pipettor, 1- or 2-μl spots of these phospholipids are placed onto nitrocellulose sheets (NitroBind: MSI) in a consistent pattern. After drying, the nitrocellulose is blocked overnight at 4° in TBS containing 3% bovine serum albumin (BSA), but no detergent. The 0.22-μm filtered ^{32}P-labeled GST–PH fusion protein is then added to TBS/3% BSA to a final concentration of approximately 0.5 μg/ml, and this mixture is used to probe the phosphoinositide-containing nitrocellulose filter for 30 min with shaking at room temperature. The filters are then washed 5 times (5 min each) with TBS (containing no detergent), dried by blotting with Whatman paper, and bound radioactivity is visualized using a PhosphorImager (Molecular Dynamics, Sunnyvale, CA). A similar approach, termed "far Western blotting," in which binding of an epitope-tagged PH domain to phosphoinositides immobilized on nitrocellulose is detected by immunoblotting, has been described by Stevenson et al.[24]

[24] J. M. Stevenson, I. Y. Perera, and W. F. Boss, *J. Biol. Chem.* **273,** 22761 (1998).

Generation of Phosphoinositide-Containing Lipid Vesicles

A simple method for producing reasonably homogeneous small unilamellar vesicles (SUVs) that can be pelleted with high efficiency by ultracentrifugation has been described by Tortorella and London.[25] The procedure uses a brominated form of PtdCho, specifically di(9,10-dibromostearoyl)-PtdCho (available from Avanti Polar Lipids, Inc.) as the background lipid to increase the density of the vesicles. Brominated PtdCho and the phosphoinositide or other phospholipid of interest are mixed at the desired ratio in 1:1 chloroform:methanol, containing 0.1% HCl. For the studies described by Klein *et al.*[20] the phosphoinositide was present at 3% (molar) in brominated PtdCho. In other studies,[9] we have used a background of 20% PtdSer:80% brominated PtdCho. The mixture of phospholipids in chloroform:methanol is largely dried under a stream of dry nitrogen gas, followed by high vacuum in a Speed-Vac (Savant, Hicksville, NY) to complete drying. The dried lipid mixture is then hydrated in HBS to give a final *total* lipid concentration of 25 mM. To generate SUVs in the hydrated lipid mixture, repeated cycles of freezing in liquid nitrogen and thawing with intensive sonication in a bath sonicator (Bransonics, Danbury, CT) are employed. Once all of the dried material is in suspension, the pH is checked using pH strips, and corrected to pH 7.2 with 1 M NaOH if residual HCl made this necessary. Multiple rounds (between 10 and 20) of freeze/thaw/sonication are then performed, until the sample becomes optically clear. We have used SUVs generated using this method for both centrifugation[9,20] and calorimetric approaches[26] for studying PH domain binding to the headgroups of phosphoinositides. Isothermal titration calorimetry (ITC) studies of the PLC-δ_1 PH domain [which forms a 1:1 complex with Ins(1,4,5)P$_3$ and PtdIns(4,5)P$_2$] indicate that SUVs formed using this method have approximately 57% of their PtdIns(4,5)P$_2$ available (on the outside of the vesicles) for PH domain binding.[26]

Centrifugation Assay for PH Domain Binding to Phosphoinositides in SUVs

To minimize the amount of phosphoinositide required for a study of lipid binding, centrifugation binding assays are performed using a Beckman Optima TLX benchtop ultracentrifuge. We use the 14-place TLA-120.1 rotor, although the 20-place TLA-100 would be more useful. Using thick-walled polycarbonate tubes (Beckman) for the TLA-120.1 rotor, we rou-

[25] D. Tortorella and E. London, *Anal. Biochem.* **217,** 176 (1994).
[26] M. A. Lemmon, K. M. Ferguson, R. O'Brien, P. B. Sigler, and J. Schlessinger, *J. Proc. Natl. Acad. Sci. U.S.A.* **92,** 10472 (1995).

tinely use assay volumes of just 80 μl. It is critical for the assay that significant quantities of DTT, glycerol, glutathione, or other compounds that will interfere with protein assays be removed. Prior to use in the centrifugation assay, Dyn1-PH (or the GST fusion protein) is therefore dialyzed exhaustively against HBS.

Centrifugation assay samples (80 μl) contain 10 μM (or 5 μM in some assays) of purified, dialyzed, PH domain (or GST-PH) plus limit-sonicated SUVs at various concentrations (of total lipid) ranging from 0 to 4 mM. This corresponds to a concentration range of the phosphoinositide of interest in the binding assay of between 0 and 60 μM (assuming that only 50% is accessible on the outside of the vesicles). Vesicle–protein mixtures are centrifuged in the TLX centrifuge for 1 hr at 25° at 85,000 rpm, which corresponds to a g_{max} of approximately 300,000g. After centrifugation, 55 μl of the supernatant is carefully removed and transferred to another tube, and the remaining supernatant gently removed and discarded. The pellet is then resuspended in 80 μl of HBS by bath sonication, and 55 μl of this suspension is transferred to a fresh tube. The Pierce (Rockford, IL) bicinchonic (BCA) protein assay kit is used to determine the concentration of protein in both the supernatant and the pellet, following the manufacturer's recommendations. To remove scattering artifacts when measuring the absorbance of the pellet-derived samples, SDS is added to a final concentration of 0.5% to dissolve lipid after incubation with the BCA reagent. A standard curve for each protein is generated in tandem in order to determine the percentage of total added protein that is pelleted by the vesicles. Molar partition coefficients (K), as defined by Peitsch and McLaughlin,[27] are then estimated by fitting the data (in ORIGIN) to Eq. (1):

$$\% \text{ protein bound} = 100 \times \frac{K[\text{lipid}]}{(1 + K[\text{lipid}])} \tag{1}$$

where [lipid] is the concentration of total available lipid (>[protein]$_{bound}$), approximated by one-half the total lipid concentration (assuming 50% is available on the SUV outer leaflet), and K is the proportionality constant between the concentration of protein bound to the outer SUV leaflet and its concentration in bulk solution. Determination of K makes no assumptions of stoichiometry, although K_D for phosphoinositide binding can be estimated as (mole ratio)/K if 1:1 binding of phosphoinositides is assumed.

Interpretation of Results

The dot-blot approach has been a valuable tool for comparing the specificity of different PH domains. It has also been useful for studies of

[27] R. M. Peitsch and S. McLaughlin, *Biochemistry* **32,** 10436 (1993).

other lipid-binding modules such as FYVE domains. However, as described by Kavran et al.,[9] good note should be taken of certain caveats. Because the different phosphoinositides have different solubilities in aqueous solution, they will tend to be removed, or washed off, from the nitrocellulose filters at different rates. $PtdIns(3,4,5)P_3$ is likely to be leached from the filters more rapidly that $PtdInsP_2$ isomers, which in turn will be washed away more rapidly that PtdInsP isomers. This was well illustrated with the PH domain from phospholipase C-γ_1, which binds PtdIns-3-P, $PtdIns(3,4)P_2$, and $PtdIns(3,4,5)P_3$ with approximately similar affinities according to studies of vesicle binding using the centrifugation assay.[9,28] Despite the similar affinities for these phosphoinositides, dot-blot studies indicated that the selectivity of the PLC-γ_1 PH domain followed the order PtdIns-3-P > $PtdIns(3,4)P_2$ > $PtdIns(3,4,5)P_3$, which is actually likely to reflect simply the amount of phospholipid that remains on the nitrocellulose. Despite this problem, we have found the dot-blot approach very useful for identifying PH or other domains that recognize one or two phosphoinositides with a high degree of specificity, and for analyzing rapidly the influence of various mutations on this specificity. The results of our dot-blot experiments were also critical in providing the initial suggestion that that dynamin's PH domain requires the avidity afforded by oligomerization to recruit its host protein to the membrane surface.

Requirement for Oligomerization in Dyn1-PH Binding to Phosphoinositides

In our own initial studies of Dyn1-PH, we could not detect significant binding to any phosphoinositide, despite the fact that the PLC-δ_1 PH domain bound strongly and specifically to $PtdIns(4,5)P_2$ in the same assays.[26] By contrast, Zheng et al.[18] and Salim et al.[19] both reported significant binding of Dyn1-PH to $PtdIns(4,5)P_2$ in different assays. Zheng et al. monitored quenching of the intrinsic tryptophan fluorescence of Dyn1-PH to study binding to phospholipids that were added from a detergent solution. Salim et al. immobilized a GST fusion protein of Dyn1-PH with an anti-GST antibody attached to a BIAcore sensor chip, and passed suspensions of phosphoinositide-containing lipid vesicles over this surface. Although we do not have a good explanation for the results of Zheng et al., the inconsistency between our initial negative findings and the binding observed by Salim et al. now appears to be the result of an avidity effect that arises because of GST dimerization. Indeed, when we began using the dot-blot approach

---

[28] M. Falasca, S. Logan, V. P. Lehto, G. Baccante, M. A. Lemmon, and J. Schlessinger, *EMBO J.* **17**, 414 (1998).

for Dyn1-PH, its phosphoinositide binding became evident in our hands. However, we were unable to displace ^{32}P-labeled GST-Dyn1PH from phosphoinositides on nitrocellulose using the monomeric Dyn1-PH that we had used to determine the X-ray crystal structure. The failure of this displacement assay indicated that the GST fusion protein binds more strongly than the monomeric protein to phosphoinositides. Because the phosphoinositides were immobilized on a solid support, and since GST itself dimerizes[29] ($K_D < 1$ μM), we reasoned that the higher affinity of GST-Dyn1PH might reflect an avidity effect resulting from the presence of the PH domains in a dimer. This would also explain why Salim *et al.* were able to detect PtdIns(4,5)P$_2$ binding in their studies.[19] Several other examples have been reported in which GST can functionally replace a native oligomerization domain,[30] or enhance apparent binding affinities through avidity effects.[31]

As a direct test of the effect of dimerization on phosphoinositide binding by Dyn1-PH, we mutated the only cysteine in Dyn1-PH (Cys-607 of dynamin-1) to serine, and introduced a unique cysteine at the amino terminus of the PH domain. By oxidizing the protein in the presence of copper(II) 1,10-phenanthroline, we were able to induce its dimerization. As assessed by sedimentation equilibrium analytical ultracentrifugation, this dimerization was reversible on addition of reducing agents such as DTT and tris (2-carboxyethylphosphine) hydrochloride (TCEP, Pierce). Using the centrifugation assay described above, we were then able to measure directly the influence of PH domain dimerization on phosphoinositide binding by measuring the partition coefficient for purified dimers of Dyn1-PH both before and after addition of 1 mM TCEP. TCEP was used for reducing the disulfide-linked dimer since it does not interfere significantly with the BCA protein assay. As seen in Fig. 1, analysis of binding to SUVs containing 3% PtdIns(4,5)P$_2$ shows that the Dyn1-PH dimer binds substantially more strongly than the monomer.[20] Effective K_D values are in the range of 9 μM (dimer) and >60 μM (monomer), suggesting that even higher order oligomerization will be required for strong membrane-targeting *in vivo*.

Importance of Phosphoinositide Binding by Dyn1-PH Studied *in Vivo*

To determine whether the oligomerization-dependent binding of Dyn1-PH to phosphoinositides is physiologically relevant, we and other groups

[29] M. A. McTigue, D. R. Williams, and J. A. Tainer, *J. Mol. Biol.* **246**, 21 (1995); L. G. Riley, G. B. Ralston, and A. S. Weiss, *Protein Eng.* **9**, 223 (1996).

[30] Y. Maru, D. E. Afar, O. N Witte, and M. Shibuya, *J. Biol. Chem.* **271**, 15353 (1996); C. Inouye, N. Dhillon, and J. Thorner, *Science* **278**, 103 (1997); H. Yan, J. T. E. Lim, L. G. Contillo, and J. J. Krolewski, *Anal. Biochem.* **231**, 455 (1995).

[31] T. Tudyka and A. Skerra, *Protein Sci.* **6**, 2180 (1997); J. E. Ladbury, M. A. Lemmon, M. Zhou, J. Green, M. C. Botfield, and J. Schlessinger, *Proc. Natl. Acad. Sci. U.S.A.* **92**, 3199 (1995).

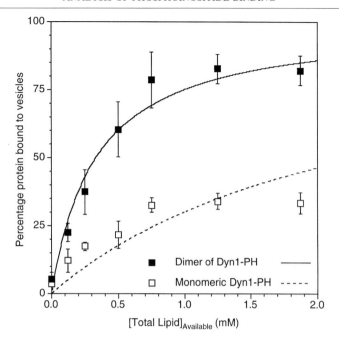

Fig. 1. Effect of dimerization on Dyn1-PH binding to SUVs containing 3% (molar) PtdIns(4,5)P$_2$, analyzed using the centrifugation assay described. Dyn1-PH with a unique cysteine at its amino terminus was oxidized to induce dimerization as described.[20] Binding to PtdIns(4,5)P$_2$-containing SUVs was then measured both in the absence of reducing agent, where the protein is dimeric, and in the presence of 1 mM TCEP, which reduces the intermolecular disulfide bond to yield monomeric Dyn1-PH. Fits to Eq. (1) are shown for the data. Error bars represent standard derivations from the mean for 3 independent experiments.

made mutated forms of dynamin-1 that contained a PH domain with impaired phosphoinositide binding.[16,17] Deletion of the PH domain from dynamin is known to abolish the ability of PtdIns(4,5)P$_2$ to increase its GTPase activity.[19] Mutation of K535, in the first interstrand loop of Dyn1-PH, to alanine or methionine abolishes binding of the isolated PH domain to PtdIns(4,5)P$_2$ in the assay of Salim *et al.*[19] The same mutations in the context of full-length dynamin abolish PtdIns(4,5)P$_2$ activation.[16] Dynamin with either of these mutations (or with the PH domain deleted) was found to act as a dominant-negative inhibitor of transferrin uptake by receptor-mediated endocytosis in COS cells.[16] Using a different PH domain mutation that prevents phosphoinositide binding by dimeric Dyn1-PH, we also found that dynamin-1 defective in PtdIns(4,5)P$_2$ binding is a dominant negative inhibitor of transferrin uptake in HeLa cells.[17] In each case, cells transiently overexpressing the mutated form of dynamin (as assessed by immunofluorescence staining) failed to internalize a fluorescently labeled or biotinylated form of transferrin. We suggest that the endogenous dynamin in these cells is impaired in PH domain-mediated membrane binding at some point in the endocytic process as a result of its co-oligomerization with exogenous dynamin that is defective in phosphoinositide binding. The avidity effect normally achieved by oligomerization is lost if other molecules in the assembly have defective PH domains, and this appears to be sufficient to inhibit dynamin function.

[49] Mapping Dynamin Interdomain Interactions with Yeast Two-Hybrid and Glutathione S-Transferase Pulldown Experiments

By Elena Smirnova, Dixie-Lee Shurland, and Alexander M. van der Bliek

Introduction

Dynamin is a 100-kDa GTPase that contributes to the formation of clathrin-coated vesicles.[1] Dynamin assembles into multimeric spirals at the necks of budding vesicles. We investigated protein interactions that in all likelihood form the basis for dynamin self-assembly using yeast two-hybrid and glutathione S-transferase (GST) pulldown techniques.[2] These techniques were chosen because they allowed us to systematically test binding between individual dynamin protein domains and they allowed us to determine the boundaries of those domains with deletion series. We discovered binding interactions between three different domains of dynamin. Our binding data support a new model that describes how dynamin and other members of the dynamin family might assemble into multimeric spirals. This chapter describes the methods that were used to identify and investigate these binding interactions.

Preparation of Yeast Two-Hybrid Expression Plasmids

We used the yeast two-hybrid system as modified by Durfee et al. to identify and initially characterize protein–protein interactions.[3] This system

---

[1] A. M. van der Bliek, *Trends Cell Biol.* **9,** 96 (1999).

[2] E. Smirnova, D. L. Shurland, E. D. Newman-Smith, B. Pishvaee, and A. M. van der Bliek, *J. Biol. Chem.* **274,** 14942 (1999).

[3] T. Durfee, K. Becherer, P. L. Chen, S. H. Yeh, Y. Yang, A. E. Kilburn, W. H. Lee, and S. J. Elledge, *Genes Dev* **7,** 555 (1993).

uses leucine and tryptophan deficiencies to select for bait and target plasmids and it allows both β-galactosidase staining and histidine selection as methods for detecting protein–protein interactions. The two-hybrid constructs were generated by recloning dynamin cDNA fragments into pAS1-CYH2 (bait) or pAct-II (prey) vectors. The cDNA fragments were generated by PCR (polymerase chain reaction). Cloning sites were included at the 5' ends of the amplification primers with an additional three to four bases (usually Gs and Cs) at the very 5' ends of the primers to increase the efficiency of the restriction digests. To reduce the number of unwanted mutations, PCR reactions were performed with Pfu polymerase (Stratagene, La Jolla, CA).

Site-directed mutations were introduced by fusion PCR as described.[4] Briefly, complementary mutagenic oligonucleotides were used in two separate PCR reactions with upstream and downstream primers. The two ensuing fragments overlapped in the sequences derived from the mutagenic oligonucleotides. A third PCR reaction was performed using only the outer (nonmutagenic) primers with a 10,000-fold dilution of the two first PCR products as template. The first two PCR products become fused in the first stages of the third PCR reaction, thereby incorporating the desired mutation in a new longer PCR product. Where possible, we incorporated additional silent mutations in the mutagenic primers to generate a new restriction site that would help identify clones that had the desired mutations. After cloning, the new plasmids were purified in large quantities with a plasmid kit from Qiagen (Valencia, CA). To ensure that no unwanted mutations had been introduced by PCR, all cloned PCR fragments were sequenced.

Transformation of Yeast

Two-hybrid constructs are transformed into the yeast strain Y190[3] using polyethylene glycol/lithium acetate, as described.[5]

Filter Assay for β-Galactosidase Activity

1. Place sterile 100-mm circular nylon filters onto selective agar plates containing synthetic dextrose (SD) medium with essential amino acids, but without tryptophan or leucine. Streak yeast from freshly grown colonies onto the nylon filters using the blunt ends of toothpicks. Allow the yeast to grow into patches by incubating the plates with filters overnight at 30°.

[4] A. M. van der Bliek, T. E. Redelmeier, H. Damke, E. J. Tisdale, E. M. Meyerowitz, and S. L. Schmid, *J. Cell Biol.* **122**, 553 (1993).
[5] D. Gietz, A. St. Jean, R. A. Woods, and R. H. Schiestl, *Nucleic Acids Res.* **20**, 1425 (1992).

2. Lift the nylon filters from the agar plates with forceps and float them with their colony side up on liquid nitrogen. Make sure the colonies are frozen by gently submerging the filters for 10 sec in the liquid nitrogen.

3. Place the nylon filters with their colony side up on filter paper and allow the colonies to thaw at room temperature.

4. Place 100-mm disks of Whatman (Clifton, NJ) 3MM filter papers in empty 100-mm petri dishes and soak these disks with staining solution [1.8 ml of Z buffer/X-Gal per filter (0.67 mg/ml X-Gal, 0.27% (v/v) 2-mercaptoethanol in 100 mM sodium phosphate, pH 7.0, 10 mM KCl, and 1 mM MgSO$_4$)].

5. Place each nylon filter, colony side up, on the Whatman filters that were presoaked with Z buffer and X-Gal. Incubate the filters at room temperature and check periodically for the appearance of blue staining (typically after 0.5–1.5 hr).

6. Quench the staining reaction by transferring the filters with their colony side up onto a set of Whatman filters that was presoaked with 1 M Na$_2$CO$_3$.

7. Remove excess salt by transferring the filters with their colony side up onto Whatman filters that were presoaked with water.

Comments. The X-Gal staining reactions should be done with fresh patches made from recently transformed colonies. Each combination of bait and target plasmids should be tested in triplicate with patches made from independent colonies. Each bait and prey construct should also be tested separately or in combination with empty vectors to ensure that the cloned DNAs do not have inherent transcription-activating properties (self-activating plasmids).

To delineate the boundaries between interacting domains of dynamin, we made shorter fragments and tested them by cotransforming and performing X-Gal staining until the binding (blue color appearance) was lost. In the case of GTPase–assembly domain interaction we found it easier to follow the binding of the deletion variant of GTPase using K44A or S45N GTPase mutant, which shows much stronger binding to the assembly domain.

In a control set of experiments we used the homologous domains of another member of the dynamin family, Mx protein. We found that Mx domains do not bind to dynamin domains, but do have analogous binding interactions between Mx domains.

Transformation of Yeast with Two-Hybrid cDNA Library

A yeast two-hybrid library containing human brain cDNA is purchased from Clontech Laboratories (Palo Alto, CA). Large quantities of plasmid

DNA are isolated with an amplification procedure described by Durfee *et al.*[3] Briefly, the plasmids are transformed into *Escherichia coli* and plated onto ten 150-mm Luria–Bertani (LB) AMP plates. The lawns of bacteria are scraped off with a razor blade and transferred to six Fernbach flasks, each containing 1 liter of Terrific broth medium with ampicillin (12 g/liter Bacto-tryptone, 24 g/liter Bacto-yeast extract, 4 ml/liter glycerol supplemented with 17 mM KH_2PO_4, 72 mM K_2HPO_4, and 100 μg/liter ampicillin). These bacterial cultures are grown for an additional 4 hr and harvested by pelleting. The plasmids are isolated with 6 Maxi-kits from Qiagen. The yield is in the range of 10 mg plasmid.

The yeast strain Y190 is transformed with the library DNA using the standard lithium acetate procedure[5] and plated on SD plates with essential amino acids without leucine, tryptophan, or histidine, but with 25 mM 3-amino-1,2,4-triazole, which enhances the histidine selection. Surviving colonies are tested with the filter assay for β-galactosidase activity (see above). Selected target plasmids are assayed for their dependence on the bait plasmid as described.[3] Plasmid DNA is isolated from the yeast clones using standard procedures.[6]

Constructs for Protein Expression with Baculovirus System

Proteins are expressed for large-scale purification with the MaxBac baculovirus system (Invitrogen, Carlsbad, CA). This system uses the pBlueBac4 vector, which allows for the identification of recombinant plaques by staining with X-Gal. The dynamin fragments are recloned from the pAS and pACT plasmids into pBlueBac4 using standard techniques. GST fusions are made by PCR amplification of GST-coding sequences from the vector pGEX (Pharmacia, Piscataway, NJ) and recloning this fragment into the pBlueBac clones. Expression constructs that did not have the GST fusion are instead tagged with hexahistidine (His_6) residues to enable purification by nickel chromatography. The His_6 tags are derived by recloning a DNA fragment from the pET21 expression vector (Novagen Inc., Madison, WI), which introduces a C-terminal His_6 tag.

Production of Recombinant Virus for Baculovirus Expression

Sf9 (*Spodoptera frugiperda* ovary) insect cells are grown in a 0.5-liter round-bottom spinner flask with 0.25 liter complete medium [Grace's insect medium with supplements (GIBCO-BRL, Grand Island, NY) and 10% fetal bovine serum (FBS)] at 27°. The culture is split every 2–3 days to maintain a density of 10–2.5 × 10^6 cell/ml as measured with a hemocytome-

---

[6] C. S. Hoffman and F. Winston, *Gene* **57**, 267 (1987).

ter. Transfections with pBlueBac constructs, determining the virus titer with a plague assay, purification of recombinant plagues, and generating a high-titer virus stocks are done as described in the MaxBac 2.0 Transfection and Expression Manual (Invitrogen, San Diego, CA). As an alternative to the spinner flask of producing high-titer virus stock described in the manual, we also use cells growing adherently in 10 tissue culture flasks of 175 cm² each containing 20×10^6 cells in 30 ml medium. This modification results in a higher titer of virus, which is useful for efficient protein expression.

Expression and Purification of Recombinant Protein

1. Infect 0.5 liter complete medium containing 1×10^6 cell/ml of Sf9 cells with the P-3 virus stock to reach an MOI (multiplicity of infection) of 5 virion/cell. Alternatively, we also use TN5 cells (kindly provided by J. H. Elder, Scripps Research Institute, La Jolla, CA), which gave better yields with our GST-fusion proteins.
2. Incubate between 48 and 60 hr at 27°.
3. Harvest the cells by centrifugation for 10 min at 6000g and 4° (Sorvall type GSA, Newtown, CT).
4. Resuspend the cells in lysis buffer (20 mM HEPES, pH 7.2, 160 mM NaCl, 2 mM MgCl$_2$, 0.5 mM EDTA, 0.5 mM DTT) with 0.1% Triton X-100 and protease inhibitors [Complete from Roche Diagnostics (Indianapolis, IN) and PMSF to 100 μg/ml]. We use 40 ml lysis buffer for 500-ml cultures.
5. Lyse the cells by passing the cell suspension 3 times through a French pressure cell (American Instrument Co, Inc., Silver Spring, MD) adjusted to 20,000 psi.
6. Draw the lysate 10 times through an 18-gauge 1 1/2 needle (Becton Dickinson, Franklin Lakes, NJ). This step releases more of the recombinant protein into the soluble fraction.
7. Pellet unlyzed cells and cell debris by centrifugation for 30 min at 30,000g (Sorvall type SS-34).
8. Mix the supernatant with 500 μl of a 50% slurry of protein purification resins. The His$_6$-tagged proteins are purified with Talon resin (Clontech, Palo Alto, CA). The GST fusions are purified with glutathione-agarose (Pharmacia Biotech Inc., Piscataway, NJ).
9. Gently rock the mixture for 30 min at 4° in a 50-ml tube mounted on a Nutator platform (Becton Dickinson, Sparks, MD).
10. Wash the resin by pelleting and resuspending it 3 times in lysis buffer.
11. Pour the slurry into a disposable column. Elute the recombinant proteins from the Talon resin with 250 mM imidazole in lysis buffer

but without Triton. Elute the proteins from the glutathione-agarose with 10 mM glutathione in 50 mM Tris-HCl, pH 8.0. Collect 4 eluate fractions of 200 μl each. Most of the recombinant protein is eluted in fractions 1 and 2.

12. Dialyze the purified proteins into lysis buffer and store single-use aliquots at $-80°$. We consistently had higher yields of the GST fusion proteins as compared to the His$_6$-tagged ones. The results of a typical protein purification procedure are shown in Fig. 1. Protein yields from a 0.5-liter culture are usually in the range of 0.1–0.5 mg, as determined by Bradford analysis.

Expression of Radiolabeled Proteins with Reticulocyte Lysates

DNAs encoding the desired fragments of dynamin were first cloned into pET21 vectors (Novagen Inc., Madison, WI), which have the T7 promoter.

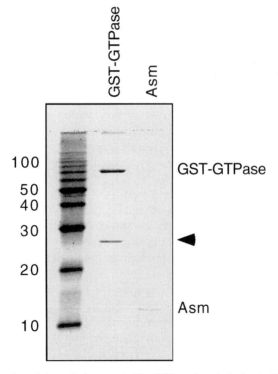

FIG. 1. Purification of dynamin fragments. The GTPase domain fusion to GST was purified using a glutathione-agarose affinity column. The assembly domain with C-terminal His$_6$ residues attached was purified on the Talon resin. The eluates from the column are shown. The arrowhead shows the band corresponding to copurifying insect cell endogenous GST.

Radiolabeled protein fragments were synthesized with the TNT T7 Coupled Transcription/Translation System (Promega Corp., Madison, WI). The methods were as follows:

1. Mix 0.5 μg DNA with 20 μl TNT T7 Quick Master Mix and 1 μl [^{35}S]methionine (Amersham, Buckinghamshire, England) in a 1.5-ml microcentrifuge tube.
2. Adjust the volume to 25 μl with nuclease-free water.
3. Unlabeled methionine was added to a final concentration of 100 μM to the synthesis of GST fusion proteins. This increases the amount of protein synthesis and decreases the specific activity by 100-fold. To maximize the detection of binding interactions, the remaining proteins were synthesized without adding cold methionine.
4. Incubate the reaction mixture for 90 min at 30°.
5. Pellet protein aggregates with 10 min of centrifugation at 16,000g and 4°. Transfer the supernatant to a new tube and use this immediately in the protein binding assays.

Protein Binding Experiments

1. Mix 1 μg unlabeled GST fusion protein produced with baculovirus and 22.5 μl of the [^{35}S]methionine-labeled target protein in a 1.5-ml microcentrifuge tube. Alternatively, use 22.5 μl of both GST-tagged and target proteins that are labeled with [^{35}S]methionine, but then the GST-tagged protein is synthesized with 100-fold excess of unlabeled methionine (see above).
2. Depending on which parameters are to be investigated, add 250 μl PBS with 10 mM EDTA or add 250 μl PBS with 5 mM MgCl$_2$ and 100 μM GTP (binding buffer). These conditions are used to investigate binding of the dynamin GTPase domain in the empty state and in the GTP-bound state.
3. Add 20 μl of a 50% glutathione-Sepharose slurry and rock the mixture on a Nutator platform (Becton Dickinson, Sparks, MD) for 1 hr at 4°.
4. Pellet the beads by centrifugation for 2 min at 16,000g and 4°.
5. Wash the glutathione-agarose beads by pelleting and resuspending them 4 times with binding buffer.
6. Elute the bound proteins by resuspending the beads in 330 μl 10 mM glutathione in 50 mM Tris-HCl, pH 8.0. Pellet the beads and transfer the supernatant to a fresh tube. Repeat this procedure two more times, collecting all three eluates in a single tube.
7. Add 260 μl 20% TCA to the combined eluates. Incubate on ice for 30 min.

8. Pellet the eluted proteins by centrifugation for 20 min at 13,000g and 4°.

9. Dissolve the pelleted proteins in SDS–PAGE sample buffer, neutralize the TCA by adding 2 μl 1 M NaOH, and size-fractionate the samples by SDS–PAGE.

10. Transfer the gel to Whatman 3MM paper. Dry the gel in a gel dryer with a moistened sheet of Whatman paper on top of plastic wrap and gel. This keeps the gel from cracking once it cools. Expose the

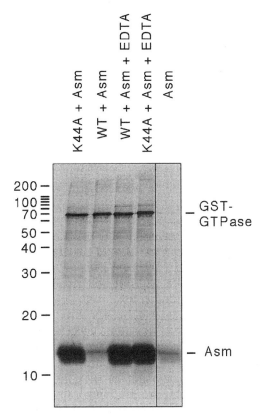

Fig. 2. *In vitro* binding between the dynamin GTPase domain and the assembly domain as detected with GST pulldown. The assembly domain (Asm) was pulled down by the wild-type (WT) GTPase or the K44A mutant fused to GST. The interaction with the wild-type GTPase domain was enhanced by EDTA, which inhibits GTP binding. The interactions with the K44A mutant GTPase domain were not influenced by EDTA, because the K44A mutant already has a GTP binding defect. Radiolabeled protein fragments were made by *in vitro* transcription/translation with [^{35}S]methionine.

gel to X-ray film or to a PhosphorImager (Molecular Dynamics, Sunnyvale, CA).

Figure 2 shows the results of the *in vitro* binding between the dynamin GTPase domain and the assembly domain as detected with GST pulldown.

Comments. We found it useful to elute bound proteins from the glutathione-agarose beads using excess of free glutathione, and then loading them on the gel (as described above) as opposed to direct elution of the proteins from the beads into SDS–PAGE sample buffer. Following the former approach we achieved higher specificity of the binding, while we had higher background using the latter one. This was especially important in case of the histidine-tagged assembly domain of dynamin, which had high unspecific

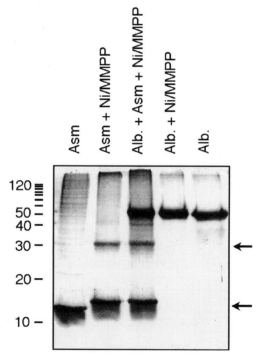

FIG. 3. *In vitro* cross-linking of the dynamin assembly domain. Dimerization of the assembly domain was tested with the cross-linking reagent magnesium monoperoxyphthalic acid (MMPP), which activates the nickel complexed with the C-terminal His tag of the assembly domain fragments. The lanes show assembly domain without the cross-linker (lane 1), assembly domain with cross-linker (lane 2), assembly domain with cross-linker and chicken albumin (Alb), which was used as a nonspecific competitor (lane 3), chicken albumin with cross-linker (lane 4), and chicken albumin without cross-linker (lane 5). The arrows indicate the sizes of monomeric and dimeric assembly domain fragments.

binding to the glutathione beads. In addition we always conducted a control reaction, omitting the GST-tagged protein in order to estimate the level of background binding.

Cross-Linking of Dynamin Assembly Domain

Dimerization of the dynamin assembly domain was tested using an *in vitro* cross-linking protocol devised by Fancy *et al.*[7] This procedure is based on the ability of nickel to catalyze cross-linking by magnesium monoperoxyphthalic acid (MMPP). Nickel is kept in association with the dynamin protein fragments using the hexahistidines that were also used for protein purification.

1. Add 1 μg of His-tagged protein to 20 μl PBS with 2 mM nickel acetate. In a separate reaction, add 3 μg carbonic anhydrase (Sigma, St. Louis, MO) or another non-His-tagged protein as a negative control.
2. Incubate the mixture for 5 min on ice.
3. Add MMPP to a final concentration of 1 mM.
4. Incubate the mixture for 15 min on ice.
5. Stop the reaction by adding SDS–PAGE sample buffer.
6. Size-fractionate the proteins by SDS–PAGE. Stain the gel with Coomassie Brilliant Blue. Dry the gel and analyze the pattern with a scanning densitometer (Molecular Dynamics). Figure 3 demonstrates dimerization of the dynamin assembly domain as detected by *in vitro* cross-linking.

Acknowledgments

We thank S. L. Schmid and members of her labarotory for sharing their experience with baculovirus and protein purification. We thank the other laboratory members for their comments on this manuscript. E.S. was supported by fellowships from the Myasthenia Gravis Foundation and the American Heart Association of Greater Los Angeles. This work was supported by grants from the American Heart Association (965084) and the NIH (GM58166) to A.M.v.d.B.

[7] D. A. Fancy, K. Melcher, S. A. Johnston, and T. Kodadek, *Chem. Biol.* **3**, 551 (1996).

[50] Interactions of Dynamin and Amphiphysin with Liposomes

By KOHJI TAKEI, VLADIMIR I. SLEPNEV, and PIETRO DE CAMILLI

Introduction

Dynamin-1 and amphiphysin-1 are two proteins highly expressed in the brain, where they are concentrated in presynaptic nerve terminals. Both proteins are implicated in clathrin-mediated endocytosis at the synapse, a process that mediates the recycling of synaptic vesicles. Dynamin-1 is a brain-specific isoform of the dynamin subfamily of GTP-binding proteins. The dynamin-1 molecule comprises a highly conserved NH_2-terminal GTPase domain, a pleckstrin homology (PH) domain, a GTPase effector domain (GED), and a COOH-terminal proline/arginine-rich domain (PRD).[1] The first piece of evidence suggesting a function of dynamin-1 in synaptic vesicle endocytosis came from the identification of the mammalian homolog of the *shibire* gene product in *Drosophila*.[2,3] Temperature-sensitive mutations in the *shibire* gene result in a temperature-sensitive block in synaptic transmission, which correlates with the disappearance of synaptic vesicles and the accumulation of endocytic pits on the presynaptic plasma membrane. These observations pointed to a role for dynamin in the process of endocytic vesicle fission.[4]

This conclusion was subsequently strengthened by the demonstration that dynamin forms rings and spirals around the neck of coated pits *in vitro* in the presence of GTPγS, a nonhydrolyzable GTP analog.[5] The latter observation led to the proposal that dynamin functions as a mechano-chemical enzyme. According to this model, dynamin oligomerizes into open rings and spirals around the neck of endocytic pits in its active, GTP-bound form, and GTP hydrolysis triggers constriction of the ring, thus mediating separation of the vesicle from the donor membrane. Consistent with the model, dynamin was shown by electron microscopy (EM) and light scattering to fragment liposomes in a GTP-dependent

[1] S. L. Schmid, M. A. McNiven, and P. De Camilli, *Curr. Opin. Cell Biol.* **10,** 504 (1998).
[2] M. S. Chen, R. A. Obar, C. C. Schroeder, T. W. Austin, C. A. Poodry, S. C. Wadsworth, and R. B. Vallee, *Nature* **351,** 583 (1991).
[3] M. van der Bliek and E. M. Meyerowitz, *Nature* **351,** 411 (1991).
[4] J. H. Koenig and K. Ikeda, *J. Neurosci.* **9,** 3844 (1989).
[5] K. Takei, P. S. McPherson, S. L. Schmid, and P. De Camilli, *Nature* **374,** 186 (1995).

manner.[6,7] The GTPase activity of dynamin can be considerably enhanced by oligomerization[8–10] and by interaction with variety of SH3 domain-containing proteins[11] and lipids,[12] in particular inositol-containing lipids.[13,14] The importance of the latter interaction, mediated by dynamin's PH domain,[15] is stressed by the dominant negative effect on clathrin-mediated endocytosis of dynamins harboring mutations in this domain.[16–18] Alternative models suggest that dynamin, like other GTPases, acts as a "molecular switch" that activates downstream effectors.[19,20] This model has found support in the observation that dynamin mutants predicted to be deficient in GTPase activity, enhance clathrin-mediated endocytosis.[20] However, no downstream effector for dynamin has been identified so far.

Amphiphysin-1 contains several functional domains. The NH_2-terminal domain is responsible for dimerization[21] and binding to membrane lipids.[22] The central domain contains binding sites for clathrin heavy chain and for the α subunit of the clathrin adaptor AP-2.[21] The COOH-terminal SH3 domain binds dynamin and another endocytic protein, the polyphosphoinositide phosphatase synaptojanin-1.[23] This structure suggests that amphiphysin acts as an adaptor that recruits dynamin (and synaptojanin-1) to clathrin

[6] K. Takei, V. Haucke, V. Slepnev, K. Farsad, M. Salazar, H. Chen, and P. De Camilli, *Cell* **94,** 131 (1998).

[7] S. M. Sweitzer and J. E. Hinshaw, *Cell* **93,** 1021 (1998).

[8] P. L. Tuma and C. A. Collins, *J. Biol. Chem.* **269,** 30842 (1994).

[9] D. E. Warnock, J. E. Hinshaw, and S. L. Schmid, *J. Biol. Chem.* **271,** 22310 (1996).

[10] E. Smirnova, D. L. Shurland, E. D. Newman-Smith, B. Pishvaee, and A. M. van der Bliek, *J. Biol. Chem.* **274,** 14942 (1999).

[11] I. Gout, R. Dhand, I. D. Hiles, M. J. Fry, G. Panayotou, P. Das, O. Truong, N. F. Totty, J. Hsuan, G. W. Booker *et al., Cell* **75,** 25 (1993).

[12] P. L. Tuma, M. C. Stachniak, and C. A. Collins, *J. Biol. Chem.* **268,** 17240 (1993).

[13] H. C. Lin, B. Barylko, M. Achiriloaie, and J. P. Albanesi, *J. Biol. Chem.* **272,** 25999 (1997).

[14] B. Barylko, D. Binns, K. M. Lin, M. A. Atkinson, D. M. Jameson, H. L. Yin, and J. P. Albanesi, *J. Biol. Chem.* **273,** 3791 (1998).

[15] K. Salim, M. J. Bottomley, E. Querfurth, M. J. Zvelebil, I. Gout, R. Scaife, R. L. Margolis, R. Gigg, C. I. Smith, P. C. Driscoll, M. D. Waterfield, and G. Panayotou, *EMBO J.* **15,** 6241 (1996).

[16] M. Achiriloaie, B. Barylko, and J. P. Albanesi, *Mol. Cell Biol.* **19,** 1410 (1999).

[17] A. Lee, D. W. Frank, M. S. Marks, and M. A. Lemmon, *Curr. Biol.* **9,** 261 (1999).

[18] Y. Vallis, P. Wigge, B. Marks, P. R. Evans, and H. T. McMahon, *Curr. Biol.* **9,** 257 (1999).

[19] P. De Camilli and K. Takei, *Neuron* **16,** 481 (1996).

[20] S. Sever, A. B. Muhlberg, and S. L. Schmid, *Nature* **398,** 481 (1999).

[21] V. I. Slepnev, G. C. Ochoa, M. H. Butler, D. Grabs, and P. De Camilli, *Science* **281,** 821 (1998).

[22] K. Takei, V. I. Slepnev, V. Haucke, and P. De Camilli, *Nature Cell Biol.* **1,** 33 (1999).

[23] Grabs, V. I. Slepnev, Z. Songyang, C. David, M. Lynch, L. C. Cantley, and P. De Camilli, *J. Biol. Chem.* **272,** 13419 (1997).

coated pits. Such a model is supported by several lines of evidence involving studies in cell-free systems[21,22] and in living cells.[21,24,25] Amphiphysin-1, like dynamin-1, evaginates liposomes into narrow tubules as a result of its lipid binding properties.[22] Amphiphysin-1 coassembles with dynamin-1 on lipid tubules and enhances the ability of dynamin-1 to fragment the tubules in the presence of GTP.[22] Thus, amphiphysin may also cooperate with dynamin in the fission of endocytic vesicles.[22]

The methodology to study the interactions of dynamin-1 and amphiphysin-1 with liposomes is described below.

Methods

Expression and Purification of Dynamin

Human dynamin-1aa[26,27] is expressed in insect cells as a glutathione *S*-transferase (GST) fusion protein using a modified Bac-to-Bac baculovirus expression system (Life Technologies) and purified by affinity chromatography on glutathione-Sepharose (Amersham Biotech) resin. GST is removed from the fusion protein by cleavage with Prescission protease (Amersham Biotech) and separated from dynamin by adsorption on glutathione-Sepharose beads. The detailed protocol follows.

The pFastBac 1 baculovirus transfer vector (Life Technologies) is modified by the insertion of the cDNA encoding GST into *Bam*HI and *Eco*RI sites of the vector. The sequence surrounding the starting ATG of GST is modified to ACAATCATGTCA corresponding to the highly expressed viral protein p10.[28] The nucleotide sequence encoding a recognition site for Prescission protease is inserted at the end of the GST cDNA. The ends of dynamin-1 cDNA cloned in pTM1 vector[27] are amplified by PCR using the following pairs of primers:

N terminus: 5'-AAAGAATTCGGCAACCGCGGCATGGA AGA and
            5'-GCCGGTGACCCTGTCGGTCTCG-3'
COOH terminus: 5'-GCTTCATGCATTCCATGGACCC-3' and
            5'-AAAACTAGTTTAGAGGTCGAAGGGGGGCCT-3'

[24] O. Shupliakov, P. Low, D. Grabs, H. Gad, H. Chen, C. David, K. Takei, P. De Camilli, and L. Brodin, *Science* **276**, 259 (1997).
[25] P. Wigge, Y. Vallis, and H. T. McMahon, *Curr. Biol.* **7**, 554 (1997).
[26] R. Urrutia, J. R. Henley, T. Cook, and M. A. McNiven, *Proc. Natl. Acad. Sci. U.S.A.* **94**, 377 (1997).
[27] H. Damke, T. Baba, D. E. Warnock, and S. L. Schmid, *J. Cell Biol.* **127**, 915 (1994).
[28] C. Richardson, D., Ed., "Baculovirus Expression Protocols: Methods in Molecular Biology," Vol 39. Humana Press, Totowa, NJ, 1995.

The fragments containing NH_2-terminal and COOH portions of the dynamin gene are digested with *Bst*EII and *Nsi*I, respectively, and ligated to the BstEII-NsiI fragment of dynamin-1.

The resulting fragment is digested with *Eco*RI and *Spe*I restriction enzymes and cloned into corresponding sites of the modified pFastBac1 vector. The transfer vector containing the dynamin cDNA is transformed into DH10Bac strain of *Escherichia coli* (Life Technologies). The transposition reaction and extraction of the viral DNA (bacmid) are performed according to manufacturer's instruction (Life Technologies). Six independent clones of bacmid DNA are transfected into Sf9 cells (ATCC) using Cellfectin reagent (Life Technologies) according to manufacturer's protocol. Seventy hours after transfection, the virus-containing medium is harvested and cells lysed in SDS–PAGE sample buffer are analyzed for expression of GST dynamin by Western blotting using polyclonal anti-dynamin-1 DG1.[23] All six tested viral clones produce similar amounts of dynamin. One of these clones is chosen for further amplification and for the production of recombinant protein. The viral stock is amplified by stepwise infection of Sf9 cells maintained in TNT-FM medium containing 5% fetal calf serum. Typically, a baculovirus-containing supernatant is harvested 72 hr after infection, sterilized by filtration through a 0.22-μm filter and used for infection of larger batches of Sf9 cells. The kinetics of dynamin expression and the optimal amount of viral stock are determined by Western blotting with anti-dynamin antibody of Sf9 and High Five (Invitrogen) cells infected with a baculovirus in 12-well plates.

For large-scale production of GST-dynamin, High Five cells (10^6 cells/ml) grown at 28° in TNT-FM medium in a 1-liter spinner flask are infected with predetermined volume of amplified virus (usually 50–100 ml). The cells are harvested by centrifugation (5 min at 1000g) 48–56 hr postinfection, frozen in dry ice, and stored at $-70°$ (longer expression time resulted in significant proteolytic degradation of dynamin). Frozen cell pellets (typically 2–3 $\times$ 10^9 cells) are thawed in ice and resuspended in 10–15 ml of TBSM buffer (250 mM NaCl, 50 mM Tris, pH 8.0, 2 mM $MgCl_2$) containing 0.5 mM GTP, 2 μg/ml of each aprotinin, leupeptin, pepstatin, antipain, and 0.5 mM Pefabloc (all Roche Biochemicals). The cells are supplemented with Triton X-100 to 1% final and sonicated for 1 min. The extract is clarified by centrifugation at 33,000g. The supernatant is desalted on a 100-ml Sephadex G-25 column (Amersham Biotech) equilibrated with TBSM buffer containing 1% Triton X-100 and mixed with 1 ml of glutathione-Sepharose 4B (Amersham Biotech) at 4° for 1 hr. The affinity resin is washed twice in batch with TBSM containing 1% Triton X-100, transferred in a 10-ml column, and further washed twice with 10 ml of the same buffer and 3 times with 10 ml of TBSM buffer. The GST–dynamin-1 is eluted with 15 mM glutathione (Sigma, St Louis, MO) in TBSM and desalted on a PD10 column equilibrated with TBS or HBS buffer (10 mM) Tris or HEPES, respectively, 150 mM NaCl, pH 7.5). GST is removed by cleavage with Precission protease (1 unit per 100 μg of fusion protein) at 4° overnight and adsorbed by a passage through a 0.5-ml glutathione-Sepharose column). The flow-through fraction containing more than 90% pure (as judged by SDS–PAGE) dynamin-1 is aliquoted and stored at −70°. The typical yield of dynamin is 1 mg per 10^9 cells.

Expression and Purification of Amphiphysin-1

The cDNA encoding human amphiphysin-1[29] is subcloned into the pGEX6-1 expression vector (Amersham Biotech) by standard procedures and transformed into the DH5a strain of *E. coli*. Transformed cells are grown at 37° in SB medium (35 g/liter tryptone, 20 g/liter yeast extract, 5 g/liter NaCl, pH 7.5) supplemented with 100 μg/ml carbenicillin to OD_{600} of 0.7–0.8 (typically 200–250 ml culture) in a rotary shaker at 300 rpm. The induction was initiated by the addition of isopropylthiogalactoside (IPTG) to 0.3 M at 25–30° for 3–4 hr. Cells are harvested by centrifugation at 5000g, resuspended in homogenization buffer H (300 mM NaCl, 50 mM Tris, pH 8.0) supplemented with 5 mM EDTA, 2 μg/ml aprotinin, 0.5 mM Pefablock, and 1% Triton X-100, and lysed by sonication (4–5 min). The homogenate is clarified by centrifugation at 33,000g in a Sorvall ultracentrifuge. The supernatant is mixed with 1–1.5 ml of glutathione-Sepharose preequilibrated with H buffer containing 1% Triton and incubated for 0.5–1 hr on a rotating wheel at 4°. The affinity resin is washed twice in batch with buffer H containing 1% Triton X-100, transferred into a 10-ml disposable column (Bio-Rad, Hercules, CA), washed twice with 10 ml of the same buffer and 3 times with 10 ml of buffer H. The fusion protein is eluted with 15 mM glutathione in buffer H, concentrated using Centriprep 10 (Millipore, Bedford, MA) and desalted into HBS buffer using PD10 column. Removal of the GST is carried out as described for dynamin. The typical yield of amphiphysin is 2–2.5 mg/200 ml of bacterial culture.

Preparation of Large Unilamellar Liposomes

Large unilamellar liposomes are prepared as described[30] with some modifications. Bovine total brain lipid extract (Folch fraction 1, Sigma) is solubilized in chloroform:methanol (1:2) at the concentration of 25 mg/ml to generate a stock solution, which can be stored at −20°. Eighty μl

---

[29] C. David, M. Solimena, and P. De Camilli, *FEBS Lett.* **351,** 73 (1994).
[30] J. P. Reeves and R. M. Dowben, *J. Cell Physiol.* **73,** 49 (1969).

of the lipid stock solution, which contains 2 mg lipid, is transferred to a 10-ml glass tube and diluted with 0.5 ml of the same organic solvent. A stream of nitrogen gas is passed through the tube to evaporate the solvent. During this process, the test tube is rotated by hand so that the lipid is evenly distributed, forming an opaque thin lipid layer on the wall. The lipid film is further dried in a lyophilyzer for 2 hr. Next the film is hydrated by passing a stream of water-saturated nitrogen (approximately 20 min) through the tube until opacity is slightly lost. After hydration, 2 ml of degassed 0.3 M sucrose in distilled water is gently added to the test tube. Nitrogen gas is flushed into the tube without agitating the solution; the tube is sealed and left undisturbed for 2 hr at 37° to allow spontaneous formation of liposomes. The tube is then gently swirled to resuspend the liposomes and large aggregates of lipids are removed by quick centrifugation. The properties of the obtained liposomes are checked by negative staining (see below). The liposomes (concentration, 1 mg/ml) are mostly unilamellar spheres with large diameters, some exceeding 1 μm (Fig. 1A). They could be stored at 4° for several days.

Incubation of Liposomes with Proteins

Liposomes, proteins, nucleotides, and a stock of cytosolic buffer ($\times$10) are mixed at room temperature in either Eppendorf tubes or glass test tubes depending on the amount of sample needed. In the case of negative staining, a total volume of less then 50 μl is sufficient for several EM grids. Larger amounts of samples (500–1000 μl) are needed for plastic embedding and thin sectioning. Final concentrations are as follows:

Liposomes: 0.1 mg/ml
Dynamin 1: 0.1 mg/ml
Amphiphysin 1: 0.1 mg/ml
GTP: 200 μM
Cytosolic buffer: 25 mM HEPES–KOH, 25 mM KCl, 2.5 mM magnesium acetate, 150 mM potassium glutamate, 10 μM Ca^{2+}, pH 7.4

The samples are incubated at 37° for 15 min. After the incubation, samples are processed for negative staining without fixation. Samples for ultrathin sectioning are fixed at the end of the incubation as described below.

Electron Microscopy

For negative staining (Figs. 1–3), liposomes are absorbed onto freshly glow-discharged Formvar- and carbon-coated EM grids by placing the grids on paddles of the samples (10–20 μl) for a couple of minutes. The grids are then passed three paddles of 0.1 M HEPES buffer over 1 min to wash out unabsorbed materials. The grids are quickly blotted by touching filter paper with their edges to remove an excess of buffer, negatively stained

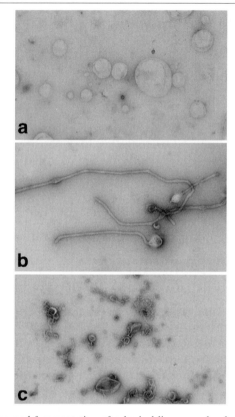

FIG. 1. Tubulation and fragmentation of spherical liposomes by dynamin. Samples were analyzed by negative staining electron microscopy. (a) Large unilamelar liposomes before incubation. (b) Liposomes incubated with dynamin in the absence of nucleotides. Long tubules coated with oligomerized dynamin (see Fig. 3a) were generated from the liposomes. (c) Liposomes incubated with dynamin in the presence of GTP have undergone massive fragmentation.

on a paddle of 2% uranyl acetate in doubly distilled H_2O for 1 min, and finally blotted and air dried.

For ultrathin sectioning, incubation mixtures are fixed in suspension by addition of an equal volume of 2× fixative (final 3% formaldehyde, 2% glutaraldehyde in 50 mM HEPES-KOH buffer, pH 7.4) for 30 min. The samples are centrifuged in a Beckman TLA100.3 rotor at 50,000 rpm for 10 min. The pellets are washed with 0.1 M Na cacodylate buffer and then postfixed with OsO_4 in the same buffer for 1 hr. Samples are dehydrated with increasing concentration of ethanol, followed by substitution with propylene oxide and embedded in Epon 812. For some samples, OsO_4

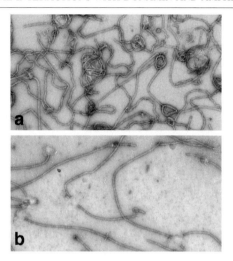

FIG. 2. Tubulation of liposomes by amphiphysin. Samples were analyzed by negative staining electron microscopy. (a) Liposomes incubated with amphiphysin alone. (b) Liposomes incubated with dynamin and amphiphysin together. [Reproduced with permission from K. Takei, V. I. Slepnev, V. Haucke, and P. De Camilli, *Nature Cell Biol.* **1,** 33 (1999).]

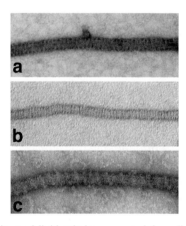

FIG. 3. High-power views of lipid tubules generated from liposomes by dynamin and amphiphysin. Tubules were generated by incubating liposomes with dynamin (a), with amphiphysin (b), or both proteins together (c). (a and b) Both dynamin and amphiphysin form tightly packed rings around the lipid tubules. However, these rings are different in size and periodicity. The combination of dynamin and amphiphysin forms thicker rings, suggesting a co-oligomerization of two proteins. [Reproduced with permission from K. Takei, V. I. Slepnev, V. Haucke, and P. De Camilli, *Nature Cell Biol.* **1,** 33 (1999).]

postfixation is followed by impregnation with 1% tannic acid in doubly distilled H_2O to enhance visualization of membrane coats.[31]

Light Scattering Assay

Light scattering at a 90° angle is measured on a Hitachi F-3010 fluorescence spectrophotometer. The excitation and emission wavelengths are set at 320 nm. Excitation and emission slit widths are set at 3 nm. Liposomes of total brain lipid mixture (Sigma) are prepared as above. The suspension of liposomes (100 μg/ml of total lipid) in cytosolic buffer (see above) is incubated in the presence of 25 μg/ml of dynamin-1, 50 μg/ml amphiphysin-1, and 1 mM GTP. Measurements began 1 min after mixing the components and data are recorded for 10 min.

[31] L. Orci, B. S. Glick, and J. E. Rothman, *Cell* **46,** 171 (1986).

[51] Activation of Dynamin GTPase Activity by Phosphoinositides and SH3 Domain-Containing Proteins

By BARBARA BARYLKO, DERK D. BINNS, and JOSEPH P. ALBANESI

Introduction

Dynamin guanosine 5'-triphosphate (GTP)ase activity is involved in membrane vesiculation *in vivo*[1] and *in vitro.*[2,3] Neuron-specific dynamin-1 and ubiquitously expressed dynamin-2 have basal activities of 1–2 and 10–20 min^{-1}, respectively (for comprehensive reviews, see Refs. 4–6). These activities can be stimulated to approximately 200–300 min^{-1} by microtubules and anionic liposomes and to about 40–50 min^{-1} by src homology 3 (SH3) domain-containing proteins and polyclonal anti-dynamin antibodies. In all cases the mechanism of GTPase activation appears to be indirect, i.e., the activators do not by themselves affect catalysis but serve to bring dynamin molecules close together. Indeed, at low ionic strength dynamin-

[1] K. Takei, P. McPherson, S. Schmid, and P. De Camilli, *Nature* **374,** 186 (1995).
[2] S. M. Sweitzer and J. E. Hinshaw, *Cell* **93,** 1021 (1998).
[3] K. Takei, V. Haucke, V. Slepnev, K. Farsad, M. Salazar, H. Chen, and P. De Camilli, *Cell* **94,** 131 (1998).
[4] S. J. McClure and P. J. Robinson, *Mol. Mem. Biol.* **13,** 189 (1996).
[5] R. B. Vallee and P. M. Okamoto, *Trends Cell Biol.* **5,** 43 (1995).
[6] S. L. Schmid, M. A. McNiven, and P. DeCamilli, *Curr. Opin. Cell Biol.* **10,** 504 (1998).

2 and, to a lesser extent, dynamin-1 display concentration-dependent increases in specific activity even in the absence of activators. Enhancement of activity correlates well with increases in dynamin self-association as monitored by light scattering and sedimentation.[7] Thus dynamin is unusual in that its specific activity is not invariant as a function of enzyme concentration. This phenomenon was explained at a molecular level when Sever *et al.*[8] identified a GTPase effector domain (GED) in the C-terminal half of dynamin that stimulates GTPase activity, presumably in an adjacent dynamin molecule.

In this chapter, we describe procedures used in our laboratory to isolate milligram quantities of dynamin-1 from bovine brain and to assay dynamin GTPase activity. Although there is nothing unique about the assay itself, the mechanism of activation by phosphoinositides and SH3 domain-containing proteins raises complexities that are also addressed below.

Purification of Dynamin

Several strategies have been reported for the purification of dynamin from rat and bovine brain. Our optimized procedure yields ~10 mg of pure dynamin from 500 g of bovine brain. The procedure is diagrammed in Fig. 1.

Fresh brain tissue is homogenized with 1–1.5 volumes of a solution containing 0.1 M 2-[N-morpholino]ethanesulfonic acid (MES), pH 7.0, 1 mM EGTA, 1 mM MgSO$_4$, 1 mM dithiothrietol (DTT), 0.2 mM phenylmethylsulfonyl fluoride (PMSF), and a battery of other protease inhibitors (10 mg of each: N^a-p-tosyl-L-arginine methyl ester, N^a-p-benzoyl-L-arginine methyl ester, N^a-p-tosyl-L-lysine chloromethyl ketone, leupeptin, and pepstatin A per liter) (buffer A). The homogenate is centrifuged at 30,000g and 4° for 40 min, then the supernatant is centrifuged again at high speed (100,000g at 4° for 1 hr). The final supernatant is immediately passed through DE52 cellulose equilibrated with buffer A, and the flow-through is loaded directly on a P11 phosphocellulose column, also equilibrated with buffer A. Dynamin is eluted with 0.5 M NaCl. The eluate is concentrated by adding (NH$_4$)$_2$SO$_4$ to 65% saturation, centrifuged (30,000g for 30 min), and the pellet resuspended in buffer A, dialyzed extensively against the same buffer, then loaded on an SP-Sepharose column equilibrated in buffer A. The column is washed with buffer A containing 0.05 M NaCl and eluted with a 0.05–0.3 M NaCl linear gradient. Fractions containing dynamin are combined and dialyzed against buffer A. The sample is next incubated with Taxol-stabilized microtubules for 15 min at 37° and then centrifuged at

[7] D. E. Warnock, T. Baba, and S. L. Schmid, *Mol. Biol. Cell* **8**, 2553 (1997).
[8] S. Sever, A. B. Muhlberg, and S. L. Schmid, *Nature* **398**, 481 (1999).

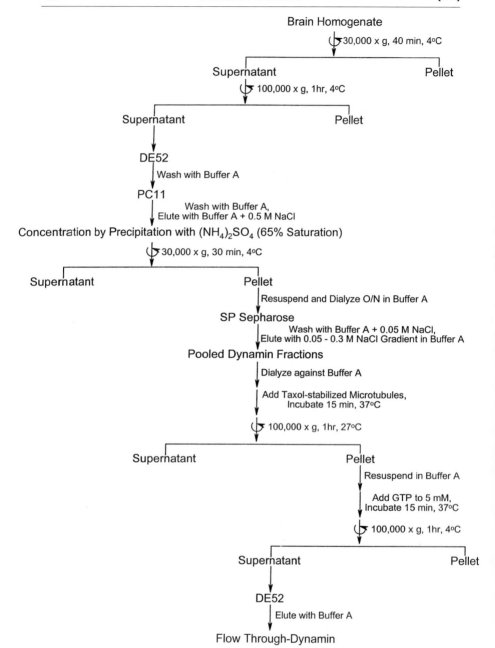

FIG. 1. Flowchart for dynamin purification.

100,000g for 1 hr at 27°. To release dynamin from microtubules, the pellet is resuspended in small (10–15 ml) volume of buffer A containing 5 mM GTP and the sample is centrifuged again at 100,000g for 1 hr at 4°. The supernatant is passed through a DE52 column to remove any traces of tubulin. At this stage, dynamin is usually about 90% pure (see Fig. 2, lane 10). If necessary, dynamin is further purified on a 5–15% sucrose gradient centrifuged at 112,000g at 4° for 16 hr. Aliquots of dynamin solution are frozen in liquid nitrogen and stored at −70°. Because dynamin has a tendency to aggregate, the samples are always centrifuged (200,000g for 15 min) directly before GTPase assays.

Enzyme Assay

Principle

The most commonly used methods to analyze GTP hydrolysis utilize radioactive nucleotide, labeled either in the α- or γ-phosphate. Production of guanosine 5′-diphosphate (GDP) can be monitored using α-labeled nucleotide and separation of GDP from GTP by thin-layer chromatography

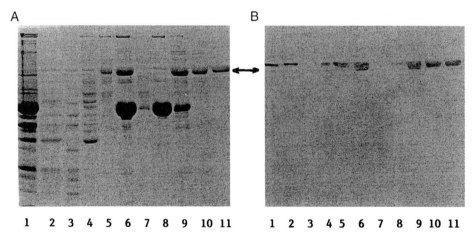

FIG. 2. Protein composition of the major steps of dynamin purification. Lane 1, high-speed supernatant; lane 2, flow-through from DE52; lane 3, flow-through from PC11; lane 4, eluate of PC11; lane 5, pooled fractions from SP-Sepharose; lanes 6 and 7, microtubule pellet (6) and supernatant (7); lanes 8 and 9, pellet (8) and supernatant (9) after incubation of microtubule pellet with 10 mM GTP; lane 10, DE52 flow-through of the supernatant (9); lane 11, dynamin purified on 5–15% sucrose gradient. (A) Coomassie blue-stained gel. (B) Corresponding blot with anti-dynamin antibodies. (Polyclonal antibodies raised against rat brain dynamin-1 were provided by Dr. Thomas Sudhof, University of Texas Southwestern Medical Center.)

(TLC) after the reaction. Phosphate release is conveniently assayed using nucleotide labeled in the γ position and, following the reaction, separating radioactive P_i from GTP and GDP. Separation is usually carried out by one of the following methods: adsorption of GTP and GDP to charcoal, which is then pelleted by brief centrifugation, leaving P_i in the supernatant; or extraction of P_i into an organic phase. Although radioactive assays are more sensitive, colorimetric assays[9] can also be used to measure the amount of released phosphate.

The procedure based on organic extraction of radioactive P_i was used in our studies and is described in detail below. It is essentially identical to a previously described procedure used to analyze myosin ATPase activity, and is based on the original method of Martin and Doty[10] with modifications by Korn et al.[11] In this method, released P_i forms a complex with molybdate and, as phosphomolybdate, is extracted quantitatively into an (upper) organic phase while GTP and GDP remain in the (lower) aqueous phase.

Reagents

Isobutanol/benzene, 1 : 1

4% Silicotungstic acid (tungstosilic acid, Sigma, St. Louis, MO) in 3 N sulfuric acid

10% Ammonium molybdate (ammonium salt tetrahydrate, Sigma) in H_2O

All three solutions can be stored at room temperature for months. However, ammonium molybdate has a tendency to precipitate and should be discarded if cloudiness is evident.

Substrate

GTP mix: nonradioactive GTP (sodium salt, Sigma) is stored frozen ($-20°$) as a solution containing 5 mM GTP, 10 mM MgSO$_4$, and either 0.1 M MES, pH 7.0, or 20 mM N-[2-hydroxyethyl]piperazine-N'-[2-ethanesulfonic acid] (HEPES), pH 7.4. GTP is normally added from a 0.1 M stock in H_2O adjusted to pH 7.0 with 1 N NaOH.

Radioactive GTP

[γ-^{32}P]GTP (Amersham Pharmacia Biotech, Piscataway, NJ) is essentially carrier free (>5000 Ci/mmol). Thus, we prepare a 5 mM stock solution consisting of 25 μl (250 μCi) [γ-^{32}P]GTP, 12.5 μl of 0.1 M nonradioactive

[9] T. D. Pollard, Methods Cell Biol. **24**, 333 (1982).
[10] J. B. Martin and D. M. Doty, Anal. Chem. **21**, 965 (1949).
[11] E. D. Korn, J. H. Collins, and H. Maruta, Methods Enzymol. **85**, 357 (1982).

GTP, and 212.5 μl H$_2$O. Final specific radioactivity is 200 μCi/μmol of GTP. This solution is kept at $-20°$ for no longer than 1 month.

Procedure

Radioactive substrate solution is made immediately prior to the assay by adding [γ^{32}-P]GTP stock to GTP mix to obtain approximately 0.6–0.8 μCi/μmol GTP (1400–2000 cpm/nmol GTP). The GTPase reaction is started by adding 20 μl of 5 mM [γ-^{32}P]GTP mix to 80 μl of dynamin solution and incubating for various times at the desired temperature. Because the K_m of dynamin for GTP is in the range of 5–50 μM depending on salt and temperature conditions, we typically used 1 mM GTP in our assays. Termination of the assay and extraction of [^{32}P]P$_i$ is accomplished by the following steps:

1. Add 1 ml of isobutanol/benzene followed by 0.25 ml of silicotungstic acid.
2. Vortex for 5 sec.
3. Add 0.1 ml ammonium molybdate.
4. Vortex for 10 sec.
5. Centrifuge at low speed (2000–4000 rpm) for 1 min at room temperature in a tabletop centrifuge to separate the organic from the aqueous phase.
6. Pipette 0.5 ml (50%) of the upper, organic phase for scintillation counting. Only a portion of the upper phase is taken to prevent contamination from the lower, aqueous phase, which contains unhydrolyzed [γ-^{32}P]GTP, usually representing >90% of the total cpm. When calculating specific activity, adjustments should be made to account for the total cpm in the entire volume of the upper phase.

The method is sensitive (specific radioactivity of the GTP mix can be increased but may give a high background) and very rapid. It requires as little dynamin as 0.5 μg per assay.

Activation by Phospholipids

At low ionic strength dynamin GTPase activity is potently stimulated by microtubules and anionic liposomes. These activators interact with the highly basic C-terminal proline/arginine-rich domain (PRD) of dynamin[5] via ionic interactions that are essentially abrogated at physiologic ionic strength. In contrast, stimulation of GTPase activity by D-5 phosphoinositides [phosphatidylinositol 4,5-bisphosphate (PI(4,5)P$_2$) and 1-O-(1,2-di-O-palmitoyl-sn-glycerol-3-benzyloxyphosphoryl)-D-myo-inositol 3,4,5-triphosphate (PI(3,4,5)P$_3$)] remains high even in the presence of 100–130 mM

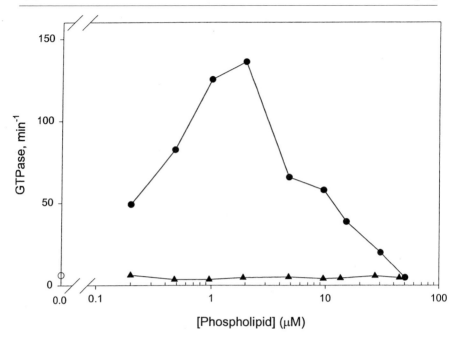

FIG. 3. Stimulation of GTPase activity of dynamin by micelles composed of 10% PI(4,5)P$_2$ and 90% PC (solid circles) or 100% PC (solid triangles) presented as a function of phospholipid concentrations. Open circle shows GTPase activity of dynamin alone. Dynamin concentration was 0.1 μM. Assay was performed in 0.1 M "nonactivating" MES.

NaCl.[12,13] These salt-resistant phosphoinositide interactions are mediated by the dynamin pleckstrin homology (PH) domain. They occur in proteolyzed dynamin-1 or truncated recombinant dynamin-2 lacking the PRD but not in dynamin-1 mutants lacking the PH domain.[12,14,15] Moreover, point mutants in which lysine-535 in the PH domain of dynamin-1 are converted to methionine[16] or alanine[17] fail to be activated by phosphoinositides and, when overexpressed, act as dominant inhibitors of endocytosis. Activation is specific for D-5 phosphoinositides since, at physiologic ionic strength,

[12] H. C. Lin, B. Barylko, M. Achiriloaie, and J. P. Albanesi, *J. Biol. Chem.* **272,** 25999 (1997).

[13] M. H. Stowell, B. Marks, P. Wigge, and H. T. McMahon, *Nature Cell Biol.* **1,** 27 (1999).

[14] K. Salim, M. J. Bottomley, E. Querfurth, M. J. Zvelebil, I. Gout, R. Scaife, R. L. Margolis, R. Gigg, C. I. E. Smith, P. C. Driscoll, M. D. Waterfield, and G. Panayotou, *EMBO J.* **15,** 6241 (1996).

[15] J. Zheng, S. M. Cahill, M. A. Lemmon, D. Fushman, J. Schlessinger, and D. Cowburn, *J. Mol. Biol.* **255,** 14 (1996).

[16] M. Achiriloaie, B. Barylko, and J. P. Albanesi, *Mol. Cell Biol.* **19,** 1410 (1999).

[17] Y. Vallis, P. Wigge, B. Marks, P. R. Evans, and H. T. McMahon, *Curr. Biol.* **9,** 257 (1999).

there is essentially no stimulation of GTPase activity by phosphatidylinositol 4-phosphate (PI4P) or 1-*O*-(1,2-di-*O*-palmityol-*sn*-glycerol-3-benzyloxy-phosphoryl)-D-*myo*-inositol 3,4-bisphosphate (PI(3,4)P$_2$).[12] There is no evidence to suggest that phosphoinositides activate dynamin by a mechanism other than facilitation of its self-association. Thus, the most important practical consideration in assaying phosphoinositide-stimulated GTPase activity is the presentation of the lipid. Mixed phosphoinositide/detergent micelles fail to activate dynamin[13] as do water-soluble short-chain phosphoinositides, unless they are incorporated into phosphatidylcholine (PC) vesicles (1998, our unpublished observations). In contrast, specific activities of up to 250 min^{-1} are achieved using large unilamellar vesicles containing 10% PI(4,5)P$_2$/90% PC[12] or "PIP$_2$ nanotubes," made in the presence of nonhydroxylated fatty-acid galactoceramides, which induce the formation of tubular rather than vesicular lipid structures.[13] The dynamin : phosphoinositide molar ratio is also critical. When using 9 : 1 PC/PIP$_2$ vesicles, if the ratio falls below approximately 1 : 40 the specific GTPase activity begins

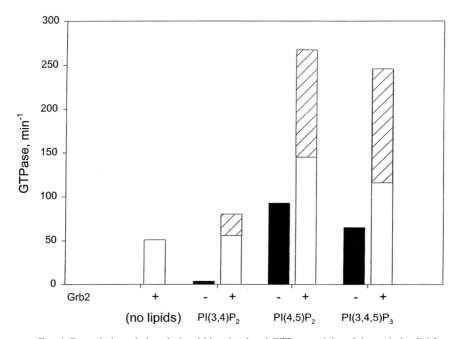

FIG. 4. Potentiation of phosphoinositide-stimulated GTPase activity of dynamin by Grb2. Cross-hatched portion of the bars represents the increase of activity above the additive activities obtained if PI(4,5)P$_2$ and Grb2 were introduced separately. Concentrations of dynamin, Grb2, and each phosphoinositide were 0.1, 4, and 2 μM, respectively. Assay was performed in 0.1 M "nonactivating" MES.

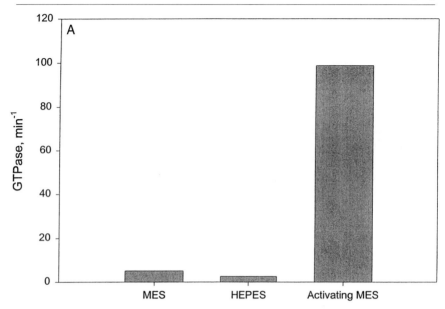

FIG. 5. Effects of buffers on the GTPase activity (A) and turbidity (B) of dynamin. (A) GTPase activity was measured in a solution contained either 20 mM HEPES, pH 7.4, or 0.1 M MES, pH 7.0, 5 mM MgCl$_2$, and 1 mM [γ-^{32}P]GTP. Dynamin concentration was 0.1 μM. (B) The turbidity was monitored at 310 nm and 20° for 10 min, with readings every 20 sec in a Beckman DU650 spectrophotometer. For each buffer tested (as in part A but without GTP), the spectrophotometer was blanked by measuring the absorbance at 310 nm of 300 μl buffer alone. After 30 sec, 100 μl of dynamin (2 μM) was introduced into the cuvette and thoroughly mixed. Nonactivating MES was from Midwest Scientific; activating MES was from Sigma.

to decline (Fig. 3), presumably because dynamin self-interactions are disrupted as dynamin molecules redistribute themselves on the lipid surface. This biphasic kinetic behavior was also observed in measurements of GTPase activation by microtubules[18] or phosphatidylserine vesicles.[12,18,19] In practice, it is worthwhile to assay enzymatic activity at a fixed lipid concentration (e.g., 5 μM PIP$_2$, 45 μM PC) and a broad range of dynamin concentrations (e.g., up to 1 μM) to determine optimal conditions.

Activation by Proteins Containing Src Homology 3 (SH3) Domains

A number of SH3 domain-containing proteins interact with dynamin and stimulate its GTPase activity *in vitro*,[20] but the physiologic significance

[18] P. Tuma and C. Collins, *J. Biol. Chem.* **269**, 30842 (1994).
[19] P. Tuma, M. Stachniak, and C. Collins, *J. Biol. Chem.* **268**, 17240 (1993).
[20] I. Gout, R. Dhand, I. Hiles, M. Fry, G. Panayotou, P. Das, O. Truong, N. Totty, J. Nsuan, G. Booker, I. Campbell, and M. Waterfield, *Cell* **75**, 25 (1993).

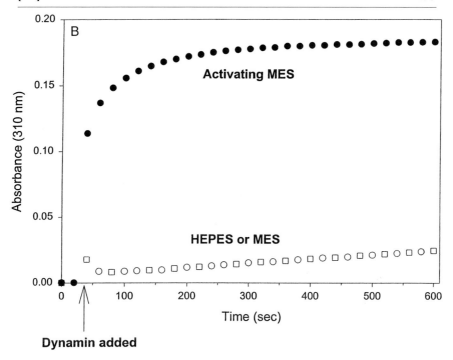

FIG. 5. (*continued*)

of this interaction is questionable. First, the level of activation is severalfold lower than that achieved in the presence of microtubules or anionic liposomes; Second, most studies have employed glutathione *S*-transferase (GST)–SH3 fusion proteins, which, given the dimeric nature of GST itself, can cross-link dynamin molecules[21]; and third, amphiphysin, the SH3 domain-containing protein most likely to interact with dynamin at the coated pit,[22] provides negligible GTPase stimulation, despite its dimeric nature.[23,24] Therefore, the interaction of dynamin with SH3 domain-containing proteins may be more important for targeting dynamin to proper locations than for stimulation of GTPase activity. This view is consistent with our recent observations that growth factor receptor binding protein (Grb2) remarkably potentiates GTPase activity of dynamin stimulated by phosphoinositides,

[21] B. Barylko, D. Binns, K. M. Lin, M. A. L. Atkinson, D. M. Jameson, H. L. Yin, and J. P. Albanesi, *J. Biol. Chem.* **273,** 3791 (1998).
[22] K. Takei, V. I. Slepnev, V. Haucke, and P. DeCamilli, *Nature Cell Biol.* **1,** 33 (1999).
[23] O. Shupliakov, P. Löw, D. Grabs, H. Gad, H. Chen, C. David, K. Takei, P. DeCamilli, and L. Brodin, *Science* **276,** 259 (1997).
[24] P. Wigge, Y. Vallis, and H. T. McMahon, *Curr. Biol.* **7,** 554 (1997).

$PI(4,5)P_2$ and $PI(3,4,5)P_3$ (Fig. 4), and the results of Rasmussen *et al.*[25] who showed a similar effect with mixed-lineage kinase 2 (MLK2). The simplest explanation for this synergy is that Grb2 clusters dynamin molecules by cross-linking their SH3 binding domains, thus increasing the effective concentration of dynamin bound to $PI(4,5)P_2$-containing liposomes. In this context it would be interesting to know whether other SH3 domain-containing proteins, particularly amphiphysin which targets dynamin to clathrin-coated pit, are able to "prime" dynamin for superactivation by phosphoinositides.

Buffer Considerations

It is generally agreed that dynamin GTPase activity will be stimulated by any factor that induces the formation of appropriately oriented dynamin polymers. Thus far, all reported GTPase activators have been macromolecules that either cross-link dynamin (e.g., GST-Grb2[21] or polyclonal anti-dynamin antibodies[26]) or provide a surface for dynamin self-assembly (e.g., microtubules or liposomes). However, we found that dynamin self-association, and consequent GTPase activation, can be promoted by yet unidentified small molecules occurring as buffer contaminants. Figure 5A shows the stimulation of dynamin-1 GTPase activity by a particular batch of MES buffer but not by HEPES or by a different source of MES. This activating MES buffer also promoted dynamin-1 self-association, assayed by measuring changes in solution turbidity (Fig. 5B). Although the contaminant in MES responsible for GTPase stimulation has not yet been identified, it appears to be a nonproteinaceous ion because it resists boiling and binds to Chelex 100 ion-exchange resin.

Acknowledgments

This research was supported by National Institutes of Health grant GM38567 (to J.P.A.) and American Heart Association, Texas Affiliate, grant 97G-111 (to B.B.).

[25] R. K. Rasmussen, J. Rusak, G. Price, P. J. Robinson, R. J. Simpson, and D. S. Dorow, *Biochem. J.* **335,** 119 (1998).
[26] D. Warnock, L. Terlecky, and S. Schmid, *EMBO J.* **14,** 1322 (1995).

Section V

Septin GTPases

[52] Expression and Analysis of Properties of Septin CDCrel-1 in Exocytosis

By Crestina L. Beites, Xiao-Rong Peng, and William S. Trimble

Introduction

The mammalian protein CDCrel-1 is a novel member of a family of highly conserved GTPases called *septins*.[1] Septins were first discovered in *Saccharomyces cerevisiae* as gene products required for the completion of budding.[1–3] In that organism, the septins CDC3, CDC10, CDC11, and CDC12 are thought to be constituents of 10-nm filamentous rings assembled at the mother bud neck at the onset of budding. Cloning revealed that each of the yeast septins contained a conserved GTP-binding domain near the amino terminus of the protein and, except for CDC10, all had an α-helical region near the carboxyl end with a high probability of forming a coiled coil. Although the function of the septins in yeast is not clear, the proteins appear to localize Bud3p, which is necessary for proper bud localization,[4] CHS4, which is required for proper chitin deposition,[5] and regulate the kinases Gin4p[6] and Hsl1p[7] during the cell cycle.

Septins are conserved throughout evolution and homologs have been found in a variety of organisms including *Drosophila, Caenorhabditis elegans, Xenopus laevis,* and in several mammalian species. It appears that at least 10 different genes encode septin isoforms in mammals, several give rise to multiple proteins through alternative splicing, and most appear to be broadly or ubiquitously expressed. The function of septins in the cell cycle may also be conserved, as mutations in the *Drosophila* septin *pnut* blocks cytokinesis in that organism, and perturbation of Nedd5 blocks cytokinesis in mammalian cells.[8] In mammals and *Drosophila*, septins accu-

[1] S. L. Sanders and C. M. Field, *Curr. Biol.* **4,** 907 (1994).

[2] M. S. Longtine, D. J. DeMarini, M. L. Valencik, O. S. Al-Awar, H. Fares, C. De Virgilio, and J. R. Pringle, *Curr. Opin. Cell. Biol.* **8,** 106 (1996).

[3] W. S. Trimble, *J. Membrane Biol.* **169,** 75 (1999).

[4] J. Chant, *Cell* **84,** 187 (1996).

[5] D. J. DeMarini, A. E. M. Adams, H. Fares, C. De Vergilio, G. Valle, J. S. Chuang, and J. R. Pringle, *J. Cell Biol.* **139,** 75 (1997).

[6] C. Carroll, R. Altman, D. Schieltz, J. Yates, and D. Kellogg, *J. Cell Biol.* **143,** 709 (1998).

[7] Y. Barral, M. Parra, S. Bidlingmaier, and M. Snyder, *Genes Dev.* **13,** 176 (1999).

[8] M. Kinoshita, S. Kumar, A. Mizoguchi, C. Ide, A. Kinoshita, T. Haraguchi, Y. Hiraoka, and M. Noda, *Genes Devel.* **11,** 1535 (1997).

mulate at the cleavage furrow that forms during late anaphase and constricts the cell membrane during telophase.

However, it is unlikely that the function of septins is limited to cell division. CDCrel-1 is a septin-like protein that is expressed almost exclusively in postmitotic neurons in the brain. It was first identified as one of several proteins that coimmunoprecipitated along with synaptophysin and the SNARE proteins VAMP, syntaxin, and SNAP-25 from human brain.[9] Monoclonal antibodies specific to this protein were used to screen a human fetal brain cDNA library and sequencing of the isolated clones revealed that CDCrel-1 belonged to the septin family.[10] CDCrel-1 is most homologous to the mammalian septin H5 with which it shares 76% sequence identity. The coding sequence of the CDCrel-1 cDNA predicts a 369 amino acid protein containing putative coiled coils near the C terminus and consensus motifs for GTP binding in the first 190 amino acids. The protein lacks transmembrane domains or other motifs associated with membrane localization. Through studies of CDCrel-1 expression, subcellular distribution, protein–protein interactions, mutagenesis, and transfection, we have found that it binds directly to the SNARE protein syntaxin and regulates secretion.[11]

Determination of Expression Patterns and Subcellular Distribution of CDCrel-1 Proteins

To examine the tissue specificity of CDCrel-1, proteins from several organs are prepared by placing tissues in homogenization buffer and homogenizing them with five strokes of a Dounce homogenizer. Then aliquots are suspended in Laemmli sample buffer and boiled followed by SDS–PAGE and Western blotting with antibody SP20.[9] CDCrel-1 is expressed predominantly in the brain, with little or no expression found in other tissues tested (spleen, heart, liver, kidney, pancreas, testis). Western blots with known amounts of recombinant CDCrel-1 electrophoresed alongside brain lysate reveal that CDCrel-1 represents about 0.1% of the total protein in the rat brain. To determine which subcellular compartment CDCrel-1 is associated with, we use a standard fractionation procedure to purify synaptosomes and synaptic vesicles (see Fig. 1).

This fractionation method[12] involves crushing 10 frozen rat brains to powder in liquid nitrogen followed by further homogenization in 100 ml of homogenization buffer [0.32 M sucrose, 20 mM HEPES, pH 7.4, 0.3

[9] W. Honer, L. Hu, and P. Davies, *Brain Res.* **609,** 9 (1993).

[10] J. Caltagarone, J. Rhodes, W. Honer, and R. Bowser, *Neuroreport* **9,** 2907 (1998).

[11] C. L. Beites, H. Xie, R. Bowser, and W. S. Trimble, *Nat. Neurosci.* **2,** 434 (1999).

[12] W. Huttner, W. Schiebler, P. Greengard, and P. De Camilli, *J. Cell Biol.* **96,** 1374 (1983).

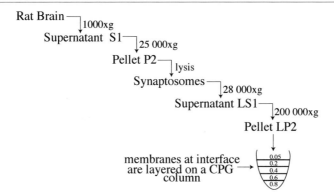

Fig. 1. Synaptic vesicle purification scheme. Rat brain homogenates are subjected to a differential centrifugation series leading to the purification of synaptic vesicles. Discarded fractions are not shown. [This method is adapted from W. Huttner, W. Schiebler, P. Greengard, and P. De Camilli, *J. Cell Biol.* **96**, 1374 (1983).]

mM phenylmethylsulfonyl fluoride (PMSF)] with five strokes of a Dounce homogenizer. The homogenate is then centrifuged at low speed (1000g, 10 min, 4°) to remove large cellular debris, unbroken cells, and nuclei. Typically, three phases are present after spinning: a small particulate fraction, which is discarded; a middle milky layer; and a clear top layer. The latter two are collected and centrifuged at 25,000g, for 10 min, 4°. The pellet fraction (P2) enriched in synaptosomes is washed once with 20 mM HEPES, pH 7.5, 140 mM NaCl, 2 mM KCl, 1 mM MgCl$_2$, and 1 mM PMSF, recentrifuged, and then resuspended in 10 volumes of ice-cold water for 15 min to lyse the synaptosomes. The synaptic vesicles released as a result of this incubation are collected in the supernatant (LS1) after a 15-min spin at 15,000 rpm in a SW28 rotor (Beckman, Palo Alto, CA). To pellet these small vesicles, LS1 is centrifuged at high speed (45,000 rpm, 50.2 Ti rotor) for 2 hr and the resulting pellet (LP2) contains the crude synaptic vesicle fraction.

Further purification of these vesicles is achieved by layering the fraction on a discontinuous 0.05–0.8 M sucrose gradient and collecting the turbid fractions at the 0.2–0.4 M interface after centrifugation for 5 hr at 25,000g and 4°. Further purification of synaptic vesicles is achieved by chromatography of this turbid material over a controlled pore glass bead column (CPG-3000, Sigma, St. Louis, MO). Although CDCrel-1 does not contain a transmembrane domain, it is more prevalent in the membrane fractions than the cytosolic fraction (LS2). Significant amounts of CDCrel-1 are also found to be insoluble when membrane fractions are dissolved with 4% Triton X-100, a common feature of cytoskeleton-associated proteins.

Expression and Purification of Wild-Type and Mutant CDCrel-1

For *in vitro* studies of CDCrel-1, we create a fusion of CDCrel-1 with *Schistosoma japonicum* glutathione *S*-transferase (GST) in the vector pGEX-KG.[13] Rat CDCrel-1 cDNA is amplified by PCR (polymerase chain reaction) using the oligonucleotides 5′-CCGGATCCATGAGCACAG-GACTGCG-3′ and CAGAATTCATGGTCCTGCATTTGC-3′ that correspond to the N and C termini of the protein and, respectively, include *Bam*HI and *Eco*RI restriction sites. The PCR product is cloned into the *Bam*HI and *Eco*RI sites of pGEX-KG for expression. Because GST retains its enzymatic activity after expression in *Escherichia coli*, CDCrel-1 can be purified from the bacterial lysates by affinity chromatography using glutathione-coated beads (details below). The Quick Change Site-Directed Mutagenesis kit (Stratagene, La Jolla, CA) is then used to make site-specific mutations in the GTP binding loop. We chose to substitute the serine at position 58 of CDCrel-1 with asparagine because we predicted that this mutation would be equivalent to the dominant negative S17N mutation in ras. In the latter case this mutation significantly reduces GTP binding and results in a permanently inactive protein that can competitively interfere with endogenous active ras proteins when introduced into cells.

The pGEX-CDCrel-1 plasmid is first denatured and the synthetic oligo-nucleotide primers 5′-CCGGCCTGGGGAAGAATACCCTTGTCAA-CAGTCTCTT-3′ and 5′-AAGAGACTGTTGACAAGGGTATTCTTC-CCCAGGCCGG-3′, in which the serine at position 58 (G1) is altered to encode asparagine, are annealed. Because this kit utilizes the *Pyrococcus furiosis* (Pfu) DNA polymerase, high fidelity in DNA synthesis is achieved and the mutant oligonucleotide primers are not displaced. Through PCR, the oligonucleotides are incorporated and the resulting nicked mutated plasmid is transformed into competent *E. coli* where it is subsequently repaired. The mutant gene is then subcloned into the pGEX vector to allow its expression.

To express recombinant CDCrel-1, *E. coli* containing the wild-type or mutant plasmid are inoculated into 3 ml of Luria–Bertani (LB) broth containing 100 μg/ml of ampicillin for 8 hr at 37° in a shaking incubator. The 3 ml is then added to 50 ml of LB broth, and incubated overnight. The next day the culture is diluted into 500 ml of LB for 1 hr or until the OD_{600} reaches 0.8–1.0. At this time, 0.1 mM of isopropyl-β-D-thiogalacto-pyranoside (IPTG) is added and the culture is switched to a 30° bacterial shaker and incubated for an additional 6 hr. We have found that inducing expression for 6 hr at 30° (rather than shorter times at 37°) decreases

---

[13] K. Guan and J. Dixon, *Biochemistry* **192,** 262 (1990).

degradation and greatly improves the yield of purified protein. The cells are then spun at 6000g at 4° for 10 min, the supernatant decanted, and the pelleted cells resuspended in 5–7 ml of cold resuspension buffer (PBS-T: 150 mM NaCl, 12.5 mM NaH_2PO_4, 90 mM Na_2HPO_4, 0.05% Tween 20), supplemented with 0.1% 2-mercaptoethanol, and the following protease inhibitors: 1 μg/ml pepstatin, 1 μg/ml leupeptin, 2 mM benzamidine, 1 mM PMSF). Using a French pressure cell press (Spectronic Instruments, Rochester, NY), the bacterial cells are lysed at 20,000 psi at a rate of about 30 drops per minute. Successful bacterial lysis is often accompanied by a darkening of the suspension. To increase the yield of soluble protein, the lysis is further incubated with 1% Triton X-100 for 30 min at 4° followed by a spin at 15,000 rpm in a JA-17 rotor (Beckman) for 15 min to remove cellular debris and insoluble fusion proteins. Even after the detergent incubation, we have found that approximately 75% of the proteins remain insoluble and are discarded.

To purify the protein from bacterial lysates, glutathione-agarose beads (Sigma, G4510) are preswollen in water, washed twice in PBS-T, and diluted to a 50% slurry. The beads are then incubated with the precleared lysate for 2 hr at 4°, with constant rotation and then washed five times with 10 ml of ice-cold PBS-T. The volume is readjusted to achieve a 50% bead slurry. Protease inhibitors (1 mM PMSF, 1 μg/ml leupeptin and pepstatin, 2 mM benzamidine) are added if the CDCrel-1 fusion protein is to be left on the bead for subsequent assays.

To remove the GST carrier from the fusion proteins, the beads are washed twice in TCB buffer (20 mM HEPES–KOH, pH 7.8, 150 mM KCl, 2.5 mM $CaCl_2$, 0.5% Tween 20 and freshly added 0.1% 2-mercaptoethanol) and incubated with 500 μl of TCB with the addition of 5 units of thrombin (Sigma, T7513). Although efficient cleavage occurs after 10 min at room temperature, some cleaved protein remains bound to the beads, possibly through aggregation of the protein at high concentration. The thrombin is then inactivated by the addition of 1 mM [4-(2-aminoethyl)benzenesulfonyl fluoride hydrochloride] (AEBSF) and the aforementioned protease inhibitors are also added. Incubation of this protein with thrombin leads to the production of two fragments, one a predominant 45-kDa species reflecting the full-length CDCrel-1 and a minor lower band possibly due to degradation of the protein from some protease sensitive site. For most studies the protein is used immediately because overnight incubation at 4° results in loss of its GTP binding capability (see below).

In spite of its low solubility, a typical yield of 1.5–2.5 mg/ml of protein is eluted from the glutathione beads per 500 ml starting culture. In our experience many unlysed bacterial pellets or precleared lysates of other GST fusion proteins can be frozen at −80° prior with purification continued at a later date, but several attempts at recovery of GST–CDCrel-1 after freezing (in the presence or absence of glycerol, or within the bacteria after induction) have been unsuccessful. Most of the protein remains in insoluble aggregates and the little that does bind to the glutathione beads can no longer bind GTP (see below).

Measurement of GTP Binding Properties of CDCrel-1

Like other GTPases, septin proteins contain several motifs needed for GTP binding and hydrolysis.[14] We examined the ability of CDCrel-1 to bind and hydrolyze GTP, and the effect of our S58N mutation on properties, using a simple assay described by Kinoshita et al. for the septin Nedd5.[8] In all GTP binding assays, freshly prepared fusion proteins must be used since GTP binding activity is greatly diminished (to mutant levels) if the protein is stored at 4° overnight. Glutathione beads to which GST–CDCrel-1 or control GST are bound are first washed twice in ice-cold binding buffer (buffer A) [20 mM Tris-HCl, pH 8.0, 5 mM MgCl$_2$, 100 mM NaCl, 0.1% 2-mercaptoethanol, 10% (v/v) glycerol, 200 μg/ml bovine serum albumin (BSA)] in a 1.5-ml Eppendorf tube in the presence of 10 mM EDTA to prevent GTP hydrolysis. To measure GTP binding, the protein is loaded with a GTP analog (Amersham, Piscataway, NJ) in which the phosphate group in the gamma position is radiolabeled. In this manner, actual GTP binding is measured and not simply the ability of the protein to bind to a guanine nucleotide, regardless of its phosphate content. The binding is allowed to continue on ice for 30 min, with rotation. The beads are then washed four times with 1 ml of binding buffer, the caps of the Eppendorf tubes are removed, and the tube is dropped into 10 ml of liquid scintillation fluid and shaken vigorously. Bound radioactivity is then measured on an automated RACKBETA (LKB) liquid scintillation counter. In a typical experiment, the wild-type protein had 1.6E6 cpm bound while the mutant protein and control GST had bound 73,000 and 2,400 cpm, respectively. The fact that the mutant protein is still able to bind a very low amount of GTP in vitro (compared to GST) may not necessarily reflect its in vivo capabilities. The mutant protein, although able to bind GTP, may prefer to bind GDP so that in vivo, the protein may never bind GTP through competition by GDP.

CDCrel-1–Syntaxin Binding Assays

The observations that CDCrel-1 fractionates along with cytosol and membrane in brain lysates, that it associates with purified synaptic vesicles,

[14] H. Bourne, D. Saunders, and F. McCormick, Nature 349, 117 (1991).

and that it does not contain a transmembrane domain suggested that it might be associating with vesicles via a transmembrane synaptic vesicle receptor. Immunoprecipitation of brain extract reveals that CDCrel-1 coimmunoprecipitates syntaxin 1A.[11] This 35-kDa membrane protein is mainly found on presynaptic membrane but is also found on synaptic vesicles. Because immunoprecipitation experiments do not prove direct interactions, recombinant binding assays are used to determine whether CDCrel-1 and syntaxin are direct binding partners. As a preliminary experiment, glutathione immobilized recombinant GST (10 μg) as a control, or GST–CDCrel-1 (10 μg) purified as described above, is incubated with recombinant syntaxin 1A (20 μg) lacking the transmembrane domain in 500 μl of buffer C [20 mM Tris-HCl, 140 mM KCl, 0.5% Triton X-100, 0.1% BSA, 1 mM PMSF, 1 mM dithiothreitol (DTT), 1 mM benzamidine, 5 mM MgCl$_2$, and 1 mM EDTA] for 2 hr at 4°.

After washing four times in ice-cold buffer C, lacking BSA and protease inhibitors, the beads are boiled in 10 μl of SDS loading buffer and subjected to SDS–PAGE and Western blotting. Syntaxin 1A binds to GST–CDCrel-1 but not to GST alone, confirming a direct interaction between the two proteins. Although the affinity for these proteins seems to be low, it is possible that the CDCrel-1–syntaxin interaction might be potentiated by the presence of nucleotides. To test this hypothesis, GST-CDCrel-1 (0.3 μM) is preincubated for 10 hr in buffer A, with 10 mM EDTA and either 1 mM GTPγS or 1 mM GDPβS. The beads are then washed in buffer A and incubated with soluble recombinant syntaxin 1A (1.5 μM) in buffer B (20 mM Tris-HCl, pH 8.0, 5 mM MgCl$_2$, 100 mM NaCl, 0.1% 2-mercaptoethanol, 0.1% BSA, 0.1% Triton X-100, 10 mM EDTA) for 1 hr at 4°. The CDCrel-1–syntaxin complexes are then washed twice in buffer B lacking BSA. Furthermore, it was reported that at least in the case of Gα proteins, tetrafluoroaluminate (AlF$_4^-$) is thought to allow a conformation mimicking the transition state of the GTPase reaction.[15] For heterotrimeric G proteins, when GDP and AlF$_4^-$ bind to the Gα subunit an active conformation ensues where the AlF$_4^-$ species binds in the position of the γ-phosphate. This transitional state may be necessary for signaling to downstream effectors.

It was also shown that not only could heterotrimeric G proteins be activated in this manner but that AlF$_4^-$ can induce stable complex formation between the GDP-bound Rho GTPase Cdc42 and its effectors,[16] thus activating the GTPase. To test the effect of this ion on the binding of CDCrel-1 to syntaxin, CDCrel-1 is preincubated as above with the GDP analog, GDPβS. The GST fused protein is then washed once in buffer A

[15] P. C. Sternweis and A. G. Gilman, *Proc. Natl. Acad. Sci. U.S.A.* **79**, 4888 (1982).
[16] G. R. Hoffman, N. Nassar, R. E. Oswald, and R. A. Cerione, *J. Biol. Chem.* **273**, 4392 (1998).

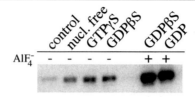

$$AlF_4^- \quad - \quad - \quad - \quad - \quad + \quad +$$

Fig. 2. Effect of nucleotide on the binding of GST–CDCrel-1 to syntaxin. Immobilized GST–CDCrel-1 or GST (control) preincubated with GTPγS, GDPβS, or buffer alone (nucl. free) were treated with (+) or without (−) AlF_4^- followed by incubation with soluble syntaxin 1A. The protein complexes were resolved on SDS–PAGE gels followed by immunoblotting with anti-syntaxin antibodies. Although atypical, some binding of syntaxin to GST is observed in this experiment.

followed by and incubation with 30 μM $AlCl_3$ and 10 mM NaF[17] for 1 hr at 4°. Unbound tetrafluoroaluminate is washed off by three consecutive washes with buffer A and cleaved recombinant syntaxin is added (as above). After another 1 hr of incubation, the beads are washed, boiled in loading buffer, and the amount of syntaxin bound to GST or GST CDCrel-1 is analyzed by Western blotting. Interestingly, although it appears that the presence of GDPβS might slightly potentiate the binding of CDCrel-1 to syntaxin, the amount of syntaxin bound to CDCrel-1 is greatly potentiated in the presence of AlF_4^- suggesting that this transitional state allows the optimal conformation for binding to syntaxin 1A (see Fig. 2).

Secretion Assay

Because syntaxin is a SNARE protein required for secretion, we tested the hypothesis that CDCrel-1 interaction with syntaxin may regulate this process. The insulin secreting pancreatic β cell line HIT-T15 contains the same isoforms of the SNARE proteins, namely, SNAP-25, VAMP2, and syntaxin-1A[18] that are found in the neuron and needed for neurotransmitter release. Moreover, the use of this cell line permits us to monitor the effect of transfection of wild-type and GTP mutant CDCrel-1 on secretion. To measure secretion only from transfected cells, we cotransfect the HIT-T15 with the CDCrel-1 containing plasmids and a plasmid pXGH5,[19] which encodes the mouse metallothionein promoter, followed by the human

[17] R. Mittal, M. R. Ahmadian, R. S. Goody, and A. Wittinghofer, *Science* **273**, 115 (1996).
[18] K. Sadoul, A. Berger, H. Niemann, U. Weller, P. A. Roche, A. Klip, W. S. Trimble, R. Regazzi, S. Catsicas, and P. A. Halban, *J. Biol. Chem.* **272**, 33023 (1997).
[19] R. Selden, K. Howie, M. Rowe, H. Goodman, and D. Moore, *Mol. Cell. Biol.* **6**, 3173 (1986).

growth hormone (hGH) gene. On depolarization of the cells, the level of hGH released into the culture medium is measured and the effect of the septin on this process determined.

To do this, HIT-T15 cells (ATCC) are grown in RPMI, supplemented with 10% fetal calf serum, l-glutamine as well as penicillin and streptomycin and plated on 24-well dishes at 2×10^5 cells per well. The following day, 3 μl of Lipofectamine reagent (Life Technologies) is premixed with 100 μl of serum-free medium. One μg of Qiagen purified pXGH5 and 2 μg of either the wild-type or mutant CDCrel-1 in the mammalian expression vector pcDNA3.1 is then diluted separately into 100 μl of serum-free medium and added to the 100 μl of diluted Lipofectamine. The DNA and Lipofectamine are incubated for 0.5 hr at room temperature to allow the formation of liposome–DNA complexes. Then 800 μl of serum free medium is added to the lipid–DNA mixture, the cells are washed twice in serum-free medium, and the lipid–DNA mixture (1 ml) is added to the cells. After an overnight incubation, the cells are washed again, medium containing serum is then added, and the cells are incubated for 48 hr at 37°, 5% CO_2.

To determine the effects of the wild-type and mutant CDCrel-1 protein on evoked secretion, the cells are then treated with a low K^+ (basal) or high K^+ (stimulated) buffer. The insulinomas are first incubated for 0.5 hr in 300 μl of low potassium buffer, which consists of 4.8 mM KCl, 129 mM NaCl, 5 mM NaHCO$_3$, 1.2 mM KH$_2$PO$_4$, 1 mM CaCl$_2$, 1.2 mM MgSO$_4$, 0.5 mM glucose, and 10 mM HEPES, pH 7.4. Culture supernatant is removed to assay for basal levels of secretion. Evoked secretion is then achieved by incubating the cells with the same buffer in which the KCl concentration has been raised to 30 mM and the NaCl reduced to 104 mM for 1 hr at 37°. The supernatants are again collected, and levels of hGH released are assayed by enzyme-linked immunosorbent assay (ELISA) (Boehringer Mannheim). Briefly, in our experiments, 100 μl of supernatant (or $E. coli$ hGH protein standard supplied with the kit for a calibration curve) is added to microtiter plates that have been precoated with anti-hGH antibody. A digoxigenin-labeled antibody to hGH is then added to the plates. The wells are then washed and an antibody to digoxigenin conjugated to peroxidase is added to the wells. After another wash, the peroxidase substrate is added. The substrate is cleaved, resulting in a colored reaction product. The plates are then read on a standard ELISA plate reader (490 nm). In control experiments, where the hGH vector alone is transfected, typically 80–90 pg/ml of hGH is secreted. However, the wild-type protein greatly reduces secreti levels to 30 pg/ml while the mutant significantly protein potentiated secretion to 120/150 pg/ml. This indicated that the septin protein inhibits secretion in a GTPase-dependent manner.

Use of C-Terminal GFP-Tagged CDCrel-1 Fusion Protein to Monitor Septin Accumulation *in vivo*

Green fluorescent protein (GFP) has been widely used to tag proteins and to study the dynamic changes of cellular processes in living cells. Septins are synthesized as cytosolic proteins, but have been shown to assemble into heterologous filaments both *in vitro* and *in vivo*. However, CDCrel-1 is primarily expressed in neurons and neuroendocrine cells and there is no evidence for the ability of this protein to form filament structures, nor is it known if GTP binding or hydrolysis is involved in this process. To address these issues, we have constructed GFP–CDCrel-1 fusion proteins for analysis in cultured cells.

To generate a fusion protein of CDCrel-1 with the N-terminal end of GFP, a cDNA fragment encoding the rat CDCrel-1 is generated by polymerase chain reaction (PCR) using the oligonucleotides 5'-CGGAA-TTCATGAGCACAGGACTGCGGTACAAG-3' and 5'-CAGAATT-CATGGTCCTGCATTTGC-3' with *Eco*RI sites built in to each end. After restriction digestion of the PCR fragment with *Eco*RI, the fragment is ligated into pBluescript KS vector (Stratagene, La Jolla, CA). The CDCrel-1 cDNA is then excised from the pBluescript KS using *Hin*dIII and *Pst*I in an orientation of 5' to 3' and ligated in frame into the *Hin*dIII/Pst1 sites of pEGFP-N1 (Clontech Laboratories, Inc, Palo Alto, CA). The S58N dominant negative mutant of CDCrel-1 is cloned into pEGFP-N1 in a similar manner except the mutant version of CDCrel-1 cDNA is used as a template for the PCR. Also, an N-terminal FLAG-tagged version of the CDCrel-1 protein is generated and this demonstrated that tagging of either end of the protein can give rise to proper localization.

Overexpression of CDCrel-1 Proteins in Mammalian Cells by Transfection

One day before the transfection, 1×10^5 HeLa cells are seeded onto coverslips in 35-mm dishes containing 2 ml Dulbecco's modified Eagle's medium (DMEM) + 10% fetal calf serum, and the transfection is performed the next day using Fugene reagent (Boehringer Mannheim). In detail, 2 μl of Fugene reagent is first incubated with 100 μl serum-free media for 5 min and the Fugene–media mixture is then added to another Eppendorf tube containing 500 ng plasmid DNA. The plasmid DNA is incubated with Fugene for 15 min and then added to the cells drop by drop. The expression of GFP–CDCrel-1 in living cells can be monitored directly by inverting the coverslips on slides in PBS or culture medium for observation using a fluorescent microscope. Alternatively, the cells are fixed 24 hr after transfec-

GFP-CDCrel-1 (wt)   GFP-CDCrel-1 (S58N)

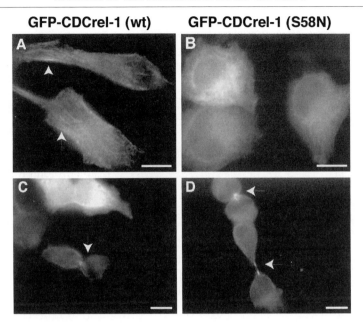

Fig. 3. Localization of CDCrel-1–GFP fusion protein in HeLa Cells. In nondividing cells that transiently expressed moderate levels of wild-type CDCrel-1–GFP protein (A), filaments were observed in the cell (arrows). In contrast to the wild-type CDCrel-1–GFP protein, the mutant CDCrel-1–GFP that does not bind GTP appears cytosolic in HeLa cells (B). However, both wild-type (C) and mutant proteins (D) accumulate at the cleavage furrow during cytokinesis as indicated by arrows. Bars: 10 μm.

tion. First, the coverslips are fixed for 15 min in 4% formaldehyde (EM Science) in PBS, pH 7.4, at room temperature, and washed with PBS once. The coverslips are then mounted on glass microscope slides in a fluorescent mounting media (DAKO Corporation, Carpinteria, CA) to reduce photobleaching.

Despite that fact that CDCrel-1 is normally only expressed in the brain, both GFP- and FLAG-tagged CDCrel-1 are able to form filaments in HeLa cells. As shown in Fig. 3A, wild-type GFP–CDCrel-1 has a filamentous appearance in cells, while GFP-CDCrel-1 (S58N) appears diffuse through the cytoplasm. In cells undergoing cytokinesis, like other septins,[20] GFP–CDCrel-1 accumulates at the cleavage furrow. Surprisingly, as shown in Fig. 3D, GFP–CDCrel-1 (S58N) also accumulates at the cleavage furrow during cytokinesis, indicating that the presence of a functional GTPase

[20] H. Xie, M. Surka, J. Howard, and W. S. Trimble, *Cell Motil. Cytoskel.* **43**, 52 (1999).

domain is not necessary for this accumulation. Similar results are obtained when FLAG-tagged CDCrel-1 is expressed in HeLa cells.

Conclusions

CDCrel-1 is a brain-specific septin that functions in regulating secretion by binding directly to syntaxin. Recombinant CDCrel-1 is relatively insoluble, yet maintains GTPase activity when freshly prepared. Extended incubation, even at 4°, or repeated freeze–thaw cycles, result in loss of this activity. When expressed in heterologous cells, it colocalizes with other septins in a filamentous, actin-associated pattern that is dependent on the presence of a functional GTPase domain. Mutant proteins defective in GTP binding accumulate in the cytoplasm. Interestingly, during cytokinesis both mutant and wild-type CDCrel-1 are recruited to the cleavage furrow. This indicates that filamentous appearance, but not cleavage furrow recruitment, required GTP binding, and that septins such as CDCrel-1, which are not normally involved in cytokinesis, are capable of cleavage furrow localization. Future studies will be aimed at identifying the cleavage furrow localization signals and the role of this and other septin proteins in a variety of cellular processes.

Author Index

Numbers in parentheses are footnote reference numbers and indicate that an author's work is referred to although the name is not cited in the text.

A

Aasland, R., 121, 132
Abagyan, R., 458
Abeliovich, H., 234
Abrahamson, D. R., 59
Achiriloaie, M., 454, 459, 467(16), 479, 492, 493(12), 494(12)
Achstetter, T., 264, 301, 302(10), 303(10), 305(10)
Adamik, R., 301, 302(6), 303(6, 14), 304(6, 14), 305, 305(14), 306, 306(14), 331, 334, 417, 425, 427(14), 428(15)
Adams, A. E. M., 499
Adams, M., 423
Adams, S. R., 7
Adari, H., 226
Adesnik, M., 165, 197
Admon, A., 307, 311(11), 312(11), 315(11)
Aelst, L. V., 146, 148
Afar, D. E., 466
Ahmadian, M. A., 327
Ahmadian, M. R., 16, 51, 506
Al-Awar, O., 249, 255(13), 343, 499
Albanesi, J. P., 454, 459, 467(16), 479, 486, 492, 493(12), 494(12), 495
Albert, S., 50, 51(3, 4), 52(3, 4), 54(3), 55(3, 4), 58(3, 4), 91
Albertsen, H., 425
Albright, C. F., 92
Alessi, D. R., 147, 148
Alexandrov, K., 4, 14, 19, 27, 31, 32(23), 33, 40, 50, 120, 133, 139, 141(19), 204, 357
Allan, V. J., 395
Altman, R., 499
Altschuler, Y., 7, 165, 166(7), 212
Altshuller, Y. M., 365
Amano, K., 121

Amano, M., 121
Amherdt, M., 372, 373, 380, 388, 389, 391, 404, 431, 438, 442(6), 443(6)
Amor, J. C., 250, 325
Anandasabapathy, N., 198
Anant, J. S., 32(24), 33
Anderson, D., 141, 142(21)
Anderson, K., 347
Andrade, J., 308, 312(19), 334, 336, 343(8), 344, 345(12), 346(12), 349(12), 351(12)
Andreev, J., 308, 318, 334, 336, 343(10), 344, 345(13), 349(13), 351(13), 458
Andres, D. A., 32, 33(22), 36(22)
Aneja, R., 347, 348(25)
Aniento, F., 187, 413
Anjelkovich, M., 148
Antonny, B., 264, 265, 265(6), 271, 271(6), 272, 275, 306, 312, 315(25), 344, 347, 388
Antony, C., 157
Aoe, T., 308
Apodaca, G., 211
Araki, K., 61, 76, 77
Araki, S., 40, 60, 61(14), 207
Aravamudan, B., 85
Aridor, M. A., 187, 431, 433(2), 435, 435(2), 437(2)
Armstrong, J., 92, 188, 190(4)
Armstrong, S. A., 32, 33, 33(22), 36(22), 139, 357
Arnold, A. D., 256, 305, 346
Aron, L. M., 225
Aronheim, A., 458
Arpin, M., 197, 198(8)
Artzt, K., 226
Arvidsson, A., 287, 289(23), 458, 463(9), 465(9)
Asakura, T., 68
Ashery, U., 84, 87(20), 281

511

F

Falasca, M., 280, 287(6), 458, 463(9), 465, 465(9)
Fales, H. M., 326, 347
Fancy, D. A., 477
Fares, H., 499
Farnsworth, C. C., 31, 32
Farquhar, M. G., 14, 186, 211
Farsad, K., 380, 389(30), 390, 479, 486
Feinberg, A. P., 50
Feingold, D. C., 382
Feingold, D. S., 412
Feliciano, E., 235
Feng, Y., 40, 165, 166, 169(6), 170(9), 174, 174(9), 175, 176, 176(2), 177, 177(6), 179(2), 183(6), 184(6), 185(6), 186(6), 187, 188
Fensome, A., 325, 344, 355, 357, 417
Feramisco, J. R., 78
Ferenz, C. R., 83
Fergestad, T., 85
Ferguson, K. M., 457, 458, 458(1, 7), 461, 463, 465(26)
Ferrans, V. J., 305, 328
Ferrige, A. G., 326, 347
Ferro-Novick, S., 120, 234, 235
Fesik, S. W., 306, 457
Feuerstein, J., 15, 56
Fiedler, K., 373, 389, 404(19)
Field, C., 100, 264, 301, 302(10), 303(10), 305(10), 438, 499
Fields, S., 146
Finazzi, D., 248, 299, 300, 307
Finger, F. P., 100, 107(4)
Fink, G. R., 321, 324(21)
Fiske, C. H., 268
Flavell, R. A., 459
Fleischer, S., 398, 399(41)
Florio, S. K., 207
Fontaine, J. J., 197, 198(8)
Forget, B. G., 407, 416(16)
Foroni, L., 353
Fossum, R. D., 39
Foster, L., 37
Franco, M., 248, 256, 265, 272, 274(8), 275, 280, 283(7), 290, 294(12), 295, 305, 313, 343, 370, 383, 388, 390, 392(33)
Frank, D. W., 459, 463(20), 466(20), 467(17, 20), 479

Frank, R., 196
Frank, S. R., 248, 256, 257, 258, 258(9), 264(8), 272, 283, 343
Franke, T. F., 147, 148
Franzuksoff, A., 264
Franzusoff, A., 290, 296, 301, 302(10), 303(10), 305(10)
Freeman, J. L., 308, 334, 336, 337(7), 344
Freundlieb, S., 259
Fried, V. A., 83
Fries, E., 370
Fritsch, E. F., 78, 149
Frohman, M. A., 365, 383
Fruman, D. A., 457
Fry, M., 494
Fry, M. H., 479
Fry, M. J., 146
Fu, H., 147
Fuerst, T. R., 8, 170, 174(11), 190, 195(11)
Fujisawa, H., 76
Fujita, Y., 77
Fukuda, M., 214
Fukui, K., 55, 60, 68
Füllekrug, K., 389
Fuller, R. S., 290
Funato, K., 15, 175
Furge, K. A., 92
Furuhjelm, J., 188
Fushman, D., 459, 465(18), 492
Futter, C. E., 173, 180, 187(20)

G

Gad, H., 480, 495
Gaffney, P. R. J., 147, 347
Gaidarov, I., 458
Gallagher, P. G., 407, 409(10)
Galli, T., 111, 118(5), 197, 200, 205(14)
Gallwitz, D., 50, 51, 51(3, 4), 52, 52(3, 4, 7, 9), 53(8), 54(3), 55, 55(3, 4, 7, 9), 58(3, 4), 91
Gamblin, S. J., 51
Garcia, P., 457
Garcia-Estefania, D., 431
Garoff, H., 195
Garrett, M., 91, 92
Garrett, M. D., 39
Garsky, V. M., 32, 39(20)

H

Hackney, D. D., 161
Haga, T., 148
Hagi, S., 76
Haguenauer-Tsapis, R., 133
Hahn, K., 7
Hahn, U., 191
Hajduk, P. J., 457
Halban, P. A., 506
Hall, A., 330
Hall, H., 148
Hamaguchi, H., 76, 78(8)
Hamajima, Y., 121
Hamamoto, S., 380, 387, 389, 389(29), 390, 438, 442(6), 443(6)
Hamilton, A. D., 39
Hamilton, T. C., 148
Hammer, J. A. III, 255
Hammer, R. E., 59, 76, 83
Hammond, S. M., 365
Hammonds-Odie, L. P., 336
Han, J.-S., 325, 424, 425(8), 426(8), 427(8)
Han, W. P., 249
Han, Y., 425
Hancock, J. F., 34, 35(35), 37
Hannai, N., 356
Hansen, G. H., 210, 211(5)
Hansen, J. C., 32(24), 33
Hansen, S. H., 248, 257, 258(9), 272
Hanson, S. H., 283
Haraguchi, T., 499, 504(8)
Hardwick, K. G., 299
Harlan, J. E., 457
Harlow, E., 42
Harris, D. F., 336
Harris, E. A. S., 306
Harris, R., 141, 142(21)
Harrison, D. H., 250, 325
Hart, M., 337, 344
Harter, C., 389, 393, 395, 402(39)
Haser, W., 146
Hata, Y., 40, 60, 61(14), 65, 84, 207
Hatfield, J. C., 257, 258, 264(8), 343
Haucke, V., 380, 389(30), 390, 479, 480(22), 486, 495
Haun, R. S., 325, 326, 329, 417, 419
Hauri, H. P., 175, 176(1), 188
Hawkins, P. T., 147, 347, 364
Hawkins, W., 226

Hayes, B., 188, 190(4)
Haystead, T. A. J., 146
Hazuka, C. D., 100
Heim, R., 7, 8
Heimberg, H., 248, 250(5)
Helenius, A., 178, 436
Hellio, R., 197, 198(8)
Helms, J. B., 299, 300, 307, 389, 391, 402, 404, 417
Hemmings, B. A., 148
Hengst, L., 55
Henley, J. R., 447, 480
Hennecke, S., 317, 373
Henry, J. P., 68
Herman, P. K., 146, 154(26)
Herrmann, C., 16
Herrmann, D., 395, 402(39)
Herskovits, J. S., 453
Heuckeroth, R. O., 273
Heuser, J., 156
Hewlett, L. J., 180, 187(20)
Hicke, L., 69, 431
Hies, I. D., 479
Hieter, P., 199, 322
Higuchi, R., 49
Hiles, I., 325, 344, 355, 417, 494
Hill, S. C., 146
Hille-Rehfeld, A., 173, 178
Hinshaw, J. E., 447, 451, 458, 479, 486
Hirano, H., 60, 67(12)
Hiraoka, Y., 499, 504(8)
Hirata, H., 457
Hirata, M., 457
Hirokawa, N., 453
Hirschberg, K., 7
Hodgkin, M. N., 344
Hoe, M. H., 388(14), 389
Hoekstra, M. F., 320
Hofer, F., 146
Hoffenberg, S., 14
Hoffman, C. S., 471
Hoffman, G. R., 505
Hoflack, B., 10, 40, 120, 132, 145, 157, 174, 176, 177(6), 183(6), 184(6), 185(6), 186(6), 187, 317, 379
Hofmann, K., 77, 84, 197
Hohl, T., 59
Holcomb, C., 34
Holden, J., 303
Holik, J. J., 256, 265, 279, 280, 281(1), 283(1,

Subject Index

A

α-Actinin
antibody preparation, 78
Rabphilin-3 assays
actin filament bundling activity effects, 80–81
interaction assay, 80
Adapter protein-1
ADP-ribosylation factor interactions in clathrin coat assembly, 379
clathrin coat assembly systems, *in vitro*
advantages, 379–380
clathrin-coated vesicle preparation, 381
clathrin coat protein preparation, 381–382
coat recruitment assay, 382–383
cytosol preparation, 381
electron microscopy morphologic studies, 383–384
Golgi membrane preparation, 382
lipid requirements for ARF and AP-1 binding, 385, 387
liposome preparation, 380
ADP-ribosylation factors
adapter protein-1-dependent clathrin coat assembly initiation, *see* Adapter protein-1
ADP-ribosylation factor domain protein 1, *see* ARD1
ADP-ribosylation factor-like proteins
ADP-ribosylation factor-like protein-1, *see* ARL1
human proteins, 425
yeast, *see also* yARL3
functions, 418
types, 418
ARF1
coat protein I, *see* Coatomer
crystal structure, 245–246, 325
functions, 307, 417
GTPase activating proteins, *see* GAP1

guanine nucleotide exchange factors, *see* ARNO1; Grp1
purification from recombinant *Escherichia coli*
ammonium sulfate precipitation, 391
anion-exchange chromatography, 391–392
cation-exchange chromatography, 392
cell growth, induction, and lysis, 391
solutions, 390–391
ARF2
ARF1 homology, 291
expression and purification of recombinant myristoylated protein from yeast, 292, 294
GTPγS binding assay, 294–296
ARF6
antibodies, 249–250
comparison with ARF1, 247–248
endocytosis studies in HeLa cells, 250–252
epitope tagging, 249
expression in mammalian cells, 248–250
functions, 248, 417
GTPγS binding assay, 276–277
guanine nucleotide exchange factors, *see* ARNO1; EFA6; Grp1
myristoylation analysis, 275–276
purification from recombinant *Escherichia coli*
myristoylated protein, 273–274
unmyristoylated protein, 273
Tac trafficking
internalization and recycling of anti-Tac antibodies, 252–256
marking of ARF6 endosomal pathway, 251
monoclonal antibody, 251–252
tryptophan fluorescence analysis of bound nucleotides, 274

R

ISBN 0-12-182230-3